W9-BKN-498

FINITE
MATHEMATICS
SECOND EDITION

FINITE
MATHEMATICS
SECOND EDITION

MARGARET L. LIAL
CHARLES D. MILLER

American River College

SCOTT, FORESMAN AND COMPANY
Glenview, Illinois

Dallas, Tex.
Oakland, N.J.
Palo Alto, Cal.
Tucker, Ga.
London, England

Library of Congress Cataloging in Publishing Data

Lial, Margaret L.
  Finite mathematics.

  Includes bibliographical references and index.
  1. Mathematics—1961–   . I. Miller, Charles
David, 1942–   . II. Title.
QA37.2.L49 1982   510   81–14441
ISBN 0–673–15536–6   AACR2

Copyright © 1982, 1977 Scott, Foresman and Company.
All Rights Reserved.
Printed in the United States of America.

Some portions of this book were previously published in *Mathematics with Applications in the Management, Natural, and Social Sciences, Second Edition,* © 1979, 1974 Scott, Foresman and Company.

2 3 4 5 6 - KPF - 86 85 84 83 82

Preface

Finite Mathematics, Second Edition, gives a solid foundation in the non-calculus portions of mathematics needed by students in business, social science, and biology. Along with sound mathematics, the text supplies numerous applications to motivate students. A course in either intermediate or college algebra is the only prerequisite.

Features of the text include the following:

Detailed treatment of linear programming The book features two complete chapters on linear programming, one on the graphical method and one on the simplex method. Additional material on duality, prepared by Nancy Shoemaker of Oakland University, and a BASIC computer program for the simplex method are included in the Instructor's Guide.

Expanded coverage of mathematics of finance The chapter on mathematics of finance has been extensively rewritten and now provides ample opportunity to use calculators.

Emphasis on mathematical models The idea of mathematical models for real-world application is emphasized throughout the book, beginning in Chapter 1. Both the strengths and the limitations of models are mentioned.

Applications Optional applications, set off with a special typeface, are introduced following appropriate sections. These applications—generally briefer than the case studies of the previous edition, but comparable to them—show the practicality of the mathematics. The concepts of the course come alive when students can see how mathematics is used at Upjohn (page 34) or at Xerox (page 297).

Extensive range of exercises Exercises range from routine drill problems to more advanced exercises that will challenge capable students.

Chapter review exercises An extensive set of review exercises has been included at the end of each chapter. These review exercises help the student reinforce and review the ideas of the chapter.

The following features allow instructors to adapt the book to the needs of their courses:

Optional review chapter Chapter 1, on linear equations, is optional. Well-prepared students can go right into Chapter 2, on matrix theory, or Chapter 6, on probability, depending on course emphasis. If Chapter 1 is not discussed in class, its presence in the text provides a convenient student reference.

Brief coverage of Markov chains and decision theory It is not unusual to finish Chapter 6, "Probability," only a few days before the end of a term, leaving time for only a brief treatment of "Markov Chains" (Chapter 7) or "Decision Theory" (Chapter 9). For this reason, these chapters have been written so that the first section of each provides a good introduction to the topic, and to the types of problems for which the topic is useful.

Chapter interdependence is as follows:

Chapter	Prerequisite
1. Linear Models	None
2. Matrix Theory	None
3. Linear Programming—The Graphical Method	Sections 1.1 and 1.2 of Chapter 1
4. Linear Programming—The Simplex Method	Chapters 2 and 3
5. Sets and Counting	None
6. Probability	Chapter 5
7. Markov Chains	Chapter 6
8. Statistics and Probability Distributions	Chapter 6
9. Decision Theory	Chapter 6
10. Mathematics of Finance	None

Additional materials have been compiled for use with this book.

A Study Guide with Computer Problems, by Margaret Lial, offers a brief introduction to the computer terminal, and to BASIC programming. Many programs for solving problems from the text are included. Numerous additional exercises are designed to provide the student with deeper insight into the topics of the course.

An Instructor's Guide contains answers to even-numbered exercises and an extensive selection of additional exercises that can be used in test preparation or as additional exercises for student use.

A Solutions Guide, included in the Instructor's Guide, prepared by Professor Joseph Buckley and Lee Witt of Western Michigan University, may be reproduced and made available for student use as desired.

We received much useful advice in the preparation of this book from Professors Erik A. Schreiner and Joseph Buckley of Western Michigan University. Valuable help was also given by Professor Tom Brylawski, University of North Carolina; Professor William G. Chinn, City College of San Francisco; Professor Raymond F. Coughlin, Temple University; Professor James A. Crenshaw, Southern Illinois University; Professor A. George Dors, Eastern Washington University; Professor Henry Howard, University of Kentucky; Professor Morris Marx, University of Oklahoma; Professor Norman Mittman, Northeastern Illinois University; Professor Donald R. Sherbert, University of Illinois; Professor Nancy Stokey, Northwestern University; and Professor Bert Waits, Ohio State University. The editorial staff of Scott, Foresman did an excellent job in helping turn a manuscript into a finished bound book.

Margaret L. Lial
Charles D. Miller

Contents

Introduction on Calculators

A calculator will come in very handy throughout this course. Calculators save much tedious computation, leaving time for more productive things such as learning the key ideas of this course.

You really don't need an expensive calculator for the great bulk of our exercises — for the most part a calculator with only the four operations of addition, subtraction, multiplication, and division will be fine. However, if you do plan to do much further work in mathematics or business, you may want to consider the purchase of a financial or scientific calculator. The prices of the basic models of these calculators are in the same range as the price of this textbook. These calculators offer the advantage of doing away with most tables; for example, it is not necessary to look up numbers for finding compound interest, or present value. This can save a lot of time, especially in the mathematics of finance portion of the course.

How sophisticated a calculator do you need? It is a waste of money to buy one with more features than you will use, but it is also a waste of money to buy one that does not have all the features that you will need. Here are some of the things to consider when looking at calculators.

Memory   A memory is an electronic scratch pad. You can store intermediate calculations in the memory and then recall them later. This feature is very helpful and usually adds only a few dollars to the cost of a calculator.

Automatic Constant   Suppose you have to find the sales tax for a great number of different items. You could do this on a calculator by entering the tax rate and the cost of each item, but an automatic constant feature would save you time. With this feature, you enter the tax rate only once. Then enter each separate price, touch "equals," and the tax for that item appears.

Rounding   If the tax rate in your area is 6.25%, then the tax on an item costing $17.95 can be found by multiplying .0625 and 17.95, getting 1.121875, or, after rounding, a tax of $1.12. A calculator with a rounding feature will automatically round the answer. In this case, it would give .0625 × 17.95 = 1.12.

Logarithms   Both common logarithms and natural logarithms occur again and again in mathematics and its applications.

Powers and Roots   These functions should prove very useful.

Statistical Functions   Some advanced machines do means, factorials, standard deviations, and produce linear regression equations. You should find considerable use for these functions, both in this course and in later work, both in school and on the job.

There are two types of logic in common use on calculators today. Both algebraic and Reverse Polish Notation (RPN) have advantages and disadvantages. Algebraic logic is the easiest to learn. For example, the problem $8 + 17$ is entered into an algebraic machine by pressing

$$8 + 17 =.$$

On a machine with Reverse Polish Notation, this same problem would be entered as

$$8 \text{ ENTER } 17 +.$$

Some people prefer Reverse Polish machines, claiming that they work advanced problems more easily than algebraic machines. Others find that algebraic machines are preferable because they are easier to use for the great bulk of common problems. It's up to you to decide which to buy. Our recommendation: one of us has a Reverse Polish machine and the other has an algebraic machine.

Significant Digits   A common mistake when working with calculators is to give the answer to more accuracy than is warranted by the original data. Almost all numbers encountered in real-world problems are approximations.

If we measure a wall to the nearest meter and say that it is 18 meters long, then we are really saying that the wall has a length between 17.5 meters and 18.5 meters. If we measure the wall more accurately and say that it is 18.3 meters long, then we know that its length is really between 18.25 meters and 18.35 meters. A measurement of 18.00 meters would indicate that the wall's length is between 17.995 and 18.005 meters. The measurement 18 meters is said to have 2 significant digits of accuracy; 18.3 has 3 significant digits and 18.00 has 4.

To find the number of **significant digits** in a measurement, count from left to right, starting at the first nonzero digit and continuing to the last digit.

The following chart shows some numbers, the number of significant digits in each number, and the range represented by each number.

Number	Number of significant digits	Range represented by number
29.6	3	29.55 to 29.65
1.39	3	1.385 to 1.395
.000096	2	.0000955 to .0000965
.03	1	.025 to .035
100.2	4	100.15 to 100.25

There is one possible place for trouble when finding significant digits. We know that the measurement 19.00 meters is a measurement to the nearest hundredth meter. What about the measurement 93,000 meters? Does it represent a measurement to the nearest meter? the nearest ten meters? hundred meters? thousand meters? We cannot tell by the way the number is written. To get around this problem, we write the number in **scientific notation,** the product of a number between 1 and 10 and a power of 10. Depending on what we know about the accuracy of the measurement, we could write 93,000 using scientific notation as follows.

Measurement to nearest . . .	Scientific notation	Number of significant digits
Meter	9.3000×10^4	5
Ten meters	9.300×10^4	4
Hundred meters	9.30×10^4	3
Thousand meters	9.3×10^4	2

When calculating with approximate data, use the following rules:

1. For *adding and subtracting,* add or subtract normally, and then round the answer so that the last digit you keep is in the right-most column in which all the numbers have significant digits.

2. When *multiplying or dividing,* round your answers to the *least* number of significant digits found in any of the given numbers.

3. For *powers and roots,* round the answer so that it has the same number of significant digits as the number whose power or root you are finding.

Calculator Errors   A calculator can store only so many digits in its memory. Because of this, numbers which have more digits than can be stored must be rounded. For example, 1/3 is not stored as the exact fraction 1/3, but rather as a decimal, perhaps .3333333333333. Since this rounded form of 1/3 is used, errors can occur in calculations. To see how this happens, use a calculator to divide 1 by 3, and then multiply the result by 3. You should get 1 (exactly), but many machines produce

$$(1 \div 3) \times 3 = \left(\frac{1}{3}\right) \times 3 = .9999999999.$$

Some machines round this result to 1, but do not treat the number internally as 1. To see this, subtract 1 from the result above on such a machine; you should get 0 but probably won't. On one expensive Texas Instruments machine,

$$(1 \div 3) \times 3 - 1 = -1 \times 10^{-12}.$$

Another calculator error results when numbers of very different sizes are used in addition. For example,

$$10^9 + 10^{-5} - 10^9 = 10^{-5}.$$

However, most calculators would give

$$10^9 + 10^{-5} - 10^9 = 0.$$

These calculator errors seldom occur in realistic problems, but if they do occur you should know what is happening.

Exercises   *The following numbers represent approximate measurements. State the range represented by each of the measures.*

1. 5 pounds   2. 8 feet   3. 9.6 tons

4. 7.8 quarts   5. 8.95 meters   6. 2.37 kilometers

7. 19.7 liters   8. 32 centimeters   9. 253.741 meters

10. 47.358 ounces

11. When the area around Mt. Everest was first surveyed, the surveyors obtained a height of 29,000 feet to the nearest foot. State the range represented by this number. (The surveyors felt that no one would believe a measurement of 29,000 feet, so they reported it as 29,002.)

12. At Denny's, a chain of restaurants, the Low-Cal Special is said to have "approximately 472 calories." What is the range of calories represented by this number? By claiming "approximately 472 calories," they are probably claiming more accuracy than is possible. In your opinion, what might be a better claim?

Give the number of significant digits in each of the following.

13. 21.8   14. 37   15. 42.08   16. 600.9

17. 31.00   18. 20,000   19. 3.9×10^7 20. 5.43×10^3

21. 2.7100×10^4 22. 3.700×10^3

Round each of the following numbers to three significant digits. Then refer to the original numbers and round each to two significant digits.

23. 768.7   24. 921.3   25. 12.53   26. 28.17

27. 9.003   28. 1.700   29. 7.125   30. 9.375

31. 11.55   32. 9.155

Use a calculator to work the following problems to the correct number of significant digits.

33. $(8.742)^2$

34. $(.98352)^2$

35. $\dfrac{.746}{.092}$

36. $\dfrac{.375}{.005792}$

37. $(.425)(89.3)(746,000)$

38. $\dfrac{1.0000}{897.62}$

39. $\dfrac{3.0000}{521.84}$

40. $\dfrac{2.000}{(74.83)(.0251)}$

41. $\dfrac{1.000}{(.0900)^2 + (3.21)^2}$

42. $\dfrac{(6.93)^2 + (21.74)^2}{(38.76)^2 - 29.4}$

43. $\dfrac{8.92}{[(3.14)^2 + 2.79]^2}$

44. $\dfrac{4.63 - (2.158)^2}{[(5.728)^2 - 33.9142]}$

45. $\sqrt{74.689}$

46. $\sqrt{215.89}$

47. $\sqrt{89,000,000}$

48. $\sqrt{253,000}$

49. $\dfrac{1.00}{\sqrt{28.6} + \sqrt{49.3}}$

50. $\dfrac{4.00}{\sqrt{59.7} - \sqrt{74.6}}$

51. $\dfrac{-5.000(2.143)}{\sqrt{.009826}}$

52. $\dfrac{78.9(258.6)}{\sqrt{.05382}}$

53. $6.0(7.4896) + 58\sqrt{79.42} - 38(489.7)$

54. $128.9(3.02) + 97.6(.0589) - \sqrt{700.9}$

55. $\left[\dfrac{89^2(25.8) + (314.2)(5.098) - \sqrt{910.593}}{258(.0972) - \sqrt{104.38} + (65.923)^2}\right]^2$

56. $\left[\dfrac{(.00900)(74)}{1.0 - (.0382)\sqrt{741.6} + \sqrt{98.32}}\right]^2$

Answers to Exercises on Calculators

1. 4.5 to 5.5   **3.** 9.55 to 9.65   **5.** 8.945 to 8.955   **7.** 19.65 to 19.75
9. 253.7405 to 253.7415   **11.** 28,999.5 to 29,000.5   **13.** 3   **15.** 4   **17.** 4
19. 2   **21.** 5   **23.** 769; 770   **25.** 12.5; 13   **27.** 9.00; 9.0   **29.** 7.13; 7.1
31. 11.6; 12   **33.** 76.42   **35.** 8.1   **37.** 28,300,000 (or 2.83×10^7)
39. 0.0057489   **41.** 0.0970   **43.** 0.0557   **45.** 8.6423   **47.** 9400 (or 9.4×10^3)
49. 0.0808   **51.** −108.1   **53.** −18,000 (or -1.8×10^4)   **55.** 2200 (or 2.2×10^3)

FINITE
MATHEMATICS
SECOND EDITION

Linear Models

To use mathematics to help in the solution of a real-world problem, it is usually necessary to set up a **mathematical model**—a mathematical description of the real-world situation. To construct a mathematical model of a given situation, we need a solid understanding of the situation to be modeled, along with a good knowledge of the possible mathematical ideas that can be used to construct the model.

Much mathematical theory has been developed over the years, with a large fraction being useful for model building. Yet the very richness and diversity of contemporary mathematics too often serves as a barrier between a person in another field and the mathematical tools that person needs. There are so many useful parts of mathematics that it is often hard to know which to pick for a particular model.

One way to get around this problem is to have a thorough understanding of the basic and most useful mathematical tools that are available for model building. In this chapter we look at the mathematics of *linear* models—those used for data whose graphs can be approximated by a straight line.

1.1 Functions

A common problem in many real-life situations is to describe relationships between quantities. For example, assuming that the number of hours a student studies each day is related to the grade he or she receives in a course, how can the relationship be expressed? One way is to set up a table showing the hours of study and the corresponding grade that resulted. Such a table might appear as follows.

Hours of study	Grade
3	A
2 1/2	B
2	C
1	D
0	F

In other relationships, a formula of some sort is often used to describe how the value of one quantity depends on the value of another. For example, if a certain bank account pays 12% interest per year, then the interest, I, that a deposit of P dollars would earn in one year is given as

$$I = .12 \times P, \quad \text{or} \quad I = .12P.$$

The formula, $I = .12P$, describes the relationship between interest and the amount of money deposited.

In this example, P, which represents the amount of money deposited, is called the **independent variable,** while I is the **dependent variable.** (The amount of interest earned *depends* on the amount of money deposited.) When a specific number, say 2000, is substituted for P, then I takes on *one* specific value—in this case, $.12 \times 2000 = 240$. The variable I is said to be a *function* of P.

A **function** is a rule which assigns to each number from one set exactly one number from another set. The first set of numbers is the set of all possible values for the independent variable; this set is called the **domain** of the function. The second set, the set of all possible values of the dependent variable, is the **range** of the function. (The two sets may or may not be the same.)

Example 1   Do the following represent functions? (Assume x represents the independent variable, an assumption we shall make throughout this book.)

(a) $y = -4x + 11$

For a given value of x, a corresponding value of y is found by multiplying x by -4 and then adding 11 to the result. This process will take a given value of x and produce exactly one value of y. (For example, if $x = -7$, then $y = -4(-7) + 11 = 39$.) Since one value of the independent variable leads to exactly one value of the dependent variable, $y = -4x + 11$ is a function.

(b) $y^2 = x$

Suppose $x = 36$. Then $y^2 = x$ becomes $y^2 = 36$, from which $y = 6$ or $y = -6$. Here one value of the independent variable leads to *two* values of the dependent variable, so that $y^2 = x$ does not represent a function.   ∎

Example 2   Find the domain and range for each of the following functions.

(a) $y = -4x + 11$

Here x may take on any value, so that the domain is the set of all real numbers. Also, y may take on any value, making the range also the set of all real numbers.

(b) $y = 2x + 7$, $x = 1, 2, 3,$ or 4

The independent variable x is restricted to the values 1, 2, 3, or 4, making the domain the set $\{1, 2, 3, 4\}$. If $x = 1$, then $y = 2x + 7$ becomes $y = 2 \cdot 1 + 7 = 9$, while if $x = 2$, then $y = 2 \cdot 2 + 7 = 11$. If $x = 3$, then $y = 13$, while $x = 4$ produces $y = 15$. The range is the set of all possible values of y, or $\{9, 11, 13, 15\}$. ■

$f(x)$ Notation   Letters such as f, g, or h are often used to name functions. For example, we might use f to name the function

$$y = 5 - 3x.$$

To show that this function is named f, and to also show that x is the independent variable, it is common to replace y with $f(x)$ (read "f of x") to get

$$f(x) = 5 - 3x.$$

If we choose 2 as a value of x, then $f(x)$ becomes $5 - 3 \cdot 2 = 5 - 6 = -1$, written

$$f(2) = -1.$$

In a similar manner,

$$f(-4) = 5 - 3(-4) = 17, \qquad f(0) = 5, \qquad f(-6) = 23,$$

and so on.

Example 3   Let $g(x) = x^2 - 4x + 5$. Find $g(3)$, $g(0)$, and $g(a)$.
To find $g(3)$, substitute 3 for x.

$$g(3) = 3^2 - 4(3) + 5 = 9 - 12 + 5 = 2$$

Find $g(0)$ and $g(a)$ in the same way.

$$g(0) = 0^2 - 4(0) + 5 = 5$$
$$g(a) = a^2 - 4a + 5 \qquad ■$$

Example 4   Suppose the sales of a small company have been estimated to be

$$S(x) = 125 + 80x,$$

where $S(x)$ represents the total sales in thousands of dollars in year x, with $x = 0$ representing 1981. Estimate the sales in each of the following years.

(a) 1981

Since $x = 0$ corresponds to 1981, we can find the sales for 1981 by finding $S(0)$; that is, by substituting 0 for x.

$$S(0) = 125 + 80(0) \qquad \text{Let } x = 0$$
$$= 125$$

We know that $S(x)$ represents sales in thousands of dollars, so that sales would be estimated as 125×1000, or \$125,000 in 1981.

(b) 1985

To estimate sales in 1985, let $x = 4$.

$$S(4) = 125 + 80(4) = 125 + 320 = 445,$$

so that sales should be about \$445,000 in 1985.   ▐

Graphs   Given a function $y = f(x)$, any given value in the domain of f produces a value for y. This pair of numbers, one for x and one for y, can be written as an **ordered pair,** (x, y).

For example, let $y = f(x) = 8 + x^2$. If $x = 1$, then $f(1) = 8 + 1^2 = 9$, producing the ordered pair $(1, 9)$. (Always write the value of the independent variable first.) If $x = -3$, then $f(-3) = 8 + (-3)^2 = 17$, giving $(-3, 17)$. Other ordered pairs for this function include $(0, 8)$, $(-1, 9)$, $(2, 12)$, and so on.

It is often useful to draw a graph of the ordered pairs produced by a function. To do this, use the perpendicular number lines of a **Cartesian coordinate system,** as shown in Figure 1. We let the horizontal number line, or **x-axis,** represent the first component of the ordered pairs of the function, while the vertical or **y-axis** represents the second component. The point where the number lines cross is the zero point on both lines; this point is called the **origin.**

To plot points on the graph corresponding to ordered pairs, proceed as in the following example. Locate the point $(-2, 4)$ by starting at the origin and counting 2 units to the left on the horizontal axis, and 4 units upward parallel to the vertical axis. This point is shown in Figure 1, along with several other sample points. The number -2 is the **x-coordinate** and 4 is the **y-coordinate** of the point $(-2, 4)$.

The x-axis and y-axis divide the graph into four parts or **quadrants.** For example, Quadrant I includes all those points whose x- and y-coordinates are both positive. The quadrants are numbered as shown in Figure 1. The points of the axes themselves belong to no quadrant. The set of points corresponding to the ordered pairs of a function is called the **graph** of the function.

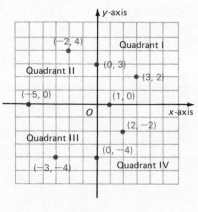

Figure 1

Example 5   Let $f(x) = 7 - 3x$, with domain $\{-2, -1, 0, 1, 2, 3, 4\}$. Graph the ordered pairs produced by this function.

If $x = -2$, then $f(-2) = 7 - 3(-2) = 13$, giving the ordered pair $(-2, 13)$. In a similar way, we can complete the following table.

x	-2	-1	0	1	2	3	4
y	13	10	7	4	1	-2	-5
Ordered pair	$(-2, 13)$	$(-1, 10)$	$(0, 7)$	$(1, 4)$	$(2, 1)$	$(3, -2)$	$(4, -5)$

These ordered pairs are graphed in Figure 2. ■

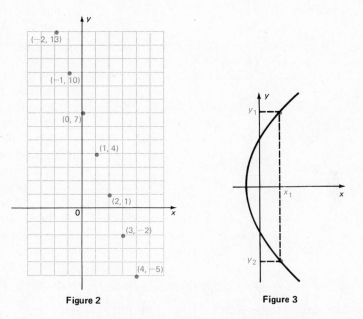

Figure 2   Figure 3

For a graph to be the graph of a function, we must be able to assign exactly one value of y to each value of x in the domain of the function. Figure 3 shows a graph. For the value $x = x_1$, the graph gives the two y-values, y_1 and y_2. Since the given x-value corresponds to two different y-values, this graph is not the graph of a function. Based on this is the **vertical line test** for a function.

> If a vertical line cuts a graph in more than one point, then the graph is not the graph of a function.

1.1 Exercises

List the ordered pairs obtained from each of the following if the domain of x for each exercise is $\{-2, -1, 0, 1, 2, 3\}$. Graph each set of ordered pairs. Give the range.

1. $y = x - 1$ **2.** $y = 2x + 3$ **3.** $y = -4x + 9$

4. $y = -6x + 12$ **5.** $y = -x - 5$ **6.** $y = -2x - 3$

7. $2x + y = 9$ **8.** $3x + y = 16$ **9.** $2y - x = 5$

10. $6x - y = -3$ **11.** $y = x(x + 1)$ **12.** $y = (x - 2)(x - 3)$

13. $y = x^2$ **14.** $y = -2x^2$ **15.** $y = 3 - 4x^2$

16. $y = 5 - x^2$ **17.** $y = \dfrac{1}{x + 3}$ **18.** $y = \dfrac{-2}{x + 4}$

19. $y = \dfrac{3x - 3}{x + 5}$ **20.** $y = \dfrac{2x + 1}{x + 3}$ **21.** $y = 4$

22. $y = -2$

Identify any of the following that represent functions.

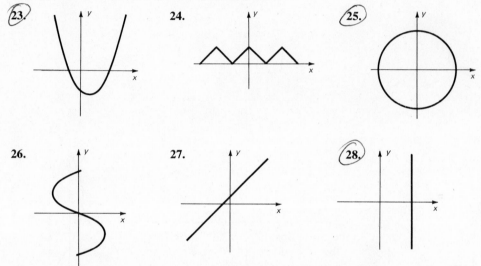

23.   **24.**   **25.**

26.   **27.**   **28.**

For each of the following functions, find (a) $f(4)$; (b) $f(-3)$; (c) $f(0)$; (d) $f(a)$.

29. $f(x) = 3x + 2$ **30.** $f(x) = 5x - 6$

31. $f(x) = -2x - 4$ **32.** $f(x) = -3x + 7$

33. $f(x) = 6$ **34.** $f(x) = 0$

35. $f(x) = 2x^2 + 4x$ **36.** $f(x) = x^2 - 2x$

37. $f(x) = -x^2 + 5x + 1$ **38.** $f(x) = -x^2 - x + 5$

39. $f(x) = (x + 1)(x + 2)$ **40.** $f(x) = (x + 3)(x - 4)$

41. Let $f(x) = 2x - 3$. Find each of the following.

   (a) $f(0)$ **(b)** $f(-1)$ **(c)** $f(-6)$ **(d)** $f(4)$ **(e)** $f(a)$

   (f) $f(-r)$ **(g)** $f(m + 3)$ **(h)** $f(p - 2)$ **(i)** $f[f(2)]$ **(j)** $f[f(-3)]$

42. Suppose the sales of a small company that sells by mail are approximated by

$$S(t) = 1000 + 50(t + 1),$$

where $S(t)$ represents sales in thousands of dollars. Here t is time in years, with $t = 0$ representing the year 1981. Find the estimated sales in each of the following years.

   (a) 1981   **(b)** 1982   **(c)** 1984   **(d)** 1986

43. A chain-saw rental firm charges $7 per day or fraction of a day to rent a saw, plus a fixed fee of $4 for resharpening the blade. Let $S(x)$ represent the cost of renting a saw for x days. Find each of the following.

(a) $S\left(\dfrac{1}{2}\right)$ (b) $S(1)$ (c) $S\left(1\dfrac{1}{4}\right)$ (d) $S\left(3\dfrac{1}{2}\right)$

(e) $S(4)$ (f) $S\left(4\dfrac{1}{10}\right)$ (g) $S\left(4\dfrac{9}{10}\right)$

(h) A portion of the graph of $y = S(x)$ is shown here. Explain how the graph could be continued.

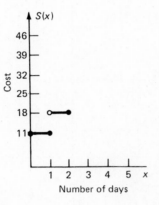

44. To rent a midsized car from Avis costs $40 per day or fraction of a day. If you pick up the car in Boston and drop it off in Utica, there is a fixed $40 charge. Let $C(x)$ represent the cost of renting the car for x days, taking it from Boston to Utica. Find each of the following.

(a) $C\left(\dfrac{3}{4}\right)$ (b) $C\left(\dfrac{9}{10}\right)$ (c) $C(1)$ (d) $C\left(1\dfrac{5}{8}\right)$

(e) $C\left(2\dfrac{1}{9}\right)$ (f) Graph the function $y = C(x)$.

1.2 Linear Functions

Any equation of the form

$$ax + by = c,$$

where a, b, and c are real numbers, with not both a and b equal to 0, is a **linear equation.** Examples of linear equations include $4x + 5y = 9$, $x - 6y = 8$, $x = 3$, $y = 5$, and so on. To see that $x = 3$ is a linear equation, rewrite it as $x + 0y = 3$.

If we are given a particular linear equation, such as $y = x + 1$, and a particular value of x, say $x = 2$, we can find the corresponding value of y:

$$y = x + 1$$
$$y = 2 + 1 \qquad \text{Let } x = 2$$
$$y = 3.$$

As this work shows, the ordered pair (2, 3) satisfies the equation $y = x + 1$. Check that $(0, 1), (4, 5), (-2, -1), (-5, -4), (-3, -2)$, among many others, are also ordered pairs which satisfy the equation.

To graph $y = x + 1$, begin by locating the ordered pairs obtained above. This graph is shown in Figure 4(a). All the points of this graph appear to be on one straight line, which can be drawn through the plotted points, as in Figure 4(b). This straight line is the graph of $y = x + 1$. Since any vertical line will cut the graph of Figure 4(b) in only one point, $y = x + 1$ is a function.

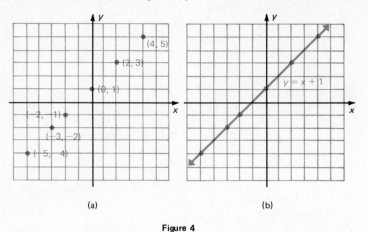

(a)   (b)

Figure 4

Example 1   Use the equation $x + 2y = 6$ to complete the ordered pairs $(-6,\ ), (-4,\ )$, $(-2,\ ), (0,\ ), (2,\ ), (4,\ )$. Graph these points and then draw a straight line through them.

To complete the ordered pair $(-6,\ )$, let $x = -6$ in the equation $x + 2y = 6$.

$$x + 2y = 6$$
$$-6 + 2y = 6 \qquad \text{Let } x = -6$$
$$2y = 12 \qquad \text{Add 6 to both sides}$$
$$y = 6$$

Ordered pair: $(-6, 6)$.

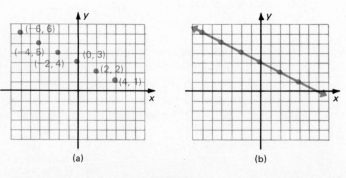

(a)   (b)

Figure 5

In the same way, if $x = -4$, then $y = 5$, giving $(-4, 5)$. Check that the remaining ordered pairs are as graphed in Figure 5(a). A line is drawn through the points in Figure 5(b).   ■

Intercepts   It can be shown that the graph of any linear equation is a straight line. Since a straight line is completely determined by any two distinct points through which it passes, we really need to locate only two distinct points to draw the graph. Two points that are often useful for this purpose are the x-intercept and the y-intercept. The **x-intercept** is the x-value (if one exists) where the graph of the equation crosses the x-axis. The **y-intercept** in turn is the y-value (if one exists) at which the graph crosses the y-axis. At the point where the graph crosses the y-axis, $x = 0$. Also, $y = 0$ at the x-intercept. (See Figure 6.)

Example 2   Use the intercepts to draw the graph of $y = -2x + 5$.
  To find the y-intercept, the point where the line crosses the y-axis, let $x = 0$.

$$y = -2x + 5$$
$$y = -2(0) + 5 \qquad \text{Let } x = 0$$
$$y = 5$$

The y-intercept is 5, leading to the ordered pair $(0, 5)$. In the same way, the x-intercept may be found by letting $y = 0$.

$$0 = -2x + 5 \qquad \text{Let } y = 0$$
$$-5 = -2x$$
$$\frac{5}{2} = x$$

The x-intercept is 5/2, or 2 1/2, with the graph going through (2 1/2, 0).   ■

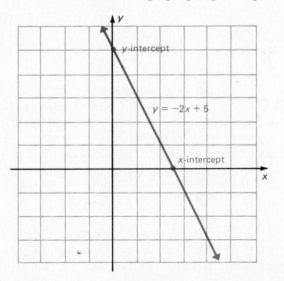

Figure 6

Using the two intercepts, we obtained the graph of Figure 6. To check the work, a third point can be found. Do this by choosing another value of x (or y) and finding the corresponding value of the other variable. Check that $(1, 3), (2, 1), (3, -1)$, and $(4, -3)$, among many other points, satisfy the equation $y = -2x + 5$ and thus also lie on the line of Figure 6. Does $y = -2x + 5$ represent a function?

In the discussion of intercepts given above, we added the phrase "if one exists" when talking about the place where a graph crosses an axis. The next example shows a graph that does not cross the x-axis, and thus has no x-intercept.

Example 3   Graph $y = -3$.

Using $y = -3$, or equivalently, $y = 0x - 3$, we always get the same y value, -3, for any value of x. Therefore, there is no value of x that will make $y = 0$, so that the graph has no x-intercept. Since $y = -3$ is a linear equation, with a straight-line graph, and since the graph cannot cross the x-axis, the line must be parallel to the x-axis. For any value of x, we have $y = -3$. Therefore, the graph is the horizontal line which is parallel to the x-axis and has y-intercept -3, as shown in Figure 7. As the vertical line test shows, the graph is the graph of a function. In general, the graph of $y = k$, where k is a real number, is the horizontal line having y-intercept k.   ∎

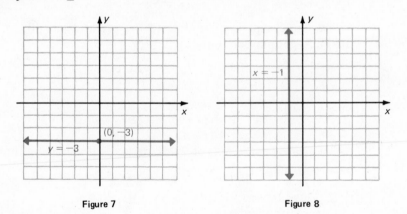

Figure 7   Figure 8

Example 4   Graph $x = -1$.

To obtain the graph of $x = -1$, we can complete some ordered pairs using the equivalent form, $x = 0y - 1$. For example, $(-1, 0), (-1, 1), (-1, 2)$, and $(-1, 4)$ are some ordered pairs satisfying $x = -1$. (The first coordinate of these ordered pairs is always -1, which is what $x = -1$ means.) Here, more than one second coordinate corresponds to the same first coordinate, -1. As the graph of Figure 8 shows, a vertical line can cut this graph in more than one point. (In fact, a vertical line cuts the graph in an infinite number of points.) For this reason, $x = -1$ is not a function.   ∎

As shown in Example 4 above, $x = -1$ is a linear equation that is not a linear function. Linear equations of this form, $x = k$, where k is a real number, are the only linear equations that are not also linear functions.

Example 5 Graph $y = -3x$.

Begin by looking for the x-intercept. If $y = 0$, then

$$y = -3x$$
$$0 = -3x \qquad \text{Let } y = 0$$
$$0 = x. \qquad \text{Multiply both sides by } -\frac{1}{3}$$

We have the ordered pair $(0, 0)$. If we start with $x = 0$, we end up with exactly the same ordered pair, $(0, 0)$. Two points are needed to determine a straight line, and the intercepts have led to only one point. To get a second point, choose some other value of x (or y). For example, if $x = 2$,

$$y = -3x = -3(2) = -6, \qquad \text{Let } x = 2$$

giving the ordered pair $(2, -6)$. The two ordered pairs that we have found, $(0, 0)$ and $(2, -6)$, were used to get the graph shown in Figure 9. ∎

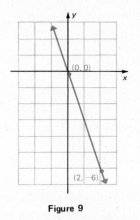

Figure 9

Linear functions can be very useful in setting up a mathematical model for a real-life situation. In almost every case, linear (or any other reasonably simple) functions provide only approximations to real-world situations. However, these can often be remarkably useful approximations.

Supply and Demand In particular, linear functions are often good choices for **supply and demand curves.** Typically, as the price of an item increases, the demand for the item decreases, while the supply increases. The following example shows this.

Example 6 Suppose that Greg Odjakjian, an economist, has studied the supply and demand for aluminum siding and has come up with the conclusion that price, p, and demand, x, in appropriate units and for an appropriate domain, are related by the linear function

$$p = 60 - \frac{3}{4}x.$$

(a) Find the demand at a price of $40.

$$p = 60 - \frac{3}{4}x$$

$$40 = 60 - \frac{3}{4}x \qquad \text{Let } p = 40$$

$$-20 = -\frac{3}{4}x \qquad \text{Add } -60 \text{ on both sides}$$

$$\frac{80}{3} = x \qquad \text{Multiply both sides by } -\frac{4}{3}$$

At a price of $40, 80/3 units will be demanded; this gives the ordered pair (80/3, 40). (It is customary to write the ordered pairs so that price comes second.)

(b) Find the price if the demand is 32 units.

$$p = 60 - \frac{3}{4}x$$

$$p = 60 - \frac{3}{4}(32) \qquad \text{Let } x = 32$$

$$p = 60 - 24$$

$$p = 36$$

With a demand of 32 units, we get a price of $36. This gives the ordered pair (32, 36).

(c) Graph $p = 60 - \frac{3}{4}x$.

Use the ordered pairs (80/3, 40) and (32, 36) to get the demand graph shown in Figure 10. Only the portion of the graph in Quadrant I is shown, since our function is only meaningful for positive values of p and x. ∎

Example 7 Suppose now that the economist of Example 6 concludes that the price and supply of siding are related by

$$p = \frac{3}{4}x,$$

where x now represents supply.

(a) Find the supply if the price is $60.

$$60 = \frac{3}{4}x \qquad \text{Let } p = 60$$

$$80 = x$$

If the price is $60, then 80 units will be supplied to the marketplace. This gives the ordered pair (80, 60).

(b) Find the price if the supply is 16 units.

$$p = \frac{3}{4}(16) = 12 \qquad \text{Let } x = 16$$

If the supply is 16 units, then the price is $12. This gives the ordered pair (16, 12).

(c) Graph $p = \dfrac{3}{4}x$.

Use the ordered pairs (80, 60) and (16, 12) to get the supply graph shown in Figure 10.   ▮

As shown in the graphs of Figure 10, both the supply and the demand functions pass through the point (40, 30). If the price of the siding is more than $30, the supply will exceed the demand. At a price less than $30, the demand will exceed the supply. Only at a price of $30 will demand and supply be equal. For this reason, $30 is called the *equilibrium price*. When the price is $30, demand and supply both equal 40 units, the *equilibrium supply* or *equilibrium demand*. In general, the **equilibrium price** of a commodity is the price found at the point where the supply and demand graphs for that commodity cross. The **equilibrium demand** is the demand at that same point; the **equilibrium supply** is the supply at that point.

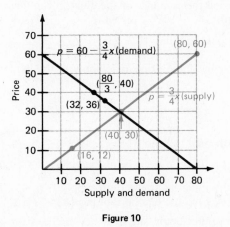

Figure 10

Example 8   Use algebra to find the equilibrium supply for the aluminum siding. (See Examples 6 and 7.)

The equilibrium supply is found when the prices from both supply and demand are equal. From Example 6 we have $p = 60 - (3/4)x$; in Example 7 we have $p = (3/4)x$. Thus,

$$60 - \frac{3}{4}x = \frac{3}{4}x$$

$$240 - 3x = 3x \qquad \text{Multiply both sides by 4}$$

$$240 = 6x \qquad \text{Add } 3x \text{ to both sides}$$

$$40 = x.$$

The equilibrium supply is 40 units, the same answer found above.   ▮

1.2 Exercises   *Graph each of the following linear equations. Identify any which are* not *linear functions.*

1. $y = 2x + 1$
2. $y = 3x - 1$
3. $y = 4x$
4. $y = x + 5$
5. $3y + 4x = 12$
6. $4y + 5x = 10$
7. $y = -2$
8. $x = 4$
9. $6x + y = 12$
10. $x + 3y = 9$
11. $x - 5y = 4$
12. $2y + 5x = 20$
13. $x + 5 = 0$
14. $y - 4 = 0$
15. $5y - 3x = 12$
16. $2x + 7y = 14$
17. $8x + 3y = 10$
18. $9y - 4x = 12$
19. $y = 2x$
20. $y = -5x$
21. $y = -4x$
22. $y = x$
23. $x + 4y = 0$
24. $x - 3y = 0$

25. Suppose that the demand and price for a certain model of electric can opener are related by

$$p = 16 - \frac{5}{4}x,$$

where p is price and x is demand, in appropriate units. Find the price for a demand of
(a) 0 units;   (b) 4 units;   (c) 8 units.

Find the demand for the electric can opener at a price of
(d) \$6;   (e) \$11;   (f) \$16.

(g) Graph $p = 16 - \frac{5}{4}x$.

Suppose the price and supply of the item above are related by

$$p = \frac{3}{4}x,$$

where x represents the supply, and p the price. Find the supply when the price is
(h) \$0;   (i) \$10;   (j) \$20.

(k) Graph $p = \frac{3}{4}x$ on the same axes used for 25 (g).

(l) Find the equilibrium supply.
(m) Find the equilibrium price.

26. Let the supply and demand functions for strawberry-flavored licorice be

$$\text{supply: } p = \frac{3}{2}x \quad \text{and} \quad \text{demand: } p = 81 - \frac{3}{4}x.$$

(a) Graph these on the same axes.
(b) Find the equilibrium demand.
(c) Find the equilibrium price.

27. Let the supply and demand functions for butter pecan ice cream be given by

$$\text{supply: } p = \frac{2}{5}x \quad \text{and} \quad \text{demand: } p = 100 - \frac{2}{5}x.$$

(a) Graph these on the same axes.
(b) Find the equilibrium demand.
(c) Find the equilibrium price.

28. Let the supply and demand functions for sugar be given by

$$\text{supply: } p = 1.4x - .6 \quad \text{and} \quad \text{demand: } p = -2x + 3.2.$$

(a) Graph these on the same axes.
(b) Find the equilibrium demand.
(c) Find the equilibrium price.

29. In a recent issue of *Business Week,* the president of Insta-Tune, a chain of franchised automobile tune-up shops, says that people who buy a franchise and open a shop pay a weekly fee of

$$y = .07x + \$135$$

to company headquarters. Here y is the fee and x is the total amount of money taken in during the week by the tune-up center. Find the weekly fee if x is
(a) \$0;   (b) \$1000;   (c) \$2000;   (d) \$3000.
(e) Graph the function.

30. In a recent issue of *The Wall Street Journal,* we are told that the relationship between the amount of money that an average family spends on food, x, and the amount of money it spends on eating out, y, is approximated by the model

$$y = .36x.$$

Find y if x is
(a) \$40;   (b) \$80;   (c) \$120.
(d) Graph the function.

1.3   Slope and the Equation of a Line

As we said in the previous section, the graph of a straight line is completely determined by two different points on the line. We can also draw the graph of a straight line knowing only *one* point on the line *if* we also know the "steepness" of the line. The number which represents the "steepness" of a line is called the *slope* of that line.

To see how slope is defined, start with Figure 11, which shows a line passing through the two different points $(x_1, y_1) = (-3, 5)$ and $(x_2, y_2) = (2, -4)$. The difference in the two x values,

$$x_2 - x_1 = 2 - (-3) = 5$$

in this example, is called the **change in x.** The symbol Δx (read "delta x") is used to represent the change in x. In the same way, Δy represents the **change in y.** In our example,

$$\Delta y = y_2 - y_1 = -4 - 5 = -9.$$

The **slope** of a line through the two different points (x_1, y_1) and (x_2, y_2) is defined as the change in y divided by the change in x or

$$\text{slope} = \frac{\text{change in } y}{\text{change in } x} = \frac{\Delta y}{\Delta x} = \frac{y_2 - y_1}{x_2 - x_1}.$$

The slope of the line in Figure 11 is

$$\text{slope} = \frac{\Delta y}{\Delta x} = \frac{-4 - 5}{2 - (-3)} = \frac{-9}{5}.$$

Using similar triangles from geometry, it can be shown that the slope is independent of the choice of points on the line. That is, the same slope will be obtained for *any* choice of two different points on the line. (See Exercise 66 below.)

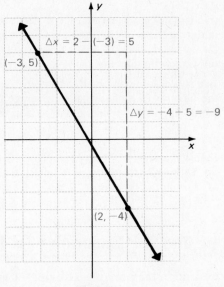

Figure 11

Example 1   Find the slope of the line through the points $(-7, 6)$ and $(4, 5)$.

Let $(x_1, y_1) = (-7, 6)$. Then $(x_2, y_2) = (4, 5)$. Use the definition of slope:

$$\text{slope} = \frac{\Delta y}{\Delta x} = \frac{5 - 6}{4 - (-7)} = \frac{-1}{11}.$$ ■

In finding the slope, we could have let $(x_1, y_1) = (4, 5)$ and $(x_2, y_2) = (-7, 6)$. In that case,

$$\text{slope} = \frac{6 - 5}{-7 - 4} = \frac{1}{-11} = \frac{-1}{11},$$

the same answer. The order in which coordinates are subtracted does not matter, as long as it is done consistently.

As we said earlier, the slope of a line is a measure of the steepness of the line. Figure 12 shows examples of lines with different slopes. Lines with positive slopes go up as we move from left to right along the x-axis, while lines with negative slopes go down as we move from left to right.

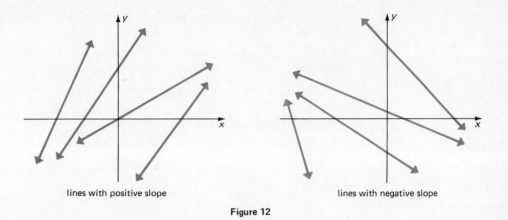

lines with positive slope   lines with negative slope

Figure 12

Example 2   Find the slope of the line $3x - 4y = 12$.

To find the slope, we need two different points on the line. We can find two such points here by finding the intercepts. First let $x = 0$ and then let $y = 0$.

If $x = 0$,   If $y = 0$,

$3x - 4y = 12$ $3x - 4y = 12$

$3(0) - 4y = 12$ Let $x = 0$ $3x - 4(0) = 12$ Let $y = 0$

$-4y = 12$ $3x = 12$

$y = -3$ $x = 4$

This gives us the ordered pairs $(0, -3)$ and $(4, 0)$. Now the slope can be found from the definition above. Use the two ordered pairs to find the slope.

$$\text{slope} = \frac{0 - (-3)}{4 - 0} = \frac{3}{4} \quad \blacksquare$$

Slope-Intercept Form   We can use a generalization of the method of Example 2 to find the equation of a line given its y-intercept and slope. Assume that a line has y-intercept b, so that it goes through $(0, b)$. Let the slope of the line be represented by m. If (x, y) is any point on the line *other* than $(0, b)$, then we can use the definition of slope with the points $(0, b)$ and (x, y) to get

$$m = \frac{y - b}{x - 0}$$

$$m = \frac{y - b}{x}$$

or

$$mx = y - b,$$

from which

$$y = mx + b.$$

This result, called the *slope-intercept form* of the equation of a line, can be summarized as follows.

> **Slope-intercept form** If a line has slope m and y-intercept b, then the equation of the line is
> $$y = mx + b.$$

Example 3   Find an equation for the line having y-intercept 7/2 and slope $-5/2$.

Here $b = 7/2$ and $m = -5/2$. Use the slope-intercept form.

$$y = mx + b$$

$$y = -\frac{5}{2}x + \frac{7}{2}$$

To get an equation without fractions, multiply both sides by 2.

$$2y = -5x + 7 \quad \blacksquare$$

The slope of a line can be found from its equation by solving the equation for y. Then the coefficient of x is the slope and the constant term is the y-intercept. (See Exercises 67 and 68.) For example, we found in Example 2 above that the slope of the line $3x - 4y = 12$ is 3/4. We can also find this slope by solving for y.

$$3x - 4y = 12$$

$$-4y = -3x + 12$$

$$y = \frac{3}{4}x - 3$$

As the coefficient of x shows, the slope is 3/4.

Example 4   Find the slope and y-intercept for each of the following lines.

(a) $5x - 3y = 1$

Solve for y:
$$5x - 3y = 1$$
$$-3y = -5x + 1$$
$$y = \frac{5}{3}x - \frac{1}{3}.$$

The slope is 5/3 and the y-intercept is $-1/3$.

(b) $-9x + 6y = 2$

Solve for y:
$$-9x + 6y = 2$$
$$6y = 9x + 2$$
$$y = \frac{3}{2}x + \frac{1}{3}.$$

The slope is 3/2 and the y-intercept is 1/3. $\blacksquare$

The slope and y-intercept of a line can be used to draw the graph of the line as shown in the next example.

Example 5   Use the slope and y-intercept to graph $3x - 2y = 2$.

Solve for y:

$$3x - 2y = 2$$
$$-2y = -3x + 2$$
$$y = \frac{3}{2}x - 1$$

The slope is 3/2 and the y-intercept is -1.

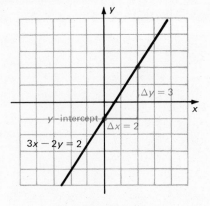

Figure 13

To draw the graph, first locate the y-intercept, as shown in Figure 13. To find a second point on the graph, use the slope. If m represents the slope, then

$$m = \frac{\Delta y}{\Delta x} = \frac{3}{2}$$

in our example. If x changes by 2 units ($\Delta x = 2$), then y will change by 3 units ($\Delta y = 3$). To find our second point, start at the y-intercept graphed in Figure 13 and move 2 units to the right and 3 units up. Once this second point is located, a line can be drawn through it and the y-intercept. ∎

Point-Slope Form   The slope-intercept form of the equation of a line involves the slope and the y-intercept. Sometimes, however, we know the slope of a line, together with one point (perhaps *not* the y-intercept) that the line goes through. The *point-slope form* of the equation of a line is used to find the equation in this case. Let (x_1, y_1) be any fixed point on the line and let (x, y) represent any other point on the line. If m is the slope of the line, then, by the definition of slope,

$$\frac{y - y_1}{x - x_1} = m,$$

or

$$y - y_1 = m(x - x_1).$$

Point-slope form If a line has slope m and passes through the point (x_1, y_1), then an equation of the line is given by

$$y - y_1 = m(x - x_1).$$

Example 6   Find an equation of the line going through the following point and having the given slope.

(a) $(-4, 1)$, $m = -3$

Since we know a point the line goes through, together with the slope of the line, we use the point-slope form. Here $x_1 = -4$, $y_1 = 1$, and $m = -3$. Substitute these values into the point-slope form.

$$y - y_1 = m(x - x_1)$$
$$y - 1 = -3[x - (-4)]$$
$$y - 1 = -3(x + 4)$$
$$y - 1 = -3x - 12$$
$$y = -3x - 11$$

(b) $(3, -7)$, $m = 5/4$

$$y - y_1 = m(x - x_1)$$
$$y - (-7) = \frac{5}{4}(x - 3) \qquad \text{Let } y_1 = -7, \ m = 5/4, \ x_1 = 3$$
$$y + 7 = \frac{5}{4}(x - 3)$$
$$4y + 28 = 5(x - 3) \qquad \text{Multiply both sides by 4}$$
$$4y + 28 = 5x - 15$$
$$4y = 5x - 43 \quad \blacksquare$$

The point-slope form can also be used to find an equation of a line if we know two different points that the line goes through. The procedure for doing this is shown in the next example.

Example 7   Find an equation of the line through $(5, 4)$ and $(-10, -2)$.

Begin by using the definition of slope to find the slope of the line which passes through the two points.

$$\text{slope} = m = \frac{-2 - 4}{-10 - 5} = \frac{-6}{-15} = \frac{2}{5}$$

Use $m = 2/5$ and either of the given points in the point-slope form. If we let $(x_1, y_1) = (5, 4)$, we get

$$y - y_1 = m(x - x_1)$$
$$y - 4 = \frac{2}{5}(x - 5) \qquad \text{Let } y_1 = 4, \ m = \frac{2}{5}, \ x_1 = 5$$
$$5y - 20 = 2(x - 5) \qquad \text{Multiply both sides by 5}$$
$$5y - 20 = 2x - 10$$
$$5y = 2x + 10.$$

Check that we get the same result if we let $(x_1, y_1) = (-10, -2)$. $\quad \blacksquare$

Example 8   Find an equation of the line through $(8, -4)$ and $(-2, -4)$.

Find the slope.

$$m = \frac{-4 - (-4)}{-2 - 8} = \frac{0}{-10} = 0$$

Choose, say, $(8, -4)$ as (x_1, y_1).

$$y - y_1 = m(x - x_1)$$
$$y - (-4) = 0(x - 8) \qquad \text{Let } y_1 = -4,\ m = 0,\ x_1 = 8$$
$$y + 4 = 0 \qquad\qquad\quad 0(x - 8) = 0$$
$$y = -4 \qquad \blacksquare$$

As we saw in the previous section, $y = -4$ represents a horizontal line, with y-intercept -4. In general, every horizontal line has a slope of 0.

Example 9   Find an equation of the line through $(4, 3)$ and $(4, -6)$.

Find the slope.

$$m = \frac{-6 - 3}{4 - 4} = \frac{-9}{0}$$

Division by 0 is impossible, so there is no slope here. If we graph the given ordered pairs $(4, 3)$ and $(4, -6)$ and draw a line through them, we find that the line is vertical. In the last section, we saw that vertical lines have equations of the form $x = k$, where k can be any real number. Since the x-coordinate of the two ordered pairs given above is 4, the desired equation is

$$x = 4. \qquad \blacksquare$$

> **A vertical line is the only line that has no slope.**

Following is a summary of the types of equations of lines that we have studied.

Equation	Description
$ax + by = c$	if $a \neq 0$ and $b \neq 0$, line has x-intercept c/a and y-intercept c/b
$x = k$	**vertical line,** x-intercept k, no y-intercept, no slope
$y = k$	**horizontal line,** y-intercept k, no x-intercept, slope 0
$y = mx + b$	**slope-intercept form,** slope m, y-intercept b
$y - y_1 = m(x - x_1)$	**point-slope form,** slope m, line passes through (x_1, y_1)

1.3 Exercises   *In each of the following exercises, find the slope, if it exists, of the line going through the given pair of points.*

1. $(-8, 6)$, $(2, 4)$

2. $(-3, 2)$, $(5, 9)$

3. $(-1, 4)$, $(2, 6)$

4. $(3, -8)$, $(4, 1)$

5. The origin and $(-4, 6)$

6. The origin and $(8, -2)$

7. $(-2, 9)$, $(-2, 11)$

8. $(7, 4)$, $(7, 12)$

9. $(3, -6)$, $(-5, -6)$

10. $(5, -11)$, $(-9, -11)$

Find the slope and y-intercept of each of the following lines.

11. $y = 3x + 4$

12. $y = -3x + 2$

13. $y + 4x = 8$

14. $y - x = 3$

15. $3x + 4y = 5$

16. $2x - 5y = 8$

17. $3x + y = 0$

18. $y - 4x = 0$

19. $2x + 5y = 0$

20. $3x - 4y = 0$

21. $y = 8$

22. $y = -4$

23. $y + 2 = 0$

24. $y - 3 = 0$

25. $x = -8$

26. $x = 3$

Graph the line going through the given point and having the given slope.

27. $(-4, 2)$, $m = 2/3$

28. $(3, -2)$, $m = 3/4$

29. $(-5, -3)$, $m = -2$

30. $(-1, 4)$, $m = 2$

31. $(8, 2)$, $m = 0$

32. $(2, -4)$, $m = 0$

33. $(6, -5)$, no slope

34. $(-8, 9)$, no slope

35. $(0, -2)$, $m = 3/4$

36. $(0, -3)$, $m = 2/5$

37. $(5, 0)$, $m = 1/4$

38. $(-9, 0)$, $m = 5/2$

Find an equation for each line having the given y-intercept and slope.

39. 4, $m = -3/4$

40. -3, $m = 2/3$

41. -2, $m = -1/2$

42. $3/2$, $m = 1/4$

43. $5/4$, $m = 3/2$

44. $-3/8$, $m = 3/4$

Find equations for each of the following lines.

45. Through $(-4, 1)$, $m = 2$

46. Through $(5, 1)$, $m = -1$

47. Through $(0, 3)$, $m = -3$

48. Through $(-2, 3)$, $m = 3/2$

49. Through $(3, 2)$, $m = 1/4$

50. Through $(0, 1)$, $m = -2/3$

51. Through $(-1, 1)$ and $(2, 5)$

52. Through $(4, -2)$ and $(6, 8)$

53. Through $(9, -6)$ and $(12, -8)$

54. Through $(-5, 2)$ and $(7, 5)$

55. Through $(-8, 4)$ and $(-8, 6)$

56. Through $(2, -5)$ and $(4, -5)$

57. Through $(-1, 3)$ and $(0, 3)$

58. Through $(2, 9)$ and $(2, -9)$

Many real-world situations can be approximately described by a straight-line graph. One way to find the equation of such a straight line is to use two typical data points from the graph and the point-slope form of the equation of a line. In each of the following problems, assume that the data can be approximated fairly closely by a straight line. Use the given information to find an equation of the line. Find the slope of each of the lines.

59. A company finds that it can make a total of 20 hot tubs for $13,900, while 10 tubs cost $7500. Let y be the total cost to produce x hot tubs.

60. The sales of a small company were $27,000 in its second year of operation and $63,000 in its fifth year. Let y represent sales in the x-th year of operation.

61. When a certain industrial pollutant is introduced into a river, the reproduction of catfish declines. In a given period of time, three tons of the pollutant results in a fish population of 37,000. Also, 12 tons of pollutant produce a fish population of 28,000. Let y be the fish population when x tons of pollutant are introduced into the river.

62. According to research done by the political scientist James March, if the Democrats win 45% of the two-party vote for the House of Representatives, they win 42.5% of the seats. If the Democrats win 55% of the vote, they win 67.5% of the seats. Let y be the percent of seats won, and x the percent of the two-party vote.

63. If the Republicans win 45% of the two-party vote, they win 32.5% of the seats (see Exercise 62.) If they win 60% of the vote, they get 70% of the seats. Let y represent the percent of the seats, and x the percent of the vote.

64. A person's tibia bone goes from ankle to knee. A male with a tibia 40 cm in length will have a height of 177 cm, while a tibia 43 cm in length corresponds to a height of 185 cm.
   (a) Write a linear equation showing how the height of a male, h, relates to the length of his tibia, t.
   (b) Estimate the height of a male having a tibia of length 38 cm; 45 cm.
   (c) Estimate the length of the tibia for a height of 190 cm.

65. The radius bone goes from the wrist to the elbow. A female whose radius bone is 24 cm long would be 167 cm tall, while a radius of 26 cm corresponds to a height of 174 cm.
   (a) Write a linear equation showing how the height of a female, h, corresponds to the length of her radius bone, r.
   (b) Estimate the height of a female having a radius of length 23 cm; 27 cm.
   (c) Estimate the length of a radius bone for a height of 170 cm.

66. Use similar triangles from geometry to show that the slope of a line is the same, no matter which two distinct points on the line are chosen to compute it.

67. Show that b is the y-intercept in the slope-intercept form $y = mx + b$.

68. Suppose that $(0, b)$ and (x_1, y_1) are distinct points on the line $y = mx + b$. Show that $(y_1 - b)/x_1$ is the slope of the line, and that $m = (y_1 - b)/x_1$.

1.4   Linear Mathematical Models

Throughout this book, we have been setting up mathematical models—mathematical descriptions of real-world situations. In this section, we discuss mathematical models in more detail; then we look at some mathematical models using the linear equations we have discussed in the last two sections.

   If we completely understand the principles causing a certain event to happen, then the mathematical model describing that event can often be very accurate.

As a rule, the mathematical models constructed in the physical sciences are excellent at predicting events. For example, if a body falls in a vacuum, then *d,* the distance in feet that the body will fall in *t* seconds, is given by

$$d = \frac{1}{2}gt^2,$$

where *g* is a constant (number) representing gravity. (As an approximation, $g = 32$ feet per second per second.) Using this equation, *d* can be predicted exactly for a known value of *t*. In the same way, astronomers have formulated very precise mathematical models of the movements of our solar system. Predictions about the future positions of heavenly bodies can be made with great accuracy. Thus, when astronomers say that Halley's comet will next appear in 1986, few dispute the statement.

The situation is different in the fields of management and social science. Mathematical models in these fields tend to be less accurate approximations, and often gross approximations at that. While an astronomer can say with almost certainty that Halley's comet will reappear in 1986, no corporation vice-president would dare predict that company sales will be $185,000 higher in 1986 than in 1985. So many variables enter into determining sales of a company that no present-day mathematical model can ever hope to produce results comparable to those produced in physical science.

In spite of the limitations of mathematical models, they have found a large and increasing acceptance in management and economic decision making. There is one main reason for this—mathematical models produce very useful results.

The former vice-president for corporate planning for Anheuser-Busch, Mr. Robert S. Weinberg, offers the following reasons for the wide use of mathematical models in management and economics.

(1) A mathematical model leads to a well-defined statement of the problem. For example, decisions about new products have often been made strictly on the intuition of a top manager. By studying past records, by surveying potential consumers, and by carefully analyzing costs at various possible levels of production, it is possible to produce a mathematical model of the factors affecting a new product. This model can then be used to make predictions such as those below.

There is a 15% chance of making a profit of $10 million.
There is a 20% chance of making a profit of $2 million.
There is a 25% chance of breaking even.
There is a 40% chance of losing $3 million.

With this information at hand, management can make a more informed decision.

(2) A model makes clear the assumptions behind the problem and its possible solutions.

(3) A model can be used to identify areas in which further research is needed.

Sales Analysis   Let us now look at some mathematical models of real-world situations.

Example 1   It is common to compare the change in sales of two companies by comparing the rates at which these sales change. If the sales of the two companies can be approximated by linear functions, we can use the work of the last section to find rates of change. For example, the chart below shows sales in two different years for two different companies.

Company	Sales in 1981	Sales in 1984
A	$10,000	$16,000
B	5000	14,000

The sales of Company A increased from $10,000 to $16,000 over this 3-year period, for a total increase of $6000. Thus, the average rate of change of sales is

$$\frac{\$6000}{3} = \$2000.$$

(a) If we assume that the sales of Company A have increased linearly (that is, that the sales can be closely approximated by a linear function), then we can find the equation describing the sales by finding the equation of the line through $(-3, 10,000)$ and $(0, 16,000)$, where $x = 0$ represents 1984 so that $x = -3$ represents 1981. The slope of the line is

$$\frac{16,000 - 10,000}{0 - (-3)} = 2000,$$

which is the same as the annual rate of change found above. Using the point-slope form of the equation of a line, we find that

$$y - 16,000 = 2000(x - 0)$$
$$y = 2000x + 16,000$$

gives the equation describing sales.

(b) Assume that the sales of Company B have also increased linearly. Find the equation giving its sales and the average rate of change of sales. The equation is

$$y = 3000x + 14,000.$$

The average rate of change of sales is $3000. ∎

As the example shows, the average rate of change is the same as the slope of the line. This is always true for data that can be modeled with a linear function.

Example 2   Suppose that a researcher has concluded that a dosage of x grams of a certain stimulant causes a rat to gain

$$y = 2x + 50$$

grams of weight, for appropriate values of x. If the researcher administers 30 grams of the stimulant, how much weight will the rat gain?

Let $x = 30$. The rat will gain

$$y = 2(30) + 50 = 110$$

grams of weight.

The average rate of change of weight gain with respect to the amount of stimulant is given by the slope of the line. The slope of $y = 2x + 50$ is 2, so that the difference in weight gain when the dose is varied by 1 gram is 2 grams. ▦

Cost Analysis In manufacturing, it is common for the cost of manufacturing an item to be made up of two parts: a **fixed cost** for designing the product, establishing a factory, training workers, and so on. Within broad limits, the fixed cost is constant for a particular product and does not change as more items are made. There is also a **variable cost** per item for labor, materials, packing, shipping, and so on. The variable cost may well be the same per item, with total variable cost increasing as the number of items increases.

Example 3 Suppose that the cost of producing clock-radios can be approximated by the linear model

$$C(x) = 12x + 100,$$

where $C(x)$ is the cost in dollars to produce x radios. The cost to produce 0 radios is

$$C(0) = 12(0) + 100 = 100,$$

or $100. This sum, $100, is the fixed cost.

Once the company has invested the fixed cost into the clock-radio project, what will then be the additional cost per radio? To find out, let's first find the cost of a total of 5 radios:

$$C(5) = 12(5) + 100 = 160,$$

or $160. The cost of 6 radios is

$$C(6) = 12(6) + 100 = 172,$$

or $172.

The sixth radio itself thus costs $172 - $160 = $12 to produce. In the same way, the 81st radio costs $C(81) - C(80) = \$1072 - \$1060 = \$12$ to produce. In fact, the $(n + 1)$st radio costs

$$C(n + 1) - C(n) = [12(n + 1) + 100] - [12n + 100] = 12$$

dollars to produce. Since each additional radio costs $12 to produce, $12 is the variable cost per radio. Note that 12 is also the slope of the cost function, $C(x) = 12x + 100$. ▦

In economics, the cost of producing an additional item is called the **marginal cost** of that item. In the clock-radio example, the marginal cost of each radio is $12.

> If a cost function is given by a linear function of the form $C(x) = mx + b$, then m represents the **variable cost** per item and b the **fixed cost.** Conversely, if the fixed cost of producing an item is b and the variable cost is m, then the **cost function,** $C(x)$, for producing x items, is given by $C(x) = mx + b$.

Example 4   In a certain city, a taxi company charges riders a fixed charge of $1.50 plus $1.80 per mile. Write a cost function, $C(x)$, which is a mathematical model for a ride of x miles.

Here the fixed cost is $b = 1.50$ dollars, with a variable cost of $m = 1.80$ dollars. The cost function, $C(x)$, is thus

$$C(x) = 1.80x + 1.50.$$

For example, a taxi ride of 4 miles will cost $C(4) = 1.80(4) + 1.50 = 870$, or $8.70. For each additional mile, the cost increases by $1.80.   ∎

Example 5   The variable cost for raising a certain type of frog for laboratory study is $12 per unit of frogs, while the cost to produce 100 units is $1500. Find the cost function, $C(x)$, if we know it is linear.

Since the cost function is linear, it can be expressed in the form $C(x) = mx + b$. We are told that the variable cost is $12 per unit, which gives the value for m in the model. The model can thus be written $C(x) = 12x + b$. To find b, we use the fact that the cost of producing 100 units of frogs is $1500, or $C(100) = 1500$. Substituting $x = 100$ and $C(x) = 1500$ into $C(x) = 12x + b$, we have

$$C(x) = 12x + b$$
$$1500 = 12 \cdot 100 + b$$
$$1500 = 1200 + b$$
$$300 = b.$$

The desired model is given by $C(x) = 12x + 300$. The fixed cost is $300.   ∎

Depreciation   Since machines and equipment wear out or become obsolete over a period of time, business firms must take into account the amount of value that the equipment has lost during each year of its useful life. This lost value, called **depreciation,** may be calculated in several ways. The simplest way is to use **straight-line,** or **linear,** depreciation, in which an item having a useful life of n years is assumed to lose $1/n$ of its value each year. Thus, a typewriter with a ten-year life would be assumed to lose $1/10$ of its value each year.

A machine may have some **scrap value** at the end of its useful life. For this reason, depreciation is calculated on **net cost**—the difference between purchase price and scrap value. To find the annual straight-line depreciation on an item having a net cost of I dollars and a useful life of n years, multiply the net cost by the fraction of the value lost each year, $1/n$. The annual straight-line depreciation is then

$$\frac{1}{n} \cdot I = \frac{I}{n}.$$

For example, in four years the depreciation will total

$$4\left(\frac{I}{n}\right) = \frac{4I}{n}.$$

The total undepreciated balance after four years will be given by the difference of the net cost and the depreciation so far, or

$$I - \frac{4I}{n} = I\left(1 - \frac{4}{n}\right).$$

To generalize this result, first note that the total depreciation at the end of year j, where j is a whole number between 1 and n inclusive, or $1 \le j \le n$, is

$$j \cdot \frac{I}{n} = \frac{jI}{n}$$

while the amount undepreciated after j years is

$$I - \frac{jI}{n} = I\left(1 - \frac{j}{n}\right).$$

Example 6   An asset has a purchase price of $100,000 and a scrap value of $40,000. The useful life of the asset is 10 years. Find each of the following for this asset.

(a) net cost
Since the net cost is the difference between purchase price and scrap value,

$$I = \text{net cost} = \$100,000 - \$40,000 = \$60,000.$$

(b) annual depreciation
The useful life of the asset is 10 years. Therefore, 1/10 of the net cost is depreciated each year. The annual depreciation by the straight-line method is

$$\frac{I}{n} = \frac{60,000}{10} = 6000,$$

or $6000.

(c) undepreciated balance after 4 years
Let $j = 4$ in the formula above.

$$I\left(1 - \frac{j}{n}\right) = 60,000\left(1 - \frac{4}{10}\right) = 60,000\left(\frac{6}{10}\right) = 36,000,$$

or $36,000. The total amount that will be depreciated over the life of the asset is $60,000. After 4 years, $36,000 of this amount has not yet been claimed as depreciation. ∎

Straight-line depreciation is the easiest method of depreciation to use, but it often does not accurately reflect the rate at which assets actually lose value. Some assets, such as new cars, lose more value annually at the beginning of their useful life than at the end. For this reason, two other methods of depreciation, the *sum-of-the-years'-digits* method, discussed in Exercises 33 and 34 below, and *double declining balance*, discussed in Section 1.5, are often used.

Break-Even Analysis   A company can make a profit only if the revenue it receives from its customers exceeds the cost of producing its goods and services. The point at which revenue just equals cost is called the **break-even point.**

Example 7   A firm producing chicken feed finds that the total cost, $C(x)$, of producing x units is given by

$$C(x) = 20x + 100.$$

The feed sells for $24 per unit, so that the revenue, $R(x)$, from selling x units is given by the product of the price per unit and the number of units sold, or

$$R(x) = 24x.$$

The firm will just break even (no profit and no loss), as long as revenue just equals cost, or $R(x) = C(x)$. This is true whenever

$R(x) = C(x)$

$24x = 20x + 100$ Substitute for $R(x)$ and $C(x)$

$4x = 100$ Add $-20x$ to both sides

$x = 25.$

Here the break-even point is at $x = 25$.

The graphs of $C(x) = 20x + 100$ and $R(x) = 24x$ are shown in Figure 14. The break-even point is shown on the graph. If the company produces more than 25 units (if $x > 25$), it makes a profit; if $x < 25$ it loses money. ∎

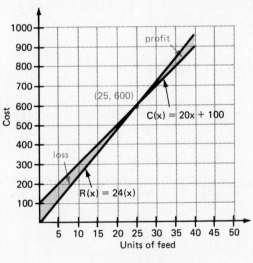

Figure 14

1.4 Exercises   **1.** Suppose the sales of a particular brand of electric guitar satisfy the relationship

$$S(x) = 300x + 2000,$$

where $S(x)$ represents the number of guitars sold in year x, with $x = 0$ corresponding to 1982. Find the sales in each of the following years.
(a) 1984   **(b)** 1985   **(c)** 1986   **(d)** 1982
(e) Find the annual rate of change of the sales.

2. If the population of ants in an anthill satisfies the relationship

$$A(x) = 1000x + 6000,$$

where $A(x)$ represents the number of ants present at the end of month x, find the number of ants present at the end of each of the following months. Let $x = 0$ represent June.
   (a) June   (b) July   (c) August   (d) December
   (e) What is the monthly rate of change of the number of ants?

3. Let $N(x) = -5x + 100$ represent the number of bacteria (in thousands) present in a certain tissue culture at time x, measured in hours, after an antibacterial spray is introduced into the environment. Find the number of bacteria present at each of the following times.
   (a) $x = 0$ (b) $x = 6$ (c) $x = 20$
   (d) What is the hourly rate of change in the number of bacteria? Interpret the negative sign in the answer.

4. Let $R(x) = -8x + 240$ represent the number of students present in a large business mathematics class, where x represents the number of hours of study required weekly. Find the number of students present at each of the following levels of required study.
   (a) $x = 0$ (b) $x = 5$ (c) $x = 10$
   (d) What is the rate of change of the number of students in the class with respect to the number of hours of study? Interpret the negative sign in the answer.
   (e) The professor in charge of the class likes to have exactly 16 students. How many hours of study must he require in order to have exactly 16 students?

5. Assume that the sales of a certain appliance dealer are approximated by a linear function. Suppose that sales were $850,000 in 1977 and $1,262,500 in 1982. Let $x = 0$ represent 1977.
   (a) Find the equation giving the dealer's yearly sales.
   (b) What were the dealer's sales in 1980?
   (c) Estimate sales in 1985.

6. Assume that the sales of a certain automobile parts company are approximated by a linear function. Suppose that sales were $200,000 in 1975, and $1,000,000 in 1982. Let $x = 0$ represent 1975 and $x = 7$ represent 1982.
   (a) Find the equation giving the company's yearly sales.
   (b) Find the sales in 1977.
   (c) Estimate the sales in 1984.

7. Suppose the number of bottles of a vitamin, $V(x)$, on hand at the beginning of the day in a health food store is given by

$$V(x) = 600 - 20x,$$

where $x = 1$ corresponds to June 1, and x is measured in days. If the store is open every day of the month, find the number of bottles on hand at the beginning of each of the following days.
   (a) June 6   (b) June 12   (c) June 24
   (d) When will the last bottle from this stock be sold?
   (e) What is the daily rate of change of this stock?

8. In psychology, the just-noticeable-difference (JND) for some stimulus is defined as the amount by which the stimulus must be increased so that a person will perceive it as having just barely been increased. For example, suppose a research study indicates

that a line 40 centimeters in length must be increased to 42 cm before a subject thinks that it is longer. In this case, the JND would be $42 - 40 = 2$ cm. In a particular experiment, the JND is given by

$$y = 0.03x,$$

where x represents the original length of the line and y the JND. Find the JND for lines having the following lengths.

(a) 10 cm   **(b)** 20 cm   **(c)** 50 cm   **(d)** 100 cm

(e) Find the rate of change in the JND with respect to the original length of the line.

Write a cost function for each of the following. Identify all variables used.

9. A chain saw rental firm charges $12 plus $1 per hour.

10. A trailer-hauling service charges $45 plus $2 per mile.

11. A parking garage charges 50¢ plus 35¢ per half-hour.

12. For a one-day rental, a car rental firm charges $44 plus 28¢ per mile.

Assume that each of the following can be expressed as a linear cost function. Find the appropriate cost function in each case.

13. Fixed cost, $100; 50 items cost $1600 to produce.

14. Fixed cost, $400; 10 items cost $650 to produce.

15. Fixed cost, $1000; 40 items cost $2000 to produce.

16. Fixed cost, $8500; 75 items cost $11,875 to produce.

17. Variable cost, $50; 80 items cost $4500 to produce.

18. Variable cost, $120; 100 items cost $15,800 to produce.

19. Variable cost, $90; 150 items cost $16,000 to produce.

20. Variable cost, $120; 700 items cost $96,500 to produce.

21. The manager of a local restaurant told us that his cost function for producing coffee is $C(x) = .097x$, where $C(x)$ is the total cost in dollars of producing x cups. (He is ignoring the cost of the coffee pot and the cost of labor.) Find the total cost of producing the following numbers of cups.

(a) 1000 cups   **(b)** 1001 cups

(c) Find the marginal cost of the 1001st cup.

(d) What is the marginal cost for *any* cup?

22. In deciding whether or not to set up a new manufacturing plant, company analysts have decided that a reasonable function for the total cost to produce x items is

$$C(x) = 500,000 + 4.75x.$$

(a) Find the total cost to produce 100,000 items.

(b) Find the marginal cost of the items to be produced in this plant.

Let $C(x)$ be the total cost to manufacture x items. Then the quotient $(C(x))/x$ is the average cost per item. Use this definition in Exercises 23 and 24.

23. $C(x) = 800 + 20x$; find the average cost per item if x is

(a) 10;   **(b)** 50;   **(c)** 200.

24. $C(x) = 500,000 + 4.75x$; find the average cost per item if x is

(a) 1000;   **(b)** 5000;   **(c)** 10,000.

For each of the assets in Exercises 25–30 find the straight-line depreciation in year four, and the amount undepreciated after 4 years.

25. Cost: $50,000; scrap value: $10,000; life: 20 years

26. Cost: $120,000; scrap value: $0; life: 10 years

27. Cost: $80,000; scrap value: $20,000; life: 30 years

28. Cost: $720,000; scrap value: $240,000; life: 12 years

29. Cost: $1,400,000; scrap value: $200,000; life: 8 years

30. Cost: $2,200,000; scrap value: $400,000; life: 12 years

31. Suppose an asset has a net cost of $80,000 and a four-year life.
    (a) Find the straight-line depreciation in each of years 1, 2, 3, and 4 of the item's life.
    (b) Find the sum of all depreciation for the four-year life.

32. A forklift truck has a net cost of $12,000, with a useful life of 5 years.
    (a) Find the straight-line depreciation in each of years 1, 2, 3, 4, and 5 of the forklift's life.
    (b) Find the sum of all depreciation for the five-year life.

Some assets, such as new cars, lose more value annually at the beginning of their useful life than at the end. By one method of depreciation for such assets, called the **sum-of-the-years'-digits** *method, the depreciation in year j, which we call A_j, is given by*

$$A_j = \frac{n - j + 1}{n(n + 1)} \cdot 2I,$$

where n is the useful life of the item, $1 \leq j \leq n$, and I is the net cost of the item. (Another approach to this method of depreciation is shown in the exercises of Section 2.2.)

33. For a certain asset, $n = 4$ and $I = \$10,000$.
    (a) Use the sum-of-the-years'-digits method to find the depreciation for each of the four years covering the useful life of the asset.
    (b) What would be the annual depreciation by the straight-line method?

34. A machine tool costs $105,000 and has a scrap value of $25,000, with a useful life of 4 years.
    (a) Use the sum-of-the-years'-digits method to find the depreciation in each of the four years of the machine tool's life.
    (b) Find the total depreciation by this method.

For each of the following assets, use the sum-of-the-years'-digits method to find the depreciation in year 1 and year 4.

35. Cost: $36,500; scrap value: $8500; life: 10 years

36. Cost: $6250; scrap value: $250; life: 5 years

37. Cost: $18,500; scrap value: $3900; life: 6 years

38. Cost: $275,000; scrap value: $25,000; life: 20 years

39. The cost to produce x units of wire is $C(x) = 50x + 5000$, while the revenue is $R(x) = 60x$. Find the break-even point and the revenue at the break-even point.

40. The cost to produce x units of squash is $C(x) = 100x + 6000$, while the revenue is $R(x) = 500x$. Find the break-even point.

You are the manager of a firm. You are considering the manufacture of a new product, so you ask the accounting department to produce cost estimates and the sales department to produce sales estimates. After you receive the data, you must decide whether or not to go ahead with production of the new product. Analyze the following data (find a break-even point) and then decide what you would do.

41. $C(x) = 85x + 900$; $R(x) = 105x$; not more than 38 units can be sold.

42. $C(x) = 105x + 6000$; $R(x) = 250x$; not more than 400 units can be sold.

43. $C(x) = 70x + 500$; $R(x) = 60x$ (Hint: what does a negative break-even point mean?)

44. $C(x) = 1000x + 5000$; $R(x) = 900x$.

45. The sales of a certain furniture company in thousands of dollars are shown in the chart below.

x year	y sales
0	48
1	59
2	66
3	75
4	80
5	90

(a) Graph this data, plotting years on the x-axis and sales on the y-axis. (Note that the data points can be closely approximated by a straight line.)

(b) Draw a line through the points (2, 66) and (5, 90). The other four points should be close to this line. (These two points were selected as "best" representing the line that could be drawn through the data points.)

(c) Use the two points of 45(b) to find an equation for the line that approximates the data.

(d) Complete the following chart.

Year	Sales (actual)	Sales (predicted from equation of 45(c))	Difference, actual minus predicted
0			
1			
2			
3			
4			
5			

(We will obtain a formula for the "best" line through the points in Section 8.6.)

(e) Use the result of (c) to predict sales in year 7.

(f) Do the same for year 9.

46. Most people are not very good at estimating the passage of time. Some people's estimations are too fast, and others, too slow. One psychologist has constructed a mathematical model for actual time as a function of estimated time: if y represents actual time and x estimated time, then

$$y = mx + b,$$

where m and b are constants that must be determined experimentally for each person.

Suppose that for a particular person, $m = 1.25$ and $b = -5$. Find y if x is
(a) 30 minutes;   (b) 60 minutes;
(c) 120 minutes;   (d) 180 minutes.

Suppose that for another person, $m = .85$ and $b = 1.2$. Find y if x is
(e) 15 minutes;   (f) 30 minutes;
(g) 60 minutes;   (h) 120 minutes.

For this same person, find x if y is
(i) 60 minutes;   (j) 90 minutes.

> *Application*   *Estimating Seed Demands — The Upjohn Company*

The Upjohn Company has a subsidiary which buys seeds from farmers and then resells them. Each spring the firm contracts with farmers to grow the seeds. The firm must decide on the number of acres that it will contract for. The problem faced by the company is that the demand for seeds is not constant, but fluctuates from year to year. Also, the number of tons of seed produced per acre varies, depending on weather and other factors. In an attempt to decide the number of acres that should be planted in order to maximize profits, a company mathematician created a model of the variables involved in determining the number of acres to plant.[1]

The analysis of this model required advanced methods that we will not go into. We can, however, give the conclusion; the number of acres that will maximize profit in the long run is found by solving the equation

$$F(AX + Q) = \frac{(S - C_p)X - C_A}{(S - C_p + C_c)X} \tag{1}$$

for A. The function $F(z)$ represents the chances that z tons of seed will be demanded by the marketplace. The variables in the equation are

$A =$ number of acres of land contracted by the company,
$X =$ quantity of seed produced per acre of land,
$Q =$ quantity of seed in inventory from previous years,
$S =$ selling price per ton of seed,
$C_p =$ variable cost (production, marketing, etc.) per ton of seed,
$C_c =$ cost to carry over one ton of seed from previous year,
$C_A =$ variable cost per acre of land.

To advise management of the number of acres of seed to contract for, the mathematician studied past records to find the values of the various variables. From these records and from predictions of future trends, it was concluded that $S = \$10,000$ per ton, $X = .1$ ton per acre (on the average), $Q = 200$ tons, $C_p = \$5000$ per ton, $C_A = \$100$ per acre, $C_c = \$3000$ per ton.

---

[1]Based on work by David P. Rutten, Senior Mathematician, The Upjohn Company, Kalamazoo, Michigan.

The function $F(z)$ is found by the same process to be approximated by

$$F(z) = \frac{z}{1000} - \frac{1}{2}, \text{ if } 500 \leq z \leq 1500 \text{ tons.} \tag{2}$$

Exercises   **1.** Evaluate $AX + Q$ using the values of X and Q given above.

2. Find $F(AX + Q)$, using equation (2) and your results from Exercise 1.

3. Solve equation (1) for A.

4. How many acres should be planted?

5. How many tons of seed will be produced?

6. Find the total revenue that will be received from the sale of the seeds.

Application   *Marginal Cost—Booz, Allen, And Hamilton*

Booz, Allen, and Hamilton is a large management consulting firm.[1] One of the services they provide to client companies is profitability studies, in which they show ways in which the client can increase profit levels. The client company requesting the analysis presented in this case is a large producer of a staple food. The company buys from farmers, and then processes the food in its mills, resulting in a finished product. The company sells both at retail under its own brands, and in bulk to other companies who use the product in the manufacture of convenience foods.

The client company has been reasonably profitable in recent years, but the management retained Booz, Allen, and Hamilton to see whether its consultants could suggest ways of increasing company profits. The management of the company had long operated with the philosophy of trying to process and sell as much of its product as possible, since, they felt, this would lower the average processing cost per unit sold. However, the consultants found that the client's fixed mill costs were quite low, and that, in fact, processing extra units made the cost per unit start to increase. (There are several reasons for this: the company must run three shifts, machines break down more often, and so on.)

In this case, we shall discuss the marginal cost of two of the company's products. The marginal cost (cost of producing an extra unit) of production for product A was found by the consultants to be approximated by the linear function

$$y = .133x + 10.09,$$

where x is the number of units produced (in millions) and y is the marginal cost.

For example, at a level of production of 3.1 million units, an additional unit of product A would cost about

$$y = .133(3.1) + 10.09 \approx \$10.50.[2]$$

---

[1] This case was supplied by John R. Dowdle of the Chicago office of Booz, Allen, and Hamilton.

[2] The symbol $\approx$ means *approximately equal to*.

At a level of production of 5.7 million units, an extra unit costs $10.85. Figure 1 shows a graph of the marginal cost function from $x = 3.1$ to $x = 5.7$, the domain over which the function above was found to apply.

The selling price for product A is $10.73 per unit, so that, as shown on the graph that follows, the company was losing money on many units of the product that it sold. Since the selling price could not be raised if the company was to remain competitive, the consultants recommended that production of product A be cut.

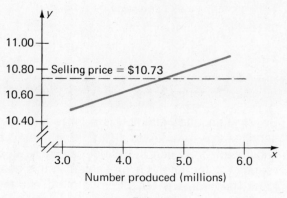

Figure 1

For product B, the Booz, Allen, and Hamilton consultants found a marginal cost function given by

$$y = .0667x + 10.29,$$

with x and y as defined above. Verify that at a production level of 3.1 million units, the marginal cost is about $10.50; while at a production level of 5.7 million units, the marginal cost is about $10.67. Since the selling price of this product is $9.65, the consultants again recommended a cutback in production.

The consultants ran similar cost analyses of other products made by the company, and then issued their recommendation to the company: The company should reduce total production by 2.1 million units. The analysts predicted that this would raise profits for the products under discussion from $8.3 million annually to $9.6 million—which is very close to what actually happened when the client took the advice.

Exercises   1. At what level of production, x, was the marginal cost of a unit of product A equal to the selling price?

2. Graph the marginal cost function for product B from $x = 3.1$ million units to $x = 5.7$ million units.

3. Find the number of units for which marginal cost equals the selling price for product B.

4. For product C, the marginal cost of production is

$$y = .133x + 9.46.$$

(a) Find the marginal cost at a level of production of 3.1 million units; of 5.7 million units.

(b) Graph the marginal cost function.

(c) For a selling price of $9.57, find the level of production for which the cost equals the selling price.

1.5   Constructing Mathematical Models (Optional)

In this section we actually construct two different mathematical models—one for double declining balance depreciation and one showing sales and prices for Volvo automobiles for the last few years.

As we develop these two models, notice the fundamental difference between them. To develop the model for depreciation, we start with the basic principles, apply some of the mathematics we have learned in this course, and come up with a model which gives the depreciation of an item *exactly*.

On the other hand, the model we develop for Volvo sales cannot go back to basic principles. (What are the basic principles for car sales? How do we write equations for them?) Rather, we construct a mathematical model for Volvo sales by gathering data on past sales and using it to predict future sales. Such methods can never give exact answers. The best we can expect is an approximation; if we are careful and lucky we can get a good approximation.

Double Declining Balance Depreciation   As we said in the last section, when a business buys an asset (such as a machine or building) it doesn't treat the total cost of the asset as an expense immediately. Instead, it *depreciates* the cost of the asset over the lifetime of the asset. For example, a machine costing $10,000 and having a useful life of 8 years, at which time it is worthless, might be depreciated at the rate of $10,000/8 = $1250 per year. As noted in the previous section, this method of depreciation is called straight-line depreciation, and assumes that the asset loses an equal amount of value during each year of its life.

This assumption of equal loss of value annually is not valid for many assets, such as new cars. A new car may lose 30% of its value during the first year. For assets that lose value quickly at first, and then less rapidly in later years, the Internal Revenue Service permits the use of alternate methods of depreciation. In Exercise Set 1.4, we discussed the sum-of-the-years'-digits method of depreciation, one alternate method.

Another common method is **double declining balance** depreciation. To find the depreciation in the first year, multiply the cost of the asset, C, by the fraction $2/n$, where n is the life of the asset in years. That is, the **depreciation in year 1** is

$$\frac{2}{n} \cdot C.$$

The number 2 in this formula is the origin of the word "double" in the name of this method.

Example 1   A forklift costs \$9600 and has a useful life of 8 years. Find the depreciation in year 1. Use the double declining balance method.

Using the formula above,

$$\text{depreciation in year } 1 = \frac{2}{n} \cdot C$$

$$= \frac{2}{8} \cdot 9600 \qquad \text{Let } n = 8 \text{ and } C = 9600$$

$$= 2400,$$

that is, \$2400.   ∎

The depreciation in later years of an asset's life can be found by multiplying the depreciation from the previous year by $1 - 2/n$. For example, an asset costing \$5000 with a life of 5 years would lead to a depreciation of $5000(2/5)$, or \$2000, during the first year of its life. To find the depreciation in year 2, multiply the depreciation in year 1 by $1 - 2/5$, as follows.

$$\text{depreciation in year } 2 = (\text{depreciation in year } 1) \times \left(1 - \frac{2}{n}\right)$$

$$= 2000\left(1 - \frac{2}{5}\right) \qquad \text{Let } n = 5$$

$$= 2000\left(\frac{3}{5}\right)$$

$$= 1200,$$

or \$1200. To find the depreciation in year 3, multiply this result by $1 - 2/5$, or $3/5$, again.

The depreciation by the double declining balance method in each of the first four years of the life of an asset is shown in the following table.

Year	1	2	3	4
Amount of depreciation	$\frac{2}{n} \cdot C$	$\frac{2}{n} \cdot C \cdot \left(1 - \frac{2}{n}\right)$	$\frac{2}{n} \cdot C \cdot \left(1 - \frac{2}{n}\right)^2$	$\frac{2}{n} \cdot C \cdot \left(1 - \frac{2}{n}\right)^3$

As the table suggests, each entry is found by multiplying the preceding entry by $1 - 2/n$. Based on this, the depreciation in year j, written a_j, is the amount

$$a_j = \frac{2}{n} \cdot C \cdot \left(1 - \frac{2}{n}\right)^{j-1}.$$

This result is a general formula for the entries in the table above. It is a mathematical model for double declining balance depreciation. If double declining balance depreciation were to be used for each year of the life of an asset, then the total depreciation would be less than the net cost of the asset. For this reason, it is permissible to switch to straight-line depreciation toward the end of the useful life of the asset.

Example 2 Oxford Typo, Inc., buys a new printing press for $39,000. The press has a life of 6 years. Find the depreciation in years 1, 2, and 3. Use the double declining balance method.

Use the mathematical model given above for a_j; replace j in turn by 1, 2, and 3.

$$a_j = \frac{2}{n} \cdot C \cdot \left(1 - \frac{2}{n}\right)^{j-1}$$

$$a_1 = \frac{2}{6} \cdot (39,000) \cdot \left(1 - \frac{2}{6}\right)^{1-1} \qquad \text{Let } j = 1, n = 6, C = 39,000$$

$$= 13,000\left(1 - \frac{2}{6}\right)^0 \qquad \frac{2}{6} \cdot 39,000 = 13,000$$

$$= 13,000(1) \qquad \left(1 - \frac{2}{6}\right)^0 = 1$$

$$a_1 = 13,000$$

Find a_2 and a_3 in the same way.

$$a_2 = \frac{2}{6} \cdot (39,000) \cdot \left(1 - \frac{2}{6}\right)^{2-1} \qquad a_3 = \frac{2}{6} \cdot (39,000) \cdot \left(1 - \frac{2}{6}\right)^{3-1}$$

$$= 13,000 \cdot \left(1 - \frac{1}{3}\right)^{2-1} \qquad = 13,000 \cdot \left(1 - \frac{1}{3}\right)^2$$

$$= 13,000 \cdot \left(\frac{2}{3}\right)^1 \qquad = 13,000 \cdot \left(\frac{2}{3}\right)^2$$

$$= 13,000 \cdot \left(\frac{2}{3}\right) \qquad = 13,000 \cdot \left(\frac{4}{9}\right)$$

$$a_2 \approx 8667 \qquad a_3 \approx 5778 \qquad \blacksquare$$

Note that we have used the symbol "$\approx$" in the last line of the example above. This symbol means "*approximately* equal to", and will be used throughout the book.

Supply and Demand   In the past few years, Volvo automobile sales have declined in the United States. Much of the cause of this decline might be attributed to price increases—in the last five years the cost of a basic Volvo sedan has increased by 78%. In the rest of this example, let us set up a mathematical model showing the relationship between sales and price. To begin, we gather data on past sales and prices. Typical price and sales data are shown in the following chart.[1] (The numbers have been rounded for simplicity.)

Total Volvo sales	Price of basic sedan
54,000	$4100
61,000	$4350
53,000	$5250
60,000	$5900
44,000	$7100

[1]Most information in this chart was supplied by Nancy Fiesler of Volvo of America.

Clearly, we cannot use this data to obtain a precise and exact mathematical model, such as the one for depreciation. There are too many variables here; the state of the economy, gasoline prices, and other factors all affect Volvo sales. For example, the chart shows that when the price increased from $5250 to $5900, sales *increased* from 53,000 to 60,000. However, a careful model can at least approximate the degree to which sales have fallen as the price has increased. A graph of the data in the chart is shown in Figure 15.

None of the curves that we have studied will fit exactly through the data points of the graph. However, a straight line provides a reasonable fit to the points. In Chapter 8 we discuss methods for finding the best possible line through these data points. For now, we can just choose two typical points and draw a line through them. Let's choose (44,000, 7100) and (61,000, 4350). A line through these points is shown in Figure 16.

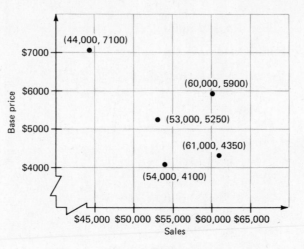

Figure 15

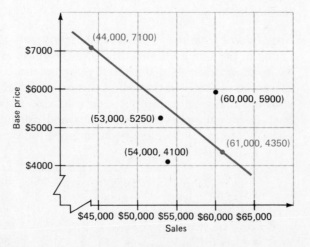

Figure 16

We can use the two points (44,000, 7100) and (61,000, 4350), along with the point-slope form of the equation of a line to find the equation of the line of Figure 16. First, find the slope. (To emphasize the fact that price is the dependent variable here, we will use p instead of y in our calculations.)

$$m = \frac{\text{change in } p}{\text{change in } x} = \frac{4350 - 7100}{61,000 - 44,000} = \frac{-2750}{17,000} \approx -.16$$

We can use $m = -.16$ and the point (44,000, 7100) for (x_1, p_1) to find the equation of the line of Figure 16.

$$\begin{aligned}
p - p_1 &= m(x - x_1) &&\text{Point-slope form} \\
p - 7100 &= -.16(x - 44,000) \\
p - 7100 &= -.16x + 7040 &&-.16(-44,000) = 7040 \\
p &= -.16x + 14,140
\end{aligned}$$

This is the mathematical model of the relationship between price, p, and demand, x.

Example 3   Suppose the factory dictates a price of $8000. Estimate the number of cars that will be sold.

Let $p = 8000$ in the model above.

$$\begin{aligned}
p &= -.16x + 14,140 \\
8000 &= -.16x + 14,140 &&\text{Let } p = 8000 \\
-6140 &= -.16x &&\text{Add } -14,140 \text{ to both sides} \\
x &= 38,375 &&\text{Divide both sides by } -.16
\end{aligned}$$

At a price of $8000, about 38,000 cars will be sold.   ∎

Example 4   Suppose the manager would like to know the price to charge so that 50,000 cars can be sold. What advice should be given?

Use the model developed above, with $x = 50,000$.

$$\begin{aligned}
p &= -.16x + 14,140 \\
p &= -.16(50,000) + 14,140 &&\text{Let } x = 50,000 \\
p &= -8000 + 14,140 \\
p &= 6140
\end{aligned}$$

To sell 50,000 cars, set the price at $6140 each.   ∎

1.5 Exercises

1. An apartment house costs $180,000. The owners decide to depreciate it over 6 years (it isn't built very well). Use the double declining balance method to find the depreciation in each of the following years. (See Examples 1 and 2.)
   (a) year 1   (b) year 2   (c) year 3   (d) year 4

2. A new airplane costs $600,000 and has a useful life of 10 years. Use the double declining balance method of depreciation to find the depreciation in each of the following years.
   (a) year 1   (b) year 2   (c) year 3   (d) year 4

Complete the following tables, which compare the three methods of depreciation. [*Recall: I is the net cost of an asset; $I = C -$ scrap value. Also, the formula for depreciation in year j, by the sum-of-the years'-digits method, is*

$$A_j = \frac{n - j + 1}{n(n + 1)}(2I).$$

3. Cost: $1400; life: 3 years; scrap value: $275

Depreciation in year	Straight-line	Double declining	Sum-of-years' digits
1			
2			
3			
Totals			

4. Cost: $55,000; life: 5 years; scrap value: $0

Depreciation in year	Straight-line	Double declining	Sum-of-years' digits
1			
2			
3			
4			
5			
Totals			

5. The following problems refer to the mathematical model developed in the text for Volvo sales. Find the price that would produce the following levels of sales.
(a) 40,000   (b) 48,000   (c) 54,000   (d) 60,000

Find the number of cars that would be sold at the following prices.
(e) $6000   (f) $7000   (g) $7400   (h) $8100

6. The following table shows the percent profit at a typical fast-food restaurant in a recent year.

Annual store sales in thousands, x	250	375	450	500	600	650
Percent pretax profit, y	9.3	14.8	14.2	15.9	19.2	21.0

(a) Draw a graph for this data that is similar to Figure 15.
(b) Pick two points that lead to a "typical" straight line through the data. Draw the line.
(c) Find the slope of your line.
(d) Find the equation of your line.
Use your line to estimate profit for sales of
(e) $300,000;   (f) $400,000;   (g) $475,000;   (h) $700,000.

When preparing a new catalog for publication, buyers at Montgomery Ward must estimate the total sales of each item in the catalog for the life of the catalog.[1] This is often done by using the actual sales of the same (or a similar) item from past catalogs. For example, the actual sales of six items in one line of goods for the 20-week life of one past catalog were as shown in the following table. (A "line of goods" is one group of items, such as men's sport shirts, plastic dinnerware, etc.)

Item	Beginning of week											
	2	3	4	5	6	7	8	9	10	11	12	13
1	20	39	80	172	251	384	558	850	1231	1697	2232	3066
2	6	34	97	182	318	480	680	998	1359	1857	2409	3108
3	4	20	50	80	122	186	273	420	584	785	1026	136,1
4	3	7	9	17	31	41	53	77	101	135	163	214
5	2	3	11	23	40	60	100	151	197	286	372	487
6	4	23	51	102	182	261	380	620	876	1223	1550	2181
Total	39	126	298	576	944	1412	2044	3116	4348	5983	7752	10,417

Item	Beginning of week							
	14	15	16	17	18	19	20	Final
1	3867	4331	4419	4468	4543	4569	4605	4646
2	3889	4209	4258	4282	4304	4317	4328	4341
3	1705	1892	1913	1927	1942	1952	1959	1961
4	267	295	307	316	324	328	329	331
5	589	660	675	686	691	694	699	700
6	2655	2861	2883	2896	2904	2911	2916	2924
Total	12,972	14,248	14,502	14,575	14,708	14,771	14,836	14,903

The numbers in this table represent *cumulative sales*. For example, 20 units of item 1 were sold by the beginning of week 2, with 39 units sold by the beginning of week 3, and 80 units sold by the beginning of week 4. A total of $80 - 39 = 41$ units were sold during week 3, for example.

After gathering the data of the table, company analysts calculate a sequence of numbers (called *terms*) as follows. Term $a_1 = 0$ and represents the fraction of total sales of all items at the beginning of week 1. To get a_2, find the total of all sales by the beginning of the second week. From the "Total" row of the table above, this number is 39. Divide this number by the total sales of all items of merchandise for the entire life of the catalog, 14,903 units. Convert this answer into a percent.

$$a_2 = 39/14{,}903 \approx .003 = .3\%$$

[1]Reprinted by permission of Philip Hartung, "A Simple Style Goods Inventory Model," *Management Science*, Vol. 19, No. 12, August 1973, copyright © 1973 The Institute of Management Sciences.

A total of .3% of all business should come before the second week begins. Also,

$$a_6 = 944/14,903 \approx .063 = 6.3\%.$$

Thus, 6.3% of all business should come before the sixth week begins.

Exercises   *Use the table above to find the number of units of item 3 sold during the following weeks:*

1. week 2   **2.** week 4   **3.** week 12   **4.** week 16

Find each of the following terms of the sequence discussed above.

5. a_4 **6.** a_8 **7.** a_{12} **8.** a_{16}

Key Words

mathematical model	supply curve
independent variable	demand curve
dependent variable	equilibrium price
domain	slope
range	change in x
ordered pair	fixed cost
Cartesian coordinate system	variable cost
x-axis	marginal cost
y-axis	depreciation
origin	straight-line depreciation
quadrant	scrap value
graph	net cost
linear equation	sum-of-the-years'-digits depreciation
x-intercept	double declining balance depreciation
y-intercept	

Chapter 1 Review Exercises

List the ordered pairs obtained from each of the following if the domain of x for each exercise is $\{-3, -2, -1, 0, 1, 2, 3\}$. *Graph each set of ordered pairs. Give the range.*

1. $2x - 5y = 10$ **2.** $3x + 7y = 21$

3. $y = (2x + 1)(x - 1)$ **4.** $y = (x + 4)(x + 3)$

5. $y = -2 + x^2$ **6.** $y = 3x^2 - 7$

7. $y = \dfrac{2}{x^2 + 1}$ **8.** $y = \dfrac{-3 + x}{x + 10}$

9. $y + 1 = 0$ **10.** $y = 3$

For each of the following functions, find (a) $f(6)$, *(b)* $f(-2)$, *(c)* $f(-4)$, *(d)* $f(r + 1)$.

11. $f(x) = 4x - 1$ **12.** $f(x) = 3 - 4x$

13. $f(x) = -x^2 + 2x - 4$ **14.** $f(x) = 8 - x - x^2$

15. Let $f(x) = 5x - 3$ and $g(x) = -x^2 + 4x$. Find each of the following.

    (a) $f(-2)$ **(b)** $g(3)$ **(c)** $g(-4)$ **(d)** $f(5)$

    (e) $g(-k)$ **(f)** $g(3m)$ **(g)** $g(k - 5)$ **(h)** $f(3 - p)$

    (i) $f[g(-1)]$ **(j)** $g[f(2)]$

16. Assume that it costs 30¢ to mail a letter weighing one ounce or less, with each additional ounce, or portion of an ounce, costing 27¢. Let $C(x)$ represent the cost to mail a letter weighing x ounces. Find the cost of mailing a letter of the following weights.

    (a) 3.4 ounces   **(b)** 1.02 ounces

    (c) 5.9 ounces   **(d)** 10 ounces

    (e) Graph C.   **(f)** Give the domain and range for C.

Graph each of the following.

17. $y = 4x + 3$ **18.** $y = 6 - 2x$

19. $3x - 5y = 15$ **20.** $2x + 7y = 14$

21. $x + 2 = 0$ **22.** $y = 1$

23. $y = 2x$ **24.** $x + 3y = 0$

25. The supply and demand for a certain commodity are related by

$$\text{supply: } p = 6x + 3; \qquad \text{demand: } p = 19 - 2x,$$

where p represents the price at a supply or demand, respectively, of x units. Find the supply and the demand when the price is

    (a) 10;   **(b)** 15;   **(c)** 18.

    (d) Graph both the supply and the demand functions on the same axes.

    (e) Find the equilibrium price.

    (f) Find the equilibrium supply; the equilibrium demand.

26. For a particular product, 72 units will be supplied at a price of 6, while 104 units will be supplied at a price of 10. Write a supply function for this product.

Find the slope for each of the following lines that have slope.

27. through $(-2, 5)$ and $(4, 7)$ **28.** through $(4, -1)$ and $(3, -3)$

29. through the origin and $(11, -2)$ **30.** through the origin and $(0, 7)$

31. $2x + 3y = 15$ **32.** $4x - y = 7$

33. $x + 4 = 9$ **34.** $3y - 1 = 14$

Find an equation for each of the following lines.

35. through $(5, -1)$, slope $2/3$ **36.** through $(8, 0)$, slope $-1/4$

37. through $(5, -2)$ and $(1, 3)$ **38.** through $(2, -3)$ and $(-3, 4)$

39. undefined slope, through $(-1, 4)$ **40.** slope 0, through $(-2, 5)$

Find each of the following linear cost functions.

41. eight units of paper cost $300; fixed cost is $60

42. fixed cost is $2000; 36 units cost $8480

43. twelve units cost $445; 50 units cost $1585

44. thirty units cost $1500; 120 units cost $5640

45. The cost of producing x units of a product is $C(x)$, where

$$C(x) = 20x + 100.$$

The product sells for $40 per unit.
(a) Find the break-even point.
(b) What revenue will the company receive if it sells just that number of units?

46. An asset costs $21,000 and has a six-year life with no scrap value. Find the first year depreciation for this asset by each of the following methods.
(a) straight line   (b) sum-of-the-years'-digits
(c) double declining balance

47. Complete the following table for an asset costing $79,000, having a scrap value of $11,000 and a four-year life.

Depreciation in year	Straight-line	Double-declining	Sum-of-the-years'-digits
1			
2			
3			
4			

48. Complete a table similar to the one of Exercise 47 for an asset costing $430,000, having a five-year life, and a scrap value of $70,000.

2

Matrix Theory

The study of matrices has been of interest to mathematicians for some time. Recently, however, the use of matrices has assumed greater importance in the fields of management, natural science, and social science because it provides a natural way to organize data. The theory of matrices is another type of mathematical model which can be used to solve many of the problems that arise in these fields, from inventory control to models of a nation's economy.

2.1 Basic Matrix Operations

A **matrix** (plural: matrices) is a rectangular array of numbers. Each number is called an **element** or entry. A matrix is indicated by enclosing the array of numbers in brackets. The matrix is referred to by a capital letter. For example,

$$A = \begin{bmatrix} 2 & -4 & 1 \\ 7 & 8 & 6 \end{bmatrix}$$

is a matrix with two (horizontal) rows and three (vertical) columns.

Example 1 demonstrates the use of matrices to organize data.

Example 1   The EZ Life Company manufactures sofas and armchairs in three models, A, B, and C. The company has regional warehouses in New York, Chicago, and San Francisco. In its August shipment, the company sends 10 model A sofas, 12 model B sofas, 5 model C sofas, 15 model A chairs, 20 model B chairs, and 8 model C chairs to each warehouse.

To organize this data, we might first list it as follows.

sofas	10 model A	12 model B	5 model C
chairs	15 model A	20 model B	8 model C

Alternatively, we might tabulate the data in a chart.

| | | Model | | |
		A	B	C
Furniture	Sofa	10	12	5
	Chair	15	20	8

With the understanding that the numbers in each row refer to the furniture type (sofa, chair) and the numbers in each column refer to the model (A, B, C), the same information can be given by a matrix, as follows.

$$M = \begin{bmatrix} 10 & 12 & 5 \\ 15 & 20 & 8 \end{bmatrix} \quad \blacksquare$$

Matrices are classified by their **order** (or **dimension**), that is, by the number of rows and columns that they contain. For example, the matrix

$$\begin{bmatrix} 2 & 7 & 5 \\ 4 & 6 & 9 \end{bmatrix}$$

has two rows and three columns. This matrix is said to be of **order** 2×3 (read "2 by 3") or **dimension** 2×3. In general, a matrix with m rows and n columns is of **order** $m \times n$. The number of rows is always given first.

Example 2   **(a)** The matrix $\begin{bmatrix} 6 & 5 \\ 3 & 4 \\ 5 & -1 \end{bmatrix}$ is of order 3×2.

(b) $\begin{bmatrix} 5 & 8 & 9 \\ 0 & 5 & -3 \\ -4 & 0 & 5 \end{bmatrix}$ is of order 3×3.

(c) $[1 \quad 6 \quad 5 \quad -2 \quad 5]$ is of order 1×5.

(d) $\begin{bmatrix} 3 \\ -5 \\ 0 \\ 2 \end{bmatrix}$ is of order 4×1. $\blacksquare$

A matrix having the same number of rows as columns is called a **square matrix**. The matrix given in Example 2(b) above is a square matrix, as are

$$\begin{bmatrix} -5 & 6 \\ 8 & 3 \end{bmatrix} \quad \text{and} \quad \begin{bmatrix} 0 & 0 & 0 & 0 \\ -2 & 4 & 1 & 3 \\ 0 & 0 & 0 & 0 \\ -5 & -4 & 1 & 8 \end{bmatrix}.$$

A matrix containing only one row is called a **row matrix** or **row vector**. The matrix in Example 2(c) is a row matrix, as are

$$[5 \quad 8], \quad [6 \quad -9 \quad 2], \quad \text{and} \quad [-4 \quad 0 \quad 0 \quad 0].$$

A matrix of only one column, such as in Example 2(d), is a **column matrix** or **column vector**.

Two matrices are **equal** if they are of the same order and if each pair of corresponding elements is equal. Using this definition, the matrices

$$\begin{bmatrix} 2 & 1 \\ 3 & -5 \end{bmatrix} \quad \text{and} \quad \begin{bmatrix} 1 & 2 \\ -5 & 3 \end{bmatrix}$$

are not equal (even though they contain the same elements and are of the same order) since the corresponding elements differ.

Example 3 (a) From the definition of equality given above, the only way that the statement

$$\begin{bmatrix} 2 & 1 \\ p & q \end{bmatrix} = \begin{bmatrix} x & y \\ -1 & 0 \end{bmatrix}$$

can be true is if $2 = x$, $1 = y$, $p = -1$, and $q = 0$.

(b) The statement

$$\begin{bmatrix} x \\ y \end{bmatrix} = \begin{bmatrix} 1 \\ 4 \\ 0 \end{bmatrix}$$

can never be true, since the two matrices are of different order. (One is 2×1 and the other is 3×1.) ∎

Addition The matrix given in Example 1,

$$M = \begin{bmatrix} 10 & 12 & 5 \\ 15 & 20 & 8 \end{bmatrix}$$

shows the August shipment from the EZ Life plant to its New York warehouse. If matrix N below gives the September shipment to the same warehouse, what is the total shipment for each item of furniture for these two months?

$$N = \begin{bmatrix} 45 & 35 & 20 \\ 65 & 40 & 35 \end{bmatrix}$$

If 10 model A sofas were shipped in August and 45 in September, then altogether 55 model A sofas were shipped in the two months. Similarly, the other corresponding entries can be added, to get a new matrix, call it Q, which represents the total shipment for the two months.

$$Q = \begin{bmatrix} 55 & 47 & 25 \\ 80 & 60 & 43 \end{bmatrix}$$

It is convenient to refer to Q as the "sum" of M and N.

The way we added the two matrices above illustrates the following definition of addition of matrices. The **sum** of two $m \times n$ matrices X and Y is the $m \times n$ matrix $X + Y$ in which each element is the sum of the corresponding elements of X and Y. It is important to remember that *only matrices with the same dimensions can be added.*

Example 4   Find each sum when possible.

(a)

$$\begin{bmatrix} 5 & -6 \\ 8 & 9 \end{bmatrix} + \begin{bmatrix} -4 & 6 \\ 8 & -3 \end{bmatrix} = \begin{bmatrix} 5 + (-4) & -6 + 6 \\ 8 + 8 & 9 + (-3) \end{bmatrix}$$

$$= \begin{bmatrix} 1 & 0 \\ 16 & 6 \end{bmatrix}$$

(b) The matrices

$$A = \begin{bmatrix} 5 & 8 \\ 6 & 2 \end{bmatrix} \quad \text{and} \quad B = \begin{bmatrix} 3 & 9 & 1 \\ 4 & 2 & 5 \end{bmatrix}$$

are of different orders. Therefore, the sum $A + B$ does not exist.   ▪

Example 5   The September shipments from the EZ Life Company to the San Francisco and Chicago warehouses are given in matrices S and C below.

$$S = \begin{bmatrix} 30 & 32 & 28 \\ 43 & 47 & 30 \end{bmatrix} \quad C = \begin{bmatrix} 22 & 25 & 38 \\ 31 & 34 & 35 \end{bmatrix}$$

What was the total amount shipped to the three warehouses in September? (See matrix N above for New York.)

The total of the September shipments is represented by the sum of the three matrices N, S, and C.

$$N + S + C = \begin{bmatrix} 45 & 35 & 20 \\ 65 & 40 & 35 \end{bmatrix} + \begin{bmatrix} 30 & 32 & 28 \\ 43 & 47 & 30 \end{bmatrix} + \begin{bmatrix} 22 & 25 & 38 \\ 31 & 34 & 35 \end{bmatrix}$$

$$= \begin{bmatrix} 97 & 92 & 86 \\ 139 & 121 & 100 \end{bmatrix}$$

From the resulting matrix above, we see, for example, that the total number of model C sofas shipped to the three warehouses in September was 86.   ▪

The **additive inverse** (or **negative**) of a matrix X is the matrix $-X$ in which each element is the additive inverse of the corresponding element of X.
If

$$A = \begin{bmatrix} 1 & 2 & 3 \\ 0 & -1 & 5 \end{bmatrix} \quad \text{and} \quad B = \begin{bmatrix} -2 & 3 & 0 \\ 1 & -7 & 2 \end{bmatrix},$$

then by the definition of the additive inverse of a matrix,

$$-A = \begin{bmatrix} -1 & -2 & -3 \\ 0 & 1 & -5 \end{bmatrix} \quad \text{and} \quad -B = \begin{bmatrix} 2 & -3 & 0 \\ -1 & 7 & -2 \end{bmatrix}.$$

By the definition of matrix addition, for each matrix X, the sum $X + (-X)$ is a **zero matrix,** O, whose elements are all zeros. There is an $m \times n$ zero matrix for each pair of values of m and n. Zero matrices have the following **identity property:** If O is an $m \times n$ zero matrix, and A is any $m \times n$ matrix, then

$$A + O = O + A = A.$$

Subtraction   We can now define **subtraction** of matrices in a manner comparable to subtraction for real numbers. That is, for matrices X and Y, the difference of X and Y, or $X - Y$, is defined as

$$X - Y = X + (-Y).$$

Using A, B, and $-B$ as defined above,

$$A - B = A + (-B) = \begin{bmatrix} 1 & 2 & 3 \\ 0 & -1 & 5 \end{bmatrix} + \begin{bmatrix} 2 & -3 & 0 \\ -1 & 7 & -2 \end{bmatrix}$$

$$= \begin{bmatrix} 3 & -1 & 3 \\ -1 & 6 & 3 \end{bmatrix}.$$

This definition means that we can perform matrix subtraction by simply subtracting corresponding elements.

Example 6 **(a)** $[8 \quad 6 \quad -4] - [3 \quad 5 \quad -8] = [5 \quad 1 \quad 4]$

(b) The matrices

$$\begin{bmatrix} -2 & 5 \\ 0 & 1 \end{bmatrix} \quad \text{and} \quad \begin{bmatrix} 3 \\ 5 \end{bmatrix}$$

have different orders and cannot be subtracted. ∎

Example 7 During September the Chicago warehouse of the EZ Life Company shipped out the following numbers of each model.

$$K = \begin{bmatrix} 5 & 10 & 8 \\ 11 & 14 & 15 \end{bmatrix}$$

What was the Chicago warehouse inventory on October 1, taking into account only the number of items received and sent out during the month?

The number of each kind of item received during September is given by matrix C from Example 5; the number of each model sent out during September is given by matrix K. Thus the October 1 inventory will be represented by the matrix $C - K$ as shown below.

$$\begin{bmatrix} 22 & 25 & 38 \\ 31 & 34 & 35 \end{bmatrix} - \begin{bmatrix} 5 & 10 & 8 \\ 11 & 14 & 15 \end{bmatrix} = \begin{bmatrix} 17 & 15 & 30 \\ 20 & 20 & 20 \end{bmatrix} \quad ∎$$

2.1 Exercises *Mark each of the following statements as* true *or* false. *If false, tell why.*

1. $\begin{bmatrix} 1 & 3 \\ 5 & 7 \end{bmatrix} = \begin{bmatrix} 1 & 5 \\ 3 & 7 \end{bmatrix}$

2. $\begin{bmatrix} 1 \\ 2 \\ 3 \end{bmatrix} = [1 \quad 2 \quad 3]$

3. $\begin{bmatrix} x \\ y \end{bmatrix} = \begin{bmatrix} 3 \\ 5 \end{bmatrix}$ if $x = 3$ and $y = 5$.

4. $\begin{bmatrix} 3 & 5 & 2 & 8 \\ 1 & -1 & 4 & 0 \end{bmatrix}$ is a 4×2 matrix.

5. $\begin{bmatrix} 1 & 9 & -4 \\ 3 & 7 & 2 \\ -1 & 1 & 0 \end{bmatrix}$ is a square matrix.

6. $\begin{bmatrix} 2 & 4 & -1 \\ 3 & 7 & 5 \\ 0 & 0 & 0 \end{bmatrix} = \begin{bmatrix} 2 & 4 & -1 \\ 3 & 7 & 5 \end{bmatrix}$

Find the order of each of the following. Identify any square, column, or row matrices.

7. $\begin{bmatrix} -4 & 8 \\ 2 & 3 \end{bmatrix}$

8. $\begin{bmatrix} -9 & 6 & 2 \\ 4 & 1 & 8 \end{bmatrix}$

9. $\begin{bmatrix} -6 & 8 & 0 & 0 \\ 4 & 1 & 9 & 2 \\ 3 & -5 & 7 & 1 \end{bmatrix}$

10. $[8 \quad -2 \quad 4 \quad 6 \quad 3]$ **11.** $\begin{bmatrix} 2 \\ 4 \end{bmatrix}$ **12.** $[-9]$

Find the values of the variables in each of the following.

13. $\begin{bmatrix} 2 & 1 \\ 4 & 8 \end{bmatrix} = \begin{bmatrix} x & 1 \\ y & z \end{bmatrix}$ **14.** $\begin{bmatrix} -5 \\ y \end{bmatrix} = \begin{bmatrix} -5 \\ 8 \end{bmatrix}$

15. $\begin{bmatrix} x+6 & y+2 \\ 8 & 3 \end{bmatrix} = \begin{bmatrix} -9 & 7 \\ 8 & k \end{bmatrix}$ **16.** $\begin{bmatrix} 9 & 7 \\ r & 0 \end{bmatrix} = \begin{bmatrix} m-3 & n+5 \\ 8 & 0 \end{bmatrix}$

17. $\begin{bmatrix} -7+z & 4r & 8s \\ 6p & 2 & 5 \end{bmatrix} + \begin{bmatrix} -9 & 8r & 3 \\ 2 & 5 & 4 \end{bmatrix} = \begin{bmatrix} 2 & 36 & 27 \\ 20 & 7 & 12a \end{bmatrix}$

18. $\begin{bmatrix} a+2 & 3z+1 & 5m \\ 4k & 0 & 3 \end{bmatrix} + \begin{bmatrix} 3a & 2z & 5m \\ 2k & 5 & 6 \end{bmatrix} = \begin{bmatrix} 10 & -14 & 80 \\ 10 & 5 & 9 \end{bmatrix}$

Perform the indicated operations where possible.

19. $\begin{bmatrix} 1 & 2 & 5 & -1 \\ 3 & 0 & 2 & -4 \end{bmatrix} + \begin{bmatrix} 8 & 10 & -5 & 3 \\ -2 & -1 & 0 & 0 \end{bmatrix}$ **20.** $\begin{bmatrix} 1 & 5 \\ 2 & -3 \\ 3 & 7 \end{bmatrix} + \begin{bmatrix} 2 & 3 \\ 8 & 5 \\ -1 & 9 \end{bmatrix}$

21. $\begin{bmatrix} 1 & 5 & 7 \\ 2 & 2 & 3 \end{bmatrix} + \begin{bmatrix} 4 & 8 & -7 \\ 1 & -1 & 5 \end{bmatrix}$ **22.** $\begin{bmatrix} 2 & 4 \\ -8 & 1 \end{bmatrix} + \begin{bmatrix} 9 & -3 \\ 8 & 5 \end{bmatrix}$

23. $\begin{bmatrix} 1 & 3 & -2 \\ 4 & 7 & 1 \end{bmatrix} + \begin{bmatrix} 3 & 0 \\ 6 & 4 \\ -5 & 2 \end{bmatrix}$ **24.** $\begin{bmatrix} 1 & 3 & -2 \\ 4 & 7 & 1 \end{bmatrix} - \begin{bmatrix} 3 & 6 & -5 \\ 0 & 4 & 2 \end{bmatrix}$

25. $\begin{bmatrix} 2 & 8 & 12 & 0 \\ 7 & 4 & -1 & 5 \\ 1 & 2 & 0 & 10 \end{bmatrix} - \begin{bmatrix} 1 & 3 & 6 & 9 \\ 2 & -3 & -3 & 4 \\ 8 & 0 & -2 & 17 \end{bmatrix}$

26. $\begin{bmatrix} 2 & 1 \\ 5 & -3 \\ -7 & 2 \\ 9 & 0 \end{bmatrix} + \begin{bmatrix} 1 & -8 & 0 \\ 5 & 3 & 2 \\ -6 & 7 & -5 \\ 2 & -1 & 0 \end{bmatrix}$

27. $\begin{bmatrix} -4x+2y & -3x+y \\ 6x-3y & 2x-5y \end{bmatrix} + \begin{bmatrix} -8x+6y & 2x \\ 3y-5x & 6x+4y \end{bmatrix}$

28. $\begin{bmatrix} 4k-8y \\ 6z-3x \\ 2k+5a \\ -4m+2n \end{bmatrix} - \begin{bmatrix} 5k+6y \\ 2z+5x \\ 4k+6a \\ 4m-2n \end{bmatrix}$

Using matrices $O = \begin{bmatrix} 0 & 0 \\ 0 & 0 \end{bmatrix}$, $P = \begin{bmatrix} m & n \\ p & q \end{bmatrix}$, $T = \begin{bmatrix} r & s \\ t & u \end{bmatrix}$, *and* $X = \begin{bmatrix} x & y \\ z & w \end{bmatrix}$, *verify that the following statements are true.*

29. $X + T$ is a 2×2 matrix (Closure property)

30. $X + T = T + X$ (Commutative property of addition of matrices)

31. $X + (T + P) = (X + T) + P$ (Associative property of addition of matrices)

32. $X + (-X) = 0$ (Inverse property of addition of matrices)

33. $P + 0 = P$ (Identity property of addition of matrices)

34. Which of the above properties are valid for matrices that are not square?

35. When John inventoried his screw collection, he found that he had 7 flathead long screws, 9 flathead medium, 8 flathead short, 2 roundhead long, no roundhead medium, and 6 roundhead short. Write this information first as a 3 × 2 matrix and then as a 2 × 3 matrix.

36. At the grocery store, Miguel bought 4 quarts of milk, 2 loaves of bread, 4 chickens, and an apple. Mary bought 2 quarts of milk, a loaf of bread, 5 chickens, and 4 apples. Write this information first as a 2 × 4 matrix and then as a 4 × 2 matrix.

37. A dietician prepares a diet specifying the amounts a patient should eat of four basic food groups: group I, meats; group II, fruits and vegetables; group III, breads and starches; group IV, milk products. Amounts are given in "exchanges" which represent 1 ounce (meat), 1/2 cup (fruits and vegetables), 1 slice (bread), 8 ounces (milk), or other suitable measurements.
    (a) The number of "exchanges" for breakfast for each of the four food groups respectively are 2, 1, 2, and 1; for lunch, 3, 2, 2, and 1; and for dinner, 4, 3, 2, and 1. Write a 3 × 4 matrix using this information.
    (b) The amounts of fat, carbohydrates, and protein (in appropriate units) in each food group respectively are as follows.

    Fat: 5, 0, 0, 10
    Carbohydrates: 0, 10, 15, 12
    Protein: 7, 1, 2, 8

    Use this information to write a 4 × 3 matrix.
    (c) There are 8 calories per exchange of fat, 4 calories per exchange of carbohydrates, and 5 calories per exchange of protein; summarize this data in a 3 × 1 matrix.

38. At the beginning of a laboratory experiment, five baby rats measured 5.6, 6.4, 6.9, 7.6, and 6.1 centimeters in length, and weighed 144, 138, 149, 152, and 146 grams respectively.
    (a) Write a 2 × 5 matrix using this information.
    (b) At the end of two weeks, their lengths were 10.2, 11.4, 11.4, 12.7, and 10.8 centimeters, and they weighed 196, 196, 225, 250, and 230 grams. Write a 2 × 5 matrix with this information.
    (c) Use matrix subtraction and the matrices found in (a) and (b) to write a matrix which gives the amount of change in length and weight for each rat.
    (d) The following week the rats gained as shown in the matrix below.

$$\begin{array}{l} \text{Length} \\ \text{Weight} \end{array} \begin{bmatrix} 1.8 & 1.5 & 2.3 & 1.8 & 2.0 \\ 25 & 22 & 29 & 33 & 20 \end{bmatrix}$$

What were their lengths and weights at the end of this week?

2.2   Multiplication of Matrices

In work with matrices, a real number is called a **scalar.** The product of a scalar k and a matrix X is the matrix kX, each of whose elements is a number equal to k times the corresponding element of X. For example,

$$(-3)\begin{bmatrix} 2 & -5 \\ 1 & 7 \end{bmatrix} = \begin{bmatrix} -6 & 15 \\ -3 & -21 \end{bmatrix}.$$

Finding the product of two matrices is more involved. However, such multiplication is important in applications which are useful in solving practical problems. To understand the reasoning behind matrix multiplication, it may be helpful to consider another example concerning the EZ Life Company discussed in Section 2.1. Sofas and chairs of the same model are often sold as sets. Matrix W below shows the number of each model set in each warehouse.

$$
\begin{array}{c}
\text{New York} \\
\text{Chicago} \\
\text{San Francisco}
\end{array}
\begin{array}{ccc}
\text{A} & \text{B} & \text{C} \\
\left[\begin{array}{ccc}
10 & 7 & 3 \\
5 & 9 & 6 \\
4 & 8 & 2
\end{array}\right] & &
\end{array} = W
$$

Suppose the selling price of a model A set is $800, of a model B set $1000, and of a model C set $1200. To find the total value of the sets in the New York warehouse, we multiply as follows.

Type	Number of sets	Price of set	Total
A	10	$800	$8000
B	7	$1000	$7000
C	3	$1200	$3600
			$18,600

(Total for New York)

Thus, the total value of the three kinds of sets in New York is $18,600.

The work done in the table above is summarized by the equation

$$10(\$800) + 7(\$1000) + 3(\$1200) = \$18,600.$$

In the same way, the Chicago sets have a total value of

$$5(\$800) + 9(\$1000) + 6(\$1200) = \$20,200,$$

and in San Francisco, the total value of the sets is

$$4(\$800) + 8(\$1000) + 2(\$1200) = \$13,600.$$

We can write the selling prices as a column matrix, $P,$ and the total value in each location as a column matrix V.

$$
\begin{bmatrix} 800 \\ 1000 \\ 1200 \end{bmatrix} = P
\qquad
\begin{bmatrix} 18,600 \\ 20,200 \\ 13,600 \end{bmatrix} = V
$$

Look at the elements of W and P; multiplying the first, second, and third elements of the first row of W by the first, second, and third elements respectively of the column matrix P and then adding these products gives the first element in V. Doing the same thing with the second row of W gives the second element of V; the third row of W leads to the third element of V. Thus it is reasonable to write the product of matrices

$$W = \begin{bmatrix} 10 & 7 & 3 \\ 5 & 9 & 6 \\ 4 & 8 & 2 \end{bmatrix} \quad \text{and} \quad P = \begin{bmatrix} 800 \\ 1000 \\ 1200 \end{bmatrix}$$

as

$$WP = \begin{bmatrix} 10 & 7 & 3 \\ 5 & 9 & 6 \\ 4 & 8 & 2 \end{bmatrix} \begin{bmatrix} 800 \\ 1000 \\ 1200 \end{bmatrix} = \begin{bmatrix} 18{,}600 \\ 20{,}200 \\ 13{,}600 \end{bmatrix} = V.$$

Thus we found the product by multiplying the elements of the *rows* of the matrix on the left and the corresponding elements of the *column* of the matrix on the right, and then finding the sum of these separate products. Notice that the product of a 3 × 3 matrix and a 3 × 1 matrix is a 3 × 1 matrix.

The **product** AB of an $m \times n$ matrix A and an $n \times k$ matrix B is found as follows. Multiply each element of the first row of A by the corresponding element of the first column of B. The sum of these n products is the first row, first column element of AB.

Similarly, the sum of the products found by multiplying the elements of the first row of A times the corresponding elements of the second column of B gives the first row, second column element of AB, and so on.

To find the ith row, jth column element of AB, multiply each element in the ith row of A by the corresponding element in the jth column of B. The sum of these products will give the elements of row i, column j of AB.

Example 1 Find the product AB given

$$A = \begin{bmatrix} 2 & 3 & -1 \\ 4 & 2 & 2 \end{bmatrix} \quad \text{and} \quad B = \begin{bmatrix} 1 \\ 8 \\ 6 \end{bmatrix}.$$

Step 1 Multiply the elements of the first row of A and the corresponding elements of the column of B.

$$\begin{bmatrix} 2 & 3 & -1 \\ 4 & 2 & 2 \end{bmatrix} \begin{bmatrix} 1 \\ 8 \\ 6 \end{bmatrix} \qquad 2(1) + 3(8) + (-1)(6) = 20$$

Therefore, 20 is the first row entry of the product matrix AB.

Step 2 Multiply the elements of the second row of A with the corresponding elements of B.

$$\begin{bmatrix} 2 & 3 & -1 \\ 4 & 2 & 2 \end{bmatrix} \begin{bmatrix} 1 \\ 8 \\ 6 \end{bmatrix} \qquad 4(1) + 2(8) + 2(6) = 32$$

The second row entry of the product is 32.

Step 3 Write the product using the two entries we found above.

$$AB = \begin{bmatrix} 2 & 3 & -1 \\ 4 & 2 & 2 \end{bmatrix} \begin{bmatrix} 1 \\ 8 \\ 6 \end{bmatrix} = \begin{bmatrix} 20 \\ 32 \end{bmatrix} \qquad \blacksquare$$

Example 2   Find the product CD given

$$C = \begin{bmatrix} -3 & 4 & 2 \\ 5 & 0 & 4 \end{bmatrix} \quad \text{and} \quad D = \begin{bmatrix} -6 & 4 \\ 2 & 3 \\ 3 & -2 \end{bmatrix}$$

Step 1

$$\begin{bmatrix} -3 & 4 & 2 \\ 5 & 0 & 4 \end{bmatrix} \begin{bmatrix} -6 & 4 \\ 2 & 3 \\ 3 & -2 \end{bmatrix} \qquad (-3)(-6) + 4(2) + 2(3) = 32$$

Step 2

$$\begin{bmatrix} -3 & 4 & 2 \\ 5 & 0 & 4 \end{bmatrix} \begin{bmatrix} -6 & 4 \\ 2 & 3 \\ 3 & -2 \end{bmatrix} \qquad (-3)(4) + 4(3) + 2(-2) = -4$$

Step 3

$$\begin{bmatrix} -3 & 4 & 2 \\ 5 & 0 & 4 \end{bmatrix} \begin{bmatrix} -6 & 4 \\ 2 & 3 \\ 3 & -2 \end{bmatrix} \qquad 5(-6) + 0(2) + 4(3) = -18$$

Step 4

$$\begin{bmatrix} -3 & 4 & 2 \\ 5 & 0 & 4 \end{bmatrix} \begin{bmatrix} -6 & 4 \\ 2 & 3 \\ 3 & -2 \end{bmatrix} \qquad 5(4) + 0(3) + 4(-2) = 12$$

Step 5   The product is

$$CD = \begin{bmatrix} -3 & 4 & 2 \\ 5 & 0 & 4 \end{bmatrix} \begin{bmatrix} -6 & 4 \\ 2 & 3 \\ 3 & -2 \end{bmatrix} = \begin{bmatrix} 32 & -4 \\ -18 & 12 \end{bmatrix}$$

Here the product of a 2×3 matrix and a 3×2 matrix is a 2×2 matrix.

As the definition of matrix multiplication shows,

> the product AB of two matrices A and B can be found only if the number of columns of A is the same as the number of rows of B.

The final product will have as many rows as A and as many columns as B.

Example 3   Suppose matrix A is 2×2 and matrix B is 2×4. Can the product AB be calculated? What is the order of the product?

The following diagram helps decide the answers to these questions.

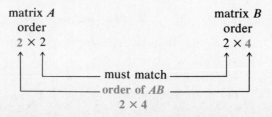

The product of A and B can be calculated because A has two columns and B has two rows. The order of the product is 2×4.   ■

Example 4   Find BA given

$$A = \begin{bmatrix} 1 & -3 \\ 7 & 2 \end{bmatrix} \quad \text{and} \quad B = \begin{bmatrix} 1 & 0 & -1 \\ 3 & 1 & 4 \end{bmatrix}.$$

Since B is a 2×3 matrix and A is a 2×2 matrix, the product BA cannot be found.   ■

Example 5   A contractor builds three kinds of houses, models A, B, and C, with a choice of two styles, Spanish or contemporary. Matrix P shows the number of each kind of house he is planning to build for a new 100-home subdivision. The amounts for each of the exterior materials he uses depend primarily on the style of the house. These amounts are shown in matrix Q. (Concrete is in cubic yards, lumber in units of 1000 board feet, brick in 1000's, and shingles in units of 100 square feet.) Matrix R gives the cost in dollars for each kind of material.

$$\begin{array}{c} \\ \text{Model A} \\ \text{Model B} \\ \text{Model C} \end{array} \begin{array}{cc} \text{Spanish} & \text{Contemporary} \\ \begin{bmatrix} 0 & 30 \\ 10 & 20 \\ 20 & 20 \end{bmatrix} \end{array} = P$$

$$\begin{array}{c} \\ \text{Spanish} \\ \text{Contemporary} \end{array} \begin{array}{cccc} \text{Concrete} & \text{Lumber} & \text{Brick} & \text{Shingles} \\ \begin{bmatrix} 10 & 2 & 0 & 2 \\ 50 & 1 & 20 & 2 \end{bmatrix} \end{array} = Q$$

$$\begin{array}{c} \\ \text{Concrete} \\ \text{Lumber} \\ \text{Brick} \\ \text{Shingles} \end{array} \begin{array}{c} \text{Cost per unit} \\ \begin{bmatrix} 20 \\ 180 \\ 60 \\ 25 \end{bmatrix} \end{array} = R$$

(a) What is the total cost for each model house?

To find the cost for each model, we must first find PQ, which will show the amount of each material needed for each model house.

$$PQ = \begin{bmatrix} 0 & 30 \\ 10 & 20 \\ 20 & 20 \end{bmatrix} \begin{bmatrix} 10 & 2 & 0 & 2 \\ 50 & 1 & 20 & 2 \end{bmatrix}$$

$$= \begin{array}{cccc} \text{Concrete} & \text{Lumber} & \text{Brick} & \text{Shingles} \\ \begin{bmatrix} 1500 & 30 & 600 & 60 \\ 1100 & 40 & 400 & 60 \\ 1200 & 60 & 400 & 80 \end{bmatrix} & \begin{array}{l} \text{Model A} \\ \text{Model B} \\ \text{Model C} \end{array} \end{array}$$

If we now multiply PQ times R, the cost matrix, we will get the total cost for each model house.

Cost

$$\begin{bmatrix} 1500 & 30 & 600 & 60 \\ 1100 & 40 & 400 & 60 \\ 1200 & 60 & 400 & 80 \end{bmatrix} \begin{bmatrix} 20 \\ 180 \\ 60 \\ 25 \end{bmatrix} = \begin{bmatrix} 72{,}900 \\ 54{,}700 \\ 60{,}800 \end{bmatrix} \begin{matrix} \text{Model A} \\ \text{Model B} \\ \text{Model C} \end{matrix}$$

(b) How much of each of the four kinds of material must he order?

    The totals of the columns of matrix PQ will give a matrix whose elements represent the total amounts of each material needed for the subdivision. Let us call this matrix T, and write it as a row matrix.

$$T = \begin{bmatrix} 3800 & 130 & 1400 & 200 \end{bmatrix}$$

(c) What is the total cost for material?

    To find the total cost of all the materials, we need the product of matrix T, the matrix showing the total amounts of each material, and matrix R, the cost matrix. (To multiply these and get a 1×1 matrix, representing total cost, we must multiply a 1×4 matrix by a 4×1 matrix. This is why T was written as a row matrix in (b) above.)

$$TR = \begin{bmatrix} 3800 & 130 & 1400 & 200 \end{bmatrix} \begin{bmatrix} 20 \\ 180 \\ 60 \\ 25 \end{bmatrix} = \begin{bmatrix} 188{,}400 \end{bmatrix}.$$

(d) Suppose the contractor builds the same number of homes in five subdivisions. Calculate the total amount of each material for each model for all five subdivisions.

    Multiply PQ by the scalar 5, as follows.

$$5 \begin{bmatrix} 1500 & 30 & 600 & 60 \\ 1100 & 40 & 400 & 60 \\ 1200 & 60 & 400 & 80 \end{bmatrix} = \begin{bmatrix} 7500 & 150 & 3000 & 300 \\ 5500 & 200 & 2000 & 300 \\ 6000 & 300 & 2000 & 400 \end{bmatrix} \quad \blacksquare$$

    We can introduce a notation to help us keep track of what quantities a matrix represents. For example, we can say that matrix P, from Example 5, represents models/styles, matrix Q represents styles/materials, and matrix R represents materials/cost. In each case we write the meaning of the rows first and the columns second. When we found the product PQ in Example 5, the rows of the matrix represented models and the columns represented materials. Therefore, we can say the matrix product PQ represents models/materials. Note that the common quantity, styles, in both P and Q was eliminated in the product PQ. Do you see that the product $(PQ)R$ represents models/cost?

    In practical problems this notation helps us decide in which order to multiply matrices so that the results are meaningful. In Example 5(c) we could have found either product RT or product TR. However, since T represents subdivisions/materials and R represents materials/cost, we multiplied T times R to get subdivisions/cost.

2.2 Exercises *In each of the following exercises, the dimensions of two matrices A and B are given. Find the dimensions of the product AB and the product BA, whenever these products exist.*

1. A is 2×2, B is 2×2 **2.** A is 3×3, B is 3×3

3. A is 4×2, B is 2×4 **4.** A is 3×1, B is 1×3

5. A is 3×5, B is 5×2 **6.** A is 4×3, B is 3×6

7. A is 4×2, B is 3×4 **8.** A is 7×3, B is 2×7

Let

$$A = \begin{bmatrix} -2 & 4 \\ 0 & 3 \end{bmatrix} \quad and \quad B = \begin{bmatrix} -6 & 2 \\ 4 & 0 \end{bmatrix}.$$

Find each of the following.

9. $2A$ **10.** $-3B$ **11.** $-4B$

12. $5A$ **13.** $-4A + 5B$ **14.** $3A - 10B$

Find each of the following matrix products where possible.

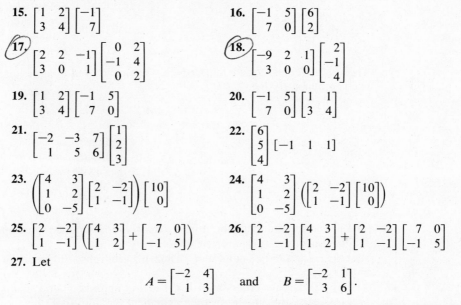

15. $\begin{bmatrix} 1 & 2 \\ 3 & 4 \end{bmatrix} \begin{bmatrix} -1 \\ 7 \end{bmatrix}$ **16.** $\begin{bmatrix} -1 & 5 \\ 7 & 0 \end{bmatrix} \begin{bmatrix} 6 \\ 2 \end{bmatrix}$

17. $\begin{bmatrix} 2 & 2 & -1 \\ 3 & 0 & 1 \end{bmatrix} \begin{bmatrix} 0 & 2 \\ -1 & 4 \\ 0 & 2 \end{bmatrix}$ **18.** $\begin{bmatrix} -9 & 2 & 1 \\ 3 & 0 & 0 \end{bmatrix} \begin{bmatrix} 2 \\ -1 \\ 4 \end{bmatrix}$

19. $\begin{bmatrix} 1 & 2 \\ 3 & 4 \end{bmatrix} \begin{bmatrix} -1 & 5 \\ 7 & 0 \end{bmatrix}$ **20.** $\begin{bmatrix} -1 & 5 \\ 7 & 0 \end{bmatrix} \begin{bmatrix} 1 & 1 \\ 3 & 4 \end{bmatrix}$

21. $\begin{bmatrix} -2 & -3 & 7 \\ 1 & 5 & 6 \end{bmatrix} \begin{bmatrix} 1 \\ 2 \\ 3 \end{bmatrix}$ **22.** $\begin{bmatrix} 6 \\ 5 \\ 4 \end{bmatrix} \begin{bmatrix} -1 & 1 & 1 \end{bmatrix}$

23. $\left(\begin{bmatrix} 4 & 3 \\ 1 & 2 \\ 0 & -5 \end{bmatrix} \begin{bmatrix} 2 & -2 \\ 1 & -1 \end{bmatrix} \right) \begin{bmatrix} 10 \\ 0 \end{bmatrix}$ **24.** $\begin{bmatrix} 4 & 3 \\ 1 & 2 \\ 0 & -5 \end{bmatrix} \left(\begin{bmatrix} 2 & -2 \\ 1 & -1 \end{bmatrix} \begin{bmatrix} 10 \\ 0 \end{bmatrix} \right)$

25. $\begin{bmatrix} 2 & -2 \\ 1 & -1 \end{bmatrix} \left(\begin{bmatrix} 4 & 3 \\ 1 & 2 \end{bmatrix} + \begin{bmatrix} 7 & 0 \\ -1 & 5 \end{bmatrix} \right)$ **26.** $\begin{bmatrix} 2 & -2 \\ 1 & -1 \end{bmatrix} \begin{bmatrix} 4 & 3 \\ 1 & 2 \end{bmatrix} + \begin{bmatrix} 2 & -2 \\ 1 & -1 \end{bmatrix} \begin{bmatrix} 7 & 0 \\ -1 & 5 \end{bmatrix}$

27. Let

$$A = \begin{bmatrix} -2 & 4 \\ 1 & 3 \end{bmatrix} \quad and \quad B = \begin{bmatrix} -2 & 1 \\ 3 & 6 \end{bmatrix}.$$

  (a) Find AB.   **(b)** Find BA.

  (c) Did you get the same answer in parts (a) and (b)? Do you think that matrix multiplication is commutative?

  (d) In general, for matrices A and B such that AB and BA both exist, does AB always equal BA?

Given matrices

$$P = \begin{bmatrix} m & n \\ p & q \end{bmatrix}, \quad X = \begin{bmatrix} x & y \\ z & w \end{bmatrix}, \quad T = \begin{bmatrix} r & s \\ t & u \end{bmatrix},$$

verify that the following statements are true.[1]

---

[1]The statements in Exercises 28–32 are valid for any matrices whenever matrix multiplication and addition can be carried out. This, of course, depends on the *order* of the matrices.

28. $(PX)T = P(XT)$ (Associative property: see Exercises 23 and 24.)

29. $P(X + T) = PX + PT$ (Distributive property: see Exercises 25 and 26.)

30. PX is a 2×2 matrix (Closure property)

31. $k(X + T) = kX + kT$ for any real number k

32. $(k + h)P = kP + hP$ for any real numbers k and h

33. Let I be the matrix $I = \begin{bmatrix} 1 & 0 \\ 0 & 1 \end{bmatrix}$, and let matrices P, X, and T be defined as above.

(a) Find IP, PI, IX.

(b) Without calculating, guess what the matrix IT might be.

(c) Suggest a reason for naming a matrix such as I an identity matrix.

34. The Bread Box, a small neighborhood bakery, sells four main items: sweet rolls, bread, cake, and pie. The amount of certain major ingredients required to make these items is given in matrix A.

$$A = \begin{array}{c} \\ \\ \\ \\ \\ \end{array} \begin{array}{ccccc} \text{Eggs} & \text{Flour[1]} & \text{Sugar[1]} & \text{Shortening[1]} & \text{Milk[1]} \\ \end{array}$$

	Eggs	Flour[1]	Sugar[1]	Shortening[1]	Milk[1]	
$A =$	1	4	$\frac{1}{4}$	$\frac{1}{4}$	1	Sweet rolls (dozen)
	0	3	0	$\frac{1}{4}$	0	Bread (loaves)
	4	3	2	1	1	Cake (1)
	0	1	0	$\frac{1}{3}$	0	Pie (1)

The cost (in cents) for each ingredient when purchased in large lots and in small lots is given by matrix B.

	Cost (in cents)		
	Large lot	Small lot	
$B =$	5	5	Eggs
	8	10	Flour[2]
	10	12	Sugar[2]
	12	15	Shortening[2]
	5	6	Milk[2]

(a) Use matrix multiplication to find a matrix representing the comparative costs per item under the two purchase options.

Suppose a day's orders consist of 20 dozen sweet rolls, 200 loaves of bread, 50 cakes, and 60 pies.

(b) Represent these orders as a 1×4 matrix and use matrix multiplication to write as a matrix the amount of each ingredient required to fill the day's orders.

(c) Use matrix multiplication to find a matrix representing the costs under the two purchase options to fill the day's orders.

35. In Exercise 37, Section 2.1, label the matrices found in parts (a), (b), and (c) respectively X, Y, and Z.

(a) Find the product matrix XY. What do the entries of this matrix represent?

(b) Find the product matrix YZ. What do the entries represent?

---

[1]Measured in cups.
[2]Cost per cup.

36. The EZ Life Company buys three machines costing $90,000, $60,000, and $120,000 respectively. They plan to depreciate the machines over a three-year period using the sum-of-the-years'-digits method, where the company assumes that the machines will have 20% of their value left at the end of three years. Thus they will compute the depreciation of 80% of the cost each year for the three-year period. (See Section 1.4, Exercises 33–38.)

(a) Write a column matrix C to represent the cost of these three assets.

(b) To get a matrix which represents 80% of the costs, let

$$P = \begin{bmatrix} .8 & 0 & 0 \\ 0 & .8 & 0 \\ 0 & 0 & .8 \end{bmatrix}.$$

Then PC is the desired matrix. Calculate PC.

(c) In the formula for the sum-of-the-years-digits method, the depreciation in year j is

$$A_j = \frac{n-j+1}{n(n+1)} \cdot 2I.$$

The coefficient of I (the net cost) in the formula is called the *rate*. Find the rate for each of the three years in the life of the machine. Use these three rates to write a row matrix R.

(d) Find $(PC)R$. This product matrix (which should be 3×3) represents the depreciation charge for each asset for each year of the three-year period.

Application *Routing*

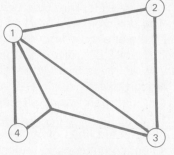

Figure 1

The diagram in Figure 1 shows the roads connecting four cities. Another way of representing this information is shown in matrix A, where the entries represent the number of roads connecting two cities without passing through another city.[1] For example, from the diagram we see that there are two roads connecting city 1

---

[1]Taken from Hugh G. Campbell, *Matrices With Applications*, © 1968, p. 50–51. Reprinted by permission of Prentice-Hall, Inc., Englewood Cliffs, N.J.

to city 4 without passing through either city 2 or 3. This information is entered in row 1, column 4 and again in row 4, column 1 of matrix A.

$$A = \begin{bmatrix} 0 & 1 & 2 & 2 \\ 1 & 0 & 1 & 0 \\ 2 & 1 & 0 & 1 \\ 2 & 0 & 1 & 0 \end{bmatrix}$$

Note that there are 0 roads connecting each city to itself. Also, there is one road connecting cities 3 and 2.

How many ways are there to go from city 1 to city 2, for example, by going through exactly one other city? Since we must go through one other city, we must go through either city 3 or city 4. On the diagram in Figure 1, we see that we can go from city 1 to city 2 through city 3 in 2 ways. We can go from city 1 to city 3 in 2 ways and then from city 3 to city 2 in one way, giving the $2 \cdot 1 = 2$ ways to get from city 1 to city 2 through city 3. It is not possible to go from city 1 to city 2 through city 4, because there is no direct route between cities 4 and 2.

Now multiply matrix A by itself, to get A^2. Let the first row, second column entry of A^2 be b_{12}. (We use a_{ij} to denote the entry in the i-th row and j-th column of matrix A.) The entry b_{12} is found as follows.

$$b_{12} = a_{11}a_{12} + a_{12}a_{22} + a_{13}a_{32} + a_{14}a_{42}$$
$$= 0 \cdot 1 + 1 \cdot 0 + 2 \cdot 1 + 2 \cdot 0$$
$$= 2.$$

The matrix A^2 gives the number of ways to travel between any two cities by passing through exactly one other city. The first product $0 \cdot 1$ in the calculations above represents the number of ways to go from city 1 to city 1 (0) and then from city 1 to city 2 (1). The 0 result indicates that such a trip does not involve a third city. The only non-zero product ($2 \cdot 1$) represents the two routes from city 1 to city 3 and the one route from city 3 to city 2 which result in the $2 \cdot 1$ or 2 routes from city 1 to city 2 by going through city 3.

Similarly, A^3 gives the number of ways to travel between any two cities by passing through exactly two cities. Also, $A + A^2$ represents the total number of ways to travel between two cities with at most one intermediate city.

The diagram can be given many other interpretations. For example, the lines could represent lines of mutual influence between people or nations; they could represent communication lines such as telephone lines.

Exercises   1. Use matrix A from the text to find A^2. Then answer the following questions.
   (a) How many ways are there to travel from city 1 to city 3 by passing through exactly one city?
   (b) How many ways are there to travel from city 2 to city 4 by passing through exactly one city?
   (c) How many ways are there to travel from city 1 to city 3 by passing through at most one city?
   (d) How many ways are there to travel from city 2 to city 4 by passing through at most one city?

2. Find A^3. Then answer the following questions.
   (a) How many ways are there to travel between cities 1 and 4 by passing through exactly two cities?
   (b) How many ways are there to travel between cities 1 and 4 by passing through at most two cities?

3. A small telephone system connects three cities. There are four lines between cities 3 and 2, three lines connecting city 3 with city 1 and two lines between cities 1 and 2.
   (a) Write a matrix B to represent this information.
   (b) Find B^2.
   (c) How many lines which connect cities 1 and 2 go through exactly one other city (city 3)?
   (d) How many lines which connect cities 1 and 2 go through at most one other city?

4. The figure shows four southern cities served by Delta Airlines.

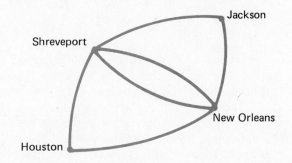

   (a) Write a matrix to represent the number of non-stop routes between cities.
   (b) Find the number of one-stop flights between Houston and Jackson.
   (c) Find the number of flights between Houston and Jackson which require at most one stop.
   (d) Find the number of one-stop flights between New Orleans and Houston.

5. The figure shows a food web. The arrows indicate the food sources of each population. For example, cats feed on rats and on mice.

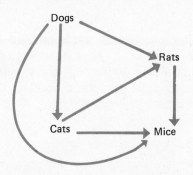

   (a) Write a matrix C in which each row and corresponding column represents a population in the food chain. Enter a one when the population in a given row feeds on the population in the given column.
   (b) Calculate and interpret C^2.

Suppose that three people have contracted a contagious disease.[1] A second group of five people may have been in contact with the three infected persons. A third group of six people may have been in contact with the second group. We can form a 3 × 5 matrix P with rows representing the first group of three and columns representing the second group of five. We enter a one in the corresponding position if a person in the first group has contact with a person in the second group. These direct contacts are called *first order contacts*. Similarly we form a 5 × 6 matrix Q representing the first order contacts between the second and third group. For example, suppose

$$P = \begin{bmatrix} 1 & 0 & 0 & 1 & 0 \\ 0 & 0 & 1 & 1 & 0 \\ 1 & 1 & 0 & 0 & 0 \end{bmatrix} \quad \text{and} \quad Q = \begin{bmatrix} 1 & 1 & 0 & 1 & 1 & 1 \\ 0 & 0 & 0 & 0 & 1 & 0 \\ 0 & 0 & 0 & 0 & 0 & 0 \\ 0 & 1 & 0 & 1 & 0 & 0 \\ 1 & 0 & 0 & 0 & 1 & 0 \end{bmatrix}.$$

From matrix P we see that the first person in the first group had contact with the first and fourth persons in the second group. Also, none of the first group had contact with the last person in the second group.

A *second order contact* is an indirect contact between persons in the first and third group through some person in the second group. The product matrix PQ indicates these contacts. Verify that the second row, fourth column entry of PQ is 1. That is, there is one second order contact between the second person in group 1 and the fourth person in group 3. Let a_{ij} denote the element in the i-th row and j-th column of the matrix PQ. By looking at the products which form a_{24} below, we see that the common contact was with the fourth individual in group 2. (The p_{ij} are entries in P, and the q_{ij} are entries in Q.)

$$a_{24} = p_{21}q_{14} + p_{22}q_{24} + p_{23}q_{34} + p_{24}q_{44} + p_{25}q_{54}$$
$$= 0 \cdot 1 + 0 \cdot 0 + 1 \cdot 0 + 1 \cdot 1 + 0 \cdot 1$$
$$= 1.$$

The second person in group 1 and the fourth person in group 3 both had contact with the fourth person in group 2.

This idea could be extended to third, fourth, and larger order contacts. It indicates a way to use matrices to trace the spread of a contagious disease. It could also pertain to the dispersal of ideas or anything that might pass from one individual to another.

Exercises

1. Find the second order contact matrix PQ mentioned in the text.

2. How many second-order contacts were there between the second contagious person and the third person in the third group?

[1] Reprinted with permission of Macmillan Publishing Co., Inc. from *Mathematics for the Biological Sciences* by Stanley L. Grossman and James E. Turner. Copyright © 1974, Macmillan Publishing Co., Inc.

3. Is there anyone in the third group who has had no contacts at all with the first group?

4. The totals of the columns in PQ give the total number of second-order contacts per person, while the column totals in P and Q give the total number of first-order contacts per person. Which person has the most contacts, counting both first- and second-order contacts?

2.3 Systems of Linear Equations

An animal feed is made from three ingredients: corn, soybeans, and cottonseed. One unit of each ingredient provides units of protein, fat, and fiber as shown in the table. How many units of each ingredient should be used to make a feed which contains 22 units of protein, 28 units of fat, and 18 units of fiber?

	Protein	Fat	Fiber
Corn	.25	.4	.3
Soybeans	.4	.2	.2
Cottonseed	.2	.3	.1

Let x represent the number of units of corn; y, the number of units of soybeans; and z, the number of units of cottonseed which are required. Since the total amount of protein is to be 22 units, we know that

$$.25x + .4y + .2z = 22.$$

Similarly, for the 28 units of fat,

$$.4x + .2y + .3z = 28,$$

and, for the 18 units of fiber,

$$.3x + .2y + .1z = 18.$$

To solve the problem, values of x, y, and z must be found that satisfy all three equations at the same time. The set of three equations is called a **system of equations.** Any values of x, y, and z that satisfy all three equations give a **solution** of the system. Verify that $x = 40$, $y = 15$, and $z = 30$ is a solution of the system, since these numbers satisfy all three equations. In fact, this is the only solution of this system. Many practical problems lead to systems of equations. In this and the next two sections, we consider methods for solving systems of first-degree equations.

A **first degree equation** in n unknowns is any equation of the form

$$a_1x_1 + a_2x_2 + \cdots + a_nx_n = k,$$

where $a_1, a_2, \cdots, a_n$ and k are all real numbers. Each of the three equations from the animal feed problem is a first degree equation in 3 unknowns. The **solution** of the first degree equation

$$a_1x_1 + a_2x_2 + \cdots + a_nx_n = k$$

is a sequence of numbers $s_1, s_2, \cdots, s_n$, such that

$$a_1 s_1 + a_2 s_2 + \cdots + a_n s_n = k.$$

The solution may be written between parentheses as $(s_1, s_2, \cdots, s_n)$. For example, $(1, 6, 2)$ is a solution of $3x_1 + 2x_2 - 4x_3 = 7$, since $3(1) + 2(6) - 4(2) = 7$.

As we saw in Section 1.2, a first-degree equation in two unknowns has a graph which is a straight line. For this reason, first-degree equations are also called **linear equations.** In this section we develop a method of solving a system of first degree equations. Although we confine our discussion to equations with only a few variables, the methods of solution can be extended to systems with many variables.

The graph of a linear equation in two variables is a straight line. Thus, there are three possibilities for the solution of a system of two linear equations in two variables.

1. The two graphs are lines intersecting at a single point. The coordinates of this point give the solution of the system. [See Figure 1(a).]

2. The graphs are distinct parallel lines. When this is the case, the system is said to be **inconsistent;** that is, there is no solution common to both equations. [See Figure 1(b).]

3. The graphs are the same line. In this case, the equations are said to be **dependent,** since any solution of one equation is also a solution of the other. There are an infinite number of solutions. [See Figure 1(c).]

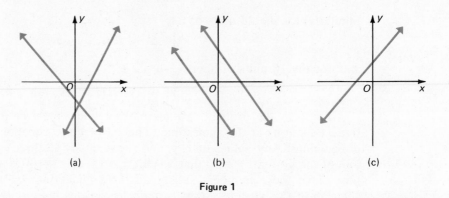

(a)   (b)   (c)

Figure 1

Transformations   To solve a linear system of equations, we use properties of algebra to change the system until a simpler equivalent system is found. An **equivalent system** is one which has the same solutions as the given system. There are three transformations that can be applied to a system to get an equivalent system.

> 1. **Exchanging any two equations.**
> 2. **Multiplying both sides of an equation by any nonzero real number.**
> 3. **Adding to any equation a multiple of some other equation.**

Use of these transformations leads to an equivalent system because each transformation can be reversed or "undone," allowing us to return to the original system.

Example 1 Solve the system

$$3x - 4y = 1 \tag{1}$$
$$2x + 3y = 12. \tag{2}$$

We want to get a system of equations which is equivalent to the given system, but simpler. If the equations are transformed so that the variable x disappears from one of them, y can be found. First, use the second transformation to multiply both sides of equation (1) by 2. This gives the equivalent system,

$$6x - 8y = 2 \tag{3}$$
$$2x + 3y = 12. \tag{2}$$

Now, using the third transformation, multiply both sides of equation (2) by -3, and add the result to equation (3).

$$(6x - 8y) + (-3)(2x + 3y) = 2 + (-3)(12)$$
$$6x - 8y - 6x - 9y = 2 - 36$$
$$-17y = -34$$
$$y = 2$$

We now have the equivalent system

$$3x - 4y = 1 \tag{1}$$
$$y = 2. \tag{4}$$

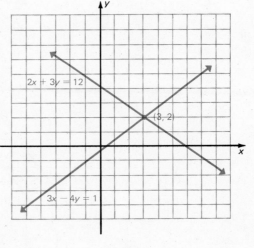

Figure 2

Substitute 2 for y in equation (1) to get

$$3x - 4(2) = 1$$
$$3x - \quad 8 = 1$$
$$x = 3.$$

The solution of the given system is (3, 2). The graphs of both equations of the system are shown in Figure 2. The graph verifies that (3, 2) satisfies both equations of the system. ∎

Since the method of solution in Example 1 results in the elimination of one variable from an equation of the system, it is called the **elimination method** for solving a system.

Example 2   Solve the system

$$3x - 2y = 4$$
$$-6x + 4y = 7.$$

We can eliminate x if we multiply the first equation by 2 and add the result to the second equation.

$$2(3x - 2y) + (-6x + 4y) = 2(4) + 7$$
$$6x - 4y - 6x + 4y = 8 + 7$$
$$0 = 15$$

The new system is

$$3x - 2y = 4$$
$$0 = 15.$$

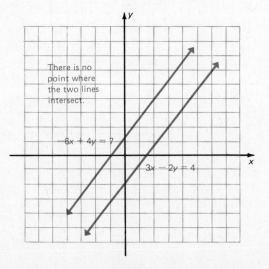

Figure 3

In the second equation, both variables are eliminated and we have a false statement. This is a signal that these two equations have no common solutions. Thus the system is inconsistent and has no solution. As we see in Figure 3, the graph of the system is two distinct parallel lines. ∎

Example 3 Solve the system

$$-4x + \; y = 2$$
$$8x - 2y = -4.$$

To eliminate x, multiply both sides of the first equation by 2 and add the result to the second equation.

$$2(-4x + y) + (8x - 2y) = 2(2) + (-4)$$
$$-8x + 2y + 8x - 2y = 4 - 4$$
$$0 = 0$$

This true statement tells us that the two equations have the same graph. When this happens, there is an infinite number of solutions for the system. In this case, all the ordered pairs that satisfy the equation $-4x + y = 2$ (or $8x - 2y = -4$) are solutions. See Figure 4. ∎

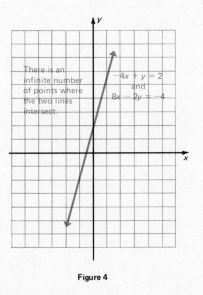

There is an infinite number of points where the two lines intersect.

$-4x + y = 2$
and
$8x - 2y = -4$

Figure 4

The Echelon Method We can extend the elimination method to solve systems with more than two equations. However, since the method then becomes more complicated, it helps to develop a systematic approach. Such an approach, together with matrix notation, makes possible computer solution of systems of equations. The next example illustrates the method.

Example 4   Solve the system

$$2x + y - z = 2 \tag{1}$$
$$x + 3y + 2z = 1 \tag{2}$$
$$x + y + z = 2. \tag{3}$$

We want to replace the given system with an equivalent system from which the solution is easily found. In the new system, the coefficient of x in the first equation should be 1. Use the first transformation to exchange equations (1) and (2). (We could have multiplied equation (1) by 1/2 instead.) The new system is

$$x + 3y + 2z = 1 \tag{2}$$
$$2x + y - z = 2 \tag{1}$$
$$x + y + z = 2. \tag{3}$$

Eliminate x in equation (1) by multiplying each term of equation (2) by -2 and adding the results to equation (1). This gives

$$-5y - 5z = 0. \tag{4}$$

Multiply both sides of equation (4) by $-1/5$ to make the coefficient of y equal to 1.

$$y + z = 0 \tag{5}$$

We now have the equivalent system

$$x + 3y + 2z = 1 \tag{2}$$
$$y + z = 0 \tag{5}$$
$$x + y + z = 2. \tag{3}$$

Eliminate x in equation (3) by multiplying both sides of equation (2) by -1 and adding the results to equation (3). The result is

$$-2y - z = 1. \tag{6}$$

Now multiply both sides of equation (5) by 2 and add to equation (6), to get

$$z = 1. \tag{7}$$

The original system has now led to the equivalent system

$$x + 3y + 2z = 1 \tag{2}$$
$$y + z = 0 \tag{5}$$
$$z = 1. \tag{7}$$

From equation (7) we know $z = 1$. Substitute 1 for z in equation (5) to get $y = -1$. Finally substitute 1 for z and -1 for y in equation (2) to get $x = 2$. Thus, the solution of the system is $(2, -1, 1)$.   ▮

This method is sometimes called the **echelon method.** In summary, to solve a linear system in n variables by the echelon method, perform the following steps using the three transformations.

> 1. Make the coefficient of the first variable equal to 1 in the first equation and 0 in the other equations.
> 2. Make the coefficient of the second variable equal to 1 in the second equation and 0 in all remaining equations.
> 3. Make the coefficient of the third variable equal to 1 in the third equation and 0 in all remaining equations.
> 4. Continue in this way until the last equation is of the form $x_n = k$, where k is a constant.

Parameters   The systems of equations discussed so far have had the same number of equations as variables. When this is the case, there is either *one* solution, *no* solution, or an *infinite number* of solutions. However, sometimes a model leads to a system of equations with fewer equations than variables. Such systems always have an infinite number of solutions or no solution.

Example 5   Solve

$$2x + 3y + 4z = 6 \qquad \text{(1)}$$
$$x - 2y + z = 9 \qquad \text{(2)}$$
$$3x - 6y + 3z = 27. \qquad \text{(3)}$$

The third equation here is a multiple of the second equation. Thus, this system really has only the two equations,

$$2x + 3y + 4z = 6 \qquad \text{(1)}$$
$$x - 2y + z = 9. \qquad \text{(2)}$$

Exchange the two equations so that the coefficient of x is 1 in the first equation of the system. This gives

$$x - 2y + z = 9 \qquad \text{(2)}$$
$$2x + 3y + 4z = 6. \qquad \text{(1)}$$

To eliminate x in equation (1), multiply both sides of equation (2) by -2 and add to equation (1). The result is

$$7y + 2z = -12.$$

Multiply both sides of this equation by $1/7$. The new system is

$$x - 2y + z = 9 \qquad \text{(2)}$$
$$y + \frac{2}{7}z = \frac{-12}{7}. \qquad \text{(3)}$$

Since we have only two equations, we can go no further. To complete the solution, solve equation (3) for y.

$$y = \frac{-2}{7}z - \frac{12}{7}$$

Now substitute the result for y in equation (2), and solve for x.

$$x - 2y + z = 9 \qquad\qquad (2)$$

$$x - 2\left(\frac{-2}{7}z - \frac{12}{7}\right) + z = 9$$

$$x + \frac{4}{7}z + \frac{24}{7} + z = 9$$

$$7x + 4z + 24 + 7z = 63$$

$$7x + 11z = 39$$

$$7x = -11z + 39$$

$$x = \frac{-11}{7}z + \frac{39}{7}$$

We now have both x and y in terms of z. Each choice of a value for z leads to values for x and y. (For this reason, z is called a *parameter*.) For example, if $z = 1$, then $x = 4$ and $y = -2$. Also,

$$\text{if } z = -6, \ x = 15 \text{ and } y = 0, \text{ and}$$

$$\text{if } z = 0, \ x = \frac{39}{7} \text{ and } y = \frac{-12}{7}.$$

There are an infinite number of solutions, since z can take an infinite number of values. This set of solutions is sometimes written as follows.

$$z \text{ arbitrary}$$

$$x = \frac{-11}{7}z + \frac{39}{7}$$

$$y = \frac{-2}{7}z - \frac{12}{7} \qquad \blacksquare$$

Example 5 discussed a system with one more variable than equations. If there are two more variables than equations, there will be two parameters, and so on. What about the case where there are fewer variables than equations? For example, consider the system

$$2x + 3y = 8$$

$$x - y = 4$$

$$5x + y = 7.$$

Since each equation has a line as its graph, the possibilities are three lines which intersect at a common point, three lines which cross at three different points, three lines of which two are the same line so that the intersection would be a point, three lines of which two are parallel so that the intersection would be two different points, three lines which are all parallel so that there would be no intersection, and so on. As in the case of n equations with n variables, the possibilities are a unique solution, no solution, or an infinite number of solutions. Verify that the point (1, 2) satisfies both the first and last equations but not the second equation of the system above, so that the system above has no solution.

Applications If the mathematical techniques you learn in this text are to be useful to you, you must be able to apply them to word problems. Always begin by reading the problem through carefully. Then note what you are to find. The unknown quantity (or quantities) should be represented by a variable. It is a good idea to *write down* exactly what quantity each variable is to represent. Then reread the problem, looking for all pertinent data. Write that down too. Finally, look for one or more sentences which can be translated into an equation (or an inequality). The next example illustrates the process.

Example 6 A bank teller has a total of 70 bills, of five-, ten- and twenty-dollar denominations. The number of fives is three times the number of tens, while the total value of the money is $960. Find the number of each type of bill.

Let x be the number of fives, y the number of tens, and z the number of twenties. Then, since the total number of bills is 70,

$$x + y + z = 70.$$

Also the number of fives, x, is three times the number of tens, y.

$$x = 3y$$

Finally the total value is $960. The value of the fives is $5x$, of the tens is $10y$, and of the twenties is $20z$. Thus,

$$5x + 10y + 20z = 960.$$

Rewriting the second equation as $x - 3y = 0$ gives the system

$$x + \quad y + \quad z = 70 \qquad \textbf{(1)}$$

$$x - \quad 3y \qquad = \ 0 \qquad \textbf{(2)}$$

$$5x + 10y + 20z = 960 \qquad \textbf{(3)}$$

To solve by the echelon method, first eliminate x from equation (2). To do this, multiply both sides of equation (1) by -1 and add to equation (2). Then multiply both sides of the result by $-1/4$. The new system is

$$x + \quad y + \quad z = 70 \qquad \textbf{(1)}$$

$$y + \frac{1}{4}z = \frac{35}{2} \qquad \textbf{(4)}$$

$$5x + 10y + 20z = 960. \qquad \textbf{(3)}$$

Use a similar procedure to eliminate x from equation (3).

$$x + y + \quad z = 70 \qquad \textbf{(1)}$$

$$y + \frac{1}{4}z = \frac{35}{2} \qquad \textbf{(4)}$$

$$5y + 15z = 610 \qquad \textbf{(5)}$$

To eliminate y from equation (5), multiply both sides of equation (4) by -5 and add to equation (5). The final equivalent system is

$$x + y + z = 70 \tag{1}$$

$$y + \frac{1}{4}z = \frac{35}{2} \tag{5}$$

$$\frac{55}{4}z = \frac{1045}{2} \tag{6}$$

From equation (6) we have $z = 38$. Substituting this result into equation (5) gives $y = 8$, and finally, from equation (1), $x = 24$. Thus, the teller had 24 fives, 8 tens, and 38 twenties. ∎

2.3 Exercises

Use the elimination method to solve each of the following systems of two equations in two unknowns. Check your answers.

1. $x + y = 9$
   $2x - y = 0$

2. $4x + y = 9$
   $3x - y = 5$

3. $5x + 3y = 7$
   $7x - 3y = -19$

4. $2x + 7y = -8$
   $-2x + 3y = -12$

5. $3x + 2y = -6$
   $5x - 2y = -10$

6. $-6x + 2y = 8$
   $5x - 2y = -8$

7. $2x - 3y = -7$
   $5x + 4y = 17$

8. $4m + 3n = -1$
   $2m + 5n = 3$

9. $5p + 7q = 6$
   $10p - 3q = 46$

10. $12s - 5t = 9$
    $3s - 8t = -18$

11. $6x + 7y = -2$
    $7x - 6y = 26$

12. $2a + 9b = 3$
    $5a + 7b = -8$

13. $3x + 2y = 5$
    $6x + 4y = 8$

14. $9x - 5y = 1$
    $-18x + 10y = 1$

15. $4x - y = 9$
    $-8x + 2y = -18$

16. $3x + 5y + 2 = 0$
    $9x + 15y + 6 = 0$

In Exercises 17–20, first multiply both sides of each equation by its common denominator to eliminate the fractions. Then use the elimination method to solve. Check your answers.

17. $\dfrac{x}{2} + \dfrac{y}{3} = 8$

    $\dfrac{2x}{3} + \dfrac{3y}{2} = 17$

18. $\dfrac{x}{5} + 3y = 31$

    $2x - \dfrac{y}{5} = 8$

19. $\dfrac{x}{2} + y = \dfrac{3}{2}$

    $\dfrac{x}{3} + y = \dfrac{1}{3}$

20. $x + \dfrac{y}{3} = -6$

    $\dfrac{x}{5} + \dfrac{y}{4} = -\dfrac{7}{4}$

Use the elimination method to solve each of the following systems of three equations in three unknowns. Check your answers.

21. $x + y + z = 2$
    $2x + y - z = 5$
    $x - y + z = -2$

22. $2x + y + z = 9$
    $-x - y + z = 1$
    $3x - y + z = 9$

23. $x + 3y + 4z = 14$
    $2x - 3y + 2z = 10$
    $3x - y + z = 9$

24. $4x - y + 3z = -2$
    $3x + 5y - z = 15$
    $-2x + y + 4z = 14$

25. $x + 2y + 3z = 8$
    $3x - y + 2z = 5$
    $-2x - 4y - 6z = 5$

26. $3x - 2y - 8z = 1$
    $9x - 6y - 24z = -2$
    $x - y + z = 1$

27. $2x - 4y + z = -4$
    $x + 2y - z = 0$
    $-x + y + z = 6$

28. $4x - 3y + z = 9$
    $3x + 2y - 2z = 4$
    $x - y + 3z = 5$

29. $x + 4y - z = 6$
    $2x - y + z = 3$
    $3x + 2y + 3z = 16$

30. $3x + y - z = 7$
    $2x - 3y + z = -7$
    $x - 4y + 3z = -6$

31. $5m + n - 3p = -6$
    $2m + 3n + p = 5$
    $-3m - 2n + 4p = 3$

32. $2r - 5s + 4t = -35$
    $5r + 3s - t = 1$
    $r + s + t = 1$

33. $a - 3b - 2c = -3$
    $3a + 2b - c = 12$
    $-a - b + 4c = 3$

34. $2x + 2y + 2z = 6$
    $3x - 3y - 4z = -1$
    $x + y + 3z = 11$

Solve each of the following systems of equations. Let x be the parameter.

35. $5x + 3y + 4z = 19$
    $3x - y + z = -4$

36. $3x + y - z = 0$
    $2x - y + 3z = -7$

37. $x + 2y + 3z = 11$
    $2x - y + z = 2$

38. $-x + y - z = -7$
    $2x + 3y + z = 7$

39. $x + y - z + 2w = -20$
    $2x - y + z + w = 11$
    $3x - 2y + z - 2w = 27$

40. $4x + 3y + z + 2w = 1$
    $-2x - y + 2z + 3w = 0$
    $x + 4y + z - w = 12$

Solve each of the following systems of equations.

41. $5x + 2y = 7$
    $-2x + y = -10$
    $x - 3y = 15$

42. $9x - 2y = -14$
    $3x + y = -4$
    $-6x - 2y = 8$

43. $x + 7y = 5$
    $4x - 3y = 2$
    $-x + 2y = 10$

44. $-3x - 2y = 11$
    $x + 2y = -14$
    $5x + y = -9$

45. $x + y = 2$
    $y + z = 4$
    $x + z = 3$
    $y - z = 8$

46. $2x + y = 7$
    $x + 3z = 5$
    $y - 2z = 6$
    $x + 4y = 10$

Write a system of equations for each of the following; then solve the system.

47. A working couple earned a total of $4352. The wife earned $64 per day; the husband earned $8 per day less. Find the number of days each worked if the total number of days worked by both was 72.

48. Midtown Manufacturing Company makes two products, plastic plates and plastic cups. Both require time on two machines: a batch of plates—one hour on machine A and two hours on machine B; a batch of cups—three hours on machine A and one hour on machine B. Both machines operate 15 hours a day. How many batches of each product can be produced in a day under these conditions?

49. A company produces two models of bicycles, model 201 and model 301. Model 201 requires 2 hours of assembly time and model 301 requires 3 hours of assembly time. The parts for model 201 cost $25 per bike and the parts for model 301 cost $30 per bike. If the company has a total of 34 hours of assembly time and $365 available per day for these two models, how many of each can be made in a day?

50. Juanita invests $10,000, received from her grandmother, in three ways. With one part, she buys mutual funds which offer a return of 8% per year. The second part, which amounts to twice the first, is used to buy government bonds at 9% per year. She puts the rest in the bank at 5% annual interest. The first year her investments bring a return of $830. How much did she invest in each way?

51. To get the necessary funds for a planned expansion, a small company took out three loans totaling $25,000. The company was able to borrow some of the money at 16%. They borrowed $2000 more than one-half the amount of the 16% loan at 20%, and the rest at 18%. The total annual interest was $4440. How much did they borrow at each rate?

52. The business analyst for Midtown Manufacturing wants to find an equation which can be used to project sales of a relatively new product. For the years 1980, 1981, and 1982 sales were $15,000, $32,000, and $123,000 respectively.
   (a) Graph the sales for the years 1980, 1981, and 1982 letting the year 1980 equal 0 on the x-axis. Let the values on the vertical axis be in thousands. [For example, the point (1981, 32,000) will be graphed as (1, 32).]
   (b) Find the equation of the straight line $ax + by = c$ through the points for 1980 and 1982.
   (c) Find the equation of the parabola $y = ax^2 + bx + c$ through the three given points.
   (d) Find the projected sales for 1985 first by using the equation from part (b) and second, by using the equation from part (c). If you were to estimate sales of the product in 1985 which result would you choose? Why?

53. For what value of k does the following system have a single solution? Find the solution.

$$\begin{aligned} 4x \qquad\quad + 8z &= 12 \\ 2y - \quad z &= -2 \\ 3x + \quad y + \quad z &= -1 \\ x + \quad y - kz &= -7 \end{aligned}$$

54. Let $A = \begin{bmatrix} 1 & 2 \\ -3 & 5 \end{bmatrix}$, $X = \begin{bmatrix} x_1 \\ x_2 \end{bmatrix}$, and $B = \begin{bmatrix} -4 \\ 12 \end{bmatrix}$.

Show that the equation $AX = B$ represents a linear system of two equations in two unknowns. Solve the system and substitute into the matrix equation to check your results.

2.4 Solution of Linear Systems by Matrices

In the last section we used the echelon method to solve linear systems of equations. We can simplify the procedure by using matrices. Since the variables are always the same, all we need to keep track of are their coefficients and the constants. For example, let's look at the system we solved in Example 4 of the previous section.

$$\begin{aligned} 2x + \quad y - \quad z &= 2 \\ x + 3y + 2z &= 1 \\ x + \quad y + \quad z &= 2 \end{aligned}$$

For a matrix solution of this system, first write the coefficients and constants as a 3×4 matrix, called the **augmented matrix** of the system.

$$\begin{bmatrix} 2 & 1 & -1 & 2 \\ 1 & 3 & 2 & 1 \\ 1 & 1 & 1 & 2 \end{bmatrix}$$

The vertical line is used to separate the constants from the coefficients.

The rows of this 3×4 matrix can be transformed in the same way as the equations of the system, since the matrix is just a shortened form of the system. The **row operations** on matrices which correspond to the transformations of systems of equations are given in the following theorem.

Theorem 2.1

> For any augmented matrix of a system of equations, the following operations produce an augmented matrix of an equivalent system:
>
> (a) interchanging any two rows;
> (b) multiplying the elements of a row by any nonzero real number;
> (c) adding a multiple of the elements of one row to the corresponding elements of some other row.

These row operations, like the transformations of systems of equations in Section 2.3, are reversible. If we use them to go from matrix A to matrix B, then it is possible to use row operations to transform B back into A. You should become expert in the use of these row operations, since they are very important in the simplex method of Chapter 4.

The first row operation allows us to change the matrix

$$\begin{bmatrix} 1 & 3 & 5 & 6 \\ 0 & 1 & 2 & 3 \\ 2 & 1 & -2 & -5 \end{bmatrix} \quad \text{to} \quad \begin{bmatrix} 0 & 1 & 2 & 3 \\ 1 & 3 & 5 & 6 \\ 2 & 1 & -2 & -5 \end{bmatrix}$$

by interchanging the first two rows. Note that row three is left unchanged.

The second row operation allows us to change

$$\begin{bmatrix} 1 & 3 & 5 & 6 \\ 0 & 1 & 2 & 3 \\ 2 & 1 & -2 & -5 \end{bmatrix} \quad \text{to} \quad \begin{bmatrix} -2 & -6 & -10 & -12 \\ 0 & 1 & 2 & 3 \\ 2 & 1 & -2 & -5 \end{bmatrix}$$

by multiplying the elements of the first row of the original matrix by -2. Note that rows two and three are left unchanged.

The third row operation allows us to change

$$\begin{bmatrix} 1 & 3 & 5 & 6 \\ 0 & 1 & 2 & 3 \\ 2 & 1 & -2 & -5 \end{bmatrix} \quad \text{to} \quad \begin{bmatrix} -1 & 2 & 7 & 11 \\ 0 & 1 & 2 & 3 \\ 2 & 1 & -2 & -5 \end{bmatrix}$$

by first multiplying each element in the third row of the original matrix by -1 and then adding the results to the corresponding elements in the first row of that matrix. Work as follows.

$$\begin{bmatrix} 1 + 2(-1) & 3 + 1(-1) & 5 + (-2)(-1) & 6 + (-5)(-1) \\ 0 & 1 & 2 & 3 \\ 2 & 1 & -2 & -5 \end{bmatrix} = \begin{bmatrix} -1 & 2 & 7 & 11 \\ 0 & 1 & 2 & 3 \\ 2 & 1 & -2 & -5 \end{bmatrix}$$

Note that rows two and three are left unchanged.

The Gauss-Jordan Method Based on Theorem 2.1, the **Gauss-Jordan method** is a method of solving systems of equations that is similar to the echelon method. The system of equations must be in proper form before using the Gauss-Jordan method. The terms with variables should be on the left and the constants on the right in each equation. The variables should be in the same order in each equation. The following example illustrates the use of the Gauss-Jordan method to solve a system of equations.

Example 1 Solve the system

$$3x - 4y = 1 \tag{1}$$
$$5x + 2y = 19 \tag{2}$$

The procedure is parallel to the echelon method we used in Section 2.3, except for the last step. First, write the matrix for the system. Here the system is already in the proper form.

Gauss-Jordan method

$$3x - 4y = 1 \tag{1}$$
$$5x + 2y = 19 \tag{2}$$

$$\begin{bmatrix} 3 & -4 & 1 \\ 5 & 2 & 19 \end{bmatrix}$$

Multiply both sides of equation (1) by 1/3 so that x has a coefficient of 1.

Using row operation (b) multiply each element of row 1 by 1/3.

$$x - \frac{4}{3}y = \frac{1}{3} \tag{3}$$
$$5x + 2y = 19 \tag{2}$$

$$\begin{bmatrix} 1 & -\dfrac{4}{3} & \dfrac{1}{3} \\ 5 & 2 & 19 \end{bmatrix}$$

Eliminate x from equation (2) by adding -5 times equation (3) to equation (2).

Using row operation (c) add -5 times the elements of row 1 to the elements of row 2.

$$x - \frac{4}{3}y = \frac{1}{3} \tag{3}$$
$$\frac{26}{3}y = \frac{52}{3} \tag{4}$$

$$\begin{bmatrix} 1 & -\dfrac{4}{3} & \dfrac{1}{3} \\ 0 & \dfrac{26}{3} & \dfrac{52}{3} \end{bmatrix}$$

Multiply both sides of equation (4) by 3/26 to get $y = 2$.

Multiply the elements of row 2 by 3/26, using row operation (b).

$$x - \frac{4}{3}y = \frac{1}{3} \tag{3}$$
$$y = 2 \tag{5}$$

$$\begin{bmatrix} 1 & -\dfrac{4}{3} & \dfrac{1}{3} \\ 0 & 1 & 2 \end{bmatrix}$$

Substitute $y = 2$ into equation (3) and solve for x to get $x = 3$.

$$x = 3 \qquad \textbf{(6)}$$
$$y = 2 \qquad \textbf{(5)}$$

The solution of the system, (3, 2), can be read directly from the last column of the final matrix. ▌

Multiply the elements of row 2 by 4/3 and add to the elements of row 1 (row operation (c)).

$$\begin{bmatrix} 1 & 0 & 3 \\ 0 & 1 & 2 \end{bmatrix}$$

Note that the last matrix has zeros above the main diagonal as well as below. When using Theorem 2.1 to transform the matrix, it is best to work column by column from left to right. For each column, the first change should produce a 1 in the proper position. Next, perform the steps that give zeros in the remainder of the column. Then proceed to the next column.

Example 2   Use the Gauss-Jordan method to solve the system.

$$x \qquad + 5z = -6 + y$$
$$3x + 3y \qquad = 10 + z$$
$$x + 3y + 2z = \ \ 5$$

The system must first be rewritten in proper form as follows.

$$x - \ \ y + 5z = -6$$
$$3x + 3y - \ \ z = 10$$
$$x + 3y + 2z = \ \ 5$$

Begin the solution by writing the matrix of the linear system.

$$\begin{bmatrix} 1 & -1 & 5 & -6 \\ 3 & 3 & -1 & 10 \\ 1 & 3 & 2 & 5 \end{bmatrix}$$

We want the final matrix to be of the form

$$\begin{bmatrix} 1 & 0 & 0 & m \\ 0 & 1 & 0 & n \\ 0 & 0 & 1 & p \end{bmatrix},$$

where m, n, and p are real numbers. From this final form of the matrix, we can read the solution: $x = m$, $y = n$, and $z = p$.

We already have 1 for the first element in column one. To get 0 for the second element in column one, multiply each element in the first row by -3 and add the results to the corresponding elements in row two (using Theorem 2.1 (c)).

$$\begin{bmatrix} 1 & -1 & 5 & -6 \\ 0 & 6 & -16 & 28 \\ 1 & 3 & 2 & 5 \end{bmatrix}$$

Now to change the last element in column one to 0, multiply each element in the first row by -1 and add the results to the corresponding elements of the third row (again using Theorem 2.1(c)).

$$\begin{bmatrix} 1 & -1 & 5 & | & -6 \\ 0 & 6 & -16 & | & 28 \\ 0 & 4 & -3 & | & 11 \end{bmatrix}$$

This transforms the first column. We transform the second and third columns similarly.

$$\begin{bmatrix} 1 & -1 & 5 & | & -6 \\ 0 & 1 & -\frac{8}{3} & | & \frac{14}{3} \\ 0 & 4 & -3 & | & 11 \end{bmatrix}$$ Second row multiplied by $\frac{1}{6}$ [Theorem 2.1(b)]

$$\begin{bmatrix} 1 & 0 & \frac{7}{3} & | & -\frac{4}{3} \\ 0 & 1 & -\frac{8}{3} & | & \frac{14}{3} \\ 0 & 4 & -3 & | & 11 \end{bmatrix}$$ Second row added to first row [Theorem 2.1(c)]

$$\begin{bmatrix} 1 & 0 & \frac{7}{3} & | & -\frac{4}{3} \\ 0 & 1 & -\frac{8}{3} & | & \frac{14}{3} \\ 0 & 0 & \frac{23}{3} & | & -\frac{23}{3} \end{bmatrix}$$ -4 times second row added to third row [Theorem 2.1(c)]

$$\begin{bmatrix} 1 & 0 & \frac{7}{3} & | & -\frac{4}{3} \\ 0 & 1 & -\frac{8}{3} & | & \frac{14}{3} \\ 0 & 0 & 1 & | & -1 \end{bmatrix}$$ Third row multiplied by $\frac{3}{23}$ [Theorem 2.1(b)]

$$\begin{bmatrix} 1 & 0 & 0 & | & 1 \\ 0 & 1 & -\frac{8}{3} & | & \frac{14}{3} \\ 0 & 0 & 1 & | & -1 \end{bmatrix}$$ $-\frac{7}{3}$ times third row added to first row [Theorem 2.1(c)]

$$\begin{bmatrix} 1 & 0 & 0 & | & 1 \\ 0 & 1 & 0 & | & 2 \\ 0 & 0 & 1 & | & -1 \end{bmatrix}$$ $\frac{8}{3}$ times third row added to second row [Theorem 2.1(c)]

The linear system associated with this last augmented matrix is

$$x \qquad\quad = 1$$
$$y \quad = 2$$
$$z = -1,$$

and the solution is $(1, 2, -1)$. ∎

Example 3   Use the Gauss-Jordan method to solve the system

$$x + y = 2$$
$$2x + 2y = 5.$$

Begin by writing the augmented matrix.

$$\begin{bmatrix} 1 & 1 & | & 2 \\ 2 & 2 & | & 5 \end{bmatrix}$$

To get a 0 for the second element in column one, multiply the numbers in row one by -2 and add the results to the corresponding elements in row two.

$$\begin{bmatrix} 1 & 1 & | & 2 \\ 0 & 0 & | & 1 \end{bmatrix}$$

The next step is to get a 1 for the second element in column two. Since this is impossible, we cannot go further.

If we put the matrix back into equation form, we have

$$x + y = 2$$
$$0x + 0y = 1.$$

Since the second equation is $0 = 1$, the system is inconsistent and has no solution. The row $[0 \quad 0 \quad 1]$ is a signal that the matrix is inconsistent.   ■

Example 4   Use the Gauss-Jordan method to solve the system

$$x + 2y - z = 0$$
$$3x - y + z = 6$$
$$-2x - 4y + 2z = 0.$$

The augmented matrix is

$$\begin{bmatrix} 1 & 2 & -1 & | & 0 \\ 3 & -1 & 1 & | & 6 \\ -2 & -4 & 2 & | & 0 \end{bmatrix}.$$

The first element in column one is a 1. Use Theorem 2.1 to get zeros in the rest of column one.

$$\begin{bmatrix} 1 & 2 & -1 & | & 0 \\ 0 & -7 & 4 & | & 6 \\ -2 & -4 & 2 & | & 0 \end{bmatrix} \quad \begin{bmatrix} 1 & 2 & -1 & | & 0 \\ 0 & -7 & 4 & | & 6 \\ 0 & 0 & 0 & | & 0 \end{bmatrix}$$

The row of all zeros in the last matrix is a signal that two of the equations (the first and last) are dependent. Continuing, multiply row 2 by $-1/7$.

$$\begin{bmatrix} 1 & 2 & -1 & | & 0 \\ 0 & 1 & -\frac{4}{7} & | & -\frac{6}{7} \\ 0 & 0 & 0 & | & 0 \end{bmatrix}$$

Finally, add -2 times row 2 to row 1.

$$\begin{bmatrix} 1 & 0 & \frac{1}{7} & | & \frac{12}{7} \\ 0 & 1 & -\frac{4}{7} & | & -\frac{6}{7} \\ 0 & 0 & 0 & | & 0 \end{bmatrix}$$

To complete the solution, let z be the parameter. The solution can then be written as

$$z \text{ arbitrary}$$

$$y = -\frac{6}{7} + \frac{4}{7}z$$

$$x = \frac{12}{7} - \frac{1}{7}z. \qquad \blacksquare$$

Although in the examples, we have used only systems with two equations and variables or three equations and variables, the Gauss-Jordan method can be used for any system with n equations and n variables. The method does become quite tedious even with 3 equations and 3 variables. However, it is very suitable for use by computers. A computer can produce the solution to a fairly large system very quickly.[1]

2.4 Exercises   *Write the augmented matrix for each of the following systems.* **Do not solve.**

1. $2x + 3y = 11$
   $x + 2y = 8$

2. $3x + 5y = -13$
   $2x + 3y = -9$

3. $x \qquad = 6 - 5y$
   $\quad y = 1$

4. $\quad 7y = 1 - 2x$
   $5x \qquad = -15$

5. $2x + y + z = 3$
   $3x - 4y + 2z = -7$
   $x + y + z = 2$

6. $4x - 2y + 3z = 4$
   $3x + 5y + z = 7$
   $5x - y + 4z = 7$

7. $\quad y \quad = 2 - x$
   $\quad 2y \quad = -4 - z$
   $\quad z = 2$

8. $x \qquad = 6$
   $\quad y \qquad = 2 - 2z$
   $x \qquad = 6 + 3z$

9. $x \qquad = 5$
   $\quad y \qquad = -2$
   $\quad z = 3$

10. $x \qquad = 8$
    $y + z = 6$
    $\quad z = 2$

Write the system of equations associated with each of the following augmented matrices. **Do not solve.**

11. $\begin{bmatrix} 1 & 0 & | & 2 \\ 0 & 1 & | & 3 \end{bmatrix}$

12. $\begin{bmatrix} 1 & 0 & | & 5 \\ 0 & 1 & | & -3 \end{bmatrix}$

13. $\begin{bmatrix} 2 & 1 & | & 1 \\ 3 & -2 & | & -9 \end{bmatrix}$

14. $\begin{bmatrix} 1 & -5 & | & -18 \\ 6 & 2 & | & 20 \end{bmatrix}$

15. $\begin{bmatrix} 1 & 0 & 0 & | & 2 \\ 0 & 1 & 0 & | & 3 \\ 0 & 0 & 1 & | & -2 \end{bmatrix}$

16. $\begin{bmatrix} 1 & 0 & 1 & | & 4 \\ 0 & 1 & 0 & | & 2 \\ 0 & 0 & 1 & | & 3 \end{bmatrix}$

Use the Gauss-Jordan method to solve each of the following systems of equations.

17. $x + y = 5$
    $x - y = -1$

18. $\quad x + 2y = 5$
    $2x + y = -2$

---

[1]See Margaret Lial, *Study Guide with Computer Problems* (Glenview, Ill., Scott, Foresman and Company, 1979).

19. $x + y = -3$
    $2x - 5y = -6$

20. $3x - 2y = 4$
    $3x + y = -2$

21. $2x = 10 + 3y$
    $2y = 5 - 2x$

22. $y = 5 - 4x$
    $2x = 3 - y$

23. $2x - 5y = 10$
    $4x - 5y = 15$

24. $4x - 2y = 3$
    $-2x + 3y = 1$

25. $2x - 3y = 2$
    $4x - 6y = 1$

26. $x + 2y = 1$
    $2x + 4y = 3$

27. $6x - 3y = 1$
    $-12x + 6y = -2$

28. $x - y = 1$
    $-x + y = -1$

29. $x + y = -1$
    $y + z = 4$
    $x + z = 1$

30. $x - z = -3$
    $y + z = 9$
    $x + z = 7$

31. $x + y - z = 6$
    $2x - y + z = -9$
    $x - 2y + 3z = 1$

32. $x + 3y - 6z = 7$
    $2x - y + 2z = 0$
    $x + y + 2z = -1$

33. $y = x - 1$
    $y = 6 + z$
    $z = -1 - x$

34. $x = 1 - y$
    $2x = z$
    $2z = -2 - y$

35. $x - 2y + z = 5$
    $2x + y - z = 2$
    $-2x + 4y - 2z = 2$

36. $3x + 5y - z = 0$
    $4x - y + 2z = 1$
    $-6x - 10y + 2z = 0$

37. $x + 2y - w = 3$
    $2x + 4z + 2w = -6$
    $x + 2y - z = 6$
    $2x - y + z + w = -3$

38. $x + 3y - 2z - w = 9$
    $2x + 4y + 2w = 10$
    $-3x - 5y + 2z - w = -15$
    $x - y - 3z + 2w = 6$

39. At rush hours, substantial traffic congestion is encountered at the traffic intersections shown in the figure. (The streets are all one way.)

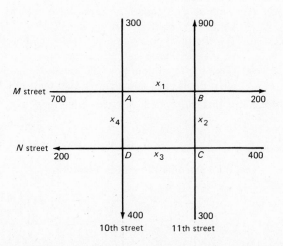

The city wishes to improve the signals at these corners so as to speed the flow of traffic. The traffic engineers first gather data. As the figure shows, 700 cars per hour come down M Street to intersection A; 300 cars per hour come to intersection A on 10th Street. A total of x_1 of these cars leave A on M Street, while x_4 cars leave A on 10th Street. The number of cars entering A must equal the number leaving, so that

$$x_1 + x_4 = 700 + 300$$

or

$$x_1 + x_4 = 1000.$$

For intersection B, x_1 cars enter B on M Street, and x_2 cars enter B on 11th Street. The figure shows that 900 cars leave B on 11th while 200 leave on M. We have

$$x_1 + x_2 = 900 + 200$$

$$x_1 + x_2 = 1100.$$

At intersection C, 400 cars enter on N Street, 300 on 11th Street, while x_2 leave on 11th Street and x_3 leave on N Street. This gives

$$x_2 + x_3 = 400 + 300$$

$$x_2 + x_3 = 700.$$

Finally, intersection D has x_3 cars entering on N and x_4 entering on 10th. There are 400 leaving D on 10th and 200 leaving on N, so that

$$x_3 + x_4 = 400 + 200$$

$$x_3 + x_4 = 600.$$

(a) Use the four equations to set up an augmented matrix, and then use the Gauss-Jordan method to solve it. (Hint: keep going until you get a row of all zeros.)

(b) Since you got a row of all zeros, the system of equations does not have a unique solution. Write three equations, corresponding to the three nonzero rows of the matrix.

(c) Solve each of the equations for x_4.

(d) One of your equations should have been $x_4 = 1000 - x_1$. What is the largest possible value of x_1 so that x_4 is not negative? What is the largest value of x_4 so that x_1 is not negative?

(e) Your second equation should have been $x_4 = x_2 - 100$. Find the smallest possible value of x_2 so that x_4 is not negative.

(f) For the third equation, $x_4 = 600 - x_3$, find the largest possible values of x_3 and x_4 so that neither variable is negative.

(g) Give the maximum possible value of x_1 so that all the equations are satisfied and all variables are nonnegative. Do the same for x_2, x_3, and x_4.

40. Show that the system of linear equations

$$\begin{align} 2x_1 + 3x_2 + x_3 &= 5 \\ x_1 - 4x_2 + 5x_3 &= 8 \end{align}$$

can be written as the matrix equation

$$\begin{bmatrix} 2 & 3 & 1 \\ 1 & -4 & 5 \end{bmatrix} \begin{bmatrix} x_1 \\ x_2 \\ x_3 \end{bmatrix} = \begin{bmatrix} 5 \\ 8 \end{bmatrix}.$$

2.5 Matrix Inverses

In Section 2.1, we defined a zero matrix which has properties similar to those of the real number zero, the identity for addition. The real number 1 is the identity element for multiplication: for any real number a, $a \cdot 1 = 1 \cdot a = a$. In this section, we define an identity matrix I which has properties similar to those of the number 1. We then use this identity matrix to find the multiplicative inverse of any square matrix which has an inverse.

We know that if I is to be the identity matrix, the products AI and IA must both equal A. This means that we can find an identity matrix only for square matrices. Otherwise, IA and AI would not both exist. The 2×2 **identity matrix** which satisfies these conditions is

$$I = \begin{bmatrix} 1 & 0 \\ 0 & 1 \end{bmatrix}.$$

To check that I, as defined above, is really the 2×2 identity matrix, let

$$A = \begin{bmatrix} a & b \\ c & d \end{bmatrix}.$$

Then AI and IA should both equal A.

$$AI = \begin{bmatrix} a & b \\ c & d \end{bmatrix} \begin{bmatrix} 1 & 0 \\ 0 & 1 \end{bmatrix} = \begin{bmatrix} a(1) + b(0) & a(0) + b(1) \\ c(1) + d(0) & c(0) + d(1) \end{bmatrix} = \begin{bmatrix} a & b \\ c & d \end{bmatrix} = A$$

$$IA = \begin{bmatrix} 1 & 0 \\ 0 & 1 \end{bmatrix} \begin{bmatrix} a & b \\ c & d \end{bmatrix} = \begin{bmatrix} 1(a) + 0(c) & 1(b) + 0(d) \\ 0(a) + 1(c) & 0(b) + 1(d) \end{bmatrix} = \begin{bmatrix} a & b \\ c & d \end{bmatrix} = A$$

This verifies that we have defined I correctly. (It can also be shown that I is the only 2×2 identity matrix.)

The identity matrices for 3×3 matrices and 4×4 matrices respectively are

$$I = \begin{bmatrix} 1 & 0 & 0 \\ 0 & 1 & 0 \\ 0 & 0 & 1 \end{bmatrix} \quad \text{and} \quad I = \begin{bmatrix} 1 & 0 & 0 & 0 \\ 0 & 1 & 0 & 0 \\ 0 & 0 & 1 & 0 \\ 0 & 0 & 0 & 1 \end{bmatrix}.$$

By generalizing, we can find an identity matrix for any $n \times n$ matrix.

Recall that the multiplicative inverse of a number a is defined to be $1/a$. The inverse $1/a$ exists for all real numbers a except 0. Because of the way matrix multiplication is defined, *only square matrices can have inverses.* The multiplicative inverse of a matrix A is written A^{-1}. (The notation A^{-1} does not mean $1/A$. We have not defined division for matrices. It is just the notation for the matrix which is the inverse of matrix A.) The matrix A^{-1} must satisfy the statements

$$AA^{-1} = I \quad \text{and} \quad A^{-1}A = I.$$

If it exists, the inverse of a matrix is unique. That is, there is only one inverse for any square matrix. The proof of this is left to Exercise 50 of this section.

The method we shall use to find the multiplicative inverse of any $n \times n$ matrix which has an inverse, depends on the row operations of Theorem 2.1.

To obtain A^{-1} for any $n \times n$ matrix A for which A^{-1} exists, we first form the augmented matrix $[A|I]$ where I is the $n \times n$ multiplicative identity matrix. Performing row operations on $[A|I]$, a matrix of the form $[I|B]$ can be obtained. Matrix B is then the desired matrix A^{-1}.

Unfortunately, it is not clear why this method should produce the desired matrix; however, in any particular case, it is easy to verify that the matrix B from $[I|B]$ is indeed A^{-1} by verifying that $AB = BA = I$.

Example 1   Find A^{-1} if $A = \begin{bmatrix} 2 & 4 \\ 1 & -1 \end{bmatrix}$.

Use the 2×2 identity matrix I to form the augmented matrix $[A|I]$.

$$[A|I] = \begin{bmatrix} 2 & 4 & | & 1 & 0 \\ 1 & -1 & | & 0 & 1 \end{bmatrix}$$

To obtain the identity matrix on the left, we must get a 1 in the upper left-hand corner. Multiply the top row of $[A|I]$ by 1/2. (We also could have exchanged the two rows.)

$$\begin{bmatrix} 1 & 2 & | & \frac{1}{2} & 0 \\ 1 & -1 & | & 0 & 1 \end{bmatrix}$$

To get a 0 for the first element in row two, multiply the new row one by -1 and add the results to row two.

$$\begin{bmatrix} 1 & 2 & | & \frac{1}{2} & 0 \\ 0 & -3 & | & -\frac{1}{2} & 1 \end{bmatrix}$$

To get a 1 for the second element in row two, multiply row two by $-1/3$.

$$\begin{bmatrix} 1 & 2 & | & \frac{1}{2} & 0 \\ 0 & 1 & | & \frac{1}{6} & -\frac{1}{3} \end{bmatrix}$$

To get a 0 for the second element in row one, multiply the new row two by -2 and add the results to row one.

$$\begin{bmatrix} 1 & 0 & | & \frac{1}{6} & \frac{2}{3} \\ 0 & 1 & | & \frac{1}{6} & -\frac{1}{3} \end{bmatrix}$$

We have now transformed $[A|I]$ into $[I|B]$. The transformation gives us

$$B = \begin{bmatrix} \frac{1}{6} & \frac{2}{3} \\ \frac{1}{6} & -\frac{1}{3} \end{bmatrix},$$

which should equal A^{-1}. To check, multiply A times B. The result should be I. We have

$$AB = \begin{bmatrix} 2 & 4 \\ 1 & -1 \end{bmatrix} \begin{bmatrix} \frac{1}{6} & \frac{2}{3} \\ \frac{1}{6} & -\frac{1}{3} \end{bmatrix} = \begin{bmatrix} \frac{1}{3} + \frac{2}{3} & \frac{4}{3} - \frac{4}{3} \\ \frac{1}{6} - \frac{1}{6} & \frac{2}{3} + \frac{1}{3} \end{bmatrix} = \begin{bmatrix} 1 & 0 \\ 0 & 1 \end{bmatrix} = I.$$

Verify that $BA = I$, also. Thus,

$$A^{-1} = \begin{bmatrix} \frac{1}{6} & \frac{2}{3} \\ \frac{1}{6} & -\frac{1}{3} \end{bmatrix}. \quad \blacksquare$$

Example 2 Find A^{-1} if $A = \begin{bmatrix} 1 & 0 & 1 \\ 2 & -2 & -1 \\ 3 & 0 & 0 \end{bmatrix}$.

Write the augmented matrix $[A|I]$.

$$[A|I] = \begin{bmatrix} 1 & 0 & 1 & 1 & 0 & 0 \\ 2 & -2 & -1 & 0 & 1 & 0 \\ 3 & 0 & 0 & 0 & 0 & 1 \end{bmatrix}$$

Since 1 is already in the upper left-hand corner as desired, we begin by selecting the row operation which will result in a 0 for the first element in row two. Multiply row one by -2 and add the result to row two. This gives

$$\begin{bmatrix} 1 & 0 & 1 & 1 & 0 & 0 \\ 0 & -2 & -3 & -2 & 1 & 0 \\ 3 & 0 & 0 & 0 & 0 & 1 \end{bmatrix}.$$

To get 0 for the first element in row three, multiply row one by -3 and add to row three. The new matrix is

$$\begin{bmatrix} 1 & 0 & 1 & 1 & 0 & 0 \\ 0 & -2 & -3 & -2 & 1 & 0 \\ 0 & 0 & -3 & -3 & 0 & 1 \end{bmatrix}.$$

To get 1 for the second element in row two, multiply row two by $-1/2$, obtaining the new matrix

$$\begin{bmatrix} 1 & 0 & 1 & 1 & 0 & 0 \\ 0 & 1 & \frac{3}{2} & 1 & -\frac{1}{2} & 0 \\ 0 & 0 & -3 & -3 & 0 & 1 \end{bmatrix}.$$

To get 1 for the third element in row three, multiply row three by $-1/3$, with the result

$$\begin{bmatrix} 1 & 0 & 1 & 1 & 0 & 0 \\ 0 & 1 & \frac{3}{2} & 1 & -\frac{1}{2} & 0 \\ 0 & 0 & 1 & 1 & 0 & -\frac{1}{3} \end{bmatrix}.$$

To get 0 for the third element in row one, multiply row three by -1 and add to row one, which gives

$$\begin{bmatrix} 1 & 0 & 0 & 0 & 0 & \frac{1}{3} \\ 0 & 1 & \frac{3}{2} & 1 & -\frac{1}{2} & 0 \\ 0 & 0 & 1 & 1 & 0 & -\frac{1}{3} \end{bmatrix}.$$

To get 0 for the third element in row two, multiply row three by $-3/2$ and add to row two.

$$\begin{bmatrix} 1 & 0 & 0 & 0 & 0 & \frac{1}{3} \\ 0 & 1 & 0 & -\frac{1}{2} & -\frac{1}{2} & \frac{1}{2} \\ 0 & 0 & 1 & 1 & 0 & -\frac{1}{3} \end{bmatrix}$$

From the last transformation, we get the desired inverse.

$$A^{-1} = \begin{bmatrix} 0 & 0 & \frac{1}{3} \\ -\frac{1}{2} & -\frac{1}{2} & \frac{1}{2} \\ 1 & 0 & -\frac{1}{3} \end{bmatrix}$$

Confirm this by forming the product $A^{-1}A$, which should equal I.   ∎

Example 3   Find A^{-1} if $A = \begin{bmatrix} 2 & -4 \\ 1 & -2 \end{bmatrix}$.

Using row operations to transform the first column of the augmented matrix

$$\begin{bmatrix} 2 & -4 & 1 & 0 \\ 1 & -2 & 0 & 1 \end{bmatrix}$$

results in the following matrices.

$$\begin{bmatrix} 1 & -2 & \frac{1}{2} & 0 \\ 1 & -2 & 0 & 1 \end{bmatrix}$$

$$\begin{bmatrix} 1 & -2 & \frac{1}{2} & 0 \\ 0 & 0 & -\frac{1}{2} & 1 \end{bmatrix}$$

At this point, we wish to change the matrix so that the second element of row two will be 1. Since that element is now 0, there is no way to complete the desired transformation.

What is wrong? Remember, near the beginning of this section we mentioned that some matrices do not have inverses. Matrix A is an example of a matrix that has no inverse. Thus, in this case there is no matrix A^{-1} such that $AA^{-1} = A^{-1}A = A$.   ∎

Solving Systems of Equations with Inverses   We used matrices to solve systems of linear equations by the Gauss-Jordan method in Section 2.4. Another way to use matrices to solve linear systems is to write the system as a matrix equation $AX = B$, where A is the matrix of the coefficients of the variables of the system, X is the matrix of the variables, and B is the matrix of the constants. Matrix A is called the **coefficient matrix**.

To solve the matrix equation $AX = B$, we first see if A^{-1} exists. If so, we then use the facts that $A^{-1}A = I$ and $IX = X$, as follows.

$$AX = B$$
$$A^{-1}(AX) = A^{-1}B \qquad \text{Multiply both sides by } A^{-1}$$
$$(A^{-1}A)X = A^{-1}B$$
$$IX = A^{-1}B$$
$$X = A^{-1}B$$

Note that when multiplying by matrices on both sides of a matrix equation, we must be careful to multiply in the same order on both sides of the equation, since (unlike multiplication of real numbers) multiplication of matrices is noncommutative.

Thus, to find X, we first find A^{-1} and then find the product $A^{-1}B$. By letting the corresponding elements of X and $A^{-1}B$ be equal, we get the solution of the system.

Example 4 Use the inverse of the coefficient matrix to solve the linear system

$$2x - 3y = 4$$
$$x + 5y = 2.$$

To represent the system as a matrix equation, we use the coefficient matrix of the system together with the matrix of variables and the matrix of constants.

$$A = \begin{bmatrix} 2 & -3 \\ 1 & 5 \end{bmatrix}, \qquad X = \begin{bmatrix} x \\ y \end{bmatrix}, \qquad \text{and} \qquad B = \begin{bmatrix} 4 \\ 2 \end{bmatrix}.$$

The system can then be written in matrix form as the equation $AX = B$ since

$$AX = \begin{bmatrix} 2 & -3 \\ 1 & 5 \end{bmatrix} \begin{bmatrix} x \\ y \end{bmatrix} = \begin{bmatrix} 2x - 3y \\ x + 5y \end{bmatrix} = \begin{bmatrix} 4 \\ 2 \end{bmatrix} = B.$$

To solve the system, we must find A^{-1}. To do this, we use row operations on matrix $[A|I]$ to get

$$\begin{bmatrix} 1 & 0 & \frac{5}{13} & \frac{3}{13} \\ 0 & 1 & -\frac{1}{13} & \frac{2}{13} \end{bmatrix}.$$

Therefore,

$$A^{-1} = \begin{bmatrix} \frac{5}{13} & \frac{3}{13} \\ -\frac{1}{13} & \frac{2}{13} \end{bmatrix}.$$

Next, find the product $A^{-1}B$.

$$A^{-1}B = \begin{bmatrix} \frac{5}{13} & \frac{3}{13} \\ -\frac{1}{13} & \frac{2}{13} \end{bmatrix} \begin{bmatrix} 4 \\ 2 \end{bmatrix} = \begin{bmatrix} 2 \\ 0 \end{bmatrix}$$

Since $X = A^{-1}B$,

$$X = \begin{bmatrix} x \\ y \end{bmatrix} = \begin{bmatrix} 2 \\ 0 \end{bmatrix}.$$

Thus, the solution of the system is $(2, 0)$. ■

Example 5 Use the inverse of the coefficient matrix to solve the system

$$-x - 2y + 2z = 9$$
$$2x + y - z = -3$$
$$3x - 2y + z = -6.$$

The matrices we need are

$$A = \begin{bmatrix} -1 & -2 & 2 \\ 2 & 1 & -1 \\ 3 & -2 & 1 \end{bmatrix}, \quad X = \begin{bmatrix} x \\ y \\ z \end{bmatrix}, \quad \text{and} \quad B = \begin{bmatrix} 9 \\ -3 \\ -6 \end{bmatrix}.$$

To find A^{-1}, start with matrix

$$[A|I] = \begin{bmatrix} -1 & -2 & 2 & | & 1 & 0 & 0 \\ 2 & 1 & -1 & | & 0 & 1 & 0 \\ 3 & -2 & 1 & | & 0 & 0 & 1 \end{bmatrix}$$

and use row operations to get $[I|A^{-1}]$, from which

$$A^{-1} = \begin{bmatrix} \frac{1}{3} & \frac{2}{3} & 0 \\ \frac{5}{3} & \frac{7}{3} & -1 \\ \frac{7}{3} & \frac{8}{3} & -1 \end{bmatrix}.$$

Now find $A^{-1} B$.

$$A^{-1} B = \begin{bmatrix} \frac{1}{3} & \frac{2}{3} & 0 \\ \frac{5}{3} & \frac{7}{3} & -1 \\ \frac{7}{3} & \frac{8}{3} & -1 \end{bmatrix} \begin{bmatrix} 9 \\ -3 \\ -6 \end{bmatrix} = \begin{bmatrix} 1 \\ 14 \\ 19 \end{bmatrix}$$

Since $X = A^{-1} B$,

$$X = \begin{bmatrix} x \\ y \\ z \end{bmatrix} = \begin{bmatrix} 1 \\ 14 \\ 19 \end{bmatrix}.$$

Then $x = 1$, $y = 14$, $z = 19$ and the solution is $(1, 14, 19)$. ∎

2.5 Exercises   *Decide whether or not the given matrices are inverses of each other. (Check to see if their product is I.)*

1. $\begin{bmatrix} 2 & 3 \\ 1 & 1 \end{bmatrix}$ and $\begin{bmatrix} -1 & 3 \\ 1 & -2 \end{bmatrix}$

2. $\begin{bmatrix} 5 & 7 \\ 2 & 3 \end{bmatrix}$ and $\begin{bmatrix} 3 & -7 \\ -2 & 5 \end{bmatrix}$

3. $\begin{bmatrix} 2 & 1 \\ 3 & 2 \end{bmatrix}$ and $\begin{bmatrix} 2 & 1 \\ -3 & 2 \end{bmatrix}$

4. $\begin{bmatrix} -1 & 2 \\ 3 & -5 \end{bmatrix}$ and $\begin{bmatrix} -5 & -2 \\ -3 & -1 \end{bmatrix}$

5. $\begin{bmatrix} 1 & 2 & 0 \\ 0 & 1 & 0 \\ 0 & 1 & 0 \end{bmatrix}$ and $\begin{bmatrix} 1 & -2 & 0 \\ 0 & 1 & 0 \\ 0 & -1 & 1 \end{bmatrix}$

6. $\begin{bmatrix} 0 & 1 & 0 \\ 0 & 0 & -2 \\ 1 & -1 & 0 \end{bmatrix}$ and $\begin{bmatrix} 1 & 0 & 1 \\ 1 & 0 & 0 \\ 0 & -1 & 0 \end{bmatrix}$

7. $\begin{bmatrix} 1 & 3 & 3 \\ 1 & 4 & 3 \\ 1 & 3 & 4 \end{bmatrix}$ and $\begin{bmatrix} 7 & -3 & -3 \\ -1 & 1 & 0 \\ -1 & 0 & 1 \end{bmatrix}$

8. $\begin{bmatrix} -1 & 0 & 2 \\ 3 & 1 & 0 \\ 0 & 2 & -3 \end{bmatrix}$ and $\begin{bmatrix} -\frac{1}{5} & \frac{4}{15} & -\frac{2}{15} \\ \frac{3}{5} & \frac{1}{5} & \frac{2}{5} \\ \frac{2}{5} & \frac{2}{15} & -\frac{1}{15} \end{bmatrix}$

Find the inverse, if it exists, for each of the following matrices.

9. $\begin{bmatrix} 1 & -1 \\ 2 & 0 \end{bmatrix}$ 10. $\begin{bmatrix} -1 & 2 \\ -2 & -1 \end{bmatrix}$

11. $\begin{bmatrix} 3 & -1 \\ -5 & 2 \end{bmatrix}$ 12. $\begin{bmatrix} -1 & -2 \\ 3 & 4 \end{bmatrix}$

13. $\begin{bmatrix} -6 & 4 \\ -3 & 2 \end{bmatrix}$ 14. $\begin{bmatrix} 5 & 10 \\ -3 & -6 \end{bmatrix}$

15. $\begin{bmatrix} 1 & 0 & 0 \\ 0 & -1 & 0 \\ 1 & 0 & 1 \end{bmatrix}$ 16. $\begin{bmatrix} 1 & 0 & 1 \\ 0 & -1 & 0 \\ 2 & 1 & 1 \end{bmatrix}$

17. $\begin{bmatrix} -1 & -1 & -1 \\ 4 & 5 & 0 \\ 0 & 1 & -3 \end{bmatrix}$ 18. $\begin{bmatrix} 2 & 0 & 4 \\ 3 & 1 & 5 \\ -1 & 1 & -2 \end{bmatrix}$

19. $\begin{bmatrix} 1 & 2 & 3 \\ -3 & -2 & -1 \\ -1 & 0 & 1 \end{bmatrix}$ 20. $\begin{bmatrix} 2 & 0 & 4 \\ 1 & 0 & -1 \\ 3 & 0 & -2 \end{bmatrix}$

21. $\begin{bmatrix} 2 & 4 & 6 \\ -1 & -4 & -3 \\ 0 & 1 & -1 \end{bmatrix}$ 22. $\begin{bmatrix} 2 & 2 & -4 \\ 2 & 6 & 0 \\ -3 & -3 & 5 \end{bmatrix}$

23. $\begin{bmatrix} 1 & -2 & 3 & 0 \\ 0 & 1 & -1 & 1 \\ -2 & 2 & -2 & 4 \\ 0 & 2 & -3 & 1 \end{bmatrix}$ 24. $\begin{bmatrix} 1 & 1 & 0 & 2 \\ 2 & -1 & 1 & -1 \\ 3 & 3 & 2 & -2 \\ 1 & 2 & 1 & 0 \end{bmatrix}$

Solve each of the following systems of equations by using the inverse of the coefficient matrix.

25. $2x + 3y = 10$
    $x - y = -5$

26. $-x + 2y = 15$
    $-2x - y = 20$

27. $2x + y = 5$
    $5x + 3y = 13$

28. $-x - 2y = 8$
    $3x + 4y = 24$

29. $-x + y = 1$
    $2x - y = 1$

30. $3x - 6y = 1$
    $-5x + 9y = -1$

31. $-x - 8y = 12$
    $3x + 24y = -36$

32. $x + 3y = -14$
    $2x - y = 7$

Solve each of the following systems of equations by using the inverse of the coefficient matrix. The inverses for the first four problems are found in Exercises 17, 18, 21, and 22 above.

33. $-x - y - z = 1$
    $4x + 5y = -2$
    $y - 3z = 3$

34. $2x + 4z = -8$
    $3x + y + 5z = 2$
    $-x + y - 2z = 4$

35. $2x + 4y + 6z = 4$
    $-x - 4y - 3z = 8$
    $y - z = -4$

36. $2x + 2y - 4z = 12$
    $2x + 6y = 16$
    $-3x - 3y + 5z = -20$

37. $x + 2y + 3z = 5$
    $2x + 3y + 2z = 2$
    $-x - 2y - 4z = -1$

38. $x + y - 3z = 4$
    $2x + 4y - 4z = 8$
    $-x + y + 4z = -3$

39. $2x - 2y = 5$
    $4y + 8z = 7$
    $x + 2z = 1$

40. $x + z = 3$
    $y + 2z = 8$
    $-x + y = 4$

Solve each of the following systems of equations by using the inverse of the coefficient matrix. The inverses were found in Exercises 23 and 24.

41.
$$\begin{aligned} x - 2y + 3z \quad\;\; &= 4 \\ y - z + w &= -8 \\ -2x + 2y - 2z + 4w &= 12 \\ 2y - 3z + w &= -4 \end{aligned}$$

42.
$$\begin{aligned} x + y \quad\;\; + 2w &= 3 \\ 2x - y + z - w &= 3 \\ 3x + 3y + 2z - 2w &= 5 \\ x + 2y + z \quad\;\; &= 3 \end{aligned}$$

Let $A = \begin{bmatrix} a & b \\ c & d \end{bmatrix}$. Show that each of the following statements is true.

43. $IA = A$ **44.** $AI = A$ **45.** $A \cdot O = O$

46. Find A^{-1}. (Assume $ad - bc \neq 0$.) Show that $AA^{-1} = I$.

47. Show that $A^{-1}A = I$.

48. Using the definitions and properties listed in this section, show that for square matrices A and B of the same order, if $AB = O$ and if A^{-1} exists, then $B = O$.

49. The Bread Box Bakery sells three types of cakes, each requiring the amounts of the basic ingredients shown in the following matrix.

Types of cake

	I	II	III
Flour (cups)	2	4	2
Sugar (cups)	2	1	2
Eggs	2	1	3

To fill its daily orders for these three kinds of cake, the bakery uses 72 cups of flour, 48 cups of sugar, and 60 eggs.

(a) Write a 3×1 matrix for the amounts used daily.

(b) Let the number of daily orders for cakes be a 3×1 matrix X with entries x_1, x_2, and x_3. Write a matrix equation which you can solve for X, using the given matrix and the matrix from part (a).

(c) Solve the equation you wrote in part (b) to find the number of daily orders for each type of cake.

50. Prove that, if it exists, the inverse of a matrix is unique. Hint: Assume there are two inverses B and C for some matrix A, so that $AB = BA = I$ and $AC = CA = I$. Multiply the first equation by C and the second by B.

> **Application**   *Code Theory*

Governments need sophisticated methods of coding and decoding messages. One example of such an advanced code uses matrix theory. Such a code takes the letters in the words and divides them into groups. (Each space between words is treated as a letter; punctuation is disregarded.) Then, numbers are assigned to the letters of the alphabet. For our purposes, let the letter *a* correspond to 1, *b* to 2, and so on. We let the number 27 correspond to a space between words.

For example, the message

mathematics is for the birds

can be divided into groups of three letters each.

mat   hem   ati   cs–   is–   for   –th   e–b   ird   s– –

Note that we used "–" to represent a space between words. We now write a column matrix for each group of three symbols using the corresponding numbers, as determined above, instead of letters. For example, the letters *mat* can be encoded as

$$\begin{bmatrix} 13 \\ 1 \\ 20 \end{bmatrix}$$

The coded message then is the set of 3×1 column matrices

$$\begin{bmatrix} 13 \\ 1 \\ 20 \end{bmatrix} \begin{bmatrix} 8 \\ 5 \\ 13 \end{bmatrix} \begin{bmatrix} 1 \\ 20 \\ 9 \end{bmatrix} \begin{bmatrix} 3 \\ 19 \\ 27 \end{bmatrix} \begin{bmatrix} 9 \\ 19 \\ 27 \end{bmatrix} \begin{bmatrix} 6 \\ 15 \\ 18 \end{bmatrix} \begin{bmatrix} 27 \\ 20 \\ 8 \end{bmatrix} \begin{bmatrix} 5 \\ 27 \\ 2 \end{bmatrix} \begin{bmatrix} 9 \\ 18 \\ 4 \end{bmatrix} \begin{bmatrix} 19 \\ 27 \\ 27 \end{bmatrix}$$

We can further complicate the code by choosing a matrix which has an inverse, in this case a 3×3 matrix, call it *M,* and find the products of this matrix and each of the above column matrices. (Note that the size of each group, the assignment of numbers to letters, and the choice of matrix *M* must all be predetermined.)

Suppose

$$M = \begin{bmatrix} 1 & 3 & 3 \\ 1 & 4 & 3 \\ 1 & 3 & 4 \end{bmatrix}$$

When the agent receives the message, he divides it into groups of numbers and forms each group into a column matrix. After multiplying each column matrix by the matrix M^{-1}, the message can be read.

Although this type of code is relatively simple, it is actually difficult to break. Many ramifications are possible. For example, a long message might be placed in groups of 20, thus requiring a 20×20 matrix for coding and decoding. Finding the inverse of such a matrix would require an impractical amount of time if calculated by hand. For this reason some of the largest computers are used by government agencies involved in coding.

If we find the products of *M* and the column matrices above, we have a new set of column matrices,

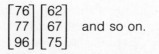

$$\begin{bmatrix} 76 \\ 77 \\ 96 \end{bmatrix} \begin{bmatrix} 62 \\ 67 \\ 75 \end{bmatrix} \text{ and so on.}$$

The entries of these matrices can then be transmitted to an agent as the message *76, 77, 96, 62, 67, 75,* and so on.

Exercises   **1.** Let $M = \begin{bmatrix} 4 & -1 \\ 2 & 6 \end{bmatrix}$

(a) Use M to encode the message: *Meet at the cave.* Use 2×1 matrices.

(b) What matrix should be used to decode the message?

2. Let $M = \begin{bmatrix} -1 & 2 \\ 2 & -5 \end{bmatrix}$.

Encode the message: *Attack at dawn unless too cold.*

3. Matrix M was used to encode the following message. Decode it.

$$\begin{bmatrix} -17 \\ 33 \end{bmatrix} \begin{bmatrix} 26 \\ -72 \end{bmatrix} \begin{bmatrix} 53 \\ -133 \end{bmatrix} \begin{bmatrix} 21 \\ -54 \end{bmatrix} \begin{bmatrix} 41 \\ -103 \end{bmatrix} \begin{bmatrix} 35 \\ -97 \end{bmatrix} \begin{bmatrix} 29 \\ -77 \end{bmatrix} \begin{bmatrix} -15 \\ 24 \end{bmatrix} \begin{bmatrix} 39 \\ -98 \end{bmatrix}$$

4. Finish encoding the message given in the text.

2.6 Input-Output Models

An interesting application of matrix theory to economics was developed by Nobel prize winner Professor Wassily Leontief. His matrix models for studying the interdependencies in an economy are called *input-output* models. In practice these models are very complicated with many variables. We shall discuss only simple examples with a few variables.

Input-output models are concerned with the production and flow of goods (and perhaps services). In an economy with n basic commodities, or sectors, the production of each commodity uses some (perhaps all) of the commodities in the economy as inputs. The amounts of each commodity used in the production of one unit of each commodity can be written as an $n \times n$ matrix A, called the **technological** or **input-output matrix** of the economy.

Example 1   Suppose a simplified economy involves just three commodity categories: agriculture, manufacturing, and transportation, all in appropriate units. Production of 1 unit of agriculture requires 1/2 unit of manufacturing and 1/4 unit of transportation. Production of 1 unit of manufacturing requires 1/4 unit of agriculture and 1/4 unit of transportation; while production of 1 unit of transportation requires 1/3 unit of agriculture and 1/4 unit of manufacturing. Find the input-output matrix of this economy.

The matrix is shown below.

$$\begin{array}{c} \text{Agriculture} \\ \text{Manufacturing} \\ \text{Transportation} \end{array} \begin{array}{ccc} \text{Agric.} & \text{Manuf.} & \text{Trans.} \end{array}$$

$$\begin{array}{c} \text{Agriculture} \\ \text{Manufacturing} \\ \text{Transportation} \end{array} \begin{bmatrix} 0 & \frac{1}{4} & \frac{1}{3} \\ \frac{1}{2} & 0 & \frac{1}{4} \\ \frac{1}{4} & \frac{1}{4} & 0 \end{bmatrix} = A$$

The first column of the input-output matrix represents the amount of each of the three commodities consumed in the production of 1 unit of agriculture. The second column gives the corresponding amounts required to produce 1 unit of manufacturing, and the last column gives the amounts needed to produce one unit of transportation. (Although it is perhaps unrealistic that production of a unit of each commodity requires none of that commodity, the simpler matrix involved is useful for our purposes.)   ∎

Another matrix used in conjunction with the input-output matrix is the matrix giving the amount of each commodity produced, called the **production matrix,** or the vector of gross output. In an economy producing n commodities, the production matrix can be represented by a column matrix X with entries x_1, $x_2, x_3, \ldots, x_n$.

Example 2   In Example 1, suppose the production matrix is

$$X = \begin{bmatrix} 60 \\ 52 \\ 48 \end{bmatrix}.$$

Then 60 units of agriculture, 52 units of manufacturing and 48 units of transportation are produced. As 1/4 unit of agriculture is used for each unit of manufacturing produced, $1/4 \times 52$ units of agriculture must be used up in the "production" of manufacturing. Similarly, $1/3 \times 48$ units of agriculture will be used up in the "production" of transportation. Thus $13 + 16 = 29$ units of agriculture are used for production in the economy. Look again at the matrices A and X. Since X gives the number of units of each commodity produced and A gives the amount (in units) of each commodity used to produce one unit of the various commodities, the matrix product AX gives the amount of each commodity used up in the production process.

$$AX = \begin{bmatrix} 0 & \frac{1}{4} & \frac{1}{3} \\ \frac{1}{2} & 0 & \frac{1}{4} \\ \frac{1}{4} & \frac{1}{4} & 0 \end{bmatrix} \begin{bmatrix} 60 \\ 52 \\ 48 \end{bmatrix} = \begin{bmatrix} 29 \\ 42 \\ 28 \end{bmatrix}$$

Thus 29 units of agriculture, 42 units of manufacturing, and 28 units of transportation are used up to produce 60 units of agriculture, 52 units of manufacturing, and 48 units of transportation.   ∎

We have seen that the matrix product AX represents the amount of each commodity used up in the production process. The remainder (if any) must be enough to satisfy the demand for the various commodities from outside the production system. In an n-commodity economy, this demand can be represented by a **demand matrix** D with entries $d_1, d_2, \ldots, d_n$. The difference between the production matrix, X, and the amount, AX, used up in the production process must equal the demand, D, or

$$D = X - AX.$$

In Example 2,

$$D = \begin{bmatrix} 60 \\ 52 \\ 48 \end{bmatrix} - \begin{bmatrix} 29 \\ 42 \\ 28 \end{bmatrix} = \begin{bmatrix} 31 \\ 10 \\ 20 \end{bmatrix}.$$

Thus production of 60 units of agriculture, 52 units of manufacturing, and 48 units of transportation would satisfy a demand of 31, 10, and 20 units of each, respectively.

Another way to state the relationship between production, X, and demand, D, is to express X as $X = D + AX$. In practice, A and D usually are known and we need to find X. That is, we need to decide what amounts of production are necessary to satisfy the required demands. We can use matrix algebra to solve the equation $D = X - AX$ for X.

$$D = X - AX$$
$$D = IX - AX$$
$$D = (I - A)X$$

If the matrix $I - A$ has an inverse, then

$$X = (I - A)^{-1}D.$$

Example 3   Suppose, in the 3-commodity economy of Examples 1 and 2, there is a demand for 516 units of agriculture, 258 units of manufacturing, and 129 units of transportation. What should production of each commodity be?

The demand matrix is

$$D = \begin{bmatrix} 516 \\ 258 \\ 129 \end{bmatrix}.$$

To find the production matrix X, first calculate $I - A$.

$$I - A = \begin{bmatrix} 1 & 0 & 0 \\ 0 & 1 & 0 \\ 0 & 0 & 1 \end{bmatrix} - \begin{bmatrix} 0 & \frac{1}{4} & \frac{1}{3} \\ \frac{1}{2} & 0 & \frac{1}{4} \\ \frac{1}{4} & \frac{1}{4} & 0 \end{bmatrix} = \begin{bmatrix} 1 & -\frac{1}{4} & -\frac{1}{3} \\ -\frac{1}{2} & 1 & -\frac{1}{4} \\ -\frac{1}{4} & -\frac{1}{4} & 1 \end{bmatrix}$$

Using row operations, we find the inverse of $I - A$ (the entries are rounded to two decimal places).

$$(I - A)^{-1} = \begin{bmatrix} 1.40 & .50 & .59 \\ .84 & 1.36 & .62 \\ .56 & .47 & 1.30 \end{bmatrix}$$

We know $X = (I - A)^{-1}D$. Thus,

$$X = \begin{bmatrix} 1.40 & .50 & .59 \\ .84 & 1.36 & .62 \\ .56 & .47 & 1.30 \end{bmatrix} \begin{bmatrix} 516 \\ 258 \\ 129 \end{bmatrix} = \begin{bmatrix} 928 \\ 864 \\ 578 \end{bmatrix}$$

(rounded to the nearest whole numbers).

From the last result, we see that production of 928 units of agriculture, 864 units of manufacturing, and 578 units of transportation is required to satisfy demands of 516, 258, and 129 units respectively.   ∎

Example 4   An economy depends on two basic products, wheat and oil. To produce 1 metric ton of wheat requires .25 metric tons of wheat and .33 metric tons of oil. Production of 1 metric ton of oil consumes .08 metric tons of wheat and .11 metric tons of oil. Find the production which will satisfy a demand of 500 metric tons of wheat and 1000 metric tons of oil.

The input-output matrix, A, and $I - A$, are

$$A = \begin{bmatrix} .25 & .08 \\ .33 & .11 \end{bmatrix} \quad \text{and} \quad I - A = \begin{bmatrix} .75 & -.08 \\ -.33 & .89 \end{bmatrix}.$$

Next, calculate $(I - A)^{-1}$.

$$(I - A)^{-1} = \begin{bmatrix} 1.39 & .13 \\ .51 & 1.17 \end{bmatrix} \quad \text{(rounded)}$$

To find the production matrix X, use the equation $X = (I - A)^{-1}D$, with

$$D = \begin{bmatrix} 500 \\ 1000 \end{bmatrix}.$$

The production matrix is

$$X = \begin{bmatrix} 1.39 & .13 \\ .51 & 1.17 \end{bmatrix} \begin{bmatrix} 500 \\ 1000 \end{bmatrix} = \begin{bmatrix} 815 \\ 1425 \end{bmatrix}.$$

Production of 815 metric tons of wheat and 1425 metric tons of oil are required to satisfy the indicated demand. ∎

The model we have discussed is referred to as an **open model,** since it allows for a surplus from the production equal to D. In the **closed model,** all the production is consumed internally in the production process so that $X = AX$. There is nothing left over to satisfy any outside demands from other parts of the economy, or from other economies. In this case, the sum of each column in the input-output matrix equals one.

To solve the equation $X = AX$ for X, first let O represent an n-row column matrix with each element equal to 0. Write $X = AX$ or $X - AX = O$: then

$$IX - AX = 0,$$
$$(I - A)X = 0.$$

The system of equations which corresponds to $(I - A)X = O$ does not have a single unique solution. However, it can be solved in terms of a parameter. As we saw in Section 2.3, this means there are infinitely many solutions.

Example 5 Use matrix A given below to find the production of each commodity in a closed model.

$$A = \begin{bmatrix} \frac{1}{2} & \frac{1}{4} & \frac{1}{3} \\ 0 & \frac{1}{4} & \frac{1}{3} \\ \frac{1}{2} & \frac{1}{2} & \frac{1}{3} \end{bmatrix}$$

Find the value of $I - A$, then set $(I - A)X = 0$ to find X.

$$I - A = \begin{bmatrix} \frac{1}{2} & -\frac{1}{4} & -\frac{1}{3} \\ 0 & \frac{3}{4} & -\frac{1}{3} \\ -\frac{1}{2} & -\frac{1}{2} & \frac{2}{3} \end{bmatrix}$$

$$(I - A)X = \begin{bmatrix} \frac{1}{2} & -\frac{1}{4} & -\frac{1}{3} \\ 0 & \frac{3}{4} & -\frac{1}{3} \\ -\frac{1}{2} & -\frac{1}{2} & \frac{2}{3} \end{bmatrix} \begin{bmatrix} x_1 \\ x_2 \\ x_3 \end{bmatrix} = \begin{bmatrix} 0 \\ 0 \\ 0 \end{bmatrix}$$

$$\begin{bmatrix} \frac{1}{2}x_1 - \frac{1}{4}x_2 - \frac{1}{3}x_3 \\ 0\,x_1 + \frac{3}{4}x_2 - \frac{1}{3}x_3 \\ -\frac{1}{2}x_1 - \frac{1}{2}x_2 + \frac{2}{3}x_3 \end{bmatrix} = \begin{bmatrix} 0 \\ 0 \\ 0 \end{bmatrix}$$

From the last matrix equation, we get the following system.

$$\tfrac{1}{2} x_1 - \tfrac{1}{4} x_2 - \tfrac{1}{3} x_3 = 0$$
$$\tfrac{3}{4} x_2 - \tfrac{1}{3} x_3 = 0$$
$$-\tfrac{1}{2} x_1 - \tfrac{1}{2} x_2 + \tfrac{2}{3} x_3 = 0$$

The solution of this system can be written as

$$x_1 = \tfrac{8}{9} x_3$$
$$x_2 = \tfrac{4}{9} x_3$$
$$x_3 \quad \text{arbitrary.}$$

If $x_3 = 9$, then $x_1 = 8$ and $x_2 = 4$. Thus the production of the three commodities is in the ratio 8: 4: 9. ▉

2.6 Exercises

Find the production matrix for the following input-output and demand matrices using the open model.

1. $A = \begin{bmatrix} .5 & .4 \\ .25 & .2 \end{bmatrix}$ $D = \begin{bmatrix} 2 \\ 4 \end{bmatrix}$ 2. $A = \begin{bmatrix} .2 & .04 \\ .6 & .05 \end{bmatrix}$ $D = \begin{bmatrix} 3 \\ 10 \end{bmatrix}$

3. $A = \begin{bmatrix} .1 & .03 \\ .07 & .6 \end{bmatrix}$ $D = \begin{bmatrix} 5 \\ 10 \end{bmatrix}$ 4. $A = \begin{bmatrix} .01 & .03 \\ .05 & .05 \end{bmatrix}$ $D = \begin{bmatrix} 100 \\ 200 \end{bmatrix}$

5. $A = \begin{bmatrix} .4 & 0 & .3 \\ 0 & .8 & .1 \\ 0 & .2 & .4 \end{bmatrix}$ $D = \begin{bmatrix} 1 \\ 3 \\ 2 \end{bmatrix}$ 6. $A = \begin{bmatrix} .1 & .5 & 0 \\ 0 & .3 & .4 \\ .1 & .2 & .1 \end{bmatrix}$ $D = \begin{bmatrix} 10 \\ 4 \\ 2 \end{bmatrix}$

In Exercises 7 and 8, refer to Example 4.

7. If the demand is changed to 690 metric tons of wheat and 920 metric tons of oil, how many units of each commodity should be produced?

8. Change the technological matrix so that production of 1 ton of wheat requires 1/5 metric ton of oil (and no wheat), and the production of 1 metric ton of oil requires 1/3 metric ton of wheat (and no oil). To satisfy the same demand matrix, how many units of each commodity should be produced?

In Exercises 9–12, refer to Example 3.

9. If the demand is changed to 516 units of each commodity, how many units of each commodity should be produced?

10. Suppose 1/3 unit of manufacturing (no agriculture or transportation) is required to produce 1 unit of agriculture, 1/4 unit of transportation is required to produce 1 unit of manufacturing, and 1/2 unit of agriculture is required to produce 1 unit of transportation. How many units of each commodity should be produced to satisfy a demand of 1000 units for each commodity?

11. Suppose 1/4 unit of manufacturing and 1/2 unit of transportation are required to produce 1 unit of agriculture, 1/2 unit of agriculture and 1/4 unit of transportation to produce 1 unit of manufacturing, and 1/4 unit of agriculture and 1/4 unit of manufacturing to produce one unit of transportation. How many units of each commodity should be produced to satisfy a demand of 1000 units for each commodity?

12. If the technological matrix is changed so that 1/4 unit of manufacturing and 1/2 unit of transportation are required to produce 1 unit of agriculture, 1/2 unit of agriculture and 1/4 unit of transportation are required to produce 1 unit of manufacturing, and 1/4 unit each of agriculture and manufacturing are required to produce 1 unit of transportation, find the number of units of each commodity which should be produced to satisfy a demand for 500 units of each commodity.

13. A primitive economy depends on two basic goods, yams and pork. Production of 1 bushel of yams requires 1/4 bushel of yams and 1/2 of a pig. To produce 1 pig requires 1/6 bushel of yams. Find the amount of each commodity which should be produced to get
    (a) 1 bushel of yams and 1 pig;
    (b) 100 bushels of yams and 70 pigs.

14. Use the input-output matrix

$$\begin{array}{c} \\ \text{yams} \\ \text{pigs} \end{array} \begin{array}{c} \text{yams} \quad \text{pigs} \\ \begin{bmatrix} 1/4 & 1/2 \\ 3/4 & 1/2 \end{bmatrix} \end{array}$$

and the closed model to find the ratios of yams and pigs produced.

Find the ratios of products A, B, *and* C, *using a closed model.*

15.
$$\begin{array}{c} \\ A \\ B \\ C \end{array} \begin{array}{c} A \quad B \quad C \\ \begin{bmatrix} .3 & .1 & .8 \\ .5 & .6 & .1 \\ .2 & .3 & .1 \end{bmatrix} \end{array}$$

16.
$$\begin{array}{c} \\ A \\ B \\ C \end{array} \begin{array}{c} A \quad B \quad C \\ \begin{bmatrix} .2 & .1 & .5 \\ .4 & .3 & .4 \\ .4 & .6 & .1 \end{bmatrix} \end{array}$$

Application   *Leontief's Model of the American Economy*

In the April 1965 issue of *Scientific American,* Leontief explained his input-output system, using the 1958 American economy as an example.[1] He divided the economy into 81 sectors, grouped into six families of related sectors. In order to keep the discussion reasonably simple, we will treat each family of sectors as a single sector and so, in effect, work with a six sector model. The sectors are listed in Table 1.

Table 1

Sector	Examples
Final nonmetal (FN)	Furniture, processed food
Final metal (FM)	Household appliances, motor vehicles
Basic metal (BM)	Machine-shop products, mining
Basic nonmetal (BN)	Agriculture, printing
Energy (E)	Petroleum, coal
Services (S)	Amusements, real estate

---

[1]From *Applied Finite Mathematics* by Robert F. Brown and Brenda W. Brown. © 1977 by Wadsworth Publishing Company, Inc. Reprinted by permission of Wadsworth Publishing Company, Belmont, California 94002.

The workings of the American economy in 1958 are described in the input-output table (Table 2) based on Leontief's figures. We will demonstrate the meaning of Table 2 by considering the first left-hand column of numbers. The numbers in this column mean that 1 unit of final nonmetal production requires the consumption of 0.170 unit of (other) final nonmetal production, 0.003 unit of final metal output, 0.025 unit of basic metal products, and so on down the column. Since the unit of measurement that Leontief used for this table is millions of dollars, we conclude that the production of $1 million worth of final nonmetal production consumes $0.170 million, or $170,000, worth of other final nonmetal products, $3000 of final metal production, $25,000 of basic metal products, and so on. Similarly, the entry in the column headed FM and opposite S of 0.074 means that $74,000 worth of output from the service industries goes into the production of $1 million worth of final metal products, and the number 0.358 in the column headed E and opposite E means that $358,000 worth of energy must be consumed to produce $1 million worth of energy.

Table 2

	FN	FM	BM	BN	E	S
FN	0.170	0.004	0	0.029	0	0.008
FM	0.003	0.295	0.018	0.002	0.004	0.016
BM	0.025	0.173	0.460	0.007	0.011	0.007
BN	0.348	0.037	0.021	0.403	0.011	0.048
E	0.007	0.001	0.039	0.025	0.358	0.025
S	0.120	0.074	0.104	0.123	0.173	0.234

By the underlying assumption of Leontief's model, the production of n units (n = any number) of final nonmetal production consumes $0.170n$ unit of final nonmetal output, $0.003n$ unit of final metal output, $0.025n$ unit of basic metal production, and so on. Thus, production of $50 million worth of products from the final nonmetal section of the 1958 American economy required $(0.170)(50) = 8.5$ units ($8.5 million) worth of final nonmetal output, $(0.003)(50) = 0.15$ unit of final metal output, $(0.025)(50) = 1.25$ units of basic metal production, and so on.

Example 1 According to the simplified input—output table for the 1958 American economy, how many dollars worth of final metal products, basic nonmetal products, and services are required to produce $120 million worth of basic metal products?

Each unit ($1 million worth) of basic metal products requires 0.018 unit of final metal products because the number in the BM column of the table opposite FM is 0.018. Thus, $120 million, or 120 units, requires $(0.018)(120) = 2.16$ units, or $2.16 million worth of final metal products. Similarly, 120 units of basic metal production uses $(0.021)(120) = 2.52$ units of basic nonmetal production and $(0.104)(120) = 12.48$ units of services, or $2.52 million and $12.48 million worth of basic nonmetal output and services, respectively.   ∎

The Leontief model also involves a *bill of demands,* that is, a list of requirements for units of output beyond that required for its inner workings as described in the input-output table. These demands represent exports, surpluses, government and individual consumption, and the like. The bill of demands for the simplified version of the 1958 American economy we have been using was (in millions)

FN	$99,640
FM	$75,548
BM	$14,444
BN	$33,501
E	$23,527
S	$263,985

We can now use the methods developed above to answer the question: how many units of output from each sector are needed in order to run the economy and fill the bill of demands? The units of output from each sector required to run the economy and fill the bill of demands is unknown, so we denote them by variables. In our example, there are six quantities which are, at the moment, unknown. The number of units of final nonmetal production required to solve the problem will be our first unknown, because this sector is represented by the first row of the input-output matrix. The unknown quantity of final nonmetal units will be represented by the symbol x_1. Following the same pattern, we represent the unknown quantities in the following manner:

x_1 = units of final nonmetal production required,

x_2 = units of final metal production required,

x_3 = units of basic metal production required,

x_4 = units of basic nonmetal production required,

x_5 = units of energy required,

x_6 = units of services required.

These six numbers are the quantities we are attempting to calculate.

To find these numbers, first let A be the 6×6 matrix corresponding to the input—output table.

$$A = \begin{bmatrix} 0.170 & 0.004 & 0 & 0.029 & 0 & 0.008 \\ 0.003 & 0.295 & 0.018 & 0.002 & 0.004 & 0.016 \\ 0.025 & 0.173 & 0.460 & 0.007 & 0.011 & 0.007 \\ 0.348 & 0.037 & 0.021 & 0.403 & 0.011 & 0.048 \\ 0.007 & 0.001 & 0.039 & 0.025 & 0.358 & 0.025 \\ 0.120 & 0.074 & 0.104 & 0.123 & 0.173 & 0.234 \end{bmatrix}$$

A is the input—output matrix. The bill of demands leads to a 6×1 demand matrix D, and X is the matrix of unknowns.

$$D = \begin{bmatrix} 99,640 \\ 75,548 \\ 14,444 \\ 33,501 \\ 23,527 \\ 263,985 \end{bmatrix} \quad \text{and} \quad X = \begin{bmatrix} x_1 \\ x_2 \\ x_3 \\ x_4 \\ x_5 \\ x_6 \end{bmatrix}$$

Now we need to find $I - A$.

$$I - A = \begin{bmatrix} 1 & 0 & 0 & 0 & 0 & 0 \\ 0 & 1 & 0 & 0 & 0 & 0 \\ 0 & 0 & 1 & 0 & 0 & 0 \\ 0 & 0 & 0 & 1 & 0 & 0 \\ 0 & 0 & 0 & 0 & 1 & 0 \\ 0 & 0 & 0 & 0 & 0 & 1 \end{bmatrix} - \begin{bmatrix} 0.170 & 0.004 & 0 & 0.029 & 0 & 0.008 \\ 0.003 & 0.295 & 0.018 & 0.002 & 0.004 & 0.016 \\ 0.025 & 0.173 & 0.460 & 0.007 & 0.011 & 0.007 \\ 0.348 & 0.037 & 0.021 & 0.403 & 0.011 & 0.048 \\ 0.007 & 0.001 & 0.039 & 0.025 & 0.358 & 0.025 \\ 0.120 & 0.074 & 0.104 & 0.123 & 0.173 & 0.234 \end{bmatrix}$$

$$= \begin{bmatrix} 0.830 & -0.004 & 0 & -0.029 & 0 & -0.008 \\ -0.003 & 0.705 & -0.018 & -0.002 & -0.004 & -0.016 \\ -0.025 & -0.173 & 0.540 & -0.007 & -0.011 & -0.007 \\ -0.348 & -0.037 & -0.021 & 0.597 & -0.011 & -0.048 \\ -0.007 & -0.001 & -0.039 & -0.025 & 0.642 & -0.025 \\ -0.120 & -0.074 & -0.104 & -0.123 & -0.173 & 0.766 \end{bmatrix}$$

By the method of Section 2.5, we find the inverse (actually an approximation),

$$(I - A)^{-1} = \begin{bmatrix} 1.234 & 0.014 & 0.006 & 0.064 & 0.007 & 0.018 \\ 0.017 & 1.436 & 0.057 & 0.012 & 0.020 & 0.032 \\ 0.071 & 0.465 & 1.877 & 0.019 & 0.045 & 0.031 \\ 0.751 & 0.134 & 0.100 & 1.740 & 0.066 & 0.124 \\ 0.060 & 0.045 & 0.130 & 0.082 & 1.578 & 0.059 \\ 0.339 & 0.236 & 0.307 & 0.312 & 0.376 & 1.349 \end{bmatrix}.$$

Therefore, $X = (I - A)^{-1}D =$

$$\begin{bmatrix} 1.234 & 0.014 & 0.006 & 0.064 & 0.007 & 0.018 \\ 0.017 & 1.436 & 0.057 & 0.012 & 0.020 & 0.032 \\ 0.071 & 0.465 & 1.877 & 0.019 & 0.045 & 0.031 \\ 0.751 & 0.134 & 0.100 & 1.740 & 0.066 & 0.124 \\ 0.060 & 0.045 & 0.130 & 0.082 & 1.578 & 0.059 \\ 0.339 & 0.236 & 0.307 & 0.312 & 0.376 & 1.349 \end{bmatrix} \begin{bmatrix} 99,640 \\ 75,548 \\ 14,444 \\ 33,501 \\ 23,527 \\ 263,985 \end{bmatrix} = \begin{bmatrix} 131,161 \\ 120,324 \\ 79,194 \\ 178,936 \\ 66,703 \\ 426,542 \end{bmatrix}.$$

From this result,

$$x_1 = 131,161$$
$$x_2 = 120,324$$
$$x_3 = 79,194$$
$$x_4 = 178,936$$
$$x_5 = 66,703$$
$$x_6 = 426,542$$

In other words, it would require 131,161 units ($131,161 million worth) of final nonmetal production, 120,324 units of final metal output, 79,194 units of basic metal products, and so on to run the 1958 American economy and completely fill the stated bill of demands.

Exercises   **1.** A much simplified version of Leontief's 42 sector analysis of the 1947 American economy divides the economy into just three sectors: agriculture, manufacturing, and the household (i.e., the sector of the economy which produces labor). It consists of the following input—output table:

	Agriculture	Manufacturing	Household
Agriculture	0.245	0.102	0.051
Manufacturing	0.099	0.291	0.279
Household	0.433	0.372	0.011

The bill of demands (in billions of dollars) is

Agriculture	2.88
Manufacturing	31.45
Household	30.91

(a) Write the input—output matrix A, the demand matrix D, and the matrix X. (b) Compute $I - A$. (c) Check that

$$(I - A)^{-1} = \begin{bmatrix} 1.454 & 0.291 & 0.157 \\ 0.533 & 1.763 & 0.525 \\ 0.837 & 0.791 & 1.278 \end{bmatrix}$$

is an approximation to the inverse of $I - A$ by calculating $(I - A)^{-1}(I - A)$. (d) Use the matrix of part (c) to compute X. (e) Explain the meaning of the numbers in X in dollars.

2. An analysis of the 1958 Israeli economy[1] is here simplified by grouping the economy into three sectors: agriculture, manufacturing, and energy. The input—output table is the following.

	Agriculture	Manufacturing	Energy
Agriculture	0.293	0	0
Manufacturing	0.014	0.207	0.017
Energy	0.044	0.010	0.216

Exports (in thousands of Israeli pounds) were

Agriculture	138,213
Manufacturing	17,597
Energy	1,786

[1] Wassily Leontief, *Input—Output Economics* (New York: Oxford University Press, 1966), pp. 54–57.

(a) Write the input—output matrix A and the demand (export) matrix D. (b) Compute $I - A$. (c) Check that

$$(I - A)^{-1} = \begin{bmatrix} 1.414 & 0 & 0 \\ 0.027 & 1.261 & 0.027 \\ 0.080 & 0.016 & 1.276 \end{bmatrix}$$

is an approximation to the inverse of $I - A$ by calculating $(I - A)^{-1}(I - A)$. (d) Use the matrix of part (c) to determine the number of Israeli pounds worth of agricultural products, manufactured goods, and energy required to run this model of the Israeli economy and export the stated value of products.

Key Words

matrix	equivalent system
element	elimination method
dimension	echelon method
order	parameter
square matrix	augmented matrix
row matrix	row operations
column matrix	Gauss-Jordan method
zero matrix	identity matrix
scalar	multiplicative inverse of a matrix
system of equations	input-output matrix
first degree equation in n unknowns	production matrix
inconsistent system	demand matrix
dependent equations	

Chapter 2 Review Exercises

For each of the following, find the order of the matrices, find the values of any variables, and identify any square, row, or column matrices.

1. $\begin{bmatrix} 2 & 3 \\ 5 & q \end{bmatrix} = \begin{bmatrix} a & b \\ c & 9 \end{bmatrix}$

2. $\begin{bmatrix} 2 & x \\ y & 6 \\ 5 & z \end{bmatrix} = \begin{bmatrix} a & -1 \\ 4 & 6 \\ p & 7 \end{bmatrix}$

3. $[m \quad 4 \quad z \quad -1] = [12 \quad k \quad -8 \quad r]$

4. $\begin{bmatrix} a+5 & 3b & 6 \\ 4c & 2+d & -3 \\ -1 & 4p & q-1 \end{bmatrix} = \begin{bmatrix} -7 & b+2 & 2k-3 \\ 3 & 2d-1 & 4l \\ m & 12 & 8 \end{bmatrix}$

5. The activities of a grazing animal can be classified roughly into three categories: grazing, moving, and resting. Suppose horses spend 8 hours grazing, 8 moving, and 8 resting; cattle spend 10 grazing, 5 moving and 9 resting; sheep spend 7 grazing,

10 moving, and 7 resting; and goats spend 8 grazing, 9 moving, and 7 resting. Write this information as a 4×3 matrix.

6. The New York Stock Exchange reports in the daily newspapers give the dividend, price to earnings ratio, sales (in hundreds of shares), last price, and change in price for each company. Write the following stock reports as a 4×5 matrix. American Telephone & Telegraph: 5, 7, 2532, 52 3/8, −1/4. General Electric: 3, 9, 1464, 56, +1/8. Gulf Oil: 2.50, 5, 4974, 41, −1 1/2. Sears: 1.36, 10, 1754, 18 7/8, +1/2.

Given the matrices

$$A = \begin{bmatrix} 4 & 10 \\ -2 & -3 \\ 6 & 9 \end{bmatrix}, \qquad B = \begin{bmatrix} 2 & 3 & -2 \\ 2 & 4 & 0 \\ 0 & 1 & 2 \end{bmatrix}, \qquad C = \begin{bmatrix} 5 & 0 \\ -1 & 3 \\ 4 & 7 \end{bmatrix},$$

$$D = \begin{bmatrix} 6 \\ 1 \\ 0 \end{bmatrix}, \qquad E = \begin{bmatrix} 1 & 3 & -4 \end{bmatrix}, \qquad F = \begin{bmatrix} -1 & 4 \\ 3 & 7 \end{bmatrix}, \qquad G = \begin{bmatrix} 2 & 5 \\ 1 & 6 \end{bmatrix},$$

find each of the following which exists.

7. $A + C$ 8. $2G - 4F$ 9. $3C + 2A$ 10. $B - A$

11. $2A - 5C$ 12. AF 13. AC 14. DE

15. ED 16. BD 17. EA 18. F^{-1}

19. B^{-1} 20. $(A + C)^{-1}$

Solve each of the following systems by the echelon method.

21. $\quad 2x + 3y = 10$
$\quad -3x + \ y = 18$

22. $\dfrac{x}{2} + \dfrac{y}{4} = 3$

$\quad \dfrac{x}{4} - \dfrac{y}{2} = 4$

23. $2x - 3y + \ z = -5$
$\quad x + 4y + 2z = 13$
$\quad 5x + 5y + 3z = 14$

24. $\quad x - \ y \quad\ = 3$
$\quad 2x + 3y + \ z = 13$
$\quad 3x \quad\quad - 2z = 21$

Write each of the following as a system of equations and solve.

25. An office supply manufacturer makes two kinds of paper clips, standard, and extra large. To make 1000 standard paper clips requires 1/4 hour on a cutting machine and 1/2 hour on a machine which shapes the clips. One thousand extra large paper clips require 1/3 hour on each machine. The manager of paper clip production has four hours per day available on the cutting machine and six hours per day on the shaping machine. How many of each kind of clip can he make?

26. Jane Schmidt plans to buy shares of two stocks. One costs $32 per share and pays dividends of $1.20 per share. The other costs $23 per share and pays dividends of $1.40 per share. She has $10,100 to spend and wants to earn dividends of $540. How many shares of each stock should she buy?

27. The Waputi Indians make woven blankets, rugs, and skirts. Each blanket requires 24 hours for spinning the yarn, 4 hours for dying the yarn, and 15 hours for weaving. Rugs require 30, 5, and 18 hours, and skirts 12, 3, and 9 hours respectively. If there

are 306, 59, and 201 hours available for spinning, dying, and weaving respectively, how many of each item can be made? (Hint: Simplify the equations you write, if possible, before solving the system.)

28. An oil refinery in Tulsa sells 50% of its production to a Chicago distributor, 20% to a Dallas distributor, and 30% to an Atlanta distributor. Another refinery in New Orleans sells 40% of its production to the Chicago distributor, 40% to the Dallas distributor, and 20% to the Atlanta distributor. A third refinery in Ardmore sells the same distributors 30%, 40%, and 30% of its production. The three distributors received 219,000, 192,000, and 144,000 gallons of oil respectively. How many gallons of oil were produced at each of the three plants?

Solve each of the following systems by the Gauss-Jordan method.

29. $\begin{aligned} 2x + 4y &= -6 \\ -3x - 5y &= 12 \end{aligned}$ 　　30. $\begin{aligned} x + 2y &= -9 \\ 4x + 9y &= 41 \end{aligned}$

31. $\begin{aligned} x - y + 3z &= 13 \\ 4x + y + 2z &= 17 \\ 3x + 2y + 2z &= 1 \end{aligned}$ 　　32. $\begin{aligned} x \quad\quad - 2z &= 5 \\ 3x + 2y \quad\quad &= 8 \\ -x \quad\quad + 2z &= 10 \end{aligned}$

33. $\begin{aligned} 3x - 6y + 9z &= 12 \\ -x + 2y - 3z &= -4 \\ x + y + 2z &= 7 \end{aligned}$

Solve each of the following systems of equations by inverses.

34. $\begin{aligned} 2x + y &= 5 \\ 3x - 2y &= 4 \end{aligned}$ 　　35. $\begin{aligned} 5x + 10y &= 80 \\ 3x - 2y &= 120 \end{aligned}$ 　　36. $\begin{aligned} x + y + z &= 1 \\ 2x + y \quad\quad &= -2 \\ 3y + z &= 2 \end{aligned}$

Find the inverse of each of the following matrices that has an inverse.

37. $\begin{bmatrix} 2 & 1 \\ 5 & 3 \end{bmatrix}$ 　　38. $\begin{bmatrix} -4 & 2 \\ 0 & 3 \end{bmatrix}$ 　　39. $\begin{bmatrix} 2 & 0 \\ -1 & 5 \end{bmatrix}$ 　　40. $\begin{bmatrix} 6 & 4 \\ 3 & 2 \end{bmatrix}$

41. $\begin{bmatrix} 2 & -1 & 0 \\ 1 & 0 & 1 \\ 1 & -2 & 0 \end{bmatrix}$ 　　42. $\begin{bmatrix} 2 & 0 & 4 \\ 1 & -1 & 0 \\ 0 & 1 & -2 \end{bmatrix}$

43. $\begin{bmatrix} 1 & 3 & 6 \\ 4 & 0 & 9 \\ 5 & 15 & 30 \end{bmatrix}$ 　　44. $\begin{bmatrix} 2 & 3 & 5 \\ -2 & -3 & -5 \\ 1 & 4 & 2 \end{bmatrix}$

Find the production matrix given the following input-output and demand matrices.

45. $A = \begin{bmatrix} .01 & .05 \\ .04 & .03 \end{bmatrix}$ 　　$D = \begin{bmatrix} 200 \\ 300 \end{bmatrix}$

46. $A = \begin{bmatrix} .2 & .1 & .3 \\ .1 & 0 & .2 \\ 0 & 0 & .4 \end{bmatrix}$ 　　$D = \begin{bmatrix} 500 \\ 200 \\ 100 \end{bmatrix}$

47. An economy depends on two commodities, goats and cheese. It takes 2/3 of a unit of goats to produce 1 unit of cheese and 1/2 unit of cheese to produce 1 unit of goats.
    (a) Write the input-output matrix for this economy.
    (b) Find the production required to satisfy a demand of 400 units of cheese and 800 units of goats.

Solve each of the following matrix equations AX = B for X.

48. $A = \begin{bmatrix} 2 & 4 \\ -1 & -3 \end{bmatrix}, \qquad B = \begin{bmatrix} 8 \\ 3 \end{bmatrix}$ **49.** $A = \begin{bmatrix} 1 & 3 \\ -2 & 4 \end{bmatrix}, \qquad B = \begin{bmatrix} 15 \\ 10 \end{bmatrix}$

50. $A = \begin{bmatrix} 1 & 0 & 2 \\ -1 & 1 & 0 \\ 3 & 0 & 4 \end{bmatrix}, \qquad B = \begin{bmatrix} 8 \\ 4 \\ -6 \end{bmatrix}$ **51.** $A = \begin{bmatrix} 2 & 4 & 0 \\ 1 & -2 & 0 \\ 0 & 0 & 3 \end{bmatrix}, \qquad B = \begin{bmatrix} 72 \\ -24 \\ 48 \end{bmatrix}$

52. A printer has three orders for pamphlets which require three kinds of paper as shown in the following matrix.

		High-grade	Medium-grade	Coated
			Paper	
Orders	I	10	12	0
	II	5	0	10
	III	8	4	5

The printer has on hand 3900 sheets of high-grade paper, 3750 sheets of medium-grade paper, and 3500 sheets of coated paper. All the paper must be used in preparing the order.

(a) Write a 3 × 1 matrix for the amounts of paper on hand.

(b) Write a matrix of variables to represent the number of pamphlets that must be printed in each of the three orders.

(c) Write a matrix equation using the given matrix and your matrices from parts (a) and (b).

(d) Solve the equation from part (c).

3

Linear Programming— The Graphical Method

While linear algebra (matrix theory) developed over several hundred years, linear programming has gained importance as a quantitative tool in management in a relatively short time. Linear programming developed from a practical problem faced by George B. Dantzig in 1947, that of allocating supplies for the United States Air Force in the least expensive way.

In this chapter, we first discuss the solution of systems of linear inequalities. We then introduce a graphical method for solving simple linear programming problems. In the next chapter, we present the simplex method, a general method for solving linear programming problems.

3.1 Systems of Linear Inequalities

Many mathematical models of real situations are best expressed as inequalities, rather than as equations. A **system of inequalities** may include some equations but must include at least one inequality. Since the solution set of an inequality is usually infinite, the clearest and most meaningful description of the solution is often its graph. The solution of a system of inequalities is shown graphically by the overlap (the part that is common to all) of the graphs of the inequalities of the system.

Let us review how to graph an inequality. A line divides a plane into three parts; the line itself, and the two regions determined by the line. Each of these two regions is called a **half-plane.** In Figure 1, line r divides the plane into half-planes P and Q. The points of r belong to neither P nor Q. Line r is the **boundary** of each half-plane.

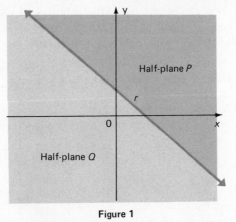

Figure 1

A **linear inequality** is one of the form $ax + by < c$, where a, b, and c are real numbers, with a and b not both equal to 0. (We could replace $<$ with $\geq$, $\leq$, or $>$.) The graph of a linear inequality is either a half-plane or a half-plane and its boundary.

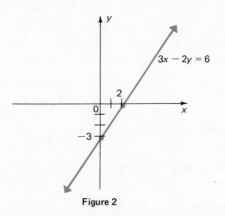

Figure 2

To graph the linear inequality $3x - 2y \leq 6$, for example, first graph the boundary, $3x - 2y = 6$, as shown in Figure 2. Since the points of the line $3x - 2y = 6$ satisfy $3x - 2y \leq 6$, this line is part of the solution. To decide which half-plane — the one above the line $3x - 2y = 6$ or the one below the line — is the rest of the solution, solve the given inequality for y.

$$3x - 2y \leq 6$$
$$-2y \leq -3x + 6$$
$$y \geq \frac{3}{2}x - 3$$

(Recall that multiplying both sides of an inequality by a negative number requires that the direction of the inequality symbol be reversed.)

For any value of x, this inequality will be satisfied by all values of y which are *greater than* or equal to $(3/2)x - 3$. These are the y-values *above* the line, as shown by the shaded region in Figure 3.

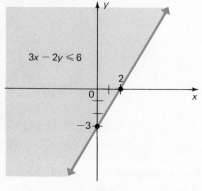

Figure 3

Example 1   Graph $x + 4y < 4$.

The boundary here is the line $x + 4y = 4$. Since the points on this line do not satisfy $x + 4y < 4$, it is customary to draw the line dashed as in Figure 4. To decide which half-plane is the solution, solve for y.

$$x + 4y < 4$$
$$4y < -x + 4$$
$$y < -\frac{1}{4}x + 1$$

Since y *is less than* $(-1/4)x + 1$, the solution is the half-plane *below* the boundary, as shown by the shaded region in Figure 4.   ■

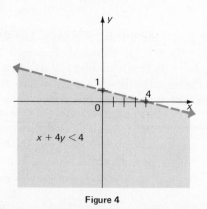

Figure 4

There is an alternate way to find the correct region to shade, or to use as a check for the method shown above. Choose as a test point any point not on the boundary line. For example, in Example 1 we could choose the point $(1, 0)$ which is not on the line $x + 4y = 4$. Substitute 1 for x and 0 for y in the given inequality.

$$x + 4y < 4$$
$$1 + 4(0) < 4$$
$$1 < 4$$

The result is a true sentence, so we shade the side of the boundary line which includes the test point. If we had a false result, we would shade the opposite side.

Example 2   Graph $x \le -1$

Recall that the graph of $x = -1$ is a vertical line. The solution includes the half plane to the left of the boundary, since we want $x \le -1$, plus the points on the boundary. See Figure 5.   ■

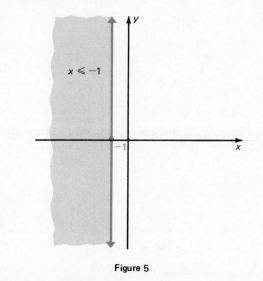

Figure 5

Systems of Inequalities   To graph the solution of a system of inequalities, graph all the inequalities on the same axes and identify, by heavy shading, the region common to all graphs. The next example shows how this is done.

Example 3   Graph the following system.

$$y < -3x + 12$$
$$x < 2y$$

The heavily shaded region of Figure 6 is the common region of the graphs of each inequality of the system. Since the points on the boundary lines are not in the solution, the lines are dashed.   ■

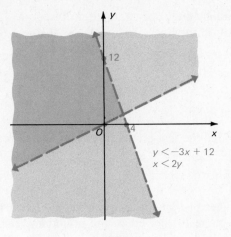

Figure 6

Example 4   Graph the system

$$2x - 5y \leq 10$$
$$x + 2y \leq \;\; 8$$
$$x \geq \;\; 0$$
$$y \geq \;\; 0.$$

Graph each inequality separately. Heavily shade the region which is common to all graphs. The solution is shown in Figure 7.   ▮

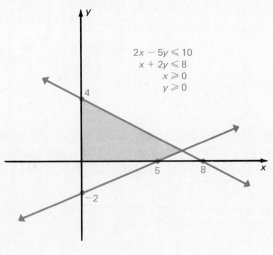

Figure 7

3.1 Exercises   *Graph each of the following linear inequalities.*

1. $x + y \leq 2$ **2.** $y \leq x + 1$ **3.** $x \geq 3 + y$

4. $y \geq x - 3$ **5.** $4x - y < 6$ **6.** $3y + x > 4$

7. $3x + y < 6$ **8.** $2x - y > 2$ **9.** $x + 3y \geq -2$

10. $2x + 3y \leq 6$ **11.** $4x + 3y > -3$ **12.** $5x + 3y > 15$

13. $2x - 4y < 3$ **14.** $4x - 3y < 12$ **15.** $x \leq 5y$

16. $2x \geq y$ **17.** $-3x < y$ **18.** $-x > 6y$

19. $x + y \leq 0$ **20.** $3x + 2y \geq 0$ **21.** $y < x$

22. $y > -2x$ **23.** $x < 4$ **24.** $y > 5$

25. $y \leq -2$ **26.** $x \geq 3$

Graph the solution of each of the following systems of inequalities.

27. $x + y \leq 1$
$x - y \geq 2$

28. $3x - 4y < 6$
$2x + 5y > 15$

29. $2x - y < 1$
$3x + y < 6$

30. $x + 3y \leq 6$
$2x + 4y \geq 7$

31. $-x - y < 5$
$2x - y < 4$

32. $6x - 4y > 8$
$3x + 2y > 4$

33. $x + y \leq 4$
$x - y \leq 5$
$4x + y \leq -4$

34. $3x - 2y \geq 6$
$x + y \leq -5$
$y \leq 4$

35. $-2 < x < 3$
$-1 \leq y \leq 5$
$2x + y < 6$

36. $-2 < x < 2$
$y > 1$
$x - y > 0$

37. $2y + x \geq -5$
$y \leq 3 + x$
$x \geq 0$
$y \geq 0$

38. $2x + 3y \leq 12$
$2x + 3y > -6$
$3x + y < 4$
$x \geq 0$
$y \geq 0$

39. $3x + 4y > 12$
$2x - 3y < 6$
$0 \leq y \leq 2$
$x \geq 0$

40. $0 \leq x \leq 9$
$x - 2y \geq 4$
$3x + 5y \leq 30$
$y \geq 0$

41. The California Almond Growers have 2400 boxes of almonds to be shipped from their plant in Sacramento to Des Moines and San Antonio. The Des Moines market needs at least 1000 boxes, while the San Antonio market must have at least 800 boxes. Let $x =$ the number of boxes to be shipped to Des Moines and $y =$ the number of boxes to be shipped to San Antonio.
   (a) Write a system of inequalities to express the conditions of the problem. Remember that the variables must be nonnegative.
   (b) Graph the system.

42. A cement manufacturer produces at least 3.2 million barrels of cement annually. He is told by the Environmental Protection Agency that his operation emits 2.5 pounds of dust for each barrel produced. The EPA has ruled that annual emissions must be reduced to 1.8 million pounds. To do this the manufacturer plans to replace the present dust collectors with two types of electronic precipitators. One type would reduce emissions to .5 pounds per barrel and would cost 16¢ per barrel. The other would reduce the dust to .3 pounds per barrel and would cost 20¢ per barrel. The manufacturer does not want to spend more than .8 million dollars on the precipitators. He needs to know how many barrels he should produce with each type.
   (a) Let $x =$ the number of barrels in millions produced with the first type and $y =$ the number of barrels in millions produced with the second type. Write inequalities to express the manufacturer's restrictions. Remember that the number of barrels cannot be a negative number.
   (b) Graph the system.

3.2 Solving Linear Programming Problems Graphically

Many mathematical models representing problems in business, biology, and economics involve optimizing (either maximizing or minimizing) a linear function subject to certain constraints. In particular, a linear programming problem either maximizes or minimizes a linear function, called the **objective function,** subject to linear **constraints** which can be expressed as linear equations or inequalities. Such problems can be solved by linear programming methods. The following examples illustrate how we can optimize an objective function by graphing a system of linear inequalities.

Example 1   Find the values of x and y which will maximize the value of the objective function $z = 2x + 5y$ subject to the constraints

$$3x + 2y \le 6$$
$$-2x + 4y \le 8$$
$$x \ge 0$$
$$y \ge 0.$$

The graph of the solution of this system is shown in Figure 8. Note that the shaded region is enclosed by boundary lines on each side. Such a region is said to be **bounded.** This enclosed region is called the **region of feasible solutions.** Any point in the interior or on the boundary of this region will satisfy all constraints.

Now we must choose one point (x, y) of the feasible region that gives the maximum possible value of z. To locate that point, let us add to the graph of Figure 8 lines which represent the objective function at possible values of z. Figure 9 shows the lines representing the objective function when $z = 0, 5, 10,$ and 15. These lines have equations

$$0 = 2x + 5y \qquad 10 = 2x + 5y$$
$$5 = 2x + 5y \qquad 15 = 2x + 5y.$$

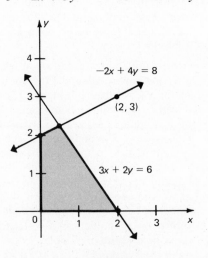

Figure 8

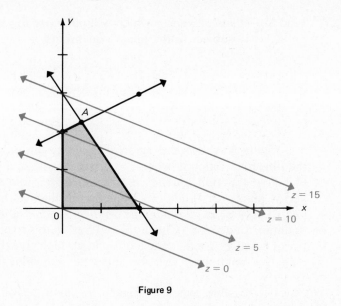

Figure 9

(Why are all the lines parallel?) From the figure, we see that z cannot be as large as 15 because the graph for $z = 15$ is outside the feasible region. The maximum possible value of z will be obtained if we draw another line, parallel to the others and between the lines representing the objective function when $z = 10$ and $z = 15$. We will maximize z and still satisfy all constraints if this intermediate line just "touches" the feasible region.

This occurs at point A, a **vertex** (or **corner**) **point** of the region of feasible solutions. To find the coordinates of this point A, note in Figure 8 that the point lies on the two lines with equations

$$3x + 2y = 6$$
$$-2x + 4y = 8.$$

By solving this system of two equations, we find that A has coordinates (1/2, 9/4). Thus, the value of z at this point is

$$z = 2x + 5y$$
$$z = 2\left(\frac{1}{2}\right) + 5\left(\frac{9}{4}\right)$$
$$z = 12\frac{1}{4}.$$

The maximum possible value of z under the given conditions is thus 12 1/4. Out of all the feasible solutions, each of which satisfies the constraints of the problem, the vertex point (1/2, 9/4) gives us the optimum value. ▌

Example 2   Find the values of x and y which will minimize the value of the objective function $z = 2x + 4y$ subject to the linear constraints

$$x + 2y \geq 10$$
$$3x + y \geq 10$$
$$x \geq 0$$
$$y \geq 0.$$

Figure 10 shows the region of feasible solutions and the lines which represent the objective function when $z = 0, 10, 20, 40,$ and 50. The line representing the objective function touches the region of feasible solutions when $z = 20$. Two vertex points $(2, 4)$ and $(10, 0)$ lie on this line. In this case, both points $(2, 4)$ and $(10, 0)$ as well as *all the points on the boundary line between them* give the same optimum value of z. (Check this.) Thus, in this example, there is an infinite number of equally "good" values of x and y which give the same minimum value of the objective function $z = 2x + 4y$. This minimum value is 20. ∎

The feasible region of Figure 10 illustrates an important point about linear programming problems: *not all problems have a solution.* For example, with this feasible region we cannot find a *maximum* value of $z = 2x + 4y$. As the graph shows, the feasible region is unbounded and goes indefinitely to the upper right. Thus, there is no point whose coordinates could maximize $z = 2x + 4y$.

In summary, in a linear programming problem, the **region of feasible solutions** is the set of all points which satisfy all linear inequalities representing the constraints of the problem. The region of feasible solutions is **bounded** if its graph is enclosed by boundary lines on all sides. *If the region of feasible solutions is bounded, the optimum value will always occur at a vertex point.* There may be

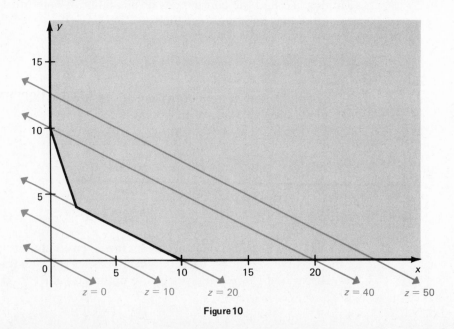

Figure 10

an infinite number of solutions which produce the same optimum value. (In this case, two of them will occur at vertex points.) If the region of feasible solutions is unbounded, there may be no solution which produces an optimum value.

Figures 9 and 10 help us see that the following theorem is true for cases with two variables. The proof of the general case is beyond the scope of this text.

Theorem 3.1

> If an optimum value (either a maximum or a minimum) of the objective function exists, it will occur at one or more of the vertex points.

Theorem 3.1 simplifies the job of finding an optimum value. All we need to do is sketch the feasible region and identify all vertex points. Then test each point in the objective function and choose the one which produces the optimum value for the objective function. For unbounded regions, we must consider whether the required optimum can be found, as illustrated in Example 2 above.

For example, we could have solved the problem in Example 1 by looking at Figure 8 and identifying the four vertex points as $(0, 0)$, $(0, 2)$, $(1/2, 9/4)$, and $(2, 0)$. To find each vertex point, identify the equations of any two lines which intersect. Solving the resulting system of two equations gives the coordinates of a vertex point. Substitute each of the four points into the objective function, $z = 2x + 5y$, and then choose the vertex point that produces the maximum value of z.

Vertex point	Value of $z = 2x + 5y$
$(0, 0)$	$2(0) + 5(0) = 0$
$(0, 2)$	$2(0) + 5(2) = 10$
$(\frac{1}{2}, \frac{9}{4})$	$2(\frac{1}{2}) + 5(\frac{9}{4}) = 12\frac{1}{4}$ (maximum)
$(2, 0)$	$2(2) + 5(0) = 4$

From these results, we see that the vertex point $(1/2, 9/4)$ yields the maximum value of 12 1/4. This is the same result that we obtained above.

Example 3 Sketch the feasible region for the following set of constraints.

$$3y - 2x \geq 0$$

$$y + 8x \leq 52$$

$$y - 2x \leq 2$$

$$x \geq 3$$

Then find the maximum and minimum values of the objective function

$$z = 5x + 2y.$$

The graph in Figure 11 shows that the feasible region is bounded. Use the vertex points from the graph to find the maximum and minimum values of the objective function.

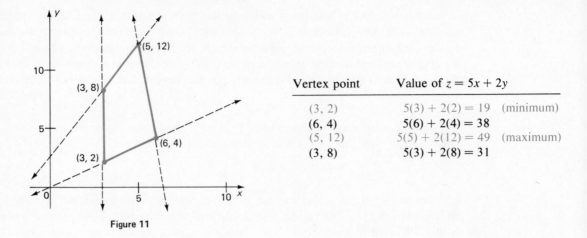

Figure 11

Vertex point	Value of $z = 5x + 2y$
(3, 2)	$5(3) + 2(2) = 19$ (minimum)
(6, 4)	$5(6) + 2(4) = 38$
(5, 12)	$5(5) + 2(12) = 49$ (maximum)
(3, 8)	$5(3) + 2(8) = 31$

The minimum value of $z = 5x + 2y$ is 19 at the vertex point (3, 2). The maximum value is 49 at (5, 12). ■

Example 4   An office manager needs to purchase new filing cabinets. He knows that Ace cabinets cost \$40 each, require 6 square feet of floor space, and hold 8 cubic feet of files. On the other hand, each Excello cabinet costs \$80, requires 8 square feet of floor space, and holds 12 cubic feet. His budget permits him to spend no more than \$560 on files, while the office has room for no more than 72 square feet of cabinets. The manager desires the greatest storage capacity within the limitations imposed by funds and space. How many of each type cabinet should he buy?

It is helpful to organize the information given in the problem as follows.

	Cost	Space required	Storage capacity
1 Ace cabinet	\$40	6 sq. ft.	8 cu. ft.
1 Excello cabinet	\$80	8 sq. ft.	12 cu. ft.

The objective function to be maximized is the amount of storage capacity provided by some combination of Ace and Excello cabinets. If we let x represent the number of Ace cabinets to be bought and y the number of Excello cabinets, then the objective function is

$$\text{storage space} = z = 8x + 12y.$$

From the chart, we see that the constraints imposed by cost and space can be expressed as follows.

$$40x + 80y \leq 560 \qquad \text{(cost)}$$
$$6x + 8y \leq 72 \qquad \text{(floor space)}$$

Also, the number of cabinets cannot be negative. Thus, $x \geq 0$ and $y \geq 0$.

The graph of the feasible region is the shaded area in Figure 12. By Theorem 3.1, we know that the maximum value of the objective function will occur at one of the vertex points or on the boundary. From the figure, we can identify three of the vertex points as (0, 0), (0, 7), and (12, 0). The coordinates of the other point, labeled Q in the figure, can be found by solving the system of equations

$$40x + 80y = 560$$

$$6x + 8y = 72.$$

By solving this system, we find that Q is the point (8, 3). We must now test the coordinates of these four points in the objective function z to determine which point maximizes it. The results are shown below.

Vertex point	Value of $z = 8x + 12y$
(0, 0)	0
(0, 7)	84
(12, 0)	96
(8, 3)	100 (maximum)

Thus, the objective function, which represents storage space, is maximized when $x = 8$ and $y = 3$. The manager should buy 8 Ace cabinets and 3 Excello cabinets. ■

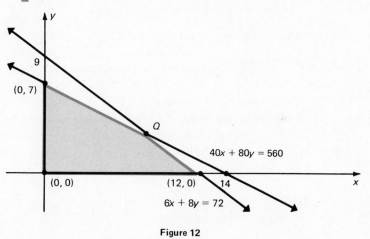

Figure 12

3.2 Exercises   *Exercises 1–4 show regions of feasible solutions. Use these regions to find maximum and minimum values of each given objective function.*

1. $z = 3x + 5y$ **2.** $z = 6x + y$

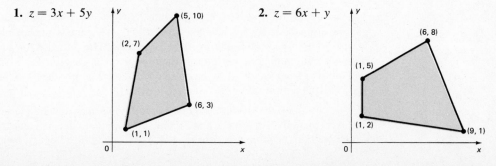

3. $z = .40x + .75y$ **4.** $z = .35x + 1.25y$

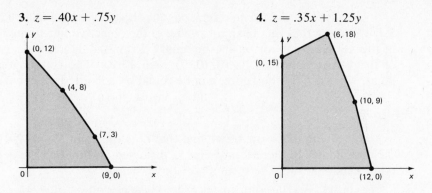

Use graphical methods to solve each of the following problems.

5. Find $x \geq 0$ and $y \geq 0$ such that
$$2x + 3y \leq 6$$
$$4x + y \leq 6$$
and $z = 5x + 2y$ is maximized.

6. Find $x \geq 0$ and $y \geq 0$ such that
$$x + y \leq 10$$
$$5x + 2y \geq 20$$
$$-x + 2y \geq 0$$
and $z = x + 3y$ is minimized.

7. Find $x \geq 2$ and $y \geq 5$ such that
$$3x - y \geq 12$$
$$x + y \leq 15$$
and $z = 2x + y$ is minimized.

8. Find $x \geq 10$ and $y \geq 20$ such that
$$2x + 3y \leq 100$$
$$5x + 4y \leq 200$$
and $z = x + 3y$ is maximized.

9. Find $x \geq 0$ and $y \geq 0$ such that
$$x - y \leq 10$$
$$5x + 3y \leq 75$$
and $z = 4x + 2y$ is maximized.

10. Find $x \geq 0$ and $y \geq 0$ such that
$$10x - 5y \leq 100$$
$$20x + 10y \geq 150$$
and $z = 4x + 5y$ is minimized.

11. Find values of $x \geq 0$ and $y \geq 0$ which maximize $z = 10x + 12y$ subject to each of the following sets of constraints.

(a) $x + y \leq 20$
$x + 3y \leq 24$

(b) $3x + y \leq 15$
$x + 2y \leq 18$

(c) $2x + 5y \geq 22$
$4x + 3y \leq 28$
$2x + 2y \leq 17$

12. Find values of $x \geq 0$ and $y \geq 0$ which minimize $z = 3x + 2y$ subject to each of the following sets of constraints.

(a) $10x + 7y \leq 42$
$4x + 10y \geq 35$

(b) $6x + 5y \geq 25$
$2x + 6y \geq 15$

(c) $x + 2y \geq 10$
$2x + y \geq 12$
$x - y \leq 8$

Solve each of the following linear programming problems by the graphical method.

13. The manufacturing process requires that oil refineries must manufacture at least two gallons of gasoline for every one of fuel oil. To meet the winter demand for fuel oil, at least 3 million gallons a day must be produced. The demand for gasoline is no more than 6.4 million gallons per day. If the refinery sells gasoline for $1.25 per gallon, and fuel oil for $1 per gallon, how much of each should be produced to maximize revenue? Find the maximum revenue.

14. Mark, who is ill, takes vitamin pills. Each day he must have at least 16 units of vitamin A, 5 units of vitamin B_1, and 20 units of vitamin C. He can choose between pill #1 which contains 8 units of A, 1 of B_1, and 2 of C, and pill #2 which contains 2 units of A,

1 of B_1, and 7 of C. Pill #1 costs 15¢ and pill #2 costs 30¢. How many of each pill should he buy in order to minimize his cost? What is the minimum cost?

15. A machine shop manufactures two types of bolts. Each can be made on any of three groups of machines, but the time required on each group differs, as shown in the table below.

		Machine groups		
		I	II	III
Bolts	Type 1	.4 hour	.5 hour	.2 hour
	Type 2	.3 hour	.2 hour	.4 hour

Production schedules are made up for one week at a time. In this period there are 1200 hours of machine time available for each machine group. Type 1 bolts sell for 10¢ and type 2 bolts for 12¢. How many of each type of bolt should be manufactured per week to maximize revenue? What is the maximum revenue?

16. Seall Manufacturing Company makes color television sets. It produces a bargain set that sells for $100 profit and a deluxe set that sells for $150 profit. On the assembly line the bargain set requires 3 hours, while the deluxe set takes 5 hours. The cabinet shop spends one hour on the cabinet for the bargain set and 3 hours on the cabinet for the deluxe set. Both sets require 2 hours of time for testing and packing. On a particular production run the Seall Company has available 3900 work hours on the assembly line, 2100 work hours in the cabinet shop, and 2200 work hours in the testing and packing department. How many sets of each type should it produce to make maximum profit? What is the maximum profit?

17. A manufacturer of refrigerators must ship at least 100 refrigerators to its two West coast warehouses. Each warehouse holds a maximum of 100 refrigerators. Warehouse A holds 25 refrigerators already, while warehouse B has 20 on hand. It costs $12 to ship a refrigerator to warehouse A and $10 to ship one to warehouse B. How many refrigerators should be shipped to each warehouse to minimize cost? What is the minimum cost?

18. In a small town in South Carolina, zoning rules require that the window space (in square feet) in a house be at least one-sixth of the space used up by solid walls. The cost to heat the house is 2¢ for each square foot of solid walls and 8¢ for each square foot of windows. Find the maximum total area (windows plus walls) if $16 is available to pay for heat.

19. A candy company has 100 kilograms of chocolate-covered nuts and 125 kilograms of chocolate-covered raisins to be sold as two different mixes. One mix will contain half nuts and half raisins and will sell for $6 per kilogram. The other mix will contain 1/3 nuts and 2/3 raisins and will sell for $4.80 per kilogram. How many kilograms of each mix should the company prepare for maximum revenue?

20. Ms. Oliveras was given the following advice. She should supplement her daily diet with at least 6000 USP units of vitamin A, at least 195 milligrams of vitamin C, and at least 600 USP units of vitamin D. Ms. Oliveras finds that Mason's Pharmacy carries Upjohn vitamin pills at 5¢ each and Lilly vitamins at 4¢ each. Each Upjohn pill contains 3000 USP units of A, 45 milligrams of C, and 75 USP units of D, while the Lilly pills contain 1000 USP units of A, 50 milligrams of C, and 200 USP units of D. What combination of vitamin pills should she buy to obtain the least possible cost? What is the least possible cost per day?

21. Sam, who is dieting, requires two food supplements, I and II. He can get these supplements from two different products, A and B, as shown in the following table.

		Supplement (grams per serving)	
		I	II
Product	A	3	2
	B	2	4

Sam's physician has recommended that he include at least 15 grams of each supplement in his daily diet. If product A costs 25¢ per serving and product B costs 40¢ per serving, how can he satisfy his requirements most economically?

22. A small country can grow only two crops for export, coffee and cocoa. The country has 500,000 hectares[1] of land available for the crops. Long-term contracts require that at least 100,000 hectares be devoted to coffee and at least 200,000 hectares to cocoa. Cocoa must be processed locally, and production bottlenecks limit cocoa to 270,000 hectares. Coffee requires two workers per hectare, with cocoa requiring five. No more than 1,750,000 people are available for these crops. Coffee produces a profit of $220 per hectare, and cocoa a profit of $310 per hectare. How many hectares should the country devote to each crop in order to maximize the profit?

3.3   The Graphical Method with Three Variables (Optional)

In this section, we extend the ideas of the last section to linear programming problems with three variables. The solution of a linear equation in three variables, such as $2x + 3y - 4z = 10$, is an **ordered triple** (x, y, z). For example, one solution to $2x + 3y - 4z = 10$ is the ordered triple $(1, 0, -2)$. To graph ordered triples, we use a three-dimensional coordinate system with axes x, y, and z as shown in Figure 13. As we shall see, it is not necessary to actually graph the system of equations in a linear programming problem to use the graphical method. The following example illustrates the method.

Example 1   A farmer has 100 acres of available land which he wishes to plant with a mixture of potatoes, corn, and cabbage. To produce an acre of potatoes costs him $400, of corn, $160, and of cabbage, $280. He has a maximum capital of $20,000 to invest. How many acres of each crop should he plant to maximize his profit, if he makes a profit of $120 per acre of potatoes, $40 per acre of corn, and $60 per acre of cabbage?

If we let x, y, and z represent the number of acres planted with potatoes, corn, and cabbage respectively, we have

$$x + y + z \leq 100.$$

---

[1] A hectare is a metric unit of land measure; one hectare is equivalent to approximately 2.47 acres.

(There are 100 acres of land available for these crops.) Also the potato crop will cost $400x$ dollars; the corn, $160y$ dollars; and the cabbage, $280z$ dollars. The total of these costs must not be more than $20,000, so

$$400x + 160y + 280z \leq 20,000.$$

All three variables must be nonnegative, so that

$$x \geq 0,\ y \geq 0,\ \text{and } z \geq 0.$$

The system formed by these five inequalities has as its graph a three-dimensional region bounded by the five planes with equations

$$x + y + z = 100 \tag{1}$$
$$400x + 160y + 280z = 20,000 \tag{2}$$
$$x = 0 \tag{3}$$
$$y = 0 \tag{4}$$
$$z = 0. \tag{5}$$

The graph of Figure 13 shows the region of feasible solutions which consists of the interior and surface points of the shaded three-dimensional figure. Theorem 3.1 in the last section applies for any number of variables. Thus, the problem is

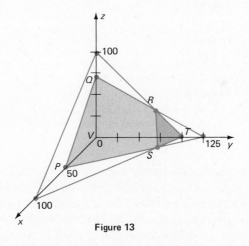

Figure 13

reduced to finding the coordinates of the vertex points. The vertex points of the region of feasible solutions in Figure 13 are labeled P, Q, R, S, T, and V.

The coordinates of the six vertex points shown in Figure 13 can be found by solving appropriate systems of three equations taken from the five equations of the planes. For example, the planes $x = 0$, $z = 0$, and $x + y + z = 100$ intersect in the point T with coordinates $(0, 100, 0)$ which is the solution of the system

$$x = 0$$
$$z = 0$$
$$x + y + z = 100.$$

Similarly, the solution of the system

$$x + y + z = 100$$
$$400x + 160y + 280z = 20{,}000$$
$$z = 0$$

is the vertex point S with coordinates (16 2/3, 83 1/3, 0).

All the vertex points can be found by solving all possible systems of three equations taken from the full system of inequalities which state the restrictions of the problem. Recall that the equations describe the planes in the graph, and it is the intersection of these planes that determine the vertex points. In general, for n variables, we must solve all possible systems of n equations in n unknowns taken from the given inequalities, and then discard the solutions that fail to satisfy the complete system.

To use this method to find all the vertex points which satisfy the given system for Example 1, begin by listing all possible combinations of three of the five equations. Such a list is shown in column I of the following table.

I Equations	II Solution (x, y, z)	III All inequalities satisfied?	IV Vertex points
1, 2, 3	$(0, 66\frac{2}{3}, 33\frac{1}{3})$	yes	$(0, 66\frac{2}{3}, 33\frac{1}{3}) = R$
1, 2, 4	$(-66\frac{2}{3}, 0, 166\frac{2}{3})$	no	
1, 2, 5	$(16\frac{2}{3}, 83\frac{1}{3}, 0)$	yes	$(16\frac{2}{3}, 83\frac{1}{3}, 0) = S$
1, 3, 4	$(0, 0, 100)$	no	
1, 3, 5	$(0, 100, 0)$	yes	$(0, 100, 0) = T$
1, 4, 5	$(100, 0, 0)$	no	
2, 3, 4	$(0, 0, 71\frac{3}{7})$	yes	$(0, 0, 71\frac{3}{7}) = Q$
2, 3, 5	$(0, 125, 0)$	no	
2, 4, 5	$(50, 0, 0)$	yes	$(50, 0, 0) = P$
3, 4, 5	$(0, 0, 0)$	yes	$(0, 0, 0) = V$

Note that it is wise to proceed through this list in some orderly way to be sure that all possible combinations are represented.[1]

Next, solve each of these systems of three equations for x, y, and z. The solutions are shown in column II of the table. (Duplicate solutions may occur.) Each solution must now be checked in all five inequalities. Only those solutions which satisfy all inequalities will be vertex points. The six vertex points of Figure 13 found in this manner are (0, 0, 0), (50, 0, 0), (0, 100, 0), (0, 0, 71 3/7), (16 2/3, 83 1/3, 0), and (0, 66 2/3, 33 1/3).

The last step is to substitute the values from each vertex point into the objective function and determine the point which leads to the maximum value. The results are shown in the following table.

---

[1]A formula for finding the number of possible combinations is given in Section 5.4.

Vertex Point	Value of $120x + 40y + 60z$
$(0, 66\frac{2}{3}, 33\frac{1}{3})$	\$4666.67
$(16\frac{2}{3}, 83\frac{1}{3}, 0)$	\$5333.33
$(0, 100, 0)$	\$4000.00
$(0, 0, 71\frac{3}{7})$	\$4285.71
$(50, 0, 0)$	\$6000.00 (maximum)
$(0, 0, 0)$	\$0.00

From the chart, we see that planting 50 acres of potatoes and none of the other crops (leaving 50 acres idle) leads to a maximum profit of \$6000. ∎

3.3 Exercises   *For each of the following systems of inequalities, find without graphing the vertex points of the region of feasible solutions. Then find the points which lead to a maximum and a minimum for the given objective function.*

1. $3x + 2y + z \leq 10$
$x + 2y \leq 8$
$x \geq 0$
$y \geq 0$
$z \geq 0$
objective function: $2x + 3y + z$

2. $x \geq 5$
$y \geq 2$
$z \geq 0$
$x + y + z \geq 15$
$2x + 3y - z \geq 10$
objective function: $3x + y + 5z$

3. $x + y + z + w \leq 12$
$2x \geq y$
$x + y \geq z$
$y \geq 0$
$w \geq 0$
objective function:
  $3x + 2y + 4z + 6w$
(Hint: There are four points.)

4. $x + y + z \geq 8$
$2m + n \leq z$
$x \geq 5$
$y \geq 3$
$m \geq 2$
$n \geq 0$
objective function:
  $2m + 5n + 3x + y + 4z$
(Hint: There are six points.)

Solve each of the following linear programming problems.

5. At most 100 kilograms of chocolate-covered caramels are to be mixed with at most 80 kilograms of chocolate-covered nuts to form three different mixtures, Deluxe, Supreme, and Economy, as shown in the table below.

	Deluxe	Supreme	Economy
Caramel	30%	50%	60%
Nuts	70%	50%	40%

How many kilograms of each mixture should be made to maximize the profit if the profit per kilogram of each mix is \$1.50, \$2, and \$1.25, respectively? Find the maximum profit.

6. Solve Exercise 5 if 90 kilograms of each type of candy is used.

7. In planning production schedules, a company must take into account three shifts with varying labor costs but with the same capacity for turning out units of the product as shown in the following table.

	Labor cost/unit	Capacity in units
First shift	$100	30
Second shift	$120	30
Third shift	$130	30

The first shift must produce at least 15 units. A particular project has been allotted labor costs of $10,000. How many units should be produced on each shift to maximize profit, if each unit produces a profit of $3? Find the maximum profit.

8. Solve Exercise 7 if the costs and capacity are as shown below, and if the first shift must produce at least 20 units.

	Labor cost/unit	Capacity
First shift	$90	40
Second shift	$100	40
Third shift	$110	40

9. If the farmer in Example 1 has 125 acres of land and $24,000 working capital, find the amounts of the three crops he should plant to maximize his profit. Find the maximum profit.

10. What amounts of each of the three crops should the farmer in Example 1 plant with the conditions in Exercise 9 if the profit on each crop is as follows: potatoes $100, corn $60, and cabbage $80?

Key Words

system of inequalities	constraints
half-plane	region of feasible solutions
boundary	vertex point
linear inequality	ordered triple
objective function	

Chapter 3 Review Exercises

Graph each of the following linear inequalities.

1. $y \geq 2x + 3$
2. $3x - y \leq 5$
3. $3x + 4y \leq 12$
4. $2x - 6y \geq 18$
5. $y \geq x$
6. $y \leq 3$

Graph the solution of each of the following systems of inequalities. Find all vertex (or corner) points.

7. $x + y \leq 6$
   $2x - y \geq 3$
8. $4x + y \geq 8$
   $2x - 3y \leq 6$
9. $-4 \leq x \leq 2$
   $-1 \leq y \leq 3$
   $x + y \leq 4$

10. $2 \leq x \leq 5$
    $1 \leq y \leq 7$
    $x - y \leq 3$

11. $x + 3y \geq 6$
    $4x - 3y \leq 12$
    $x \geq 0$
    $y \geq 0$

12. $x + 2y \leq 4$
    $2x - 3y \leq 6$
    $x \geq 0$
    $y \geq 0$

Set up a system of inequalities for each of the following problems; then graph the solution of the system.

13. A bakery makes both cakes and cookies. Each batch of cakes requires two hours in the oven and three hours in the decorating room. Each batch of cookies needs one and a half hours in the oven and two thirds of an hour in the decorating room. The oven is available no more than 15 hours a day, while the decorating room can be used no more than 13 hours a day.

14. A company makes two kinds of pizza, basic and plain. Basic contains cheese and beef, while plain contains onions and beef. The company sells at least three units a day of basic, and at least two units of plain. The beef costs $5 per unit for basic, and $4 per unit for plain. They can spend no more than $50 per day on beef. Dough for basic is $2 per unit, while dough for plain is $1 per unit. The company can spend no more than $16 per day on dough.

Use the given regions to find the maximum and minimum values of the objective function $z = 2x + 4y$.

15.
16.

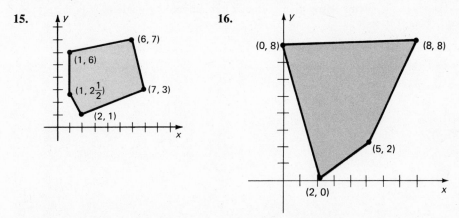

Use the graphical method to solve the following problems.

17. Find $x \geq 0$ and $y \geq 0$ such that

    $$3x + 2y \leq 12$$
    $$5x + y \geq 5$$

    and $2x + 4y$ is maximized.

18. Find $x \geq 0$ and $y \geq 0$ such that

    $$8x + 9y \geq 72$$
    $$6x + 8y \geq 72$$

    and $3x + 2y$ is minimized.

19. Find $x \geq 0$ and $y \geq 0$ such that
    $$x + y \leq 50$$
    $$2x + y \geq 20$$
    $$x + 2y \geq 30$$
    and $4x + 2y$ is minimized.

20. How many units of each kind of pizza should the company of Exercise 14 make in order to maximize profits if basic sells for $20 per unit and plain for $15 per unit?

21. How many batches of cakes and cookies should the bakery of Exercise 13 make in order to maximize profits if cookies produce a profit of $20 per batch and cakes produce a profit of $30 per batch?

For each of the following, without graphing, find the points which lead to a maximum and a minimum for the objective function $10x + 5y + 8z$.

22. $3x + 4y + 2z \leq 12$
$x + y \leq 8$
$x \geq 0$
$y \geq 0$
$z \geq 0$

23. $x \geq 10$
$y \geq 6$
$z \geq 0$
$x + y + z \geq 20$
$x + 2y + 3z \geq 30$

24. A manufacturer of outdoor equipment makes three products: sleeping bags, down jackets, and back packs. The amount of material (in yards) and number of hours of labor required for each item is given below. The profit per item is also shown.

	Material	Labor	Profit
Sleeping bags	4	3	$20
Jackets	2	4	$15
Back packs	1.5	5	$18

There are 100 yards of material and 40 hours of labor available each day for these items. How many of each item should be made to maximize profit?

4

Linear Programming— The Simplex Method

In the previous chapter, we solved linear programming problems using the graphical method. The graphical method illustrates the basic ideas of linear programming, but is practical only for problems with two variables. For problems with more than two variables, or with two variables and many constraints, we use the simplex method.

The **simplex method** involves going systematically from one feasible solution (on a boundary) to another in such a way that the objective function is automatically improved (or at least stays the same) with each new feasible solution. Finally, an optimum solution is reached or it can be seen that none exists.

The simplex method requires a number of steps. We have divided the presentation of these steps into two parts. We show how to set up the problem and begin the method in Section 4.1, and how to complete the method in Section 4.2. Then, in Section 4.3, we consider minimizing problems and more complicated maximizing problems.

4.1 Slack Variables and the Pivot

Since the simplex method is used for problems with a large number of variables it is not always possible to use letters such as x, y, z, or w as variable names. We use instead the symbols x_1 (read "x-sub-one"), x_2, x_3, and so on, as variables. These variable names lend themselves easily to use on the computer.[1]

In a linear programming problem, we must have a linear function which we want to optimize (the objective function) and a set of constraints which are ex-

---

[1] A BASIC computer program for linear programming by the simplex method is given in *Study Guide with Computer Problems,* by Margaret L. Lial, (Glenview, Ill.: Scott, Foresman and Company, 1979).

pressed as linear equations or inequalities. To use the simplex method, all constraints must be expressed in the linear form

$$a_1x_1 + a_2x_2 + a_3x_3 + \cdots \leq b$$

where $x_1, x_2, x_3, \ldots$ are variables and $a_1, a_2, \ldots, b$ are constants. That is, we want all terms with variables on the left and the constant terms on the right. We restrict our discussion at this point to problems which meet the following conditions.

1. **All variables are nonnegative. ($x_i \geq 0$)**
2. **The constant terms of the constraints are nonnegative. ($b \geq 0$)**
3. **The problems are maximization problems in which all constraints are $\leq$ inequalities.**

(In Section 4.3 we discuss problems which do not meet all of these conditions.)
We can express a linear programming problem in the matrix notation of Chapter 2. Let X represent the column matrix with entries $x_1, x_2, \ldots, x_n$, A represent the matrix of coefficients of the variables in the linear inequalities, and B represent the matrix of constants. Then the constraints can be written as

$$AX \leq B.$$

Then if C is the row matrix of coefficients of the objective function, the problem is stated as follows:

$$\text{maximize} \qquad z = CX$$
$$\text{subject to} \qquad AX \leq B$$
$$X \geq 0.$$

Slack Variables   We begin the simplex method by converting the constraints, which are linear inequalities, into linear equations. (We then have a system of linear equations which can be solved by the matrix methods of Chapter 2.) The linear inequalities are converted into linear equations by adding a nonnegative variable, called a **slack variable,** to each constraint. For example, we can convert the inequality $x_1 + x_2 \leq 10$ into an equation by adding the slack variable x_3, to get

$$x_1 + x_2 + x_3 = 10, \qquad \text{where } x_3 \geq 0.$$

The inequality $x_1 + x_2 \leq 10$ tells us that the sum $x_1 + x_2$ is less than or perhaps equal to 10. The variable x_3 "takes up any slack" and represents the amount by which $x_1 + x_2$ fails to equal 10. If $x_1 + x_2 = 10$, the value of x_3 is 0.
In the following example, we rework Example 1 of Section 3.3, using the simplex method.

Example 1   A farmer has 100 acres of available land which he wishes to plant with a mixture of potatoes, corn, and cabbage. It costs him $400 to produce an acre of potatoes, $160 to produce an acre of corn, and $280 to produce an acre of cabbage. He has a maximum of $20,000 to spend. He makes a profit of $120 per acre of potatoes, $40 per acre of corn, and $60 per acre of cabbage. How many acres of each crop should he plant to maximize his profit?

We can summarize the information in the problem as follows.

	Cost	Profit
1 acre of potatoes	$400	$120
1 acre of corn	160	40
1 acre of cabbage	280	60

If the number of acres allotted to each of the three crops is represented by x_1, x_2, and x_3, respectively, then the constraints of the example can be expressed as

$$x_1 + \quad x_2 + \quad x_3 \leq \quad 100 \quad \text{(number of acres)},$$
$$400x_1 + 160x_2 + 280x_3 \leq 20000 \quad \text{(production costs)},$$

where x_1, x_2, and x_3 are all nonnegative. The first of these constraints tells us that $x_1 + x_2 + x_3$ is less than or perhaps equal to 100. We use x_4 to represent any slack, and express $x_1 + x_2 + x_3 \leq 100$ as the equation

$$x_1 + x_2 + x_3 + x_4 = 100.$$

Here the slack variable x_4 represents the amount of the farmer's 100 acres that will not be used. (x_4 may be 0 or any value up to 100.)

In the same way, the constraint $400x_1 + 160x_2 + 280x_3 \leq 20000$ can be converted into an equation by adding a slack variable, x_5:

$$400x_1 + 160x_2 + 280x_3 + x_5 = 20000.$$

The slack variable x_5 represents any unused portion of the farmer's $20,000 in capital. (Again, x_5 may be any value from 0 to 20,000.)

The objective function here represents the profit. The farmer wants to maximize

$$z = 120x_1 + 40x_2 + 60x_3.$$

We can now restate this problem as follows. Find $x_1 \geq 0, x_2 \geq 0, x_3 \geq 0, x_4 \geq 0$, and $x_5 \geq 0$ such that

$$x_1 + \quad x_2 + \quad x_3 + x_4 = \quad 100$$
$$400x_1 + 160x_2 + 280x_3 + x_5 = 20000,$$

and $z = 120x_1 + 40x_2 + 60x_3 + 0x_4 + 0x_5$ is maximized.

Here we added slack variables x_4 and x_5. Note that each constraint requires a separate slack variable. We used coefficients of 0 for x_4 and x_5 in the objective function because there is no profit in unused land or capital. The problem is now in a form suitable for solution by the simplex method. We return to this example later.   ■

Example 2   Restate the following linear programming problem in a form suitable for solution by the simplex method. Maximize $z = 3x_1 + 2x_2 + x_3$ subject to the constraints

$$2x_1 + \quad x_2 + \quad x_3 \leq 150$$
$$2x_1 + 2x_2 + 8x_3 \leq 200$$
$$2x_1 + 3x_2 + \quad x_3 \leq 320$$

and $x_1 \geq 0, x_2 \geq 0, x_3 \geq 0$.

We must restate each of these constraints as an equation. To do this, we need three nonnegative slack variables, x_4, x_5, and x_6, one for each constraint. By adding slack variables we can restate the problem as follows: Find $x_1 \geq 0$, $x_2 \geq 0$, $x_3 \geq 0$, $x_4 \geq 0$, $x_5 \geq 0$, and $x_6 \geq 0$ such that

$$
\begin{aligned}
2x_1 + x_2 + x_3 + x_4 &= 150 \\
2x_1 + 2x_2 + 8x_3 + x_5 &= 200 \\
2x_1 + 3x_2 + x_3 + x_6 &= 320
\end{aligned}
$$

and $z = 3x_1 + 2x_2 + x_3$ is maximized.   ■

Basic Feasible Solutions   One solution for the system of linear equations from Example 2 is $x_1 = 0$, $x_2 = 0$, $x_3 = 0$, $x_4 = 150$, $x_5 = 200$, and $x_6 = 320$. We write this solution as (0, 0, 0, 150, 200, 320). We can identify such a solution more readily if we write the system of equations as an augmented matrix. For example, the system of equations above leads to the matrix

$$
\begin{array}{cccccc}
x_1 & x_2 & x_3 & x_4 & x_5 & x_6
\end{array}
$$
$$
\begin{bmatrix}
2 & 1 & 1 & 1 & 0 & 0 & | & 150 \\
2 & 2 & 8 & 0 & 1 & 0 & | & 200 \\
2 & 3 & 1 & 0 & 0 & 1 & | & 320
\end{bmatrix}.
$$

The three columns headed x_4, x_5, and x_6 form a 3×3 identity matrix. If all the other variables have the value 0, the system will reduce to

$$
\begin{aligned}
1 \cdot x_4 + 0 \cdot x_5 + 0 \cdot x_6 &= 150 \\
0 \cdot x_4 + 1 \cdot x_5 + 0 \cdot x_6 &= 200 \\
0 \cdot x_4 + 0 \cdot x_5 + 1 \cdot x_6 &= 320,
\end{aligned}
$$

or simply $x_4 = 150$, $x_5 = 200$, and $x_6 = 320$. These results lead to the same solution (0, 0, 0, 150, 200, 320) that we found above. Thus, by identifying the columns which form an identity matrix, the nonzero values of a solution can be read directly from the matrix.

The simplex method uses this procedure of identifying the columns like those of an identity matrix to determine the nonzero values of a solution. All variables associated with other columns will be given the value 0 in that solution.

Example 3   Read a solution from the matrix shown below.

$$
\begin{array}{ccccc}
x_1 & x_2 & x_3 & x_4 & x_5
\end{array}
$$
$$
\begin{bmatrix}
2 & 1 & 8 & 5 & 0 & | & 27 \\
9 & 0 & 3 & 12 & 1 & | & 45
\end{bmatrix}
$$

Here the columns of the identity matrix are under the variables x_2 and x_5. Thus we let the other three variables equal zero in the solution. Since the column of the identity matrix under x_2 has a 1 in the *first* row, then $x_2 = 27$, where 27 is the constant in the *first* row. Similarly, $x_5 = 45$. The solution is (0, 27, 0, 0, 45).   ■

In the discussion of Example 2, we arbitrarily let $x_1 = 0$, $x_2 = 0$, and $x_3 = 0$. This seems to contradict common sense. These three variables came from the given linear programming problem, while the variables x_4, x_5, and x_6 were merely introduced to help find the final answer to the given problem. However, what we are doing does have a purpose. With the simplex method we will be able to go systematically from one vertex of the region of feasible solutions to another in such a way that the value of the objective function is improved, or at least remains the same, each time. To get started, we choose the simplest solution, the one in which the original variables are all 0.

The solutions which are read from the matrix include values for both the variables of the given problem and the slack variables which are introduced to change the constraints to equations. These solutions are represented graphically (in problems where this is possible) by the vertices of the region of feasible solutions. In Example 2, setting x_1, x_2, and $x_3 = 0$ gives the solution which corresponds to the vertex (0, 0, 0) of the region of feasible solutions.

Example 4   Suppose we want to maximize $z = 6x_1 + 7x_2$, given

$$4x_1 + 5x_2 \le 20$$
$$3x_1 + 2x_2 \le 12.$$

If x_3 and x_4 are the slack variables, the inequalities become the equations

$$4x_1 + 5x_2 + x_3 \quad\quad = 20$$
$$3x_1 + 2x_2 \quad\quad + x_4 = 12,$$

with solutions written (x_1, x_2, x_3, x_4). The graph in Figure 1 shows the region of feasible solutions, which has vertices (0, 0), (0, 4), (20/7, 12/7), and (4, 0). The corresponding feasible solutions and values of z are listed below.

Vertex	Feasible solution	z	
(0, 0)	(0, 0, 20, 12)	0	
(0, 4)	(0, 4, 0, 4)	28	
$(\frac{20}{7}, \frac{12}{7})$	$(\frac{20}{7}, \frac{12}{7}, 0, 0)$	$29\frac{1}{7}$	maximum
(4, 0)	(4, 0, 4, 0)	24	

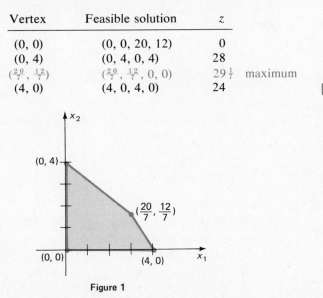

Figure 1

We have seen that to start the simplex method the original variables are set equal to 0. The following theorem tells us that there will be the same number of 0 values in the optimum solution.

Theorem 4.1

> Suppose an objective function has k variables subject to m linear inequalities, and suppose m slack variables are introduced. Then, if an optimum solution exists, there is always one optimum solution for which at least k of the $m + k$ variables are 0.

Because of this theorem, we need only consider **basic solutions,** those for which at least k variables are 0. A basic solution in which all the variables are nonnegative is called a **basic feasible solution.** It can be shown that it corresponds to a vertex of the region of feasible solutions.

The feasible solutions listed in Example 4 are all basic feasible solutions and each one corresponds to a vertex. By Theorem 4.1, $k = 2$ and $m = 2$. The only optimum solution is (20/7, 12/7, 0, 0) which has at least 2 of the 4 variables equal to 0 as predicted.

Example 5   In Example 1 the objective function

$$z = 120x_1 + 40x_2 + 60x_3$$

has three variables. Thus, by Theorem 4.1, we have $k = 3$. This objective function is subject to two constraints (number of acres and production costs). We added two slack variables to change the inequalities into equations. Thus, $m = 2$ and $m + k = 2 + 3 = 5$. By Theorem 4.1, if there is an optimal solution, it will be one in which at least three of the five variables are 0.   ■

The solution that we read directly from the augmented matrix seldom leads to a maximum or minimum. It is necessary to proceed to other solutions until an optimum is finally reached. To do this, we use row transformations to change the augmented matrix so that one of the zero variables takes on a nonzero value. Making a zero variable become nonzero forces one of the nonzero variables to become 0 (Theorem 4.1).

The Pivot   In Example 2 we had the matrix

$$\begin{array}{cccccc} x_1 & x_2 & x_3 & x_4 & x_5 & x_6 \end{array}$$
$$\begin{bmatrix} 2 & 1 & 1 & 1 & 0 & 0 & | & 150 \\ 2 & 2 & 8 & 0 & 1 & 0 & | & 200 \\ 2 & 3 & 1 & 0 & 0 & 1 & | & 320 \end{bmatrix} \begin{array}{c} x_4 \\ x_5 \\ x_6. \end{array}$$

(The variables in the far right column are placed in the row that corresponds to the equation in which the variable was introduced.) From this matrix we read the basic feasible solution (0, 0, 0, 150, 200, 320).

Now, suppose we want x_2 to be nonzero and x_5 to be zero in the solution. We say that x_2 is the **entering variable** and x_5 the **departing variable.** We show this using the matrix below by having one arrow point to row 2 to indicate that x_5

departs and another arrow point to column 2 to indicate that x_2 is to enter. The number in both row 2 and column 2 is called the **pivot.** Here the pivot, 2, is circled.

$$
\begin{array}{cccccc}
x_1 & x_2 & x_3 & x_4 & x_5 & x_6
\end{array}
$$

$$
\begin{bmatrix}
2 & 1 & 1 & 1 & 0 & 0 & | & 150 \\
2 & \textcircled{2} & 8 & 0 & 1 & 0 & | & 200 \\
2 & 3 & 1 & 0 & 0 & 1 & | & 320
\end{bmatrix}
\begin{array}{l}
x_4 \\
x_5 \quad \leftarrow \\
x_6
\end{array}
$$

We now use matrix row operations (see Chapter 2) to obtain a new matrix where the x_2 column, instead of the x_5 column, will become

$$
\begin{array}{c}
0 \\
1 \\
0.
\end{array}
$$

To change the pivot from a 2 to a 1, multiply row 2 by the reciprocal of the pivot, 1/2, which gives the following matrix.

$$
\begin{array}{cccccc}
x_1 & x_2 & x_3 & x_4 & x_5 & x_6
\end{array}
$$

$$
\begin{bmatrix}
2 & 1 & 1 & 1 & 0 & 0 & | & 150 \\
1 & 1 & 4 & 0 & \frac{1}{2} & 0 & | & 100 \\
2 & 3 & 1 & 0 & 0 & 1 & | & 320
\end{bmatrix}
$$

Now we want to get a 0 both above and below this 1 in the x_2 column. To get a 0 above, multiply row 2 of the new matrix by -1 and add the result to row 1.

$$
\begin{array}{cccccc}
x_1 & x_2 & x_3 & x_4 & x_5 & x_6
\end{array}
$$

$$
\begin{bmatrix}
1 & 0 & -3 & 1 & -\frac{1}{2} & 0 & | & 50 \\
1 & 1 & 4 & 0 & \frac{1}{2} & 0 & | & 100 \\
2 & 3 & 1 & 0 & 0 & 1 & | & 320
\end{bmatrix}
$$

Now multiply row 2 by -3 and add the result to row 3 to get a new row 3.

$$
\begin{array}{cccccc}
x_1 & x_2 & x_3 & x_4 & x_5 & x_6
\end{array}
$$

$$
\begin{bmatrix}
1 & 0 & -3 & 1 & -\frac{1}{2} & 0 & | & 50 \\
1 & 1 & 4 & 0 & \frac{1}{2} & 0 & | & 100 \\
-1 & 0 & -11 & 0 & -\frac{3}{2} & 1 & | & 20
\end{bmatrix}
\begin{array}{l}
x_4 \\
x_2 \\
x_6
\end{array}
$$

From this matrix we can read the basic feasible solution $(0, 100, 0, 50, 0, 20)$.

When we use row operations to change the matrix in this way, we are simply replacing the original system of equations with an equivalent system of equations. This is what we did when we used the row operations to solve systems of equations in Section 2.4.

In the simplex method, this process is repeated until an optimum solution is found, if one exists. In the next section, we discuss how to decide which variables to use as entering and departing variables in order to improve the value of the objective function. We also show how to tell when an optimum solution has been reached, or that one does not exist.

4.1 Exercises   *Convert each of the following inequalities into equations by adding a slack variable.*

1. $x_1 + 2x_2 \leq 6$ **2.** $3x_1 + 5x_2 \leq 100$

3. $2x_1 + 4x_2 + 3x_3 \leq 100$ **4.** $8x_1 + 6x_2 + 5x_3 \leq 250$

For each of the following problems,
(*a*) *determine the number of slack variables needed;*
(*b*) *name them;*
(*c*) *use slack variables to convert each constraint into a linear equation.*

5. Maximize $z = 10x_1 + 12x_2$ subject to
$$4x_1 + 2x_2 \leq 20$$
$$5x_1 + x_2 \leq 50$$
$$2x_1 + 3x_2 \leq 25$$
and $x_1 \geq 0$, $x_2 \geq 0$.

6. Maximize $z = 1.2x_1 + 3.5x_2$ subject to
$$2.4x_1 + 1.5x_2 \leq 10$$
$$1.7x_1 + 1.9x_2 \leq 15$$
and $x_1 \geq 0$, $x_2 \geq 0$.

7. Maximize $z = 8x_1 + 3x_2 + x_3$ subject to
$$7x_1 + 6x_2 + 8x_3 \leq 118$$
$$4x_1 + 5x_2 + 10x_3 \leq 220$$
and $x_1 \geq 0$, $x_2 \geq 0$, $x_3 \geq 0$.

8. Maximize $z = 12x_1 + 15x_2 + 10x_3$ subject to
$$2x_1 + 2x_2 + x_3 \leq 8$$
$$x_1 + 4x_2 + 3x_3 \leq 12$$
and $x_1 \geq 0$, $x_2 \geq 0$, $x_3 \geq 0$.

Write the basic feasible solution that can be read from each of the following.

9. x_1 x_2 x_3 x_4 x_5

$$\begin{bmatrix} 2 & 2 & 0 & 3 & 1 & | & 15 \\ 3 & 4 & 1 & 6 & 0 & | & 20 \end{bmatrix}$$

10. x_1 x_2 x_3 x_4 x_5

$$\begin{bmatrix} 0 & 2 & 1 & 1 & 3 & | & 5 \\ 1 & 5 & 0 & 1 & 2 & | & 8 \end{bmatrix}$$

11. x_1 x_2 x_3 x_4 x_5 x_6

$$\begin{bmatrix} 6 & 2 & 1 & 3 & 0 & 0 & | & 8 \\ 2 & 2 & 0 & 1 & 0 & 1 & | & 7 \\ 2 & 1 & 0 & 3 & 1 & 0 & | & 6 \end{bmatrix}$$

12. x_1 x_2 x_3 x_4 x_5 x_6

$$\begin{bmatrix} 0 & 2 & 0 & 1 & 2 & 2 & | & 3 \\ 0 & 3 & 1 & 0 & 1 & 2 & | & 2 \\ 1 & 4 & 0 & 0 & 3 & 5 & | & 5 \end{bmatrix}$$

For each of the following matrices, the entering and departing variables are indicated by arrows.
(*a*) *Determine the pivot and perform the indicated transformation.*
(*b*) *State the resulting basic feasible solution.*

13. x_1 x_2 x_3 x_4 x_5

$$\begin{bmatrix} 1 & 2 & 4 & 1 & 0 & | & 56 \\ 2 & 2 & 1 & 0 & 1 & | & 40 \end{bmatrix} \leftarrow$$
$\uparrow$

14. x_1 x_2 x_3 x_4 x_5

$$\begin{bmatrix} 5 & 4 & 1 & 1 & 0 & | & 50 \\ 3 & 3 & 2 & 0 & 1 & | & 40 \end{bmatrix} \begin{matrix} \leftarrow \\ \end{matrix}$$
$\uparrow$

15. x_1 x_2 x_3 x_4 x_5 x_6

$$\begin{bmatrix} 2 & 2 & 1 & 1 & 0 & 0 & | & 12 \\ 1 & 2 & 3 & 0 & 1 & 0 & | & 45 \\ 3 & 1 & 1 & 0 & 0 & 1 & | & 20 \end{bmatrix} \begin{matrix} \leftarrow \\ \\ \end{matrix}$$
$\uparrow$

16. x_1 x_2 x_3 x_4 x_5 x_6

$$\begin{bmatrix} 4 & 2 & 3 & 1 & 0 & 0 & | & 22 \\ 2 & 2 & 5 & 0 & 1 & 0 & | & 28 \\ 1 & 3 & 2 & 0 & 0 & 1 & | & 45 \end{bmatrix} \begin{matrix} \\ \leftarrow \\ \end{matrix}$$
$\uparrow$

17. $\quad x_1 \quad x_2 \quad x_3 \quad x_4 \quad x_5 \quad x_6$

$$\begin{bmatrix} 1 & 1 & 1 & 1 & 1 & 0 & | & 50 \\ 3 & 1 & 2 & 1 & 0 & 1 & | & 100 \end{bmatrix} \leftarrow$$
$\qquad\qquad\quad \uparrow$

18. $\quad x_1 \quad x_2 \quad x_3 \quad x_4 \quad x_5 \quad x_6 \quad x_7$

$$\begin{bmatrix} 1 & 2 & 3 & 1 & 1 & 0 & 0 & | & 115 \\ 2 & 1 & 8 & 5 & 0 & 1 & 0 & | & 200 \\ 1 & 0 & 1 & 0 & 0 & 0 & 1 & | & 50 \end{bmatrix} \leftarrow$$
$\quad \uparrow$

Restate the following linear programming problems, adding slack variables as necessary, and then write the resulting systems of equations as augmented matrices.

19. Find $x_1 \geq 0$ and $x_2 \geq 0$ such that
$$2x_1 + 3x_2 \leq 6$$
$$4x_1 + x_2 \leq 6$$
and $z = 5x_1 + x_2$ is maximized.

20. Find $x_1 \geq 50$, $x_2 \geq 50$ such that
$$2x_1 + 3x_2 \leq 100$$
$$5x_1 + 4x_2 \leq 200$$
and $z = x_1 + 3x_2$ is maximized.

21. Find $x_1 \geq 0$, $x_2 \geq 0$ such that
$$x_1 + x_2 \leq 10$$
$$5x_1 + 2x_2 \leq 20$$
$$x_1 + 2x_2 \leq 36$$
and $z = x_1 + 3x_2$ is maximized.

22. Find $x_1 \geq 0$, $x_2 \geq 0$ such that
$$x_1 + x_2 \leq 10$$
$$5x_1 + 3x_2 \leq 75$$
and $z = 4x_1 + 2x_2$ is maximized.

23. Find $x_1 \geq 0$, $x_2 \geq 0$ such that
$$3x_1 + x_2 \leq 12$$
$$x_1 + x_2 \leq 15$$
and $z = 2x_1 + x_2$ is maximized.

24. Find $x_1 \geq 0$, $x_2 \geq 0$ such that
$$10x_1 + 4x_2 \leq 100$$
$$20x_1 + 10x_2 \leq 150$$
and $z = 4x_1 + 5x_2$ is maximized.

Set up each of the following problems for solution by the simplex method: that is, express the linear constraints and objective function, add slack variables, and set up the matrix.

25. A candy company has 100 kilograms of chocolate-covered nuts and 125 kilograms of chocolate-covered raisins to be sold as two different mixtures. One mix will contain half nuts and half raisins and will sell for $6 per kilogram. The other mix will contain 1/3 nuts and 2/3 raisins, and will sell for $4.80 per kilogram. How many kilograms of each mix should the company prepare for maximum revenue? (This is Exercise 19, Section 3.2.)

26. Seall Manufacturing Company makes color television sets. It produces a bargain set that sells for $100 profit and a deluxe set that sells for $150 profit. On the assembly line the bargain set requires 3 hours' work, while the deluxe set takes 5 hours. The cabinet shop spends one hour on the cabinet for the bargain set and 3 hours on the cabinet for the deluxe set. Both sets require 2 hours of time for testing and packing. On a particular production run the Seall Company has available 3900 work hours on the assembly line, 2100 work hours in the cabinet shop, and 2200 work hours in the testing and packing department. How many sets of each type should it produce to make maximum profit? What is the maximum profit? (See Exercise 16, Section 3.2.)

27. A small boat manufacturer builds three types of fiberglass boats: prams, runabouts, and trimarans. The pram sells at a profit of $75, the runabout at a profit of $90, and the trimaran at a profit of $100. The factory is divided into two sections. Section A does the molding and construction work, while section B does the painting, finishing and equipping. The pram takes 1 hour in section A and 2 hours in section B. The runabout takes 2 hours in A and 5 hours in B. The trimaran takes 3 hours in A and 4 hours in B. Section A has a total of 6240 hours available and section B has 10,800 hours available for the year. The manufacturer has ordered a supply of fiberglass that will build at most 3000 boats, figuring the average amount used per boat. How many of each type of boat should be made to produce maximum profit? What is the maximum profit?

28. Caroline's Quality Candy Confectionery is famous for fudge, chocolate cremes, and pralines. Its candy-making equipment is set up to make 100-pound batches at a time. Currently there is a chocolate shortage and the company can get only 120 pounds of chocolate in the next shipment. On a week's run, the confectionery's cooking and processing equipment is available for a total of 42 machine hours. During the same period the employees have a total of 56 work hours available for packaging. A batch of fudge requires 20 pounds of chocolate while a batch of cremes uses 25 pounds of chocolate. The cooking and processing take 120 minutes for fudge, 150 minutes for chocolate cremes, and 200 minutes for pralines. The packaging time measured in minutes per one pound box are 1, 2, and 3 respectively, for fudge, cremes, and pralines. Determine how many batches of each type of candy the confectionery should make, assuming that the profit per pound box is 50¢ on fudge, 40¢ on chocolate cremes, and 45¢ on pralines.

29. A cat breeder has the following amounts of cat food: 90 units of tuna, 80 units of liver, and 50 units of chicken. To raise a Siamese cat, the breeder must use 2 units of tuna, 1 of liver, and 1 of chicken per day, while raising a Persian cat requires 1, 2, and 1 units respectively per day. If a Siamese cat sells for $12, while a Persian cat sells for $10, how many of each should be raised in order to obtain maximum gross income? What is the maximum gross income?

30. Banal, Inc. produces art for motel rooms. Its painters can turn out mountain scenes, seascapes, and pictures of clowns. Each painting is worked on by three different artists, T, D, and H. Artist T works only 25 hours per week, while D and H work 45 and 40 hours per week, respectively. Artist T spends 1 hour on a mountain scene, 2 hours on a seascape, and 1 hour on a clown. Corresponding times for D and H are 3, 2, and 2 hours, and 2, 1, and 4 hours, respectively. Banal makes $20 on a mountain scene, $18 on a seascape, and $22 from a clown. The head painting packer can't stand clowns, so that no more than 4 clown paintings may be done in a week. Find the number of each type of painting that should be made weekly in order to maximize profit. Find the maximum possible profit.

4.2   Maximization Problems

We have learned how to prepare a linear programming problem for solution by first converting the constraints to linear equations with slack variables. We then wrote the coefficients of the variables from the linear equations as an augmented matrix. Finally, we used the pivot to go from one vertex of the region of feasible solutions to another.

Now we are ready to put all this together and produce an optimum value for the objective function. To see how this is done, let us finally complete the example about the farmer. (Recall Example 1 from Section 4.1.) In this example, we are to maximize the profit

$$z = 120x_1 + 40x_2 + 60x_3$$

subject to the constraints

$$x_1 + x_2 + x_3 \leq 100$$
$$400x_1 + 160x_2 + 280x_3 \leq 20000,$$

with $x_1 \geq 0$, $x_2 \geq 0$, and $x_3 \geq 0$.

We first added slack variables x_4 and x_5, where $x_4 \geq 0$ and $x_5 \geq 0$, getting the equations

$$x_1 + \quad x_2 + \quad x_3 + x_4 \quad = \quad 100 \tag{1}$$

and

$$400x_1 + 160x_2 + 280x_3 \quad + x_5 = 20000. \tag{2}$$

The objective function was then written as

$$z = 120x_1 + 40x_2 + 60x_3 + 0x_4 + 0x_5.$$

As mentioned before, all equations used in the simplex method should be in linear form with variable terms on the left and a constant term on the right. Furthermore, that constant term and all variables must be nonnegative. Therefore, for the simplex method, the objective function should be rewritten as

$$-120x_1 - 40x_2 - 60x_3 - 0x_4 - 0x_5 + z = 0. \tag{3}$$

To begin the simplex method, form an augmented matrix using the coefficients and constants of equations (1), (2), and (3).

$$
\begin{array}{cccccc}
x_1 & x_2 & x_3 & x_4 & x_5 & z
\end{array}
$$

$$
\left[
\begin{array}{cccccc|c}
1 & 1 & 1 & 1 & 0 & 0 & 100 \\
400 & 160 & 280 & 0 & 1 & 0 & 20000 \\
\hline
-120 & -40 & -60 & 0 & 0 & 1 & 0
\end{array}
\right]
\begin{array}{l}
x_4 \\
x_5 \\
z
\end{array}
$$

We use a horizontal line to separate the objective function from the constraints. This matrix is called a **simplex tableau.** The z column in the tableau above will never change because of the two zeros and the fact that pivots are never located in the bottom row. Therefore, it will be left off later tableaus.

From the tableau, we read the solution $(0, 0, 0, 100, 20000)$. Notice that we have ignored the z column. However, from the lower right corner of the tableau, we see that this solution leads to a value of $0 for z. A profit of $0 for the farmer is certainly not an optimum, so we try to improve this profit.

Since the coefficients of x_1, x_2, and x_3 of the objective function are nonzero, we can improve the profit by making any one of these variables take on a nonzero value in a solution. To decide which variable to use, look in the bottom row in the above matrix. The coefficient of $x_1 (-120)$ is the most negative in this row. This means that x_1 has the largest coefficient in the objective function, and we can most reasonably increase profit by increasing x_1.

To make x_1 nonzero, either x_4 or x_5 must become 0. (Recall Theorem 4.1.) To decide whether x_4 or x_5 should be used, solve equations (1) and (2) above for x_4 and x_5, respectively.

$$x_4 = 100 - x_1 - x_2 - x_3$$
$$x_5 = 20000 - 400x_1 - 160x_2 - 280x_3$$

We are changing x_1 to a nonzero value. Thus, x_2 and x_3 still equal 0. Replacing x_2 and x_3 with 0 gives

$$x_4 = 100 - x_1$$
$$x_5 = 20000 - 400x_1.$$

Since both x_4 and x_5 must remain nonnegative, there is a limit to how much we can increase x_1. From the equation $x_4 = 100 - x_1$ we see that x_1 cannot exceed 100/1 or 100. In the second equation, $x_5 = 20000 - 400x_1$, we see that x_1 cannot exceed 20000/400 or 50. To satisfy both these conditions, x_1 cannot exceed 50 (the smaller of 50 and 100). If we let $x_1 = 50$, $x_2 = 0$, $x_3 = 0$, and $x_5 = 0$, we find that $x_4 = 100 - x_1 = 50$, leading to the basic feasible solution (50, 0, 0, 50, 0).

This solution gives a profit of

$$z = 120x_1 + 40x_2 + 60x_3 + 0x_4 + 0x_5$$
$$z = 120(50) + 40(0) + 60(0) + 0(50) + 0(0)$$
$$z = 6000.$$

We could have found the same result by using row operations on the first simplex tableau given above. First, choose the most negative number in the bottom row of the tableau. (If there is no negative number, we cannot further improve the value of the objective function.) The most negative number identifies the entering variable. This variable (x_1 here) is the one whose value is to be increased in the next solution. The entering variable is shown with an arrow on the tableau below.

To find the departing variable, the variable which will become 0 in place of x_1, compute the quotients that we found above. This is done automatically by dividing each number from the right-most column of the tableau by the corresponding coefficient of x_1 (the entering variable).

Quotients

$$
\begin{array}{c}
\\
100/1 = 100 \\
20000/400 = 50
\end{array}
\begin{array}{ccccc}
x_1 & x_2 & x_3 & x_4 & x_5 \\
\left[\begin{array}{ccccc|c}
1 & 1 & 1 & 1 & 0 & 100 \\
400 & 160 & 280 & 0 & 1 & 20000 \\
\hline
-120 & -40 & -60 & 0 & 0 & 0
\end{array}\right]
\begin{array}{c}
x_4 \\
x_5 \leftarrow \\
z
\end{array}
\end{array}
$$

The smaller quotient is 50. This quotient identifies x_5 as the departing variable. (See the arrow at the right above.) The entering and departing variables locate 400 as the pivot. Now use row transformations to obtain the second simplex tableau. First multiply the second row by 1/400. Then add -1 times the new second row to the old first row to get the new first row. (How do we obtain the new third row?)

$$
\begin{array}{ccccc}
x_1 & x_2 & x_3 & x_4 & x_5 \\
\left[\begin{array}{ccccc|c}
0 & .6 & .3 & 1 & -.0025 & 50 \\
1 & .4 & .7 & 0 & .0025 & 50 \\
\hline
0 & 8 & 24 & 0 & .3 & 6000
\end{array}\right]
\begin{array}{c}
x_4 \\
x_1 \\
z
\end{array}
\end{array}
$$

We can now read the new solution from this tableau. Above the horizontal line, the columns that form the 2×2 identity matrix (x_4 and x_1 here) identify the variables which have nonzero values. From the first row, we get $x_4 = 50$ and from the second row, we get $x_1 = 50$; the other variables are all zero. Thus, the solution from this tableau is (50, 0, 0, 50, 0), as we found above. The entry 6000 (in color) in the lower right hand corner of the tableau gives the value of the objective function for this solution:

$$z = \$6000.$$

Since none of the entries in the last row of the tableau are negative, we cannot improve the value of z beyond \$6000. Remember that the last row represents the coefficients of the objective function. If we add the coefficient of 1 for the z column which was dropped, we have

$$0x_1 + 8x_2 + 24x_3 + 0x_4 + .3x_5 + z = 6000.$$

Solving for z, we get

$$z = 6000 - 0x_1 - 8x_2 - 24x_3 - 0x_4 - .3x_5.$$

Since x_2, x_3, and x_5 are zero, $z = 6000$, but if any of these variables were to increase, z would decrease. Thus, we have the optimal solution as soon as no coefficients in the last row of the tableau are negative.

If the last row contained a negative number, we could improve the solution. We would repeat the entire procedure, locating new entering and departing variables and a pivot. We would then transform the matrix again. This process can be repeated as long as negative numbers appear in the last row.

We can finally state the solution to the problem about the farmer: the optimum value of z is 6000, where $x_1 = 50$, $x_2 = 0$, $x_3 = 0$, $x_4 = 50$, and $x_5 = 0$. That is, the farmer will make a maximum profit of \$6000 by planting 50 acres of potatoes. Another 50 acres should be left unplanted. (This is the same solution we found in Section 3.3.) It may seem strange that leaving assets unused can produce a maximum profit, but such results occur often in practical applications.

Example 1 Given the simplex tableau below, find the entering and departing variables and the pivot.

$$
\begin{array}{ccccc}
x_1 & x_2 & x_3 & x_4 & x_5 \\
\end{array}
$$

$$
\left[
\begin{array}{ccccc|c}
1 & -2 & 1 & 0 & 0 & 100 \\
3 & 4 & 0 & 1 & 0 & 200 \\
5 & 0 & 0 & 0 & 1 & 150 \\
\hline
-10 & -25 & 0 & 0 & 0 & 0 \\
\end{array}
\right]
\begin{array}{c}
x_3 \\
x_4 \\
x_5 \\
z \\
\end{array}
$$

To find the entering variable, look for the most negative number in the last row, which is -25. Thus, x_2 is the entering variable. To find the departing variable, we must find the quotients formed by the entries in the right-most column and those in the x_2 column. Here the three quotients are $100/-2$, $200/4$, and $150/0$. We can't divide by 0, so we must disregard the last quotient. This quotient comes from the equation

$$5x_1 + 0x_2 + 0x_3 + 0x_4 + x_5 = 150.$$

If $x_1 = 0$, the equation reduces to

$$0x_2 + x_5 = 150 \quad \text{or} \quad x_5 = 150 - 0x_2.$$

Since x_2 is to be the entering variable, x_5 cannot be 0 and thus cannot be the departing variable. In general, we should always disregard quotients with 0 denominators.

Since the quotients predict the value of the entering variable in the new basic feasible solution, they cannot be negative. The equation which corresponds to the quotient $100/-2 = -50$ is

$$x_1 - 2x_2 + x_3 + 0x_4 + 0x_5 = 100.$$

For $x_1 = 0$, this becomes

$$x_3 = 100 + 2x_2.$$

If x_3 is to be nonnegative, then

$$100 + 2x_2 \geq 0$$
$$x_2 \geq -50.$$

Since x_2 must be nonnegative anyway, this equation tells us nothing new. Thus we can also disregard negative quotients.

The only usable quotient is $200/4 = 50$. Therefore, x_4 is the departing variable and 4 is the pivot. *If all the quotients are either negative or have zero denominators, there will be no optimal solution.* The quotients, then, determine whether or not an optimal solution exists. ∎

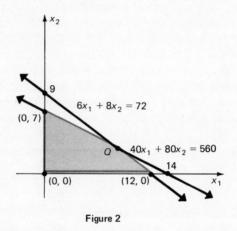

Figure 2

Example 2   To compare the simplex method with the geometric method, let's use the simplex method to solve the problem of Example 4, Section 3.2. The graph is shown again in Figure 2. The objective function to be maximized was

$$z = 8x_1 + 12x_2. \qquad \text{(storage space)}$$

(Since we are using the simplex method, we use x_1 and x_2 as variables instead of x and y.) The constraints were as follows:

$$40x_1 + 80x_2 \leq 560 \qquad \text{(cost)}$$
$$6x_1 + 8x_2 \leq 72 \qquad \text{(floor space)}$$
$$x_1 \geq 0, \, x_2 \geq 0.$$

We must add a slack variable to each constraint.

$$40x_1 + 80x_2 + x_3 = 560$$
$$6x_1 + 8x_2 + x_4 = 72.$$

The first tableau can now be written.

$$
\begin{array}{cccc}
x_1 & x_2 & x_3 & x_4 \\
\end{array}
$$

$$
\begin{bmatrix}
40 & 80 & 1 & 0 & 560 \\
6 & 8 & 0 & 1 & 72 \\
-8 & -12 & 0 & 0 & 0
\end{bmatrix}
\begin{array}{l}
x_3 \\
x_4 \\
z
\end{array}
$$

At this point the solution is $(0, 0, 560, 72)$, which corresponds to the vertex at the origin in Figure 2. The value of the objective function is $z = 0$, as in Section 3.2. Because -12 is the most negative number in the bottom row of the tableau, the entering variable will be x_2. The quotients are

$$
\frac{560}{80} = 7 \quad \text{and} \quad \frac{72}{8} = 9.
$$

Choose the smallest quotient. Since $7 < 9$, x_3 is the departing variable, and the pivot is 80. The next tableau is as follows.

$$
\begin{bmatrix}
\frac{1}{2} & 1 & \frac{1}{80} & 0 & 7 \\
2 & 0 & -\frac{1}{10} & 1 & 16 \\
-2 & 0 & \frac{3}{20} & 0 & 84
\end{bmatrix}
$$

The solution from this tableau is $(0, 7, 0, 16)$, which corresponds to another vertex of the region of feasible solutions, $(0, 7)$. This solution leads to $z = 84$, which agrees with the result in Section 3.2.

The next tableau is the final one shown below.

$$
\begin{bmatrix}
0 & 1 & \frac{3}{80} & -\frac{1}{4} & 3 \\
1 & 0 & -\frac{1}{20} & \frac{1}{2} & 8 \\
0 & 0 & \frac{1}{20} & 1 & 100
\end{bmatrix}
$$

This tableau gives the solution $(8, 3, 0, 0)$, which makes $z = 100$, the same result found by the geometric method. We know it is a maximum because none of the numbers in the last row of the tableau are negative. ∎

The simplex method tested the vertices of the region of feasible solutions in a systematic manner, starting with the vertex at the origin. We did not need to test the last vertex, $(12, 0)$, in Figure 2, because we found the maximum value of z before we reached that vertex.

Example 3 The matrix given below is the first simplex tableau of a linear programming problem. Complete the solution.

$$
\begin{array}{ccccc}
x_1 & x_2 & x_3 & x_4 & x_5 \\
\end{array}
$$

$$
\begin{bmatrix}
1 & 2 & 4 & 1 & 0 & 38 \\
2 & 1 & 1 & 0 & 1 & 16 \\
-2 & -3 & -1 & 0 & 0 & 0
\end{bmatrix}
\begin{array}{l}
x_4 \\
x_5
\end{array}
$$

Since the most negative number in the last row is -3, x_2 is the entering variable. The smaller of the two quotients, shown below on the left of the tableau, is 16, which indicates that x_5 is the departing variable. Thus, the pivot is 1.

$$
\begin{array}{cc}
 & \begin{array}{ccccc} x_1 & x_2 & x_3 & x_4 & x_5 \end{array} \\
\begin{array}{c} 38/2 = 19 \\ 16/1 = 16 \end{array} &
\left[\begin{array}{ccccc|c}
1 & 2 & 4 & 1 & 0 & 38 \\
2 & ① & 1 & 0 & 1 & 16 \\
\hline
-2 & -3 & -1 & 0 & 0 & 0
\end{array}\right]
\begin{array}{c} x_4 \\ x_5 \leftarrow \\ \\ \end{array}
\end{array}
$$

Performing row transformations, we get the second tableau.

$$
\begin{array}{c}
\begin{array}{ccccc} x_1 & x_2 & x_3 & x_4 & x_5 \end{array} \\
\left[\begin{array}{ccccc|c}
-3 & 0 & 2 & 1 & -2 & 6 \\
2 & 1 & 1 & 0 & 1 & 16 \\
\hline
4 & 0 & 2 & 0 & 3 & 48
\end{array}\right]
\begin{array}{c} x_4 \\ x_2 \\ z \end{array}
\end{array}
$$

All numbers in the last row are nonnegative. Therefore, the solution cannot be improved. The optimum value of 48 occurs when $x_1 = 0$, $x_2 = 16$, $x_3 = 0$, $x_4 = 6$, and $x_5 = 0$. ∎

We can now summarize the steps involved in the simplex method for finding the maximum value of an objective function.

1. Determine the objective function.
2. Write all necessary constraints.
3. Convert each constraint into an equation by adding the necessary slack variables.
4. Set up the first simplex tableau.
5. Locate the most negative number in the bottom row. This number establishes the entering variable.
6. Form the necessary quotients to find the departing variable. Disregard any negative quotients or quotients with a 0 denominator. The smallest non-negative quotient indicates the departing variable. If all quotients must be disregarded, no maximum solution exists.[1]
7. Locate the pivot by using the entering and departing variables.
8. Transform the tableau so that the pivot becomes 1 and all other numbers in that column become 0.
9. If the numbers in the bottom row are all positive or zero, you are through. If not, go back to step 5 above and transform the latest tableau. Keep doing this until a tableau is obtained with no negative numbers in the bottom row.
10. The maximum value of the objective function is given by the number in the lower right corner of the final tableau.

Although linear programming problems with more than a few variables would seem to be very complex, it turns out that in practical applications a large fraction of the entries in the simplex tableau are zeros.

---

[1]Some special circumstances are noted at the end of Section 4.3.

4.2 Exercises

Use the simplex method to solve the following first tableaus of linear programming problems.

1. x_1 x_2 x_3 x_4 x_5

$$
\begin{bmatrix}
1 & 2 & 4 & 1 & 0 & 8 \\
2 & 2 & 1 & 0 & 1 & 10 \\
-2 & -5 & -1 & 0 & 0 & 0
\end{bmatrix}
\begin{matrix} x_4 \\ x_5 \\ z \end{matrix}
$$

2. x_1 x_2 x_3 x_4 x_5

$$
\begin{bmatrix}
2 & 2 & 1 & 1 & 0 & 10 \\
1 & 2 & 3 & 0 & 1 & 15 \\
-3 & -2 & -1 & 0 & 0 & 0
\end{bmatrix}
\begin{matrix} x_4 \\ x_5 \\ z \end{matrix}
$$

3. x_1 x_2 x_3 x_4 x_5

$$
\begin{bmatrix}
1 & 3 & 1 & 0 & 0 & 12 \\
2 & 1 & 0 & 1 & 0 & 10 \\
1 & 1 & 0 & 0 & 1 & 4 \\
-2 & -1 & 0 & 0 & 0 & 0
\end{bmatrix}
\begin{matrix} x_3 \\ x_4 \\ x_5 \\ z \end{matrix}
$$

4. x_1 x_2 x_3 x_4 x_5 x_6

$$
\begin{bmatrix}
2 & 2 & 1 & 1 & 0 & 0 & 50 \\
1 & 1 & 3 & 0 & 1 & 0 & 40 \\
4 & 2 & 5 & 0 & 0 & 1 & 80 \\
-2 & -3 & -5 & 0 & 0 & 0 & 0
\end{bmatrix}
\begin{matrix} x_4 \\ x_5 \\ x_6 \\ z \end{matrix}
$$

5. x_1 x_2 x_3 x_4 x_5 x_6

$$
\begin{bmatrix}
2 & 2 & 8 & 1 & 0 & 0 & 40 \\
4 & -5 & 6 & 0 & 1 & 0 & 60 \\
2 & -2 & 6 & 0 & 0 & 1 & 24 \\
-14 & -10 & -12 & 0 & 0 & 0 & 0
\end{bmatrix}
\begin{matrix} x_4 \\ x_5 \\ x_6 \\ z \end{matrix}
$$

6. x_1 x_2 x_3 x_4 x_5

$$
\begin{bmatrix}
3 & 2 & 4 & 1 & 0 & 18 \\
2 & 1 & 5 & 0 & 1 & 8 \\
-1 & -4 & -2 & 0 & 0 & 0
\end{bmatrix}
\begin{matrix} x_4 \\ x_5 \\ z \end{matrix}
$$

Use the simplex method to solve each of the following problems.

7. Maximize $z = 10x_1 + 12x_2$ subject to
$$
4x_1 + 2x_2 \le 20
$$
$$
5x_1 + x_2 \le 50
$$
$$
2x_1 + 2x_2 \le 24
$$
and $x_1 \ge 0$, $x_2 \ge 0$.

8. Maximize $z = 1.2x_1 + 3.5x_2$ subject to
$$
2.4x_1 + 1.5x_2 \le 10
$$
$$
1.7x_1 + 1.9x_2 \le 15
$$
and $x_1 \ge 0$, $x_2 \ge 0$.

9. Maximize $z = 8x_1 + 3x_2 + x_3$ subject to
$$
x_1 + 6x_2 + 8x_3 \le 118
$$
$$
x_1 + 5x_2 + 10x_3 \le 220
$$
and $x_1 \ge 0$, $x_2 \ge 0$, $x_3 \ge 0$.

10. Maximize $z = 12x_1 + 15x_2 + 5x_3$ subject to
$$
2x_1 + 2x_2 + x_3 \le 8
$$
$$
x_1 + 4x_2 + 3x_3 \le 12
$$
and $x_1 \ge 0$, $x_2 \ge 0$, $x_3 \ge 0$.

11. Maximize $z = x_1 + 2x_2 + x_3 + 5x_4$ subject to
$$
x_1 + 2x_2 + x_3 + x_4 \le 50
$$
$$
3x_1 + x_2 + 2x_3 + x_4 \le 100
$$
and $x_1 \ge 0$, $x_2 \ge 0$, $x_3 \ge 0$, $x_4 \ge 0$.

12. Maximize $z = x_1 + x_2 + 4x_3 + 5x_4$ subject to
$$
x_1 + 2x_2 + 3x_3 + x_4 \le 115
$$
$$
2x_1 + x_2 + 8x_3 + 5x_4 \le 200
$$
$$
x_1 + x_3 \le 50
$$
and $x_1 \ge 0$, $x_2 \ge 0$, $x_3 \ge 0$, $x_4 \ge 0$.

Solve the following problems using the simplex method.

13. A biologist has 500 kilograms of nutrient A, 600 kilograms of nutrient B, and 300 kilograms of nutrient C. These nutrients will be used to make 4 types of food, whose contents (in percent of nutrient per kilogram of food) and whose "growth values" are as shown below.

Food	Nutrient (%) A	B	C	Growth value
P	0	0	100	90
Q	0	75	25	70
R	37.5	50	12.5	60
S	62.5	37.5	0	50

How many kilograms of each food should be produced in order to maximize total growth value? Find the maximum growth value.

14. A baker has 150 units of flour, 90 units of sugar, and 150 of raisins. A loaf of raisin bread requires 1 unit of flour, 1 of sugar and 2 of raisins, while a raisin cake needs 5, 2, and 1 units, respectively. If raisin bread sells for 35¢ a loaf and raisin cake for 80¢ each, how many of each should be baked so that gross income is maximized? What is the maximum gross income?

15. A candy company has 100 kilograms of chocolate-covered nuts and 125 kilograms of chocolate-covered raisins to be sold as two different mixtures. One mix will contain half nuts and half raisins and will sell for $6 per kilogram. The other mix will contain 1/3 nuts and 2/3 raisins, and will sell for $4.80 per kilogram. How many kilograms of each mix should the company prepare for maximum revenue? (See Exercise 25, Section 4.1.) Find the maximum revenue.

16. Caroline's Quality Candy Confectionery is famous for fudge, chocolate cremes, and pralines. Its candy-making equipment is set up to make 100-pound batches at a time. Currently there is a chocolate shortage and the company can get only 120 pounds of chocolate in the next shipment. On a week's run, the confectionery's cooking and processing equipment is available for a total of 42 machine hours. During the same period the employees have a total of 56 work hours available for packaging. A batch of fudge requires 20 pounds of chocolate while a batch of cremes uses 25 pounds of chocolate. The cooking and processing take 120 minutes for fudge, 150 minutes for chocolate cremes, and 200 minutes for pralines. The packaging times measured in minutes per one pound box are 1, 2, and 3, respectively for fudge, cremes and pralines. Determine how many batches of each type of candy the confectionery should make, assuming that the profit per pound box is 50¢ on fudge, 40¢ on chocolate cremes, and 45¢ on pralines. Also, find the maximum profit for the week. (See Exercise 28, Section 4.1.)

17. A manufacturer of bicycles builds one-, three-, and ten-speed models. The bicycles need both aluminum and steel. The company has available 91,800 units of steel and 42,000 units of aluminum. The one-, three-, and ten-speed models need respectively 17, 27, and 34 units of steel, and 12, 21, and 15 units of aluminum. How many of each type of bicycle should be made in order to maximize profit if the company makes $8 per one-speed bike, $12 per three-speed, and $22 per ten-speed? What is the maximum possible profit?

18. A political party is planning a half-hour television show. The show will have 3 minutes of direct requests for money from viewers. Three of the party's politicians will be on

the show — a senator, a congresswoman, and a governor. The senator, a party "elder statesman," demands that he be on at least twice as long as the governor. The total time taken by the senator and the governor must be at least twice the time taken by the congresswoman. Based on a pre-show survey, it is believed that 40, 60, and 50 (in thousands) viewers will watch the program for each minute the senator, congresswoman, and governor, respectively are on the air. Find the time that should be alloted to each politician in order to get the maximum number of viewers. Find the maximum number of viewers.

4.3   Problems With Both $\leq$ and $\geq$ Constraints and Minimization

Up to this point we have discussed the simplex method only for problems which met the following conditions.

1. All variables are nonnegative.
2. The constant terms are nonnegative.
3. All constraints are $\leq$ inequalities.
4. The problem is a maximization problem.

We shall refer to problems which meet all of these conditions as **standard**. In this section, we see how to handle problems which are nonstandard. The only condition we will keep is the first: all variables must be nonnegative.

Problems with $\leq$ and $\geq$ Constraints   Suppose a new constraint is added to the farmer problem from Example 1, Section 4.1. To satisfy orders from regular buyers, the farmer must plant a total of at least 60 acres of the three crops. This introduces the new inequality

$$x_1 + x_2 + x_3 \geq 60.$$

We must rewrite this inequality as an equation in which the variables all represent nonnegative numbers. The inequality $x_1 + x_2 + x_3 \geq 60$ means that

$$x_1 + x_2 + x_3 - x_6 = 60$$

for some nonnegative variable x_6. (Remember that x_4 and x_5 are the slack variables in the problem.)

The new variable, x_6, is called a **surplus variable.** It represents the excess number of acres (over 60) which may be planted. Thus, x_6 can vary from 0 to 40, since the total number of acres planted is to be less than 100 but greater than 60.

We must now solve the following system of equations,

$$
\begin{array}{rcrcrcrcrcrcr}
x_1 & + & x_2 & + & x_3 & + & x_4 & & & & & = & 100 \\
400x_1 & + & 160x_2 & + & 280x_3 & & & + & x_5 & & & = & 20000 \\
x_1 & + & x_2 & + & x_3 & & & & & - & x_6 & = & 60 \\
-120x_1 & - & 40x_2 & - & 60x_3 & & & & & & & + z = & 0
\end{array}
$$

with x_1, x_2, x_3, x_4, x_5, and x_6 all nonnegative.

Set up the simplex tableau. (The z column is omitted as before.)

$$\begin{array}{cccccc} x_1 & x_2 & x_3 & x_4 & x_5 & x_6 \end{array}$$

$$\left[\begin{array}{cccccc|c} 1 & 1 & 1 & 1 & 0 & 0 & 100 \\ 400 & 160 & 280 & 0 & 1 & 0 & 20000 \\ 1 & 1 & 1 & 0 & 0 & -1 & 60 \\ \hline -120 & -40 & -60 & 0 & 0 & 0 & 0 \end{array}\right]$$

The basic solution is

$$x_1 = 0,\ x_2 = 0,\ x_3 = 0,\ x_4 = 100,\ x_5 = 20000,\ x_6 = -60.$$

But this is not a feasible solution, since x_6 is negative. All the variables in any feasible solution must be nonnegative.

When this happens, we use row operations to transform the matrix until we can get a basic solution in which all variables are nonnegative. The difficulty is caused by the -1 in row three of the matrix. We do not have the third column of the usual 3×3 identity matrix. To fix this, we use row transformations to change a column which has nonzero entries (such as the x_1, x_2, or x_3 columns) to one in which the third row entry is 1 and the other entries are 0. The choice of a column is arbitrary. Let's choose the x_2 column. If we still are unable to get a basic feasible solution, we can try one of the other choices.

The third row entry in the x_2 column is already 1. Using row transformations to get 0's in the rest of the column gives the following tableau.

$$\begin{array}{cccccc} x_1 & x_2 & x_3 & x_4 & x_5 & x_6 \end{array}$$

$$\left[\begin{array}{cccccc|c} 0 & 0 & 0 & 1 & 0 & 1 & 40 \\ 240 & 0 & 120 & 0 & 1 & 160 & 10400 \\ 1 & 1 & 1 & 0 & 0 & -1 & 60 \\ \hline -80 & 0 & -20 & 0 & 0 & -40 & 2400 \end{array}\right]$$

The basic solution is (0, 60, 0, 40, 10400, 0) which is feasible. We now proceed with the simplex method. The entering variable is x_1, the departing variable is x_5, and the pivot is 240.

$$\begin{array}{cccccc} x_1 & x_2 & x_3 & x_4 & x_5 & x_6 \end{array}$$

$$\left[\begin{array}{cccccc|c} 0 & 0 & 0 & 1 & 0 & 1 & 40 \\ \boxed{240} & 0 & 120 & 0 & 1 & 160 & 10400 \\ 1 & 1 & 1 & 0 & 0 & -1 & 60 \\ \hline -80 & 0 & -20 & 0 & 0 & -40 & 2400 \end{array}\right] \leftarrow$$

$$\begin{array}{cccccc} x_1 & x_2 & x_3 & x_4 & x_5 & x_6 \end{array}$$

$$\left[\begin{array}{cccccc|c} 0 & 0 & 0 & 1 & 0 & 1 & 40 \\ 1 & 0 & .5 & 0 & .004 & .667 & 43.3 \\ 0 & 1 & .5 & 0 & -.004 & -1.667 & 16.7 \\ \hline 0 & 0 & 20 & 0 & .32 & 13.4 & 5864 \end{array}\right] \text{ (rounded)}$$

The second matrix above has been obtained from the first by standard row operations; some numbers in it have been rounded.

From the last matrix, the solution is (43.3, 16.7, 0, 40, 0, 0). Thus the farmer should plant 43.3 acres of potatoes, 16.7 acres of corn, and no cabbage. Forty acres of the 100 available should not be planted. The profit will be $5864, less than the $6000 profit if he planted only 50 acres of potatoes. Because of the additional constraint that he must plant at least 60 acres, his profit is reduced.

Example 1 Maximize $z = 10x_1 + 8x_2$ subject to

$$4x_1 + 4x_2 \geq 60$$
$$2x_1 + 5x_2 \leq 120$$

and $x_1 \geq 0$, $x_2 \geq 0$.

Add slack or surplus variables to the constraints and write the first simplex tableau.

$$4x_1 + 4x_2 - x_3 \qquad\qquad = 60$$
$$2x_1 + 5x_2 \qquad + x_4 \qquad = 120$$
$$-10x_1 - 8x_2 \qquad\qquad + z = \quad 0$$

$$
\begin{array}{cccc}
x_1 & x_2 & x_3 & x_4 \\
\end{array}
$$

$$
\left[
\begin{array}{cccc|c}
4 & 4 & -1 & 0 & 60 \\
2 & 5 & 0 & 1 & 120 \\
\hline
-10 & -8 & 0 & 0 & 0 \\
\end{array}
\right]
$$

Since the basic solution $(0, 0, -60, 120)$ is not feasible, it can't be used to start the simplex method. We need a column with 1 in the first row and zeros in the rest of the columns. Let's choose the x_1 column. Use row transformations. Multiply the first row by 1/4 to get 1 in the top row of the column. Then use row transformations to get zeros in the other rows of that column.

$$
\begin{array}{cccc}
x_1 & x_2 & x_3 & x_4 \\
\end{array}
$$

$$
\left[
\begin{array}{cccc|c}
1 & 1 & -\frac{1}{4} & 0 & 15 \\
0 & 3 & \frac{1}{2} & 1 & 90 \\
\hline
0 & 2 & -\frac{5}{2} & 0 & 150 \\
\end{array}
\right]
$$

The basic solution is now $(15, 0, 0, 90)$ which is feasible. Complete the solution in the usual way. The pivot is 1/2. The next tableau is

$$
\begin{array}{cccc}
x_1 & x_2 & x_3 & x_4 \\
\end{array}
$$

$$
\left[
\begin{array}{cccc|c}
1 & \frac{5}{2} & 0 & \frac{1}{2} & 60 \\
0 & 6 & 1 & 2 & 180 \\
\hline
0 & 17 & 0 & 5 & 600 \\
\end{array}
\right]
$$

Since all entries in the bottom row are nonnegative, we have the feasible solution, $(60, 0, 180, 0)$. Thus, the maximum is $z = 600$ when $x_1 = 60$ and $x_2 = 0$. ∎

The approach we have discussed above allows us to solve problems where the constraints are mixed ≤ and ≥ inequalities. It can also be used to solve a problem where the constant term is negative. (This is not likely to happen in an application, however.)

When one of the constraints is an equality, we simply add an **artificial variable** to get a basic feasible solution from the simplex tableau. As its name implies, an artificial variable has nothing to do with the actual problem. It is added to preserve the tableau form so that we can get an initial basic feasible solution. For example, if the constraint is

$$2x_1 + 3x_2 + 4x_3 = 75,$$

we would add an artificial variable, say x_4, to get

$$2x_1 + 3x_2 + 4x_3 + x_4 = 75.$$

The problem is then set up and solved by the same methods we have discussed above.

Minimization Problems   Now we consider minimization problems. These are handled by maximizing the negative of the objective function. The next example shows how this works.

Example 2   Minimize $w = 3x_1 + 2x_2$ subject to

$$x_1 + 3x_2 \geq 6$$
$$2x_1 + x_2 \geq 3$$

and $x_1 \geq 0$, $x_2 \geq 0$.

Change this to a maximization problem by letting $z = -w$ and maximizing z.

$$z = -w = -3x_1 - 2x_2$$

The problem can now be stated as follows.
Maximize $z = -3x_1 - 2x_2$ subject to

$$x_1 + 3x_2 \geq 6$$
$$2x_1 + x_2 \geq 3$$

and $x_1 \geq 0$, $x_2 \geq 0$.

To solve, add surplus variables and set up the first tableau.

$$\begin{array}{cccc} x_1 & x_2 & x_3 & x_4 \end{array}$$
$$\begin{bmatrix} 1 & 3 & -1 & 0 & | & 6 \\ 2 & 1 & 0 & -1 & | & 3 \\ \hline 3 & 2 & 0 & 0 & | & 0 \end{bmatrix}$$

The basic solution $(0, 0, -6, -3)$ contains negatives. Use row transformations to get a tableau with an acceptable basic solution. Let's use the x_1 column since it already has a 1 in the first row. First, get a 0 in the second row, and then a 0 in the third row.

$$\begin{array}{cccc} x_1 & x_2 & x_3 & x_4 \end{array}$$
$$\begin{bmatrix} 1 & 3 & -1 & 0 & | & 6 \\ 0 & -5 & 2 & -1 & | & -9 \\ \hline 0 & -7 & 3 & 0 & | & -18 \end{bmatrix}$$

The basic solution, (6, 0, 0, 9), is feasible. Now complete the solution as usual by the simplex method. The pivot is -5.

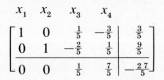

$$
\begin{array}{cccc}
x_1 & x_2 & x_3 & x_4 \\
\end{array}
$$

$$
\left[
\begin{array}{cccc|c}
1 & 0 & \frac{1}{5} & -\frac{3}{5} & \frac{3}{5} \\
0 & 1 & -\frac{2}{5} & \frac{1}{5} & \frac{9}{5} \\
\hline
0 & 0 & \frac{1}{5} & \frac{7}{5} & -\frac{27}{5}
\end{array}
\right]
$$

Now the solution is (3/5, 9/5, 0, 0) which is feasible and the last row has no negative entries to the left of the bar. Since $z = -27/5$ and $z = -w$, $w = 27/5$ is the minimum value, and the minimum is obtained when $x_1 = 3/5$ and $x_2 = 9/5$. ∎

In general, to solve a linear programming problem where all variables are nonnegative ($x_i \geq 0$), follow these steps.

---

1. **Determine the objective function.**
   (a) **In a maximization problem, rewrite the objective function so that one side equals 0. Remember to keep z positive.**
   (b) **In a minimization problem with objective function w, let $z = -w$. Then write as in step 1(a).**

2. **Write all constraints as equations, adding slack variables (or artificial variables) or subtracting surplus variables as needed.**

3. **Set up the first simplex tableau and check the basic solution.**
   (a) **If the basic solution has no negative entries, complete the solution by the simplex method.**
   (b) **If the basic solution has a negative entry, transform the matrix until you have a basic feasible solution, with no negative entries.**

In this chapter, we certainly have not covered all the possible complications that can arise in using the simplex method. Some of the difficulties (which may have occurred to you) include the following.

1. The minimum quotient may be zero in which case the entering variable retains the value zero (rather than taking on a nonzero value) in the new basic solution. In this case, there is no immediate improvement in the value of z.

2. There may be two or more equal quotients which are smallest. This leads to a basic solution in which one or more of the basic variables is zero. Each "tie" produces a zero basic variable in the row of the tying quotient.

3. Occasionally, a transformation will cycle—that is, produce a "new" solution which was an earlier solution in the process. These situations are known as degeneracies and special methods are available for handling them.

4. It may not be possible to convert a nonfeasible basic solution to a feasible basic solution. In that case, there is no solution which satisfies all the constraints. Graphically, this means there is no region of feasible solutions.

Two linear programming models in actual use, one by Boeing, the other by Upjohn, are presented after the exercises of this section. These models illustrate the usefulness of linear programming. In most real applications, the number of variables is so large that these problems could not be solved without the use of a method, like the simplex method, which can be computerized.

4.3 Exercises   *Rewrite each system of inequalities adding slack variables or subtracting surplus variables as necessary.*

1. $2x_1 + 3x_2 \leq 8$
   $x_1 + 4x_2 \geq 7$

2. $5x_1 + 8x_2 \leq 10$
   $6x_1 + 2x_2 \geq 7$

3. $x_1 + x_2 + x_3 \leq 100$
   $x_1 + x_2 + x_3 \geq 75$
   $x_1 + x_2 \quad\quad \geq 27$

4. $2x_1 + x_3 \leq 40$
   $x_1 + x_2 \geq 18$
   $x_1 + x_3 \geq 20$

Convert the following problems into maximization problems.

5. Minimize $w = 4x_1 + 3x_2 + 2x_3$ subject to
   $$x_1 + x_2 + x_3 \geq 5$$
   $$x_1 + x_2 \quad\quad \geq 4$$
   $$2x_1 + x_2 + 3x_3 \geq 15$$
   and $x_1 \geq 0, x_2 \geq 0, x_3 \geq 0.$

6. Minimize $w = 8x_1 + 3x_2 + x_3$ subject to
   $$7x_1 + 6x_2 + 8x_3 \geq 18$$
   $$4x_1 + 5x_2 + 10x_3 \geq 20$$
   and $x_1 \geq 0, x_2 \geq 0, x_3 \geq 0.$

7. Minimize $w = x_1 + 2x_2 + x_3 + 5x_4$ subject to
   $$x_1 + x_2 + x_3 + x_4 \geq 50$$
   $$3x_1 + x_2 + 2x_3 + x_4 \geq 100$$
   and $x_1 \geq 0, x_2 \geq 0, x_3 \geq 0, x_4 \geq 0.$

8. Minimize $w = x_1 + x_2 + 4x_3$ subject to
   $$x_1 + 2x_2 + 3x_3 \geq 115$$
   $$2x_1 + x_2 + x_3 \leq 200$$
   $$x_1 \quad\quad + x_3 \geq 50$$
   and $x_1 \geq 0, x_2 \geq 0, x_3 \geq 0.$

Use the simplex method to solve the following problems.

9. Find $x_1 \geq 0, x_2 \geq 0$ such that
   $$x_1 + 2x_2 \geq 24$$
   $$x_1 + x_2 \leq 40$$
   and $z = 12x_1 + 10x_2$ is maximized.

10. Find $x_1 \geq 0, x_2 \geq 0$ such that
    $$3x_1 + 4x_2 \geq 48$$
    $$2x_1 + 4x_2 \leq 60$$
    and $z = 6x_1 + 8x_2$ is maximized.

11. Find $x_1 \geq 0, x_2 \geq 0, x_3 \geq 0$ such that
    $$x_1 + x_2 + x_3 \leq 150$$
    $$x_1 + x_2 + x_3 \geq 100$$
    and $z = 2x_1 + 5x_2 + 3x_3$ is maximized.

12. Find $x_1 \geq 0, x_2 \geq 0, x_3 \geq 0$ such that
    $$x_1 + x_2 + 2x_3 \leq 38$$
    $$2x_1 + x_2 + x_3 \geq 24$$
    and $z = 3x_1 + 2x_2 + 2x_3$ is maximized.

13. Find $x_1 \geq 0, x_2 \geq 0$ such that
    $$x_1 + x_2 \leq 100$$
    $$x_1 + x_2 \geq 50$$
    $$2x_1 + x_2 \leq 110$$
    and $z = 2x_1 + 3x_2$ is maximized.

14. Find $x_1 \geq 0, x_2 \geq 0$ such that
    $$x_1 + 2x_2 \leq 18$$
    $$x_1 + 3x_2 \geq 12$$
    $$2x_1 + 2x_2 \leq 24$$
    and $z = 5x_1 + 10x_2$ is maximized.

15. Find $y_1 \geq 0$, $y_2 \geq 0$ such that
$$10y_1 + 5y_2 \geq 100$$
$$20y_1 + 10y_2 \geq 150$$
and $w = 4y_1 + 5y_2$ is minimized.

16. Minimize $w = 3y_1 + 2y_2$ subject to
$$2y_1 + 3y_2 \geq 60$$
$$y_1 + 4y_2 \geq 40$$
and $y_1 \geq 0$, $y_2 \geq 0$.

17. Minimize $w = 2y_1 + y_2 + 3y_3$ subject to
$$y_1 + y_2 + y_3 \geq 100$$
$$2y_1 + y_2 \qquad \geq 50$$
and $y_1 \geq 0$, $y_2 \geq 0$, $y_3 \geq 0$.

18. Minimize $w = 3y_1 + 2y_2$ subject to
$$y_1 + 2y_2 \geq 10$$
$$y_1 + y_2 \geq 8$$
$$2y_1 + y_2 \geq 12$$
and $y_1 \geq 0$, $y_2 \geq 0$.

Use the simplex method to solve each of the following problems.

19. Brand X Canners produce canned whole tomatoes and tomato sauce. This season, they have available 3,000,000 kilograms of tomatoes for these two products. To meet the demands of regular customers, they must produce at least 80,000 kilograms of sauce and 800,000 kilograms of whole tomatoes. The cost per kilogram is $4 to produce canned whole tomatoes and $3.25 to produce tomato sauce. How many kilograms of tomatoes should they use for each product to minimize cost?

20. Sam, who is dieting, requires two food supplements, I and II. He can get these supplements from two different products, A and B, as shown in the following table.

Supplement
(grams per serving)

		I	II
Product	A	3	2
	B	2	4

Sam's physician has recommended that he include at least 15 grams of supplement I but no more than 12 grams of II in his daily diet. If product A costs 25¢ per serving and product B costs 40¢ per serving, how can he satisfy his requirements most economically? (See Exercise 21, Section 3.2.)

21. Mark, who is ill, takes vitamin pills. Each day he must have at least 16 units of vitamin A, 5 units of vitamin B_1, and 20 units of vitamin C. He can choose between pill #1 which costs 10 cents and contains 8 units of A, 1 of B_1, and 2 of C, and pill #2 which costs 20 cents and contains 2 units of A, 1 of B_1, and 7 of C. How many of each pill should he buy in order to minimize his cost?

22. A brewery produces regular beer and a lower-carbohydrate "light" beer. Steady customers of the brewery buy 12 units of regular beer and 10 units of light beer. While setting up the brewery to produce the beers, the management decides to produce extra beer, beyond that needed to satisfy the steady customers. The cost per unit of regular beer is $36,000 and the cost per unit of light beer is $48,000. The number of units of light beer should not exceed twice the number of units of regular beer. At least twenty additional units of beer can be sold. How much of each type beer should be made so as to minimize total production costs?

23. The chemistry department at a local college decides to stock at least 800 small test tubes and 500 large test tubes. It wants to buy at least 1500 test tubes to take advantage of a special price. Since the small tubes are broken twice as often as the larger, the department will order at least twice as many small tubes as large. If the small test tubes cost 15¢ each and the large ones 12¢ each, how many of each size should they order to minimize cost?

24. Topgrade Turf lawn seed mixtures contain three types of seeds: bluegrass, rye, and bermuda. The costs per pound of the three types of seed are 20¢, 15¢, and 5¢. In each mixture there must be at least 20% bluegrass seed and the amount of bermuda must be no more than the amount of rye. To fill current orders, the company must make at least 5000 pounds of the mixture. How much of each kind of seed should be used to minimize cost?

25. A biologist must make a nutrient for her algae. The nutrient must contain the three basic elements D, E, and F, and must contain at least 10 kilograms of D, 12 kilograms of E, and 20 kilograms of F. The nutrient is made from three ingredients, I, II, and III. The quantity of D, E, and F in one unit of each of the ingredients is as given in the following chart.

One unit of ingredient	Contains the following elements in kilograms			Cost of one unit of ingredient
	D	E	F	
I	4	3	0	4
II	1	2	4	7
III	10	1	5	5

How many units of each ingredient are required to meet her needs at minimum cost?

Application   Airline Fleet Assignment—Boeing

Boeing Commercial Airplane Company has developed a mathematical model to help an airline decide on its fleet's needs.[1] This model helps airlines choose the best mix of small, medium, and large jets for the routes that it serves. To begin, the airline estimates the potential revenue from passengers and freight from one city to all the others that can be reached from that city.

For example, Figure 1 shows that 10 units of revenue ($10,000) is earned going from city A to city B. Also, $89,000 is received from traffic that goes from A to C through B, while $111,000 is received from traffic going from A directly to C. Thus, a total of $89,000 + $111,000 = $200,000 is received from traffic going from A to C. Using the information given in the figure above, a *demand matrix* can be set up. (See Exercise 1 below.)

After the expected demand has been decided upon, the mathematical model for optimum fleet utilization can be discussed. The model includes three types of variables—demand variables, itinerary variables, and fleet variables. A demand variable represents the demand from customers for flights from one city to another, including nonstop flights, through flights (not nonstop, but no change of planes), and connecting flights (which involve a change of plane). Itinerary variables show how the airplanes owned by the company can be applied to the

[1]From "Applications of an Airline Fleet Assignment Model" by A. Maimon and R. M. Peterson. Reprinted by permission of Boeing Commercial Airplane Company, Seattle, Washington.

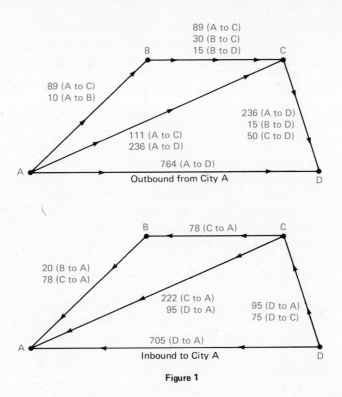

Figure 1

airline system. Finally, the fleet variables give the number of each type of plane that the airline should have.

The mathematical model is designed so that linear programming, using the simplex method, can be used to maximize profit. Profit (the objective function) is given by

Profit = P = passenger revenue + cargo revenue − flight operating costs
        − aircraft ownership costs − costs of any new aircraft.

The profit function must be maximized subject to many constraints. For example, the number of aircraft services for a city pair must be greater than some minimum marketing requirement. This constraint is of the form

$$\sum_{i=1}^{r} \sum_{j=1}^{a} N_{ij}^{pq} \geq ms_{pq}$$

where ms_{pq} is the minimum service requirement for the city pair (p, q) and N_{ij}^{pq} is the number of frequencies of aircraft type j on those routes i that serve the city pair (p, q).

System fuel usage must not exceed the total amount of fuel available.

$$\sum_{i=1}^{r} \sum_{j=1}^{a} FB_{ij} \cdot N_{ij} \leq FA$$

Here FB_{ij} is the amount of fuel burned by an aircraft of type j on route i. Also, FA represents the total amount of fuel available.

The "load factor" (the fraction of available seats that are occupied) must not exceed some preassigned value.

$$\sum_{j=1}^{a} LF_i \cdot S_j \cdot N_{ji} - \sum t \geq 0.$$

In this constraint, LF_i is the maximum load factor on a given route i ($LF_i < 1$), S_j is the number of seats on an aircraft of type j, and N_{ji} the number of aircraft of type j on route i. Also, Σt represents the total number of passengers on the route.

Other constraints relate to the number of seats available on a route, the frequency of service to a city, and so on. When a given airline wishes to use this model, the values of all necessary variables must be found. (Normally, this is the most time-consuming part of the process.) These values are entered into a large computer, which sets up a simplex tableau, goes through the steps of the simplex method, and produces a conclusion.

The conclusions are quite lengthy, since they must specify the number of flights made by several different types of aircraft between many different pairs of cities. The following table shows the size of the simplex tableau for the mathematical model of this case for several different types of airlines.

Airline type — domestic or international	Passenger or freight or combination	Number of types of aircraft	Number of itineraries	Number of city pairs	Size of simplex tableau rows	columns	Computer running time (seconds)
Domestic	Passenger	7	46	89	353	384	100
International	Combination	8	80	46	590	1283	190
Domestic	Passenger	8	203	287	1373	2008	1800
Domestic	Passenger	3	169	736	1638	3594	2700
International	Combination	7	259	151	2293	4233	1560
International	Combination	3	66	112	649	1139	210
Domestic	Freight	5	44	177	401	864	190
International	Passenger	4	140	204	604	1213	150

The size of these simplex tableaus shows why a large computer is usually necessary for realistic problems when mathematical models involve linear programming.

Exercises **1.** Complete the following demand matrix using the information in the figure at the beginning of this section.

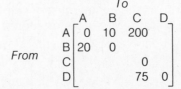

Find the total number of entries in the simplex tableaus for each of the following airlines in the table above. (Hint: find the number of rows and columns in each tableau.)

2. the first airline

3. the second airline

4. the third airline

5. the fourth airline

6. the last airline

(In practice, the number of entries is not as important as the fraction of the numbers that are not zero. In a realistic use of this model only 2% of the entries are not zero.)

Application   *Merit Pay—The Upjohn Company*

Individuals doing the same job within the management of a company often receive different salaries. These salaries may differ because of length of service, productivity of an individual worker, and so on. However, for each job there is usually an established minimum and maximum salary.

Many companies make annual reviews of the salary of each of their management employees. At these reviews, an employee may receive a general cost of living increase, an increase based on merit, both, or neither. (*The Wall Street Journal* tells us that in a recent year, about 99.97% of all employees of the Department of Agriculture who were eligible for merit increases actually received them.)

In this case, we look at a mathematical model for distributing merit increases in an optimum way.[1] An individual who is due for salary review may be described as shown in Figure 1 below. Here i represents the number of the employee whose salary is being reviewed.

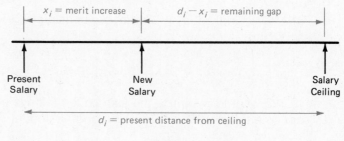

Figure 1

Here the salary ceiling is the maximum salary for the job classification, x_i is the merit increase to be awarded to the individual ($x_i \geq 0$), d_i is the present distance of the current salary from the salary ceiling, and the difference, $d_i - x_i$, is the remaining gap.

---

[1]Based on part of a paper by Jack Northam, Head, Mathematical Services Department, The Upjohn Company, Kalamazoo, Michigan.

We let w_i be a measure of the relative worth of the individual to the company. This is the most difficult variable of the model to actually calculate. One way to evaluate w_i is to give a rating sheet to a number of co-workers and supervisors of employee i. An average rating can be obtained and then divided by the highest rating received by an employee with that same job. This number then gives the worth of employee i relative to all other employees with that job.

The best way to allocate money available for merit pay increases is to minimize

$$\sum_{i=1}^{n} w_i(d_i - x_i).$$

This sum is found by multiplying the relative worth of employee i and the distance of employee i from the salary gap, after any merit increase. Here n represents the total number of employees who have this job. The one constraint here is that the total of all merit increases cannot exceed P, the total amount available for merit increases. That is,

$$\sum_{i=1}^{n} x_i \leq P.$$

Also, the increases for an employee must not put that employee over the maximum salary. That is, for employee i,

$$x_i \leq d_i.$$

We can simplify the objective function, using rules from algebra.

$$\sum_{i=1}^{n} w_i(d_i - x_i) = \sum_{i=1}^{n} (w_i d_i - w_i x_i)$$

$$= \sum_{i=1}^{n} w_i d_i - \sum_{i=1}^{n} w_i x_i$$

For a given individual, w_i and d_i are constant. Therefore, $w_i d_i$ is some constant, say Z, and

$$\sum_{i=1}^{n} w_i(d_i - x_i) = Z - \sum_{i=1}^{n} w_i x_i.$$

We want to minimize the sum on the left; we can do so by *maximizing* the sum on the right. (Why?) Thus, the original model simplifies to maximizing

$$\sum_{i=1}^{n} w_i x_i,$$

subject to

$$\sum_{i=1}^{n} x_i \leq P \qquad \text{and} \qquad x_i \leq d_i$$

for each i.

Exercises   *Here are current salary information and job evaluation averages for six employees who have the same job. The salary ceiling is $1700 per month.*

Employee number	Evaluation average	Current salary
1	570	$1600
2	500	$1550
3	450	$1500
4	600	$1610
5	520	$1530
6	565	$1420

1. Find w_i for each employee by dividing that employee's evaluation average by the highest evaluation average.

2. Use the simplex method to find the merit increase for each employee. Assume that P is 400.

Key Words

simplex method	departing variable
slack variable	pivot
basic feasible solution	simplex tableau
entering variable	surplus variable

Chapter 4   Review Exercises

For each of the following problems, (a) select appropriate variables, (b) write the objective function, (c) write the constraints as inequalities.

1. Roberta Hernandez sells three items, A, B, and C, in her gift shop. Each unit of A costs her $5 to buy, $1 to sell and $2 to deliver. For each unit of B, the costs are $3, $2 and $1 respectively, and for each unit of C the costs are $6, $2, and $5 respectively. The profit on A is $4, on B it is $3, and on C, $3. How many of each should she get to maximize her profit if she can spend $1200 to buy, $800 on selling costs, and $500 on delivery costs?

2. An investor is considering three types of investment: a high risk venture into oil leases with a potential return of 15%, a medium risk investment in bonds with a 9% return, and a relatively safe stock investment with a 5% return. He has $50,000 to invest. Because of the risk, he will limit his investment in oil leases and bonds to 30% and his investment in oil leases and stock to 50%. How much should he invest in each to maximize his return assuming investment returns are as expected?

3. The Aged Wood Winery makes two white wines, Fruity and Crystal, from two kinds of grapes and sugar. The wines require the following amounts of each ingredient per gallon and produce a profit per gallon as shown below.

	Grape A (bushels)	Grape B (bushels)	Sugar (pounds)	Profit (dollars)
Fruity	2	2	2	12
Crystal	1	3	1	15

The winery has available 110 bushels of grape A, 125 bushels of grape B, and 90 pounds of sugar. How much of each wine should be made to maximize profit?

4. A company makes three sizes of plastic bags: 5 gallon, 10 gallon and 20 gallon. The production time in hours for cutting, sealing, and packaging a unit of each size is shown below.

size	cutting	sealing	packaging
5 gallon	1	1	2
10 gallon	1.1	1.2	3
20 gallon	1.5	1.3	4

There are at most 8 hours available each day for each of the three operations. If the profit on a unit of 5-gallon bags is $1, 10-gallon bags is $.90, and 20-gallon bags is $.95, how many of each size should be made per day?

For each of the following problems, (a) add slack variables or subtract surplus variables, and (b) set up the first simplex tableau.

5. Maximize $z = 5x_1 + 3x_2$ subject to
$$2x_1 + 5x_2 \le 50$$
$$x_1 + 3x_2 \le 25$$
$$4x_1 + x_2 \le 18$$
$$x_1 + x_2 \le 12$$
and $x_1 \ge 0, x_2 \ge 0$.

6. Maximize $z = 25x_1 + 30x_2$ subject to
$$3x_1 + 5x_2 \le 47$$
$$x_1 + x_2 \le 25$$
$$5x_1 + 2x_2 \le 35$$
$$2x_1 + x_2 \le 30$$
and $x_1 \ge 0, x_2 \ge 0$.

7. Maximize $z = 5x_1 + 8x_2 + 6x_3$ subject to
$$x_1 + x_2 + x_3 \le 90$$
$$2x_1 + 5x_2 + x_3 \le 120$$
$$x_1 + 3x_2 \ge 80$$
and $x_1 \ge 0, x_2 \ge 0, x_3 \ge 0$.

8. Maximize $z = 2x_1 + 3x_2 + 4x_3$ subject to
$$x_1 + x_2 + x_3 \ge 100$$
$$2x_1 + 3x_2 \le 500$$
$$x_1 + 2x_3 \le 350$$
and $x_1 \ge 0, x_2 \ge 0, x_3 \ge 0$.

For each of the following, use the simplex method to solve the maximizing linear programming problem given the first tableau.

9. $x_1 \quad x_2 \quad x_3 \quad x_4 \quad x_5$

$$\begin{bmatrix} 1 & 2 & 3 & 1 & 0 & 28 \\ 2 & 4 & 1 & 0 & 1 & 32 \\ -5 & -2 & -3 & 0 & 0 & 0 \end{bmatrix}$$

10. $x_1 \quad x_2 \quad x_3 \quad x_4$

$$\begin{bmatrix} 2 & 1 & 1 & 0 & 10 \\ 1 & 3 & 0 & 1 & 16 \\ -2 & -3 & 0 & 0 & 0 \end{bmatrix}$$

11. $\begin{array}{cccccc} x_1 & x_2 & x_3 & x_4 & x_5 & x_6 \end{array}$

$$\begin{bmatrix} 1 & 2 & 2 & 1 & 0 & 0 & | & 50 \\ 3 & 1 & 0 & 0 & 1 & 0 & | & 20 \\ 1 & 0 & 2 & 0 & 0 & -1 & | & 15 \\ \hline -5 & -3 & -2 & 0 & 0 & 0 & | & 0 \end{bmatrix}$$

12. $\begin{array}{ccccc} x_1 & x_2 & x_3 & x_4 & x_5 \end{array}$

$$\begin{bmatrix} 3 & 6 & -1 & 0 & 0 & | & 28 \\ 1 & 1 & 0 & 1 & 0 & | & 12 \\ 2 & 1 & 0 & 0 & 1 & | & 16 \\ \hline -1 & -2 & 0 & 0 & 0 & | & 0 \end{bmatrix}$$

Convert the following problems into maximization problems.

13. Minimize $w = 10x_1 + 15x_2$ subject to
$$\begin{aligned} x_1 + x_2 &\geq 17 \\ 5x_1 + 8x_2 &\geq 42 \end{aligned}$$
and $x_1 \geq 0, x_2 \geq 0$.

14. Minimize $w = 20x_1 + 15x_2 + 18x_3$ subject to
$$\begin{aligned} 2x_1 + x_2 + x_3 &\leq 112 \\ x_1 + x_2 + x_3 &\geq 80 \\ x_1 + x_2 \quad\;\; &\geq 45 \end{aligned}$$
and $x_1 \geq 0, x_2 \geq 0, x_3 \geq 0$.

15. Minimize $w = 7x_1 + 2x_2 + 3x_3$ subject to
$$\begin{aligned} x_1 + x_2 + 2x_3 &\leq 48 \\ x_1 + x_2 \quad\;\; &\geq 12 \\ x_3 &\geq 10 \\ 3x_1 \quad\;\; + x_3 &\leq 30 \end{aligned}$$
and $x_1 \geq 0, x_2 \geq 0, x_3 \geq 0$.

The following tableaus are the final tableaus of minimizing problems. State the solution and the minimum value of the objective function for each problem.

16. $\begin{array}{cccccc} x_1 & x_2 & x_3 & x_4 & x_5 & x_6 \end{array}$

$$\begin{bmatrix} 1 & 0 & 0 & 3 & 1 & 2 & | & 12 \\ 0 & 0 & 1 & 4 & 5 & 3 & | & 5 \\ 0 & 1 & 0 & -2 & 7 & -6 & | & 8 \\ \hline 0 & 0 & 0 & 5 & 7 & 3 & | & -172 \end{bmatrix}$$

17. $\begin{array}{cccccc} x_1 & x_2 & x_3 & x_4 & x_5 & x_6 \end{array}$

$$\begin{bmatrix} 0 & 0 & 3 & 0 & 1 & 1 & | & 2 \\ 1 & 0 & -2 & 0 & 2 & 0 & | & 8 \\ 0 & 1 & 7 & 0 & 0 & 0 & | & 12 \\ 0 & 0 & 1 & 1 & -4 & 0 & | & 1 \\ \hline 0 & 0 & 5 & 0 & 8 & 0 & | & -62 \end{bmatrix}$$

18. $\begin{array}{ccccc} x_1 & x_2 & x_3 & x_4 & x_5 \end{array}$

$$\begin{bmatrix} 5 & 1 & 0 & 7 & -1 & | & 100 \\ -2 & 0 & 1 & 1 & 3 & | & 27 \\ \hline 12 & 0 & 0 & 7 & 2 & | & -640 \end{bmatrix}$$

19. Solve Exercise 1.

20. Solve Exercise 2.

21. Solve Exercise 3.

22. Solve Exercise 4.

5

Sets and Counting

The terminology and concepts of sets have proved to be very useful. Sets, which are collections of objects, help to clarify and classify many mathematical ideas, making them easier to understand. Sets are particularly useful in presenting the topics of probability. The principles of counting, discussed later in this chapter, will also be needed to find probabilities.

5.1 Sets

Think of a **set** as a collection of objects. We could form a set containing one of each type of coin now put out by the government. Another set might be made up of all the students in your class. In mathematics, sets are often made up of numbers. The set consisting of the numbers 3, 4, and 5 is written

$$\{3, 4, 5\},$$

with set braces, { }, to enclose the numbers belonging to the set. The numbers 3, 4, and 5 are called the **elements** or **members** of this set. To show that 4 is an element of the set $\{3, 4, 5\}$, we use the symbol $\in$ and write

$$4 \in \{3, 4, 5\}.$$

Also, $5 \in \{3, 4, 5\}$. To show that 8 is *not* an element of this set, place a slash through the symbol.

$$8 \notin \{3, 4, 5\}$$

We often name sets with capital letters, so that if

$$B = \{5, 6, 7\},$$

we have, for example, $6 \in B$ and $10 \notin B$.

Two sets are **equal** if they contain exactly the same elements. The sets {5, 6, 7}, {7, 6, 5}, and {6, 5, 7} all contain exactly the same elements and are equal. In symbols,

$$\{5, 6, 7\} = \{7, 6, 5\} = \{6, 5, 7\}.$$

Sets which do not contain exactly the same elements are *not equal.* For example, the sets {5, 6, 7} and {7, 8, 9} do not contain exactly the same elements and are not equal. This is written as follows:

$$\{5, 6, 7\} \neq \{7, 8, 9\}.$$

Sometimes we are more interested in a common property of a set rather than in a list of the elements in the set. We can express this common property using **set-builder notation.** We write

$$\{x | x \text{ has property } P\}$$

to represent the set of all elements having some property P.

Example 1   Write the elements belonging to each of the following sets.

(a) $\{x | x$ is a counting number less than 5$\}$
The counting numbers less than 5 make up the set {1, 2, 3, 4}.

(b) $\{x | x$ is a state that touches Florida$\}$ = {Alabama, Georgia}   ■

When discussing a particular situation or problem, we can usually identify a **universal set** (whether expressed or implied) which contains all the elements appearing in any set used in the given problem. The letter U is used to represent the universal set.

For example, when discussing the set of company employees who favor a pension proposal, we might choose the universal set to be the set of all company employees. In discussing the types of species found by Charles Darwin on the Galápagos Islands, the universal set might be the set of all species on all Pacific islands. The choice of a universal set is often arbitrary and depends on the problem under discussion.

Sometimes every element of one set also belongs to another set. For example, if

$$A = \{3, 4, 5, 6\}$$

and $\qquad\qquad\qquad B = \{2, 3, 4, 5, 6, 7, 8\},$

then every element of A is also an element of B. This means that A is a **subset** of B, written $A \subset B$. For example, the set of all presidents of corporations is a subset of the set of all executives of corporations.

Example 2   Decide whether the following statements are true or false.

(a) {3, 4, 5, 6} = {4, 6, 3, 5}
Both sets contain exactly the same elements; the sets are equal. The statement is true. (The fact that the elements are in a different order doesn't matter.)

(b) $\{5, 6, 9, 10\} \subset \{5, 6, 7, 8, 9, 10, 11\}$

Every element of the first set is also an element of the second. This statement is also true. ▮

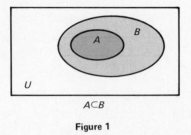

$A \subset B$

Figure 1

Figure 1 shows a drawing which represents a set A which is a subset of set B. The rectangle of the drawing represents the universal set, U. Diagrams like this are called **Venn diagrams.** We use Venn diagrams as an aid in clarifying and discussing the relationships among sets.

By the definition of subset, the **empty set** (which contains no elements) is a subset of every set. That is, if A is a set, and we use $\varnothing$ to represent the empty set, then

$$\varnothing \subset A.$$

Also the definition of subset means that any set A is a subset of itself. That is, for any set A,

$$A \subset A.$$

Example 3 List all possible subsets for each of the following sets.

(a) $\{7, 8\}$

There are four subsets of $\{7, 8\}$:

$$\varnothing, \quad \{7\}, \quad \{8\}, \quad \{7, 8\}.$$

(b) $\{a, b, c\}$

There are eight subsets of $\{a, b, c\}$:

$$\varnothing, \quad \{a\}, \quad \{b\}, \quad \{c\}, \quad \{a, b\}, \quad \{a, c\}, \quad \{b, c\}, \quad \{a, b, c\}. \quad ▮$$

The **cardinal number** of a set is the number of distinct elements in the set. The cardinal number of the set A is written $n(A)$. For example, set

$$A = \{a, b, c, d, e\}$$

has cardinal number 5, written

$$n(A) = 5.$$

Since the empty set has no elements, its cardinal number is 0, by definition. Thus, $n(\varnothing) = 0$.

Example 4 Give the cardinal number of each of the following sets.

(a) $B = \{2, 5, 7, 9, 10, 12, 15\}$.
This set has seven elements, so $n(B) = 7$.

(b) $C = \{x, y, z\}$.
Since set C has three elements, $n(C) = 3$. ■

In Example 3, we found all subsets of $\{7, 8\}$ and all subsets of $\{a, b, c\}$ by trial and error. An alternate method uses a **tree diagram.** A tree diagram is a systematic way of listing all the subsets of a given set. Figures 2(a) and (b) show tree diagrams for finding the subsets of $\{7, 8\}$ and $\{a, b, c\}$.

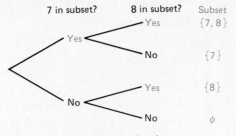

(a) Find the subsets of $\{7, 8\}$

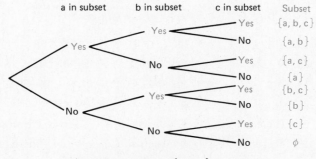

(b) Find the subsets of $\{a, b, c\}$

Figure 2

By studying examples and tree diagrams similar to the ones above, we are led to the following rule, which we prove in Section 5.5.

A set of n distinct elements has 2^n subsets.

Example 5 Find the number of subsets for each of the following sets.

(a) $\{3, 4, 5, 6, 7\}$
This set has five elements; thus it has 2^5 or 32 subsets.

(b) $\{-1, 2, 3, 4, 5, 6, 12, 14\}$
This set has 8 elements and therefore has 2^8 or 256 subsets. ■

5.1 Exercises

Write true *or* false *for each of the following.*

1. $3 \in \{2, 5, 7, 9, 10\}$

2. $6 \in \{-2, 6, 9, 5\}$

3. $1 \in \{3, 4, 5, 1, 11\}$

4. $12 \in \{19, 17, 14, 13, 12\}$

5. $9 \notin \{2, 1, 5, 8\}$

6. $3 \notin \{7, 6, 5, 4\}$

7. $\{2, 5, 8, 9\} = \{2, 5, 9, 8\}$

8. $\{3, 0, 9, 6, 2\} = \{2, 9, 0, 3, 6\}$

9. $\{5, 8, 9\} = \{5, 8, 9, 0\}$

10. $\{3, 7, 12, 14\} = \{3, 7, 12, 14, 0\}$

11. {all counting numbers less than 6} = $\{1, 2, 3, 4, 5, 6\}$

12. {all whole numbers greater than 7 and less than 10} = $\{8, 9\}$

13. {all whole numbers not greater than 4} = $\{0, 1, 2, 3\}$

14. {all counting numbers not greater than 3} = $\{0, 1, 2\}$

15. $\{x \mid x$ is a whole number, $x \leq 5\} = \{0, 1, 2, 3, 4, 5\}$

16. $\{x \mid x$ is an integer, $-3 \leq x < 4\} = \{-3, -2, -1, 0, 1, 2, 3, 4\}$

17. $\{x \mid x$ is an odd integer, $6 \leq x \leq 18\} = \{7, 9, 11, 15, 17\}$

18. $\{x \mid x$ is an even counting number, $x \leq 9\} = \{0, 2, 4, 6, 8\}$

Let

$$A = \{2, 4, 6, 8, 10, 12\} \qquad D = \{2, 10\}$$
$$B = \{2, 4, 8, 10\} \qquad\qquad U = \{2, 4, 6, 8, 10, 12, 14\}$$
$$C = \{4, 10, 12\}$$

Write true *or* false *for each of the following.*

19. $A \subset U$

20. $C \subset U$

21. $D \subset B$

22. $D \subset A$

23. $A \subset B$

24. $B \subset C$

25. $\varnothing \subset A$

26. $\varnothing \subset \varnothing$

27. $\{4, 8, 10\} \subset B$

28. $\{0, 2\} \subset D$

29. $D \not\subset B$

30. $A \not\subset C$

31. There are exactly 32 subsets of A.

32. There are exactly 16 subsets of B.

33. There are exactly 6 subsets of C.

34. There are exactly 4 subsets of D.

Find the number of subsets for each of the following sets.

35. $\{4, 5, 6\}$

36. $\{3, 7, 9, 10\}$

37. $\{5, 9, 10, 15, 17\}$

38. $\{6, 9, 1, 4, 3, 2\}$

39. $\varnothing$

40. $\{0\}$

41. $\{x \mid x$ is a counting number between 6 and 12$\}$

42. $\{x \mid x$ is a whole number between 8 and 12$\}$

Give the cardinal number of each of the following sets.

43. $\{m, p, q, n\}$

44. $\{a, b, c, d, e, f\}$

45. $\{1, 2, 3\}$

46. $\{0\}$

47. $\varnothing$

48. $\{0, \varnothing\}$

49. A Hershey bar of a certain size contains 220 calories. Suppose you eat two of these candy bars and then decide to exercise and get rid of the calories. A list of possible exercises shows the following information.

Exercise	Abbreviation	Calories per hour
Sitting around	s	100
Light exercise	l	170
Moderate exercise	m	300
Severe exercise	e	450
Very severe exercise	u	600

The universal set here is $U = \{s, l, m, e, u\}$. Find all subsets of U (with no element listed twice) that will burn off the calories from the candy bars in
(a) one hour;   **(b)** two hours.

50. The list below includes the producers of most (93%) of the hazardous wastes in the United States.

Industry	% (in 1977)
Inorganic Chemicals (1)	11
Organic Chemicals (2)	34
Electroplating (3)	12
Petroleum Refining (4)	5
Smelting and Refining (5)	26
Textiles Dyeing and Finishing (6)	5

Let x be the number associated with an industry in the table above. The universal set is $U = \{1, 2, 3, 4, 5, 6\}$. List the elements of the following subsets of U.
(a) $\{x \mid$ the industry produces more than 15% of the wastes$\}$.
(b) $\{x \mid$ the industry produces less than 5% of the wastes$\}$.
(c) Find all subsets of U containing industries which produce a total of at least 50% of the waste.

5.2 Set Operations

If we are given a set A, and a universal set U, we can form the set of all elements of U which do *not* belong to A. This set is called the **complement** of set A. For example, if set A is the set of all the female students in your class, and U is the set of all students in the class, then the complement of A would be the set of all male students in the class. The complement of set A is written A'. (Read: "A-prime.") The Venn diagram of Figure 3 shows a set B. Its complement, B', is shown in color.

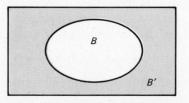

Figure 3

Example 1   Let $U = \{1, 2, 3, 4, 5, 6, 7\}$, $A = \{1, 3, 5, 7\}$ and $B = \{3, 4, 6\}$. Find each of the following sets.

(a) A'

Set A' contains the elements of U that are not in A.

$$A' = \{2, 4, 6\}$$

(b) $B' = \{1, 2, 5, 7\}$

(c) $\emptyset' = U$ and $U' = \emptyset$ ∎

Given two sets A and B, the set of all elements belonging to both set A and set B is called the **intersection** of the two sets, written $A \cap B$. For example, the elements that belong to both $A = \{1, 2, 4, 5, 7\}$ and $B = \{2, 4, 5, 7, 9, 11\}$ are 2, 4, 5, and 7, so that

$$A \cap B = \{1, 2, 4, 5, 7\} \cap \{2, 4, 5, 7, 9, 11\} = \{2, 4, 5, 7\}.$$

The Venn diagram of Figure 4 shows two sets A and B; their intersection, $A \cap B$, is shown in color.

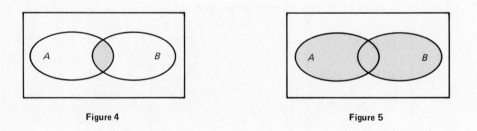

Figure 4   Figure 5

Example 2   **(a)** $\{9, 15, 25, 36\} \cap \{15, 20, 25, 30, 35\} = \{15, 25\}$

The elements 15 and 25 are the only ones belonging to both sets.

(b) $\{-2, -3, -4, -5, -6\} \cap \{-4, -3, -2, -1, 0, 1, 2\} = \{-4, -3, -2\}$ ∎

Two sets that have no elements in common are called **disjoint sets**. For example, there are no elements common to both $\{50, 51, 54\}$ and $\{52, 53, 55, 56\}$, so that these two sets are disjoint, and

$$\{50, 51, 54\} \cap \{52, 53, 55, 56\} = \emptyset.$$

The result of this example can be generalized: for any sets A and B, if A and B are disjoint sets then $A \cap B = \emptyset$.

The set of all elements belonging to either set A or set B or both is called the **union** of the two sets, written $A \cup B$. For example,

$$\{1, 3, 5\} \cup \{3, 5, 7, 9\} = \{1, 3, 5, 7, 9\}.$$

The Venn diagram of Figure 5 shows two sets A and B; their union, $A \cup B$, is shown in color.

Example 3   **(a)** Find the union of $\{1, 2, 5, 9, 14\}$ and $\{1, 3, 4, 8\}$.

Begin by listing the elements of the first set, $\{1, 2, 5, 9, 14\}$. Then include any elements from the second set that are not already listed. Doing this gives

$$\{1, 2, 5, 9, 14\} \cup \{1, 3, 4, 8\} = \{1, 2, 3, 4, 5, 8, 9, 14\}.$$

(b) $\{1, 3, 5, 7\} \cup \{2, 4, 6\} = \{1, 2, 3, 4, 5, 6, 7\}.$ ∎

Finding the complement of a set, the intersection of two sets, or the union of two sets are examples of **set operations**. These are similar to operations on numbers, such as addition, subtraction, multiplication, and division.

Example 4   The table below gives the current dividend, price to earnings ratio (the quotient of the price per share and the annual earnings per share), and price change at the end of a day for six companies, as listed on the New York Stock Exchange.

Stock	Dividend	Price to earnings ratio	Price change
ATT	5	6	$+\frac{1}{8}$
GE	3	9	0
Hershey	1.4	6	$+\frac{3}{8}$
IBM	3.44	12	$-\frac{3}{8}$
Mobil	3.40	6	$-1\frac{5}{8}$
RCA	1.80	7	$-\frac{1}{4}$

Let set A include all stocks with a dividend greater than \$3, B include all stocks with a price to earnings ratio of at least 10, and C include all stocks with a positive price change. Find the following.

(a) A'

Set A' contains all the listed stocks outside set A, or $A' = \{$GE, Hershey, RCA$\}$.

(b) $A \cap B$

The intersection of A and B will contain those stocks that offer a dividend greater than \$3 *and* have a price to earnings ratio of at least 10.

$$A \cap B = \{\text{IBM}\}.$$

(c) $A \cup C$

We want the set of all stocks with a dividend greater than \$3 *or* a positive price change (or both).

$$A \cup C = \{\text{ATT, Hershey, IBM, Mobil}\}.$$ ∎

5.2 Exercises   *Write* true *or* false *for each of the following.*

1. $\{5, 7, 9, 19\} \cap \{7, 9, 11, 15\} = \{7, 9\}$
2. $\{8, 11, 15\} \cap \{8, 11, 19, 20\} = \{8, 11\}$
3. $\{2, 1, 7\} \cup \{1, 5, 9\} = \{1\}$
4. $\{6, 12, 14, 16\} \cup \{6, 14, 19\} = \{6, 14\}$

5. $\{3, 2, 5, 9\} \cap \{2, 7, 8, 10\} = \{2\}$ **6.** $\{8, 9, 6\} \cup \{9, 8, 6\} = \{8, 9\}$

7. $\{3, 5, 9, 10\} \cap \varnothing = \{3, 5, 9, 10\}$ **8.** $\{3, 5, 9, 10\} \cup \varnothing = \{3, 5, 9, 10\}$

9. $\{1, 2, 4\} \cup \{1, 2, 4\} = \{1, 2, 4\}$ **10.** $\{1, 2, 4\} \cap \{1, 2, 4\} = \varnothing$

11. $\varnothing \cup \varnothing = \varnothing$ **12.** $\varnothing \cap \varnothing = \varnothing$

Let $U = \{2, 3, 4, 5, 7, 9\}$, $X = \{2, 3, 4, 5\}$, $Y = \{3, 5, 7, 9\}$, *and* $Z = \{2, 4, 5, 7, 9\}$. *Find each of the following sets.*

13. $X \cap Y$ **14.** $X \cup Y$ **15.** $Y \cup Z$ **16.** $Y \cap Z$

17. $X \cup U$ **18.** $Y \cap U$ **19.** X' **20.** Y'

21. $X' \cap Y'$ **22.** $X' \cap Z$ **23.** $Z' \cap \varnothing$ **24.** $Y' \cup \varnothing$

25. $X \cup (Y \cap Z)$ **26.** $Y \cap (X \cup Z)$

Let $U = \{$all students in this school$\}$
$M = \{$all students taking this course$\}$
$N = \{$all students taking accounting$\}$
$P = \{$all students taking zoology$\}$

Describe each of the following sets in words.

27. M' **28.** $M \cup N$ **29.** $N \cap P$

30. $N' \cap P'$ **31.** $M \cup P$ **32.** $P' \cup M'$

Given $U = \{1, 2, 3, 4, 5, 6, 7, 8, 9, 10\}$, $P = \{2, 4, 6, 8, 10\}$, $Q = \{4, 5, 6\}$, *and* $R = \{4\}$. *Find the cardinal number of each of the following sets.*

33. $P \cup Q$ **34.** $P \cap Q$ **35.** P' **36.** Q'

37. $P' \cap Q$ **38.** $P \cup R'$ **39.** $P \cup (R \cap Q)$ **40.** $P \cap (R \cup Q)$

Refer to Example 4 in the text. Describe each of the following sets in words. Then list the elements of each set.

41. B' **42.** C' **43.** $B \cap C$ **44.** $A \cup B$

45. $(A \cap B)'$ **46.** $(A \cup C)'$

47. The lists below show some symptoms of an overactive thyroid and an underactive thyroid.

Underactive thyroid	Overactive thyroid
Sleepiness, s	Insomnia, i
Dry hands, d	Moist hands, m
Intolerance of cold, c	Intolerance of heat, h
Goiter, g	Goiter, g

(a) Find the smallest possible universal set U that includes all the symptoms listed.

Let N be the set of symptoms for an underactive thyroid, and let O be the set of symptoms for an overactive thyroid. Find each of the following sets.
(b) O' (c) N' (d) $N \cap O$ (e) $N \cup O$ (f) $N \cap O'$

48. Let A and B be sets with cardinal numbers $n(A) = a$ and $n(B) = b$, respectively. Answer *true* or *false* for the following statements:
(a) $n(A \cup B) = n(A) + n(B)$;   (b) $n(A \cup B) = n(A) + n(B) - n(A \cap B)$.

5.3   Venn Diagrams

We used Venn diagrams in the last section to help in understanding set union and intersection. The rectangular region in a Venn diagram represents the universal set, U. If we include only a single set A inside the universal set, as in Figure 6, we divide the total region of U into two regions. Region 1 represents those elements outside of set A, while region 2 represents those elements belonging to set A. (Our numbering of the regions is arbitrary.)

If we include two sets A and B inside U, we get the Venn diagram of Figure 7. Two sets divide the universal set into four regions. As labeled in Figure 7, region 1 includes those elements outside of both set A and set B. Region 2 includes those elements belonging to A and not to B. Region 3 includes those elements belonging to both A and B. Which elements belong to region 4? (Again, the labeling is arbitrary.)

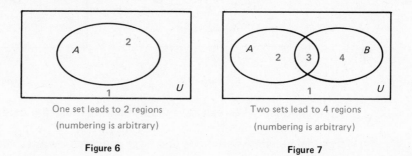

One set leads to 2 regions
(numbering is arbitrary)

Figure 6

Two sets lead to 4 regions
(numbering is arbitrary)

Figure 7

Example 1   Draw Venn diagrams similar to Figure 7 and shade the regions representing the following sets.

(a) $A' \cap B$

Set A' contains all the elements outside of set A. As labeled in Figure 7, A' is made up of regions 1 and 4. Set B is made up of the elements in regions 3 and 4. The intersection of sets A' and B, the set $A' \cap B$, is made up of the elements in the region common to regions 1 and 4 and regions 3 and 4. The result, region 4, is shaded in Figure 8.

(b) $A' \cup B'$

Again, set A' is represented by regions 1 and 4, while B' is made up of regions 1 and 2. To find $A' \cup B'$, we need the elements belonging to either regions 1 and 4 or to regions 1 and 2. The result, regions 1, 2, and 4, is shaded in Figure 9.   ■

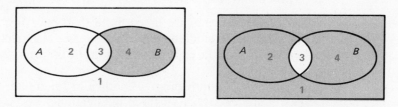

Figure 8

Figure 9

We can draw Venn diagrams with three sets inside U. Three sets divide the universal set into eight regions, which can be numbered as in Figure 10.

Example 2   Shade $A' \cup (B \cap C')$ on a Venn diagram.

We first find $B \cap C'$. Set B is made up of regions 3, 4, 7, and 8, while C' is made up of regions 1, 2, 3, and 8. The overlap of these regions, the set $B \cap C'$, is made up of regions 3 and 8. Set A' is made up of regions 1, 6, 7, and 8. The union of regions 3 and 8 and regions 1, 6, 7, 8 is regions 1, 3, 6, 7 and 8, which are shaded in Figure 11.   ▮

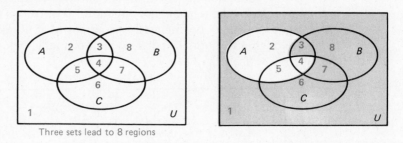

Three sets lead to 8 regions

Figure 10   **Figure 11**

Applications   We can use Venn diagrams and the cardinal number of a set to solve counting problems in surveying groups of people. Suppose a group of 60 freshman business students at a large university was surveyed, with the following results.

19 of the students read *Business Week;*

18 read *The Wall Street Journal;*

50 read *Fortune;*

13 read *Business Week* and *The Journal;*

11 read *The Journal* and *Fortune;*

13 read *Business Week* and *Fortune;*

 9 read all three.

Let us use this data to help answer the following questions.

(a) How many students read none of the publications?

(b) How many read only *Fortune?*

(c) How many read *Business Week* and *The Journal,* but not *Fortune?*

Many of the students are listed more than once in the data above. For example, some of the 50 students who read *Fortune* also read *Business Week.* The 9 students who read all three are counted in the 13 who read *Business Week* and *Fortune,* and so on.

We can use a Venn diagram, as shown in Figure 12, to better illustrate this data. Since 9 students read all three publications, we begin by placing 9 in the area that belongs to all three regions, as shown in Figure 13. We know that 13 students read *Business Week* and *Fortune*. However, 9 of these 13 also read *The Journal*. Therefore, only $13 - 9 = 4$ read just *Business Week* and *Fortune*. Place the number 4 in the area of Figure 13 common to *Business Week* and *Fortune* readers. In the same way, place 4 in the region common only to *Business Week* and *The Journal,* and 2 in the region common only to *Fortune* and *The Journal*.

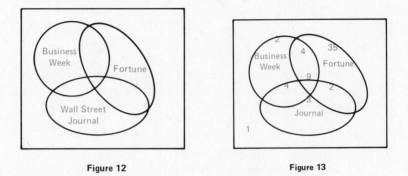

Figure 12   Figure 13

   We know 19 students read *Business Week*. However, we have already placed $4 + 9 + 4 = 17$ readers in the region representing *Business Week*. Thus, the rest of this region will contain only $19 - 17 = 2$ students. These 2 students read *Business Week* only—not *Fortune* and not *The Journal*. In the same way, 3 students read only *The Journal* and 35 read only *Fortune*.

   We have placed $2 + 4 + 3 + 4 + 9 + 2 + 35 = 59$ students in the three regions of Figure 13. We know that 60 students were surveyed; thus, $60 - 59 = 1$ student reads none of these three publications and so is placed outside all three regions.

   We can now use Figure 13 to answer the questions asked above.

(a) Only 1 student reads none of the three publications.

(b) From Figure 13, 35 students read only *Fortune*.

(c) The overlap of the regions representing *Business Week* and *The Journal* shows that 4 students read *Business Week* and *The Journal* but not *Fortune*.

Example 3 Jeff Friedman is a section chief for an electric utility company. The employees in his section cut down tall trees, climb poles, and splice wire. Friedman reported the following information to the management of the utility.

   Out of the 100 employees in my section,

   45 can cut tall trees;

   50 can climb poles;

   57 can splice wire;

   28 can cut trees and climb poles;

   20 can climb poles and splice wire;

   25 can cut trees and splice wire;

   11 can do all three;

    9 can't do any of the three (management trainees).

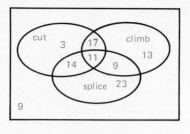

Figure 14

From the data supplied by Friedman we can find the numbers shown in Figure 14. By adding all the numbers from the regions, we find the total number of Friedman's employees to be

$$9 + 3 + 14 + 23 + 11 + 9 + 17 + 13 = 99.$$

Friedman claimed to have 100 employees, but his data indicates only 99. The management decided that Friedman didn't qualify as a section chief, and reassigned him as a nightshift meter reader in Guam. (Moral: he should have taken this course.)   ▌

Note that in both problems we worked above, we started in the innermost region, which is the intersection of the three categories. This is usually the best way to begin with these problems.

5.3 Exercises   *Use a Venn diagram similar to Figure 7 to show each of the following sets.*

1. $B \cap A'$

2. $A \cup B'$

3. $A' \cup B$

4. $A' \cap B'$

5. $B' \cup (A' \cap B')$

6. $(A \cap B) \cup B'$

7. U'

8. $\varnothing'$

Use a Venn diagram similar to Figure 10 to show each of the following sets.

9. $(A \cap B) \cap C$

10. $(A \cap C') \cup B$

11. $A \cap (B \cup C')$

12. $A' \cap (B \cap C)$

13. $(A' \cap B') \cap C$

14. $(A \cap B') \cup C$

15. $(A \cap B') \cap C$

16. $A' \cap (B' \cup C)$

17. If $n(A) = 5$, $n(B) = 8$, and $n(A \cap B) = 4$, what is $n(A \cup B)$?

18. If $n(A) = 12$, $n(B) = 27$, and $n(A \cup B) = 30$, what is $n(A \cap B)$?

19. Suppose $n(B) = 7$, $n(A \cap B) = 3$, and $n(A \cup B) = 20$. What is $n(A)$?

20. Suppose $n(A \cap B) = 5$, $n(A \cup B) = 35$, and $n(A) = 13$. What is $n(B)$?

Use Venn diagrams to answer the following questions.

21. Jeff Friedman, of Example 3 in the text, was again reassigned, this time to the home economics department of the electric utility. He interviewed 140 people in a suburban shopping center to find out some of their cooking habits. He obtained the following results. Should he be reassigned yet one more time?

> 58 use microwave ovens;
>
> 63 use electric ranges;
>
> 58 use gas ranges;
>
> 19 use microwave ovens and electric ranges;
>
> 17 use microwave ovens and gas ranges;
>
> 4 use both gas and electric ranges;
>
> 1 uses all three;
>
> 2 cook only with solar energy.

22. Toward the middle of the harvesting season, peaches for canning come in three types: earlies, lates, and extra lates, depending on the expected date of ripening. During a certain week, the following data was recorded at a fruit delivery station.

> 34 trucks went out carrying early peaches;
>
> 61 had late peaches;
>
> 50 had extra lates;
>
> 25 had earlies and lates;
>
> 30 had lates and extra lates;
>
> 8 had earlies and extra lates;
>
> 6 had all three;
>
> 9 had only figs (no peaches at all).

(a) How many trucks had only late variety peaches?

(b) How many had only extra lates?

(c) How many had only one type of peaches?

(d) How many trucks in all went out during the week?

23. A chicken farmer surveyed his flock with the following results. The farmer had

> 9 fat red roosters;
>
> 2 fat red hens;
>
> 37 fat chickens;
>
> 26 fat roosters;
>
> 7 thin brown hens;
>
> 18 thin brown roosters;
>
> 6 thin red roosters;
>
> 5 thin red hens.

Answer the following questions about the flock. Hint: you need a Venn diagram with regions for fat, for male (a rooster is a male, a hen is a female), and for red (assume that brown and red are opposites in the chicken world). How many chickens were

(a) fat?   (b) red?   (c) male?   (d) fat, but not male?

(e) brown, but not fat?   (f) red and fat?

24. Country-western songs emphasize three basic themes: love, prison, and trucks. A survey of the local country-western radio station produced the following data.

12 songs were about a truck driver who was in love while in prison;

13 about a prisoner in love;

28 about a person in love;

18 about a truck driver in love;

3 about a truck driver in prison who was not in love;

2 about a prisoner who was not in love and did not drive a truck;

8 about a person out of jail who was not in love, and did not drive a truck;

16 about truck drivers who were not in prison.

(a) How many songs were surveyed?

Find the number of songs about
(b) truck drivers;   **(c)** prisoners;   **(d)** truck drivers in prison;
(e) people not in prison;   **(f)** people not in love.

25. After a genetics experiment, the number of pea plants having certain characteristics was tallied, with the results as follows.

22 were tall;

25 had green peas;

39 had smooth peas;

9 were tall and had green peas;

17 were tall and had smooth peas;

20 had green peas and smooth peas;

6 had all three characteristics;

4 had none of the characteristics.

(a) Find the total number of plants counted.
(b) How many plants were tall and had peas which were neither smooth nor green?
(c) How many plants were not tall but had peas which were smooth and green?

26. Human blood can contain either no antigens, the A antigen, the B antigen, or both the A and B antigens. A third antigen, called the Rh antigen, is important in human reproduction, and again may or may not be present in an individual. Blood is called type A-positive if the individual has the A and Rh, but not the B antigen. A person having only the A and B antigens is said to have type AB-negative blood. A person having only the Rh antigen has type O-positive blood. Other blood types are defined in a similar manner. Identify the blood type of the individuals in regions (a)—(g) of the Venn diagram.

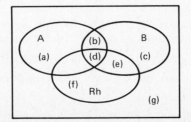

27. (Use the diagram from Exercise 26.) In a certain hospital, the following data was recorded.

> 25 patients had the A antigen;
>
> 17 had the A and B antigens;
>
> 27 had the B antigen;
>
> 22 had the B and Rh antigens;
>
> 30 had the Rh antigen;
>
> 12 had none of the antigens;
>
> 16 had the A and Rh antigens;
>
> 15 had all three antigens.

How many patients
(a) were represented?   (b) had exactly one antigen?
(c) had exactly two antigens?   (d) had O-positive blood?
(e) had AB-positive blood?   (f) had B-negative blood?
(g) had O-negative blood?   (h) had A-positive blood?

28. A survey of 80 sophomores at a western college showed that

> 36 take English;
>
> 32 take history;
>
> 32 take political science;
>
> 16 take political science and history;
>
> 16 take history and English;
>
> 14 take political science and English;
>
> 6 take all three.

How many students:
(a) take English and neither of the other two?
(b) take none of the three courses?
(c) take history, but neither of the other two?
(d) take political science and history, but not English?
(e) do not take political science?

29. The following table shows the number of people in a certain small town in Georgia who fit in the given categories.

Age	Drink vodka (V)	Drink bourbon (B)	Drink gin (G)	Totals
20–25 (Y)	40	15	15	70
26–35 (M)	30	30	20	80
over 35 (O)	10	50	10	70
Totals	80	95	45	220

Using the letters given in the table, find the number of people in each of the following sets.
(a) $Y \cap V$ (b) $M \cap B$ (c) $M \cup (B \cap Y)$
(d) $Y' \cap (B \cup G)$ (e) $O' \cup G$ (f) $M' \cap (V' \cap G')$

30. The following table shows the results of a survey in a medium-sized town in Tennessee. The survey asked questions about the investment habits of local citizens.

Age	Stocks (S)	Bonds (B)	Savings accounts (A)	Totals
18–29 (Y)	6	2	15	23
30–49 (M)	14	5	14	33
50 or over (O)	32	20	12	64
Totals	52	27	41	120

Using the letters given in the table, find the number of people in each of the following sets.

(a) $Y \cap B$ **(b)** $M \cup A$ **(c)** $Y \cap (S \cup B)$
(d) $O' \cup (S \cup A)$ **(e)** $(M' \cup O') \cap B$

For the statements of Exercises 31–34, draw Venn diagrams for the sets on each side of the equals sign. Show that the Venn diagrams are the same.[1]

31. $(A \cup B)' = A' \cap B'$ **32.** $(A \cap B)' = A' \cup B'$

33. $A \cap (B \cup C) = (A \cap B) \cup (A \cap C)$ **34.** $A \cup (B \cap C) = (A \cup B) \cap (A \cup C)$

35. Let $n(A)$ represent the number of elements in set A. Verify that $n(A \cup B) = n(A) + n(B) - n(A \cap B)$ for sets $A = \{1, 2, 3, 4, 5\}$ and $B = \{3, 5, 7\}$.

36. Do you think the statement in Exercise 35 is true for any sets A and B?

5.4 Permutations and Combinations

After making do with your old V-8 automobile for several years, you finally decide to replace it with a new small super-economy model. You drive over to Ned's New Car Emporium to choose the car that's just right for you. Once there, you find that you can select from 5 models, each with 4 power options, a choice of 8 exterior color combinations and 3 interior colors. How many different new cars are available to you? Problems of this sort are best solved by means of the counting principles which are discussed in this section. These counting methods are very useful in probability.

Let us begin with a simpler example. If there are three roads from town A to town B and two roads from town B to town C, how many ways can we travel from A to C by way of B? For each of the three roads from A there are two different routes leading from B to C; hence there are $3 \cdot 2 = 6$ different ways to make the trip, as shown in Figure 15. This example illustrates a general principle of counting, sometimes called the **multiplication axiom:**

> If an event can occur in m ways and a second event can occur in n ways, the first event followed by the second event can occur in mn ways, assuming the second event is in no way influenced by the first.

[1]The statements of Exercises 31 and 32 are known as De Morgan's Laws. They are named for the English Mathematician Augustus De Morgan (1806–71).

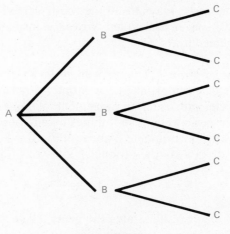

Figure 15

The multiplication axiom can be extended to any number of events, provided that no one event influences another. Thus we can apply it to our problem of determining how many choices we actually have in selecting a new car at Ned's. Applying the axiom, we find that

$$5 \cdot 4 \cdot 8 \cdot 3 = 480$$

different new cars are available from this one dealer.

Example 1   A teacher has 5 different books which he wishes to arrange side by side. How many different arrangements are possible?

There are 5 decisions (events) to be made, one for each space which will hold a book. To select a book for the first space, the teacher has 5 choices, for the second space, 4 choices (one book has already been put in the first space), for the third space, 3 choices, and so on. By the multiplication axiom, we see that the number of different arrangements is $5 \cdot 4 \cdot 3 \cdot 2 \cdot 1 = 120$. ∎

In using the multiplication axiom, we frequently encounter such products as $5 \cdot 4 \cdot 3 \cdot 2 \cdot 1$. For convenience, the symbol $n!$ (read **"*n* factorial"**) is used to denote such products. For any counting number n, we define

$$n! = n(n - 1)(n - 2)(n - 3) \cdots (2)(1).$$

Thus, we may write $5 \cdot 4 \cdot 3 \cdot 2 \cdot 1$ as $5!$. Similarly $3! = 3 \cdot 2 \cdot 1 = 6$. By this definition of $n!$, we see that $n[(n - 1)!] = n!$ for all natural numbers $n \geq 2$. It is convenient to have this relation hold also for $n = 1$, so we define

$$0! = 1.$$

Example 2   Suppose the teacher in Example 1 wishes to place only 3 of the 5 books on his desk. How many arrangements of 3 books are possible?

The teacher again has 5 ways to fill the first space, 4 ways to fill the second space, and 3 ways to fill the third. Since he wants to use only 3 books, there are only 3 spaces to be filled (3 events) instead of 5. Thus, there are $5 \cdot 4 \cdot 3 = 60$ arrangements. ∎

Permutations   The answer 60 in Example 2 is called the number of permutations of 5 things taken 3 at a time. A **permutation** of r ($r \geq 1$) elements from a set of n elements is any arrangement, *without repetition,* of the r elements. The number of permutations of n things taken r at a time ($r \leq n$) is written $P(n, r)$. Based on the work in Example 2 above, $P(5, 3) = 5 \cdot 4 \cdot 3 = 60$. We can use factorial notation to express the product $5 \cdot 4 \cdot 3$ as follows.

$$5 \cdot 4 \cdot 3 = 5 \cdot 4 \cdot 3 \cdot \frac{2 \cdot 1}{2 \cdot 1} = \frac{5 \cdot 4 \cdot 3 \cdot 2 \cdot 1}{2 \cdot 1} = \frac{5!}{2!} = \frac{5!}{(5 - 3)!}$$

This example illustrates the general rule stated in Theorem 5.1, which is a direct consequence of the multiplication axiom.

Theorem 5.1

> If $P(n, r)$ (where $r \leq n$) denotes the number of permutations of n elements taken r at a time, then
>
> $$P(n, r) = \frac{n!}{(n - r)!}.$$

The proof of the theorem follows the discussion of Example 2 above. There are n ways to choose the first of the r elements, $n - 1$ ways to choose the second and so on. We have

$$P(n, r) = n(n - 1)(n - 2) \cdots (n - r + 1)$$
$$= n(n - 1)(n - 2) \cdots (n - r + 1) \cdot \frac{(n - r)!}{(n - r)!}$$
$$= \frac{n(n - 1)(n - 2) \cdots (n - r + 1)(n - r)!}{(n - r)!}$$
$$P(n, r) = \frac{n!}{(n - r)!}.$$

To find $P(n, r)$, either Theorem 5.1 or direct application of the multiplication axiom may be used, as the following example demonstrates.

Example 3   Find the number of permutations of 8 elements taken 3 at a time.

Using the multiplication axiom, we note that there are 3 choices to be made, so that $P(8, 3) = 8 \cdot 7 \cdot 6 = 336$. However, we can also use the formula given above for $P(n, r)$.

$$P(8, 3) = \frac{8!}{(8 - 3)!} = \frac{8!}{5!} = \frac{8 \cdot 7 \cdot 6 \cdot 5!}{5!} = 8 \cdot 7 \cdot 6 = 336 \quad ∎$$

Combinations   In Example 2 above, we found that there are 60 ways that a teacher can arrange 3 of 5 different books on his desk. That is, there are 60 permutations of 5 things taken 3 at a time. Suppose now that the teacher does not wish to arrange the books on his desk, but rather wishes to choose, at random, any 3 of the 5 books to give to a book sale to raise money for his school. In how many ways can he do this?

At first glance, we might say 60 again, but this is incorrect. The number 60 counts all possible *arrangements* of 3 books chosen from 5. However, the following arrangements would all lead to the same set of three books being given to the book sale.

mystery-biography-textbook	biography-textbook-mystery
mystery-textbook-biography	textbook-biography-mystery
biography-mystery-textbook	textbook-mystery-biography

We have here 6 different *arrangements* of 3 books, but only one *set* of 3 books for the book sale. A subset of items selected *without worrying about order*, is called a **combination.** The number of combinations of 5 things taken 3 at a time is written $\binom{5}{3}$.

To evaluate $\binom{5}{3}$, we note first that there are $5 \cdot 4 \cdot 3$ *permutations* of 5 things taken 3 at a time. However, we don't care about order, and each subset of 3 items from the set of 5 items can have its elements rearranged in $3 \cdot 2 \cdot 1 = 3!$ ways. Therefore, $\binom{5}{3}$ can be found by dividing the number of permutations by 3!, or

$$\binom{5}{3} = \frac{5 \cdot 4 \cdot 3}{3!} = \frac{5 \cdot 4 \cdot 3}{3 \cdot 2 \cdot 1} = 10.$$

There are 10 ways that the teacher can choose 3 books at random for the book sale. The general formula for the number of combinations of n elements taken r at a time is

$$\binom{n}{r} = \frac{P(n, r)}{r!}.$$

We can rewrite this formula as follows.

$$\binom{n}{r} = \frac{P(n, r)}{r!}$$

$$= \frac{n!}{(n - r)!} \cdot \frac{1}{r!}$$

$$= \frac{n!}{(n - r)!r!}$$

This last form is the most useful for calculation. The steps above lead to the following theorem.

Theorem 5.2

> If $\binom{n}{r}$ denotes the number of combinations of n elements taken r at a time, then
>
> $$\binom{n}{r} = \frac{n!}{(n - r)!r!}.$$

Table 1 in the Appendix gives the values of $\binom{n}{r}$ for $n \leq 20$.

Example 4   How many committees of 3 people can be formed from a group of 8 people?

A committee is an unordered set, so we want $\binom{8}{3}$. Using Theorem 5.2, we have

$$\binom{8}{3} = \frac{8!}{5!3!} = \frac{8 \cdot 7 \cdot 6 \cdot 5 \cdot 4 \cdot 3 \cdot 2 \cdot 1}{5 \cdot 4 \cdot 3 \cdot 2 \cdot 1 \cdot 3 \cdot 2 \cdot 1} = \frac{8 \cdot 7 \cdot 6}{3 \cdot 2 \cdot 1} = 56. \quad \blacksquare$$

Example 5   From a group of 30 employees, 3 are to be selected to work on a special project.

(a) In how many different ways can the employees be selected?

Here we wish to know how many 3-element combinations can be formed from a set of 30 elements. (We want combinations, not permutations, since order within the group of 3 is irrelevant.)

$$\binom{30}{3} = \frac{30!}{27!3!} = \frac{30 \cdot 29 \cdot 28 \cdot 27!}{27! \cdot 3 \cdot 2 \cdot 1}$$

$$= \frac{30 \cdot 29 \cdot 28}{3 \cdot 2 \cdot 1}$$

$$= 4060$$

There are 4060 ways to select the project group.

(b) In how many ways can the group of three be selected if it has been decided that a particular man must work on the project?

Since one man has already been selected for the project, the problem is reduced to selecting two more people from the remaining 29 employees.

$$\binom{29}{2} = \frac{29!}{27!2!} = \frac{29 \cdot 28 \cdot 27!}{27! \cdot 2 \cdot 1} = \frac{29 \cdot 28}{2 \cdot 1} = 29 \cdot 14 = 406$$

In this case, the project group can be selected in 406 ways. $\blacksquare$

The formulas for permutations and combinations given in this section will be very useful in solving probability problems in the next chapter. Any difficulty in using these formulas usually comes from being unable to differentiate between them. In the next examples, we concentrate on recognizing which formulas should be applied.

Example 6   A manager must select 4 employees for promotion: 12 employees are eligible.

(a) In how many ways can the four be chosen?

Since there is no reason to differentiate among the 4 who are selected, use combinations.

$$\binom{12}{4} = \frac{12!}{4!8!} = 495$$

(b) In how many ways can 4 employees be chosen (from 12) to be placed in 4 different jobs?

In this case, once a group of 4 is selected, they can be assigned in many different ways (or arrangements) to the 4 jobs. Therefore, this problem requires the permutations formula.

$$P(12, 4) = \frac{12!}{8!} = 11880 \quad \blacksquare$$

Example 7 In how many ways can a full house of aces and eights (3 aces and 2 eights) be dealt in five card poker?

Here, we are not interested in the arrangement of the three aces or the two eights. Use combinations and the multiplication axiom. There are $\binom{4}{3}$ ways to get 3 aces from the 4 aces in the deck, and $\binom{4}{2}$ ways to get 2 eights. The number of ways to get 3 aces and 2 eights is

$$\binom{4}{3} \cdot \binom{4}{2} = 4 \cdot 6 = 24. \quad \blacksquare$$

Example 8 In how many ways can a flush be dealt in five-card poker? (A flush is a five-card hand of the same suit.)

The total number of ways that 5 cards of a particular suit of 13 cards can be dealt is $\binom{13}{5}$. The arrangement of the five cards is not important, so we use combinations. Since there are four different suits, by the multiplication axiom, there are

$$4 \cdot \binom{13}{5} = 4 \cdot 1287 = 5148$$

ways to deal a 5-card flush. $\quad \blacksquare$

The following table outlines the similarities of permutations and combinations as well as their differences.

Permutations	Combinations
Number of ways of selecting r items out of n items	
Repetitions are not allowed	
Order is important	Order is not important
Arrangements of r items from a set of n items	Subsets of r items from a set of n items
$P(n, r) = \dfrac{n!}{(n-r)!}$	$\dbinom{n}{r} = \dfrac{n!}{(n-r)!\,r!}$

It should be stressed that not all counting problems lend themselves to either of these techniques. Whenever a tree diagram or the product rule can be used directly, then use them.

5.4 Exercises

Evaluate the following factorials, permutations, and combinations.

1. $P(4, 2)$ **2.** $3!$ **3.** $\binom{8}{3}$ **4.** $7!$

5. $P(8, 1)$ **6.** $\binom{8}{1}$ **7.** $4!$ **8.** $P(4, 4)$

9. $\binom{12}{5}$ **10.** $\binom{10}{8}$ **11.** $P(13, 2)$ **12.** $P(12, 3)$

13. How many different two-card hands can be dealt from an ordinary deck (52 cards)?

14. How many different thirteen-card bridge hands can be dealt from an ordinary deck?

15. Five cards are marked with the numbers 1, 2, 3, 4, and 5, then shuffled, and two cards are drawn. How many different two-card combinations are possible?

16. Marbles are drawn without replacement from a bag containing 15 marbles.
  (a) How many samples of 2 marbles can be drawn?
  (b) How many samples of 4 marbles can be drawn?
  (c) If the bag contains 3 yellow, 4 white, and 8 blue marbles, how many samples of 2 marbles can be drawn in which both marbles are blue?

17. Use the multiplication axiom to decide how many 7-digit telephone numbers are possible if the first digit cannot be zero and
  (a) only odd digits may be used;
  (b) the telephone number must be a multiple of 10 (that is, it must end in zero);
  (c) the telephone number must be a multiple of 100;
  (d) the first three digits are 481;
  (e) no repetitions are allowed?

18. How many different license numbers consisting of 3 letters followed by 3 digits are possible?

19. In a club with 8 men and 11 women members, how many 5-member committees can be chosen that have **(a)** all men; **(b)** all women; **(c)** 3 men and 2 women.

20. 5 cards are drawn from an ordinary deck. In how many ways is it possible to draw
  (a) all queens;
  (b) all face cards (face cards are the Jack, Queen, and King);
  (c) no face card;
  (d) exactly 2 face cards;
  (e) 1 heart, 2 diamonds, and 2 clubs.

21. If a baseball coach has 5 good hitters and 4 poor hitters on the bench and chooses 3 players at random, in how many ways can he choose at least 2 good hitters?

22. If your college offers 400 courses, 20 of which are in mathematics, and your counselor arranges your schedule of 4 courses by random selection, how many schedules are possible that do not include a math course?

23. A bag contains 5 black, 1 red, and 3 yellow jelly beans; you reach in and take 3 at random. How many samples are possible in which the jelly beans are
  (a) all black;   **(b)** all red;
  (c) all yellow;   **(d)** 2 black, 1 red;
  (e) 2 black, 1 yellow;   **(f)** 2 yellow, 1 black;
  (g) 2 red, 1 yellow.

24. There are 5 rotten apples in a crate of 25 apples.
  (a) How many samples of 3 apples can be drawn from the crate?
  (b) How many samples of 3 could be drawn in which all 3 are rotten?
  (c) How many samples of 3 could be drawn in which there are two good apples and one rotten one?

25. How many different types of homes are available if a builder offers a choice of 5 basic plans, 3 roof styles, and 2 exterior finishes?

26. An auto manufacturer produces 7 models, each available in 6 different colors, with 4 different upholstery fabrics, and 5 interior colors. How many varieties of the auto are available?

27. How many different 4-letter radio station call letters can be made
  (a) if the first letter must be K or W and no letter may be repeated?
  (b) if repeats are allowed (but the first letter is K or W)?
  (c) How many 4-letter call letters (starting with K or W) with no repeats end in R?

28. A business school gives courses in typing, shorthand, transcription, business English, technical writing, and accounting. In how many ways can a student arrange a schedule if 3 courses are taken?

29. In how many ways can an employer select 2 new employees from a group of 4 applicants?

30. Hal's Hamburger Hamlet sells hamburgers with cheese, relish, lettuce, tomato, mustard, or catsup. How many different kinds of hamburgers can be made using any three of the extras?

31. A group of 7 workers decides to send a delegation of 2 to their supervisor to discuss their grievances.
  (a) How many delegations are possible?
  (b) If it is decided that a particular employee must be in the delegation, how many different delegations are possible?
  (c) If there are 2 women and 5 men in the group, how many delegations would include at least 1 woman?

32. In how many ways can 7 of 10 monkeys be arranged in a row for a genetics experiment?

33. A group of 3 students is to be selected from a group of 12 students to take part in a special class in cell biology.
  (a) In how many ways can this be done?
  (b) In how many ways can the group which will *not* take part be chosen?

34. In an experiment on plant hardiness, a researcher gathers 6 wheat plants, 3 barley plants, and 2 rye plants. She wishes to select 4 plants at random.
  (a) In how many ways can this be done?
  (b) In how many ways can this be done if 2 wheat plants must be included?

35. In an experiment on social interaction, 6 people will sit in 6 seats in a row. In how many ways can this be done?

36. A couple has narrowed down the choice of a name for their new baby to 3 first names and 5 middle names. How many different first and middle name arrangements are possible?

37. A concert to raise money for an economics prize is to consist of 5 works: 2 overtures, 2 sonatas, and a piano concerto. In how many ways can the program be arranged?

38. How many different license plate numbers can be formed using 3 letters followed by 3 digits if no repeats are allowed?

39. How many license plate numbers (see Exercise 38) are possible if there are no repeats and either numbers or letters can come first?

40. An economics club has 30 members. If a committee of 4 is to be selected, in how many ways can it be done?

41. A city council is composed of 5 liberals and 4 conservatives. A delegation of 3 is to be selected to attend a convention.
    (a) How many delegations are possible?
    (b) How many delegations could have all liberals?
    (c) How many delegations could have 2 liberals and 1 conservative?
    (d) If one member of the council serves as mayor, how many delegations which include the mayor are possible?

42. The coach of the Morton Valley Softball Team has 6 good hitters and 8 poor hitters. He chooses three hitters at random.
    (a) In how many ways can he choose 2 good hitters and 1 poor hitter?
    (b) In how many ways can he choose all good hitters?

5.5 The Binomial Theorem (Optional)

If we evaluate the expression $(x + y)^n$ for positive integer values of n, we get a family of expressions, called **expansions**, which are important in the study of mathematics generally, and in particular in the study of probability. For example,

$$(x + y)^1 = x + y$$
$$(x + y)^2 = x^2 + 2xy + y^2$$
$$(x + y)^3 = x^3 + 3x^2y + 3xy^2 + y^3$$
$$(x + y)^4 = x^4 + 4x^3y + 6x^2y^2 + 4xy^3 + y^4$$
$$(x + y)^5 = x^5 + 5x^4y + 10x^3y^2 + 10x^2y^3 + 5xy^4 + y^5.$$

From inspection, we see that these expansions follow a pattern. Let us try to identify the pattern so that we can write a general expression for $(x + y)^n$.

First, notice that each expansion begins with x raised to the same power as the binomial itself. That is, the expansion of $(x + y)^1$ has the first term x^1, $(x + y)^2$ has the first term x^2, $(x + y)^3$ has the first term x^3, and so on. Also, the last term in each expansion is y raised to the same power as the binomial. We can see that the expansion of $(x + y)^n$ should begin with the term x^n and end with the term y^n.

Further, note that the exponents on x decrease by 1 in each term after the first, while the exponents on y, beginning with y in the second term, increase by 1 in each succeeding term. Thus, the *variables,* in the expansion of $(x + y)^n$ should have the following pattern:

$$x^n, \; x^{n-1}y, \; x^{n-2}y^2, \; x^{n-3}y^3, \; . \; . \; . , \; x^2y^{n-2}, \; xy^{n-1}, \; y^n.$$

From this pattern, it can be seen that the sum of the exponents on x and y in each term is n. For example, in the third term above, the variable is $x^{n-2}y^2$, and the sum of the exponents, $n - 2 + 2$, is n.

Now let us try to find a pattern for the *coefficients* in the terms of the expansions shown above. In the product

$$(x + y)^5 = (x + y)(x + y)(x + y)(x + y)(x + y), \qquad (*)$$

the variable x occurs 5 times, once in each factor. To get the first term of the expansion, we form the product of these 5 x's to get x^5. The product x^5 can occur in just one way, by taking an x from each factor in equation (*), so that the coefficient of x^5 is 1. We can get the term with x^4y in more than one way. For example, we can take the x's from the first four factors and the y from the last factor, or we may take the x's from the last four factors and the y from the first, and so on. Since there are five factors in equation (*), from which we wish to select four x's for the term x^4y, there are $\binom{5}{4}$ ways in which this can be done. Thus the term x^4y occurs in $\binom{5}{4}$ or 5 ways and hence has coefficient 5. Continuing in this manner, we can use combinations to find the coefficients for each term of the expansion:

$$(x + y)^5 = x^5 + \binom{5}{4} x^4y + \binom{5}{3} x^3y^2 + \binom{5}{2} x^2y^3 + \binom{5}{1} xy^4 + y^5.$$

The coefficient 1 of the first and last terms could be written $\binom{5}{5}$ or $\binom{5}{0}$ to complete the pattern.

Generalizing from this special case, the coefficient for any term of $(x + y)^n$ in which the variable is $x^{n-r}y^r$ is $\binom{n}{n-r}$. The **binomial theorem** gives the general binomial expansion.

Theorem 5.3

(**Binomial Theorem**) For any positive integer n

$$(x + y)^n = \binom{n}{n}x^n + \binom{n}{n-1}x^{n-1}y + \binom{n}{n-2}x^{n-2}y^2$$
$$+ \binom{n}{n-3}x^{n-3}y^3 + \cdots + \binom{n}{1}xy^{n-1} + \binom{n}{0}y^n.$$

Example 1   Write out the binomial expansion of $(a + b)^7$.
Use the binomial theorem.

$$(a + b)^7 = a^7 + \binom{7}{6} a^6b + \binom{7}{5} a^5b^2 + \binom{7}{4} a^4b^3 + \binom{7}{3} a^3b^4$$
$$+ \binom{7}{2} a^2b^5 + \binom{7}{1} ab^6 + b^7$$
$$= a^7 + 7a^6b + 21a^5b^2 + 35a^4b^3 + 35a^3b^4$$
$$+ 21a^2b^5 + 7ab^6 + b^7 \quad \blacksquare$$

Example 2   Expand $\left(a - \dfrac{b}{2}\right)^4$

We use the binomial theorem to write

$$\left(a - \frac{b}{2}\right)^4 = \left(a + \left(-\frac{b}{2}\right)\right)^4$$

$$= a^4 + \binom{4}{3}a^3\left(-\frac{b}{2}\right) + \binom{4}{2}a^2\left(-\frac{b}{2}\right)^2 + \binom{4}{1}a\left(-\frac{b}{2}\right)^3 + \left(-\frac{b}{2}\right)^4$$

$$= a^4 + 4a^3\left(-\frac{b}{2}\right) + 6a^2\left(\frac{b^2}{4}\right) + 4a\left(-\frac{b^3}{8}\right) + \frac{b^4}{16}$$

$$= a^4 - 2a^3b + \frac{3}{2}a^2b^2 - \frac{1}{2}ab^3 + \frac{1}{16}b^4. \quad \blacksquare$$

Pascal's Triangle   Another method which can be used to find the coefficients of the terms in a binomial expansion is to use **Pascal's Triangle**, shown below. The nth row in the triangle gives the coefficients for the expansion of $(x + y)^n$. To see this, compare the numbers in the rows shown below with the coefficients of the expansions given at the beginning of this section. Each number in the triangle is found by adding the two numbers directly above it. Two illustrations of this are shown in color on the triangle below. A disadvantage of this method of finding coefficients is that the entire triangle must be produced down to the row which gives the desired coefficients.

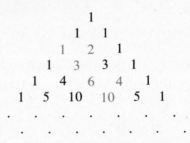

The binomial theorem can be used to prove the theorem which we used in Section 5.1.

Theorem 5.4

A set of n distinct elements has 2^n subsets.

We will illustrate the proof for $n = 6$. Subsets of a set of 6 elements can be chosen as follows: there are $\binom{6}{6}$ subsets with 6 elements, $\binom{6}{5}$ subsets with 5 elements, $\binom{6}{4}$ subsets with 4 elements and so on. Altogether there are

$$\binom{6}{6} + \binom{6}{5} + \binom{6}{4} + \binom{6}{3} + \binom{6}{2} + \binom{6}{1} + \binom{6}{0}$$

subsets. By the binomial theorem,

$$(x + y)^6 = \binom{6}{6}x^6 + \binom{6}{5}x^5y + \binom{6}{4}x^4y^2 + \binom{6}{3}x^3y^3 + \binom{6}{2}x^2y^4 + \binom{6}{1}xy^5 + \binom{6}{0}y^6.$$

If $x = 1$ and $y = 1$,

$$(1 + 1)^6 = \binom{6}{6} \cdot 1^6 + \binom{6}{5} \cdot 1^5 \cdot 1 + \binom{6}{4}1^4 \cdot 1^2 + \binom{6}{3}1^3 \cdot 1^3$$
$$+ \binom{6}{2} \cdot 1^2 \cdot 1^4 + \binom{6}{1} \cdot 1 \cdot 1^5 + \binom{6}{0} \cdot 1^6$$

or $\qquad\qquad 2^6 = \binom{6}{6} + \binom{6}{5} + \binom{6}{4} + \binom{6}{3} + \binom{6}{2} + \binom{6}{1} + \binom{6}{0}.$

Thus the total number of subsets of a set of 6 elements is $2^6 = 64$. In the general case, using the binomial theorem in the same way,

$$2^n = \binom{n}{n} + \binom{n}{n-1} + \binom{n}{n-2} + \cdots + \binom{n}{0},$$

so the total number of subsets is 2^n.

Example 3   The Yummy Yogurt Shoppe offers either chocolate or vanilla yogurt with a choice of 3 fruit toppings, chocolate topping, and chopped nuts. How many different servings are possible with one flavor of yogurt and any combination of toppings?

Use the multiplication axiom first. There are really two basic choices — which flavor of yogurt and then which combination of toppings to choose. There are two ways to select a flavor. The four toppings plus nuts form a set of five elements. The number of different subsets which can be selected from a set of five elements is 2^5. Thus, the number of different servings is

$$2 \cdot 2^5 = 2^6 = 64. \quad \blacksquare$$

5.5 Exercises   *Write out the binomial expansion and simplify the terms.*

1. $(m + n)^4$ 　　　 2. $(p - q)^5$ 　　　 3. $(3x - 2y)^6$

4. $(2x + t^3)^4$ 　　 5. $\left(\dfrac{m}{2} - 3n\right)^5$ 　　 6. $\left(2p + \dfrac{q}{3}\right)^3$

Write out the first four terms of the binomial expansion and simplify.

7. $(p + q)^{10}$ 　　 8. $(r + 5)^9$ 　　 9. $(a + 2b)^{15}$ 　　 10. $(3c + d)^{12}$

11. How many different subsets can be chosen from a set of ten elements?

12. How many different subsets can be chosen from a set of eight elements?

13. How many different pizzas can be chosen if you can have cheese, beef, sausage, pepper, tomatoes, and onion?

14. How many different vanilla sundaes can be chosen if the available trimmings are chocolate, strawberry, apricot, pineapple, peanuts, walnuts, and pecan crunch?

15. How many different school programs can be selected from 20 course offerings if at least two courses and no more than six courses can be selected?

16. How many different committees can be selected from a group of 16 people if the committee must have between two and five people (inclusive)?

17. A buffet offers four kinds of salad to any of which can be added sliced beets, bean sprouts, chopped egg, and sliced mushrooms. How many different salads are possible?

18. The buffet in Exercise 17 offers three meat and two fish entrees. How many different entree combinations are possible?

Key Words

set	intersection
element	disjoint sets
member	union
set-builder notation	set operations
universal set	multiplication axiom
subset	factorial
Venn diagrams	permutations
empty set	combinations
tree diagram	expansion
complement	binomial theorem

Chapter 5   Review Exercises

Write true or false for each of the following.

1. $9 \in \{8, 4, -3, -9, 6\}$

2. $4 \notin \{3, 9, 7\}$

3. $2 \notin \{0, 1, 2, 3, 4\}$

4. $0 \in \{0, 1, 2, 3, 4\}$

5. $\{3, 4, 5\} \subset \{2, 3, 4, 5, 6\}$

6. $\{1, 2, 5, 8\} \subset \{1, 2, 5, 10, 11\}$

7. $\{3, 6, 9, 10\} \subset \{3, 9, 11, 13\}$

8. $\varnothing \subset \{1\}$

9. $\{2, 8\} \not\subset \{2, 4, 6, 8\}$

10. $0 \subset \varnothing$

List the elements in the following sets. Give the cardinal number of each set.

11. $\{x \mid x$ is a counting number more than 5 and less than 8$\}$

12. $\{x \mid x$ is an integer, $-3 \leq x < 1\}$

13. {all counting numbers less than five}

14. {all whole numbers not greater than 2}

Let $U = \{a, b, c, d, e, f, g\}$, $K = \{c, d, f, g\}$, and $R = \{a, c, d, e, g\}$.
Find the following.

15. the number of subsets of K

16. the number of subsets of R

17. K'

18. R'

19. $K \cap R$

20. $K \cup R$

21. $(K \cap R)'$

22. $(K \cup R)'$

23. $\varnothing'$

24. U'

Let $U = $ {all employees of the K.O. Brown Company}
  $A = $ {employees in the accounting department}
  $B = $ {employees in the sales department}
  $C = $ {employees with at least 10 years in the company}
  $D = $ {employees with an MBA degree}

Describe the following sets in words.

25. $A \cap C$

26. $B \cap D$

27. $A \cup D$

28. $A' \cap D$

29. $B' \cap C'$

30. $(B \cup C)'$

Draw Venn diagrams for Exercises 31–34.

31. $A \cup B'$ **32.** $A' \cap B$ **33.** $(A \cap B) \cup C$ **34.** $(A \cup B)' \cap C$

A telephone survey of television viewers revealed the following information.

> 20 *watch situation comedies*
> 19 *watch game shows*
> 27 *watch movies*
> 5 *watch both situation comedies and game shows*
> 8 *watch both game shows and movies*
> 10 *watch both situation comedies and movies*
> 3 *watch all three*
> 6 *watch none of these*

35. How many viewers were interviewed?

36. How many viewers watch comedies and movies but not game shows?

37. How many viewers watch only movies?

38. How many viewers watch comedies and game shows but not movies?

39. In how many ways can 6 business tycoons line up their golf carts at the country club?

40. In how many ways can a sample of three oranges be taken from a bag of a dozen oranges?

41. In how many ways can a selection of two pictures from a group of 5 pictures be arranged in a row on a wall?

42. In how many ways can the pictures of Exercise 41 be arranged if one must be first?

43. In a Chinese restaurant the menu lists eight items in column A and six items in column B. To order a dinner, the diner is told to select three from column A and two from column B. How many dinners are possible?

44. A spokesperson is to be selected from each of three departments in a small college. If there are seven people in the first department, five in the second department, and four in the third department, how many different groups of three representatives are possible?

45. Write out the binomial expansion of $(2m + n)^5$.

46. Write out the first three terms of the binomial expansion of $(a + b)^{16}$.

47. Write out the first 4 terms of the binomial expansion of $(x - y/2)^{20}$.

48. How many different sums can be formed from the numbers 2, 5, 10, and 13?

49. How many different collections can be formed from a set of 8 old coins, if each collection must contain at least 2 coins?

50. How many sets of three or more books can be formed from a collection of six books?

6

Probability

If you go out to a nearby supermarket and buy five pounds of peaches at 54¢ per pound, you can easily find the *exact* price of your purchase: $2.70. Such a purchase is an example of a **deterministic phenomenon**: a phenomenon where the result can be found *exactly*.

On the other hand, the produce manager of the market is faced with the problem of ordering peaches. The manager may have a good estimate of the number of pounds of peaches that will be sold during the day, but there is no way to know exactly. The number of items that customers will purchase during a day is an example of a **random phenomenon**, a phenomenon which cannot be predicted exactly.

A great many problems that come up in applications of mathematics are random phenomena—those for which exact prediction is impossible. The best we can do is construct a mathematical model which gives the *probability* of certain events. The basics of probability are discussed in this chapter, with applications of probability discussed in succeeding chapters.

6.1 Sample Spaces

In probability, each repetition of an experiment is called a **trial**. The possible results of each trial are called **outcomes**. An example of a probability experiment is the tossing of a coin. On each trial of the experiment (each toss), there are two possible outcomes, heads and tails, abbreviated *h* and *t,* respectively. If the two outcomes, *h* and *t,* are equally likely to occur, then the coin is not "loaded" to favor one side over the other. Such a coin is called **fair**. For a coin that is not loaded, this "equally likely" assumption is made for each trial.

Since there are *two* equally likely outcomes possible, *h* and *t,* and just *one* of them is heads, we would expect that a coin tossed many, many times would come up heads approximately 1/2 of the time. We would also expect that the more times the coin was tossed, the closer the occurrence of heads should be to 1/2.

Suppose an experiment could be repeated again and again under unchanging conditions. Suppose the experiment is repeated n times, and that a certain outcome happens m times. The ratio m/n is called the **relative frequency** of the outcome after n trials.

If this ratio m/n approaches closer and closer to some fixed number p as n gets larger and larger, then p is called the **probability** of the outcome. If p exists, then

$$p \approx \frac{m}{n}$$

as n gets larger and larger.

This approach to probability is consistent with most people's intuitive feeling of what probability is—a way of measuring how likely a certain outcome is to occur. For example, since 1/4 of the cards in an ordinary deck are diamonds, we would assume that if we drew a card from a well-shuffled deck, kept track of whether it was a diamond or not, and then replaced the card in the deck, after a large number of repetitions of the experiment we would find about 1/4 of the cards drawn had been diamonds.

The definition of probability, on the other hand, has the disadvantage of not being precise, since it uses phrases such as "approaches closer and closer" and "gets larger and larger." One of the things done in the first few sections of this chapter is to give a more precise meaning to the terms associated with probability.

The probability of heads on a single toss of a fair coin is 1/2. This is written

$$P(h) = \frac{1}{2}.$$

Also, $P(t) = 1/2$.

Example 1 Suppose we spin the spinner of Figure 1.

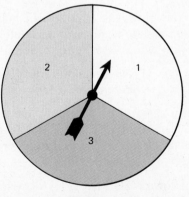

Figure 1

(a) Find the probability that it will point to 1.

A natural assumption is that if this spinner were spun many, many times, it would point to 1 about 1/3 of the time, so that

$$P(1) = \frac{1}{3}.$$

(b) Find the probability that it will point to 2.

Since getting the number 2 is one of three possible outcomes, then $P(2) = 1/3$. ▌

An ordinary die is a cube whose six faces show the numbers 1, 2, 3, 4, 5, and 6. If the die is not "loaded" to favor certain faces over others, then any one of the six faces is equally likely to come up when the die is rolled.

Example 2 (a) If a single fair die is rolled, find the probability of rolling the number 4.

Since one out of six faces is a 4, $P(4) = 1/6$.

(b) Using the same die, find the probability of rolling the number 6.

Since one out of six faces is a 6, $P(6) = 1/6$. ▌

Sample Space   Sometimes we are interested in a result that is satisfied by more than one of the possible outcomes. To find the probability that the spinner in Example 1 will point to an odd number, we notice that two of the three possible outcomes are odd numbers, 1 and 3. Therefore,

$$P(\text{odd}) = \frac{2}{3}.$$

The set of all possible outcomes for an experiment is the **sample space** for the experiment. A sample space for the experiment of tossing a coin is made up of the two outcomes, heads and tails. If S represents this sample space, then

$$S = \{h, t\}.$$

In the same way, the sample space for tossing a single fair die is

$$\{1, 2, 3, 4, 5, 6\}.$$

Example 3 (a) For the purposes of a certain public opinion poll, people are classified as young, middle-aged, or older, and as male or female. The sample space for this poll would be a set of ordered pairs:

{(young, male), (young, female), (middle-aged, male), (middle-aged, female), (older, male), (older, female)}.

(b) A firm can run its assembly line at a low, medium, or high rate. With each speed, the firm may find 1%, 2%, or 3% of the items from the line defective. Placing the line speed first in an ordered pair, and the rate of defectives second, gives the sample space

{(low, 1%), (low, 2%), (low, 3%), (medium, 1%), (medium, 2%), (medium, 3%), (high, 1%), (high, 2%), (high, 3%)}.

(c) A manufacturer tests automobile tires by running a tire until it fails or until tread depth reaches a certain unsafe level. The number of miles that the tire lasts, m, is recorded. At least in theory, m can be any nonnegative real number, so that the sample space is

$$\{m|m \geq 0\}.$$

However, for a particular type of tire, there would be practical limits on m, so that the sample space might then be

$$\{m|15{,}000 \leq m \leq 50{,}000\}. \quad \blacksquare$$

An **event** is a subset of a sample space. If the sample space for tossing a coin is $S = \{h, t\}$, then one event is $E = \{h\}$, which represents the outcome "heads." For the sample space of tossing a single fair die, $\{1, 2, 3, 4, 5, 6\}$, one event is $\{2, 4, 6\}$, or "the number showing on top is even."

Example 4   One experiment consists of studying all possible families having exactly three children. Let b represent "boy" and g represent "girl."
(a) Write a sample space for the experiment.
    A family can have three boys, written bbb, three girls, ggg, or various combinations, such as bgg. The sample space is made up of all such outcomes (there are eight).

$$S = \{\text{bbb, bbg, bgb, gbb, bgg, gbg, ggb, ggg}\}.$$

(b) Write event H, "the family has exactly two girls."
    Families can have exactly two girls with either bgg, gbg, or ggb, so that event H is

$$H = \{\text{bgg, gbg, ggb}\}.$$

(c) Write the event J, "the family has three girls."
    Only ggg satisfies this condition, so

$$J = \{\text{ggg}\}. \quad \blacksquare$$

In Example 4(c), event J had only one possible outcome, ggg. Such an event, with only one possible outcome, is a **simple event.** If event E equals the sample space S, then E is a **certain event.** If event $E = \varnothing$, then E is an **impossible event.**

Example 5   Suppose a die is rolled. As we have seen, the sample space is $\{1, 2, 3, 4, 5, 6\}$.

(a) The event "the die has a four on top" is a simple event, $\{4\}$. The event has only one possible outcome.

(b) The event "the number on top is less than ten" equals the sample space, $S = \{1, 2, 3, 4, 5, 6\}$. This event is a certain event; if a die is rolled the number on top (either 1, 2, 3, 4, 5, or 6), must be less than ten.

(c) The event "the die has 7 on top" is the empty set, $\varnothing$; this event is impossible. $\blacksquare$

Events are sets, so the union, intersection, and complement of events can be formed.

Example 6   A die is tossed; let E be the event "the number showing on top is more than 3", and let F be the event "the number showing on top is even." Then

$$E = \{4, 5, 6\} \quad \text{and} \quad F = \{2, 4, 6\}.$$

(a) $E \cap F$ is the event "the number showing on top is more than 3 *and* is even", the outcomes common to *both* E and F.

$$E \cap F = \{4, 6\}.$$

(b) $E \cup F$ is the event "the number showing on top is more than 3 *or* is even", the outcomes of E or F, or both.

$$E \cup F = \{2, 4, 5, 6\}.$$

(c) Event E' is the event "the number showing on top is *not* more than 3"; the elements of the sample space that are *not* in E. (The event E' is called the *complement* of event E.)

$$E' = \{1, 2, 3\}.$$

The sketches of Figure 2 show the events $E \cap F$, $E \cup F$, and E'.   ▌

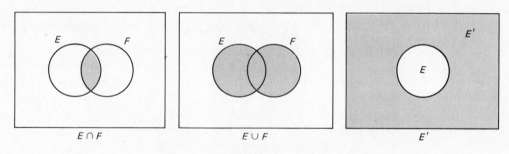

Figure 2

Two events that cannot both occur at the same time, such as getting both a head and a tail on the same toss of a coin, are **mutually exclusive events.** Thus, events A and B are mutually exclusive events if $A \cap B = \varnothing$.

Example 7   Let $S = \{1, 2, 3, 4, 5, 6\}$, the sample space for tossing a die. Let $E = \{4, 5, 6\}$, and let $G = \{1, 2\}$. Then E and G are mutually exclusive events since they have no outcomes in common; $E \cap G = \varnothing$. See Figure 3.   ▌

In summary, let E and F be events for a sample space S. Then

$E \cap F$ occurs when both E and F occur
$E \cup F$ occurs when either E or F or both occur
E' occurs when E does not
E and F are mutually exclusive if $E \cap F = \varnothing$

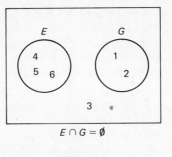

$$E \cap G = \emptyset$$

Figure 3

6.1 Exercises   *Write sample spaces for the following experiments. (The sample spaces in Exercises 11–12 are infinite.)*

1. Choose a month of the year.

2. Pick a day in April.

3. Ask a student how many points they earned on a recent 80 point test.

4. Ask a person how many hours (to the nearest hour) they watched television yesterday.

5. The management of an oil company must decide whether to go ahead with a new oil shale plant, or cancel it.

6. A record is kept for three days about whether a particular stock goes up or down.

7. The length of life of a light bulb is measured to the nearest hour; the bulbs never last more than 5000 hours.

8. Toss a coin and roll a die.

9. Toss a coin four times.

10. A box contains five balls, numbered 1, 2, 3, 4, and 5. A ball is drawn at random, the number on it recorded, and the ball replaced. After shaking the box, a second ball is drawn, and its number recorded.

11. A coin is tossed until a head appears.

12. A die is rolled, but only "odd" (1, 3, 5) or "even" (2, 4, 6) is recorded. Roll the die until odd appears.

13. A die is tossed twice, with the tosses recorded as ordered pairs. Write the
    (a) sample space;
    (b) event F, the first die is 3;
    (c) event G, the sum of the dice is 8;
    (d) event H, the sum of the dice is 13.

14. One urn contains four balls, labeled 1, 2, 3, and 4. A second urn contains five balls, labeled 1, 2, 3, 4, and 5. An experiment consists of taking one ball from the first urn, and then taking a ball from the second urn. Find the
    (a) sample space;
    (b) event E, the first ball is even;
    (c) event F, the second ball is even;
    (d) event G, the sum of the numbers on the two balls is 5;
    (e) event H, the sum of the numbers on the two balls is 1.

15. A coin is tossed until two heads appear, or until the coin is tossed five times, whichever comes first. Write the
   (a) sample space;
   (b) event E, the coin is tossed exactly two times;
   (c) event F, the coin is tossed exactly three times;
   (d) event G, the coin is tossed exactly five times, without getting two heads.

16. A committee of two people is selected from five executives, Abbot, Babbit, Coats, Dickson, and Ellsberg. Write the
   (a) sample space;
   (b) event E, Coats is on the committee;
   (c) event F, Dickson and Ellsberg are not both on the committee;
   (d) event G, both Abbot and Coats are on the committee.

17. The management of a firm wishes to check on the opinions of its assembly line workers. To do this, the workers are first divided into various categories. Define events E, F, and G as follows.

   $E:$ worker is female
   $F:$ worker has worked less than five years
   $G:$ worker contributes to a voluntary retirement plan

   Describe each of the following events in words.
   (a) E' **(b)** F' **(c)** $E \cap F$
   (d) $F \cup G$ **(e)** $E \cup G'$ **(f)** $F' \cap G'$

18. For a medical experiment, people are classified as to whether they smoke, have a family history of heart disease, and are overweight. Define events E, F, and G as follows.

   $E:$ person smokes
   $F:$ person has a family history of heart disease
   $G:$ person is overweight

   Describe each of the following events in words.
   (a) G' **(b)** $E \cup F$ **(c)** $F \cap G$
   (d) $E' \cap F$ **(e)** $E \cup G'$ **(f)** $F' \cup G'$

Decide if the following events are mutually exclusive.

19. Tossing heads and tails on a coin.

20. Wearing glasses and wearing sandals.

21. Being married and being over 30 years old.

22. Being a teen-ager and being over 30 years old.

23. Rolling a die and getting a 4 and an odd number.

24. A person being both male and a postal worker.

25. Let E be an event. Must E and E' be mutually exclusive?

26. Let E and F be mutually exclusive events. Let F and G be mutually exclusive events. Must E and G be mutually exclusive events?

27. Suppose E and F are mutually exclusive events. Must E' and F' be mutually exclusive events?

28. Suppose E and F are mutually exclusive events. Must E and $(E \cup F)'$ be mutually exclusive events?

29. A person is taking accounting and business law. Let E be the event that the student passes accounting, and let F be the event that the student passes business law. Write each of the following using $E, F, \cup, \cap$, and $'$, as needed. The student

 (a) fails accounting;
 (b) passes both;
 (c) fails both;
 (d) passes acounting but not business law;
 (e) passes exactly one course;
 (f) passes at least one course.

30. Let E and F be events. Write the following, using $\cap, \cup$, or $'$, as needed.

 (a) E does not happen
 (b) neither happens
 (c) both happen
 (d) E happens, but not F
 (e) one or the other happens, but not both

31. Let $S = \{r, s, t\}$. Write all events associated with S.

32. Let $S = \{1, 2, 3, 4, \cdots, n\}$, where n is a positive integer. How many different events could be obtained from S?

6.2 Basics of Probability

Given a sample space S, we now need to assign to each event that can be obtained from S a number, called the **probability of the event.** This number will indicate the relative likelihood of the various events. In the remainder of this chapter, attention is restricted to sample spaces with a *finite* number of elements.

For events that are *equally likely,* the probability of the event can be found from the following **basic probability principle:**

> Let a sample space S have n possible equally likely outcomes.
> Let event E contain m of these outcomes, all distinct. Then the **probability** that event E occurs, written $P(E)$, is
>
> $$P(E) = \frac{m}{n}.$$

This same result can also be given in terms of the cardinal number of a set. (Recall from Chapter 5 that $n(E)$ represents the number of elements in a finite set E.) With the same assumptions given above,

$$P(E) = \frac{n(E)}{n(S)}.$$

Example 1 Suppose a single fair die is rolled. The sample space is $S = \{1, 2, 3, 4, 5, 6\}$. Set S contains 6 outcomes, all of which are equally likely. Thus, in the basic probability principle, $n = 6$. Find the probability of the following outcomes.

(a) $E = \{1, 2\}$.

Event E contains two elements, so

$$P(E) = \frac{2}{6} = \frac{1}{3}.$$

Thus, if a single die is rolled, we could expect the numbers 1 or 2 to show up about 1/3 of the time.

(b) An even number is rolled.

Let event $F = \{2, 4, 6\}$ be the event "an even number is rolled." Event F contains three elements, so

$$P(F) = \frac{3}{6} = \frac{1}{2}.$$

(c) The die shows a 7.

A die can never show 7. If G is this event, then $G = \varnothing$, and

$$P(G) = \frac{0}{6} = 0.$$

This event is impossible. ∎

Example 2   If a single playing card is drawn at random from an ordinary 52-card bridge deck, find the probability of each of the following events.

(a) An ace is drawn.

There are four aces in the deck, out of 52 cards, so

$$P(\text{ace}) = \frac{4}{52} = \frac{1}{13}.$$

(b) A face card is drawn.

Since there are 12 face cards,

$$P(\text{face card}) = \frac{12}{52} = \frac{3}{13}.$$

(c) A spade is drawn.

The deck contains 13 spades, so

$$P(\text{spade}) = \frac{13}{52} = \frac{1}{4}.$$

(d) A spade or a heart is drawn.

Besides the 13 spades, the deck contains 13 hearts, so

$$P(\text{spade or heart}) = \frac{26}{52} = \frac{1}{2}.$$ ∎

Example 3   The manager of a department store has decided to make a study on the size of purchases made by people coming into the store. To begin, he chooses a day that seems fairly typical and gathers the following data. (Purchases have been rounded to the nearest dollar, with sales tax ignored.)

Amount of purchase	Number of customers
$0	158
$1–$5	94
$6–$9	203
$10–$19	126
$20–$49	47
$50–$99	38
$100 and over	53

First, the manager might add the numbers of customers to find that 719 people came into the store that day. Of these 719 people, $126/719 \approx .175$ made a purchase of at least $10 but no more than $19. Also, $53/719 \approx .074$ of the customers spent $100 or more. Thus, the probability (on this given day) that a customer entering the store will spend from $10 to $19 is .175, and the probability that the customer will spend $100 or more is .074. Probabilities for the various purchase amounts can be assigned in the same way, giving the results of the following table.

Size of purchase	Probability
$0	.220
$1–$5	.131
$6–$9	.282
$10–$19	.175
$20–$49	.065
$50–$99	.053
$100 and over	.074
Total	1.000

From the table, .282 of the customers spend from $6 to $9, inclusive—over a quarter of the customers. Since this price range attracts so many customers, perhaps the store's advertising should emphasize items in, or near, this price range.

The manager should use this table of probabilities to help in predicting the results on other days only if the manager is reasonably sure that the day when the measurements were made is fairly typical of the other days the store is open— for example, on the last few days before Christmas the probabilities might be quite different. ▮

Probability Distributions In Example 3 the outcomes were various purchase amounts, and a probability was assigned to each outcome. By this process, we set up a **probability distribution;** that is, for each possible outcome of an experiment, a number, called the probability of that outcome, is assigned.

As we have seen, one way to think of these probabilities is as relative frequencies; that is, for a large number of sales days of the type measured in the table, approximately 13% of all purchases would be in the $1–$5 range, approximately 28% in the $6–$9 range, and so on.

Example 4   Set up a probability distribution for the number of girls in a family with three children.

Start by writing the sample space, which shows the possible number of boys and girls in a family of three children, $S = \{bbb, bbg, bgb, gbb, bgg, gbg, ggb, ggg\}$. Let event E_0 be "the family has no girls;" from the sample space, $E_0 = \{bbb\}$.

In a similar way, let $E_1 = \{bbg, bgb, gbb\}$, $E_2 = \{bgg, gbg, ggb\}$, and $E_3 = \{ggg\}$. Sample space S has eight possible outcomes, and event E_1, for example, has three elements, giving $P(E_1) = 3/8$. Doing the same thing for the other events gives the following probability distribution.

Number of girls	Probability
0	1/8
1	3/8
2	3/8
3	1/8
Total	1

(Here we assume that the probability of having a girl baby and a boy baby is equally likely. This assumption is not quite exact in actual fact, but the correct fraction is not far from 1/2.) ∎

The probability distributions that were set up above suggest the following **properties of probability.**

Let $S = \{s_1, s_2, s_3, \cdots, s_n\}$ be a sample space containing the n distinct simple events $s_1, s_2, s_3, \cdots, s_n$. Let the simple events in S have associated probabilities $p_1, p_2, p_3, \cdots, p_n$. Then
(a) $0 \le p_1 \le 1, 0 \le p_2 \le 1, \cdots, 0 \le p_n \le 1$
   (all probabilities are between 0 and 1, inclusive);
(b) $p_1 + p_2 + p_3 + \cdots + p_n = 1$;
   (the sum of all probabilities for a sample space is 1);
(c) $P(S) = 1$;
(d) $P(\varnothing) = 0$.

The Addition Principle   Suppose event E is the union of several *simple* events, say

$$E = \{s_1, s_2, s_3\} = \{s_1\} \cup \{s_2\} \cup \{s_3\}.$$

To find $P(E)$, the probability of event E, add the probabilities for each of the simple events making up E. For $E = \{s_1, s_2, s_3\}$,

$$P(E) = P(\{s_1\}) + P(\{s_2\}) + P(\{s_3\}).$$

The generalization of this result is called the *addition principle:*

Addition Principle   Suppose $E = \{s_1, s_2, s_3, \cdots, s_m\}$, where $\{s_1\}, \{s_2\}, \{s_3\}, \cdots, \{s_m\}$ are distinct simple events. Then

$$P(E) = P(\{s_1\}) + P(\{s_2\}) + P(\{s_3\}) + \cdots + P(\{s_m\}).$$

It is important to note that the addition rule *does not necessarily apply* to the addition of probabilities of events that are not simple. For example, the sum of the probability of getting at least 4 on a single roll of a die, and the probability of getting an even number on a single roll is *not* equal to the probability of getting at least 4 or an even number on a single roll. That is, P(at least 4) $= P(4, 5,$ or $6) =$ $1/2$, P(even number) $= P(2, 4,$ or $6) = 1/2$, and P(at least 4 or even) $= P(2, 4, 5,$ or $6) = 2/3$, with P(at least 4) $+ P$(even number) $\neq P$(at least 4 or even).

Example 5   Refer back to Example 3 and find the probability that a customer spends at least $6 but less than $50.

This event is the union of three simple events, spending from $6 to $9, spending from $10 to $19, or spending from $20-$49. The probability of spending at least $6 but less than $50 can thus be found by the addition principle.

P(spending at least $6 but less than $50)

$$= P(\text{spending } \$6\text{-}\$9) + P(\text{spending } \$10\text{-}\$19) + P(\text{spending } \$20\text{-}\$49)$$

$$= .282 + .175 + .065 = .522. \quad \blacksquare$$

Let us now extend the addition rule to events that are not necessarily simple events, but that are mutually exclusive events. Suppose that E and F are mutually exclusive events, with $E = \{s_1, s_2, \cdots, s_n\}$, and $F = \{t_1, t_2, \cdots, t_m\}$, where $\{s_1\}$, $\{s_2\}, \cdots, \{s_n\}$ and $\{t_1\}, \{t_2\}, \cdots, \{t_m\}$ are simple events. Then

$$P(E) + P(F) = P(\{s_1, s_2, \cdots, s_n\}) + P(\{t_1, t_2, \cdots, t_m\})$$

$$= P(\{s_1\}) + P(\{s_2\}) + \cdots + P(\{s_n\})$$

$$+ P(\{t_1\}) + P(\{t_2\}) + \cdots + P(\{t_m\})$$

$$= P(\{s_1, s_2, \cdots, s_n, t_1, t_2, \cdots, t_m\})$$

$$= P(E \cup F)$$

For *mutually exclusive* events E and F, $P(E \cup F) = P(E) + P(F)$.

Example 6   Use the probability distribution of Example 4 to find the probability that a family with three children has at least two girls.

Event E, "the family has at least two girls," is the union of two mutually exclusive events, "the family has two girls," and "the family has three girls." Thus,

$$P(E) = P(\text{at least two girls}) = P(2 \text{ girls}) + P(3 \text{ girls})$$

$$= \frac{3}{8} + \frac{1}{8} = \frac{1}{2}. \quad \blacksquare$$

Recall earlier discussions of set theory. The set of all outcomes in a sample space that do not belong to an event E is called the *complement* of E, written E'. For example, in the experiment of drawing a single card from a well-shuffled deck of 52 cards, let E be the event "the card is an ace." Then E' is the event "the card is not an ace." Using this definition of E', for any event E from a sample space S,

$$E \cup E' = S \quad \text{and} \quad E \cap E' = \emptyset.$$

Since $E \cap E' = \emptyset$, events E and E' are mutually exclusive, so that

$$P(E \cup E') = P(E) + P(E').$$

However, $E \cup E' = S$, the sample space, and $P(S) = 1$. Thus

$$P(E \cup E') = P(E) + P(E') = 1,$$

giving two alternate and useful results:

$$P(E) = 1 - P(E') \quad \text{and} \quad P(E') = 1 - P(E).$$

Example 7   In a particular experiment, $P(E) = 2/7$. Find $P(E')$.

$$P(E') = 1 - P(E) = 1 - \frac{2}{7} = \frac{5}{7} \quad \blacksquare$$

The next example shows that it is sometimes easier to find $P(E)$ by first finding $P(E')$, and then $1 - P(E')$.

Example 8   Refer to Example 3 above, and find the probability that a customer spends less than \$100.

Let E be the event "a customer spends less than \$100". Then from the table of Example 3,

$$P(E) = .220 + .131 + .282 + .175 + .065 + .053 = .926.$$

We can also find $P(E)$ by noting that if E is the event "a customer spends less than \$100," then E' is the event "a customer spends \$100 and over." From the table, $P(E') = .074$, and

$$P(E) = 1 - P(E') = 1 - .074 = .926$$

Here $P(E)$ is easier to calculate as $1 - P(E')$. $\blacksquare$

Odds   Sometimes probability statements are given in terms of *odds,* a comparison of $P(E)$ with $P(E')$:

The **odds in favor** of an event E is defined as the ratio of $P(E)$ to $P(E')$, or

$$\frac{P(E)}{P(E')}.$$

Example 9   Suppose the weather forecaster says that the probability of rain tomorrow is 1/3. Find the odds in favor of rain tomorrow.

Let E be the event "rain tomorrow." Then E' is the event "no rain tomorrow." Since $P(E) = 1/3$, we have $P(E') = 2/3$. By the definition of odds,

$$\text{odds in favor of rain} = \frac{\frac{1}{3}}{\frac{2}{3}} = \frac{1}{2}, \quad \text{written} \quad 1 \text{ to } 2, \quad \text{or} \quad 1:2.$$

On the other hand, the odds that it will *not* rain are

$$\frac{\frac{2}{3}}{\frac{1}{3}} = \frac{2}{1}, \quad \text{written} \quad 2 \text{ to } 1, \quad \text{or} \quad 2:1. \quad \blacksquare$$

If we know that the odds in favor of an event are, say, 3 to 5, then the probability of the event is 3/8, while the probability of the complement of the event is 5/8. (Odds of 3 to 5 indicate 3 outcomes in favor of the event out of a total of 8 outcomes.) In general, if the odds favoring event E are m to n, then

$$P(E) = \frac{m}{m + n} \quad \text{and} \quad P(E') = \frac{n}{m + n}.$$

Example 10   The odds that a particular bid will be the low bid are 4 to 5. Find the probability that the bid will be the low bid.

Odds of 4 to 5 show 4 favorable chances out of $4 + 5 = 9$ chances altogether. Thus,

$$P(\text{bid will be low bid}) = \frac{4}{4 + 5} = \frac{4}{9}.$$

There is a 5/9 chance that the bid will *not* be the low bid. $\blacksquare$

Subjective Probabilities   The formulas above let us find the probability of an event that can be repeated exactly, again and again. However, we would also like to be able to assign probabilities to the many occurrences which cannot be exactly repeated. For example, we might want to state the probability of rain on the day planned for the company picnic, or the probability that the earnings of a firm will increase by 30%, or the probability that a given number of pounds of peaches will be sold on a given day (as mentioned in the introduction to this chapter). The probabilities for these events are known as **subjective probabilities**, and must be assigned, if at all, on the basis of personal judgment. A sales manager may use past experience to assign a probability to the success of the current July sale, but since this year's sale can never be exactly like that of past or future Julys, it is still a subjective assignment of probability.

In one method of weather forecasting, the forecaster compares current atmospheric conditions with similar conditions in the past, and predicts the weather *now* based on what happened *then* under such conditions. For example, if, in the past, under the kind of weather conditions we have today, rain fell 4 times out of 5, the forecaster will announce an 80% chance of rain.

In many cases, no information is available, and probability must be assigned on the basis of hunches or expectations. In fact, sometimes it is necessary to assign probabilities to events that will happen only once.

One difficulty with subjective probability is that different people may assign different probabilities to the same event. Nevertheless, subjective probabilities can be assigned to many occurrences where the objective approach to probability cannot be used. In Chapter 9 we illustrate the use of subjective probability in more detail.

6.2 Exercises   *A single fair die is rolled. Find the probability of the following events.*

1. a 2

2. an odd number

3. a number less than 5

4. a number greater than 2

A card is drawn from a well-shuffled deck of 52 cards. Find the probability of drawing

5. a 9;

6. a black card;

7. a black 9;

8. a heart;

9. the 9 of hearts;

10. a face card.

A single fair die is rolled. Find the odds in favor of rolling

11. the number 5;

12. 3, 4, or 5;

13. 1, 2, 3, or 4;

14. some number less than 2.

List the simple events whose union forms each of the following events.

15. getting an even number on a roll of a die

16. drawing a card that is both a heart and a face card from an ordinary deck

17. a record is made of whether a company's annual profit goes up, down, or stays the same

18. a batch of 7 items was checked, and the number of defectives recorded

An experiment is conducted for which the sample space is $S = \{s_1, s_2, s_3, s_4, s_5\}$. Which of the following probability distributions is feasible for this experiment? If a distribution is not feasible, tell why.

19.

Outcomes	s_1	s_2	s_3	s_4	s_5
Probability	.09	.32	.21	.25	.13

20.

Outcomes	s_1	s_2	s_3	s_4	s_5
Probability	.92	.03	0	.02	.03

21.

Outcomes	s_1	s_2	s_3	s_4	s_5
Probability	$\frac{1}{3}$	$\frac{1}{4}$	$\frac{1}{6}$	$\frac{1}{8}$	$\frac{1}{10}$

22.

Outcomes	s_1	s_2	s_3	s_4	s_5
Probability	$\frac{1}{5}$	$\frac{1}{3}$	$\frac{1}{4}$	$\frac{1}{5}$	$\frac{1}{10}$

23.

Outcomes	s_1	s_2	s_3	s_4	s_5
Probability	.64	−.08	.30	.12	.02

24.

Outcomes	s_1	s_2	s_3	s_4	s_5
Probability	.05	.35	.5	.2	−.3

The table below gives a certain golfer's probabilities of scoring in various ranges on a par-70 course.

Range	Probability
below 60	.01
60 – 64	.08
65 – 69	.15
70 – 74	.28
75 – 79	.22
80 – 84	.08
85 – 89	.06
90 – 94	.04
95 – 99	.02
100 or more	.06

In a given round, find the probability that the golfer's score will be

25. 90 or higher;   **26.** below par of 70;

27. in the 70's;   **28.** in the 90's;

29. not in the 60's;   **30.** not in the 60's or 70's.

31. Find the odds in favor of the golfer shooting below par.

32. Find the odds against the golfer shooting in the 70s.

Fransisco has set up the following probability distribution for the number of hours it will take him to finish his homework.

Hours	1	2	3	4	5	6
Probability	.05	.10	.20	.40	.10	.15

Find the probability that his homework will take

33. fewer than 3 hours;   **34.** 3 hours or less;

35. more than 2 hours;   **36.** at least 2 hours;

37. more than 1 hour and less than 5 hours;   **38.** 8 hours.

A marble is drawn from a box containing 3 yellow, 4 white, and 8 blue marbles. Find the probability of drawing

39. a yellow marble;   **40.** a blue marble;   **41.** a white marble.

For this same marble experiment, find the odds in favor of drawing a

42. yellow marble;   **43.** blue marble;   **44.** white marble.

45. Find the odds of not drawing a white marble.

46. The probability that a company will make a profit this year is .74. Find the odds against the company making money.

47. If the odds that it will rain are 4 to 7, what is the probability of rain?

48. If the odds that a given candidate will win an election are 3 to 2, what is the probability that the candidate will lose?

The probability distribution for a given experiment having sample space $S = \{s_1, s_2, s_3, s_4, s_5, s_6\}$ is shown here.

Outcomes	s_1	s_2	s_3	s_4	s_5	s_6
Probability	.17	.03	.09	.46	.21	.04

Let $E = \{s_1, s_2, s_5\}$, and let $F = \{s_4, s_5\}$. Find each of the following probabilities.

49. $P(E)$ **50.** $P(F)$ **51.** $P(E \cap F)$

52. $P(E \cup F)$ **53.** $P(E' \cup F')$ **54.** $P(E' \cap F)$

Which of the following are examples of subjective probability?

55. the probability of heads on five consecutive tosses of a coin

56. the probability that a freshman entering college will graduate with a degree

57. the probability that a person is allergic to penicillin

58. the probability of drawing an ace from a standard deck of 52 cards

59. the probability that a person will get lung cancer from smoking cigarettes

60. a weather forecaster predicts a 70% chance of rain tomorrow

61. a gambler claims that on a roll of a fair die, $P(\text{even}) = 1/2$

62. a surgeon gives a patient a 90% chance of a full recovery

63. a bridge player has a 1/4 chance of being dealt a diamond

64. a forest ranger states that the probability of a short fire season this year is only 3 in 10

65. On page 134 of Roger Staubach's autobiography, *First Down, Lifetime to Go,* Staubach makes the following statement regarding his experience in Vietnam: "Odds against a direct hit are very low but when your life is in danger, you don't worry too much about the odds." Is this wording consistent with our definition of odds, for and against? How could it have been said so as to be technically correct?

> **Application   *Making a First Down***

A first down is desirable in football — it guarantees four more plays by the team making it, assuming no score or turnover occurs in the plays. After getting a first down, a team can get another by advancing the ball at least ten yards. During the four plays given by a first down, a team's position will be indicated by a phrase such as "third and 4," which means that the team has already had two of its four plays, and that 4 more yards are needed to get the 10 yards necessary for another first down. An article in a management journal[1] offers the following results for 189 games of a recent National Football League season. "Trials"

---

[1]Reprinted by permission of Virgil Carter and Robert Machols, "Optimal Strategies on Fourth Down," *Management Science,* Vol. 24, No. 16, December 1978, copyright © 1978 The Institute of Management Sciences.

represents the number of times a team tried to make a first down, given that it was currently playing either a third or a fourth down. Here n represents the number of yards still needed for a first down.

n	Trials	Successes	Probability of making first down with n yards to go
1	543	388	
2	327	186	
3	356	146	
4	302	97	
5	336	91	

Exercises   **1.** Complete the table.

2. Why is the sum of the answers in Exercise 1 not equal to 1?

6.3 Extending the Addition Rule

We saw in the previous section that for mutually exclusive events E and F,

$$P(E \cup F) = P(E) + P(F). \tag{1}$$

We shall now extend this result to *any* two events E and F.

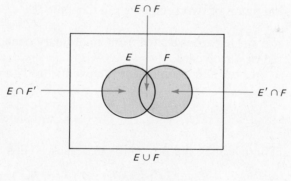

Figure 4

To obtain this more general result, take the region $E \cup F$ of Figure 4 and break it into three disjoint regions, that part of E not including F ($E \cap F'$), the intersection, $E \cap F$, and that part of F not including E ($E' \cap F$). Thus,

$$E \cup F = (E \cap F') \cup (E \cap F) \cup (E' \cap F).$$

Since all three of the sets on the right are disjoint, we can use a slight extension of property (1) above to get

$$P(E \cup F) = P(E \cap F') + P(E \cap F) + P(E' \cap F). \tag{2}$$

As illustrated by Figure 4,

$$E = (E \cap F') \cup (E \cap F) \quad \text{and} \quad F = (E' \cap F) \cup (E \cap F)$$

so that, by (1) above,

$$P(E) = P(E \cap F') + P(E \cap F) \tag{3}$$

and

$$P(F) = P(E' \cap F) + P(E \cap F). \tag{4}$$

From (3),

$$P(E \cap F') = P(E) - P(E \cap F) \tag{5}$$

and from (4),

$$P(E' \cap F) = P(F) - P(E \cap F). \tag{6}$$

Substituting from equations (5) and (6) into equation (2) gives

$$P(E \cup F) = P(E) + P(F) - P(E \cap F). \tag{7}$$

This result is called the *extended addition principle*.

Extended Addition Principle For any two events E and F from a sample space S,

$$P(E \cup F) = P(E) + P(F) - P(E \cap F).$$

Notice the similarity of this result to the formula for the number of elements in the union of two sets, given in Chapter 5.

Example 1 If a single card is drawn from an ordinary deck, find the probability that it will be red or a face card.

Let R and F represent the events "red" and "face card" respectively. Then

$$P(R) = \frac{26}{52}, \quad P(F) = \frac{12}{52}, \quad \text{and} \quad P(R \cap F) = \frac{6}{52}.$$

(There are six red face cards in a deck.) Thus, by equation (7) above,

$$P(R \cup F) = P(R) + P(F) - P(R \cap F)$$

$$= \frac{26}{52} + \frac{12}{52} - \frac{6}{52}$$

$$= \frac{32}{52} = \frac{8}{13}. \quad \blacksquare$$

Example 2 Suppose two fair dice are rolled. Find each of the following probabilities.
(a) The first die shows a 2 or the sum is 6 or 7

The sample space for the throw of two dice is shown in Figure 5. The two events are labeled A and B. From the diagram,

$$P(A) = \frac{6}{36}, \qquad P(B) = \frac{11}{36}, \qquad \text{and} \qquad P(A \cap B) = \frac{2}{36}.$$

By the extended addition principle,

$$P(A \cup B) = P(A) + P(B) - P(A \cap B),$$

$$P(A \cup B) = \frac{6}{36} + \frac{11}{36} - \frac{2}{36} = \frac{15}{36} = \frac{5}{12}.$$

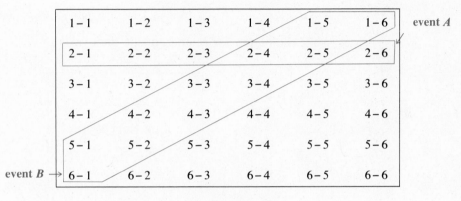

Figure 5

(b) the sum is 11 or the second die is 5
$P(\text{sum is 11}) = 2/36$, $P(\text{second die is 5}) = 6/36$, and $P(\text{sum is 11 and second die is 5}) = 1/36$, so

$$P(\text{sum is 11 or second die is 5}) = \frac{2}{36} + \frac{6}{36} - \frac{1}{36} = \frac{7}{36}. \qquad ■$$

Example 3   The personnel director at a medium sized manufacturing company has received 20 applications from people applying for a job as plant manager. Of these 20 people, 8 have MBA degrees, 9 have previous related experience, and 5 have both MBA degrees and experience. If the personnel director chooses a candidate at random, find the probability that the candidate has an MBA degree or previous related experience.

Use M for "has degree" and E for "has experience." As stated, $P(M) = 8/20$, $P(E) = 9/20$, and $P(M \cap E) = 5/20$, with

$$P(M \cup E) = \frac{8}{20} + \frac{9}{20} - \frac{5}{20} = \frac{12}{20} = \frac{3}{5}. \qquad ■$$

Example 4   Susan is a college student who receives heavy sweaters from her aunt at the first sign of cold weather. The probability that a sweater is the wrong size is .47, the probability that it is a loud color is .59, and the probability that it is both the wrong size and a loud color is .31.

To use this information to solve probability problems, we can use a Venn diagram, such as the one of Figure 6. Here W represents the event, "wrong size," while L represents "loud color." Start by placing .31, the probability that the sweater is both the wrong size and a loud color, inside the intersection of the regions for W and L.

Event W has probability .47. We have already placed .31 inside the intersection of W and L, so that

$$.47 - .31 = .16$$

goes inside region W, but outside the intersection of W and L. In the same way,

$$.59 - .31 = .28$$

goes inside the region for L, and outside the overlap.

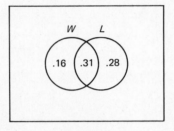

Figure 6

(a) Find the probability that the sweater is the correct size and not a loud color. Using our regions, W and L, this event becomes $W' \cap L'$. From the Venn diagram of Figure 6, the regions that are labeled have total probability

$$.16 + .31 + .28 = .75.$$

The entire region of the Venn diagram is assumed to have probability 1. The region outside W and L, or $W' \cap L'$, has probability

$$1 - .75 = .25.$$

The probability is .25 that the sweater is the correct size and not a loud color.

(b) Find the probability that the sweater is the correct size or is not loud. The region $W' \cup L'$ has probability

$$.25 + .16 + .28 = .69.$$ ■

6.3 Exercises   *Two dice are rolled. Find the probability of rolling each of the following sums.*

1. 2	**2.** 4	**3.** 5	**4.** 6
5. 8	**6.** 9	**7.** 10	**8.** 13

9. 9 or more   **10.** less than 7   **11.** between 5 and 8

12. not more than 5   **13.** not less than 8   **14.** between 3 and 7

One card is drawn from an ordinary deck of 52 cards. Find the probability of drawing

15. a 9 or a 10;

16. red or a 3;

17. a 9 or black 10;

18. a heart or black;

19. less than 4 (count aces as 1s);

20. a diamond or a 7;

21. a black card or an ace;

22. a heart or a jack.

Ms. Elliott invites ten relatives to a party: her mother, two aunts, three uncles, two brothers, one male cousin, and one female cousin. If the chances of any one guest arriving first are equally likely, find the probability that the first guest to arrive is

23. a brother or uncle;

24. a brother or cousin;

25. a brother or her mother;

26. an uncle or a cousin;

27. a male or a cousin;

28. a female or a cousin.

The numbers 1, 2, 3, 4, *and* 5 *are written on slips of paper and two slips are drawn at random without replacement. Find each of the following probabilities.*

29. the sum of the numbers is 9

30. both numbers are even

31. the sum of the numbers is 5 or less

32. one of the numbers is even or greater than 3

33. the first number is 2 or the sum is 6

34. the sum is 5 or the second number is 2

35. A student feels that her probability of passing accounting is .74, of passing mathematics is .39, and that her probability of passing both courses is .25. Find the probability that the student passes at least one course.

36. Suppose that 8% of a certain batch of calculators have a defective case, and that 11% have defective batteries. Also, 3% have both a defective case and defective batteries. A calculator is selected from the batch at random. Find the probability that the calculator has a good case and good batteries.

The table below shows the probability that a customer of a department store will make a purchase in the indicated range.

Cost	Probability
below $2	.07
$2–$4.99	.18
$5–$9.99	.21
$10–$19.99	.16
$20–$39.99	.11
$40–$69.99	.09
$70–$99.99	.07
$100–$149.99	.08
$150 or over	.03

Find the probability that a customer makes a purchase which is

37. less than $5;

38. $10 to $69.99;

39. $20 or more;

40. more than $4.99;

41. less than $100;

42. $100 or more.

Suppose $P(E) = .26$, $P(F) = .41$, and $P(E \cap F) = .17$. Use a Venn diagram to help find each of the following.

43. $P(E \cup F)$ **44.** $P(E' \cap F)$ **45.** $P(E \cap F')$ **46.** $P(E' \cup F')$

Let $P(Z) = .42$, $P(Y) = .38$, and $P(Z \cup Y) = .61$. Find each of the following probabilities.

47. $P(Z' \cap Y')$ **48.** $P(Z' \cup Y')$ **49.** $P(Z' \cup Y)$ **50.** $P(Z \cap Y')$

Color blindness is an inherited characteristic which is sex-linked, so that it is more common in males than in females. If M represents male and C represents red-green color blindness, we use the relative frequencies of the incidence of males and of red-green color blindness as probabilities to get $P(C) = .049$, $P(M \cap C) = .042$, $P(M \cup C) = .534$. Find the following.

51. $P(C')$ **52.** $P(M)$ **53.** $P(M')$

54. $P(M' \cap C')$ **55.** $P(C \cap M')$ **56.** $P(C \cup M')$

Gregor Mendel, an Austrian monk, was the first to use probability in the study of genetics. In an effort to understand the mechanism of character transmittal from one generation to the next in plants, he counted the number of occurrences of various characteristics. Mendel found that the flower color in certain pea plants obeyed this scheme:

Pure red crossed with pure white produces red.

The red offspring received from its parents genes for both red (R) and white (W) but in this case red is dominant *and white* recessive, *so the offspring exhibits the color red. However, the offspring still carries both genes, and when two such offspring are crossed, several things can happen in the third generation. The table below, which is called a* Punnet square, *shows the possibilities.*

		2nd parent	
		R	*W*
1st parent	*R*	*RR*	*RW*
	W	*WR*	*WW*

Use the fact that red is dominant over white to find

57. $P(\text{red})$;   **58.** $P(\text{white})$.

Mendel found no dominance in snapdragons, with one red gene and one white gene producing pink-flowered offspring. These second generation pinks, however, still carry one red and one white gene, and when they are crossed, the next generation still yields the Punnet square above. Find

59. $P(\text{red})$;   **60.** $P(\text{pink})$;   **61.** $P(\text{white})$.

(Mendel verified these probability ratios experimentally and did the same for many character units other than flower color. His work, published in 1866, was not recognized until 1890.)

In most animals and plants, it is very unusual for the number of main parts of the organism (arms, legs, toes, flower petals, etc.) to vary from generation to generation. Some species, however, have meristic variability, *in which the number of certain body parts varies from generation to generation. One researcher studied the front feet of certain guinea pigs and produced the following probabilities.[1]*

---

[1]From "An Analysis of Variability in Guinea Pigs" by J. R. Wright in *Genetics* 19, pp. 506–536. Reprinted by permission.

$$P(\text{only four toes, all perfect}) = .77$$

$$P(\text{one imperfect toe and four good ones}) = .13$$

$$P(\text{exactly five good toes}) = .10$$

Find the probability of each of the following events.

62. no more than four good toes   **63.** five toes, whether perfect or not

64. Let E, F, and G be events from a sample space S. Show that

$$P(E \cup F \cup G) = P(E) + P(F) + P(G) - P(E \cap F) - P(E \cap G)$$

$$- P(F \cap G) + P(E \cap F \cap G).$$

(Hint: let $H = E \cup F$ and use equation (7).)

65. A group of three people, each with a different type of hat, decides to exchange hats. To do so, they toss their hats into a pile. Each person then takes a hat at random. Find the probability that at least one person gets his own hat.

66. Prove equation (2) in the text above.

6.4 Applications of Counting

In Chapter 5 we studied permutations and combinations—ways of counting the number of outcomes for various kinds of experiments. In this section we use these methods of counting to help solve problems in probability. The solution to many of these problems depends on the basic probability principle of Section 6.2:

> Let a sample space S have n possible equally likely outcomes, and let event E contain m of these outcomes, all distinct elements. The probability that event E occurs is
>
> $$P(E) = \frac{m}{n}.$$

Example 1   From a group of 22 employees, 4 are to be selected to present a list of grievances to management.

(a) In how many ways can this be done?

We must select 4 employees from a group of 22; this can be done in $\binom{22}{4}$ ways. From Section 5.4,

$$\binom{22}{4} = \frac{22!}{4!18!} = \frac{22(21)(20)(19)}{4(3)(2)(1)} = 7315.$$

There are 7315 ways to choose 4 people from 22.

(b) One of the 22 employees is Judy Lewis: the group agrees that she *must* be one of the four people chosen. If the groups of four people are chosen at random, find the probability that Lewis will be among those chosen.

If Lewis must be one of the four people, we are reduced to the problem of finding the number of ways that the additional three employees can be chosen. The 3 are chosen from 21 employees; this can be done in

$$\binom{21}{3} = \frac{21!}{3!18!} = \frac{21(20)(19)}{3(2)(1)} = 1330$$

ways. The probability that Lewis will be one of the four employees chosen is thus

$$P(\text{Lewis is chosen}) = \frac{1330}{7315} \approx .182$$

The probability that she will *not* be chosen is $1 - .182 = .818$. ∎

Example 2   When shipping diesel engines abroad, it is common to pack 12 engines in one container which is then loaded on a rail car and sent to a port. Suppose that a company has received complaints from its customers that many of the engines arrive in nonworking condition. To help solve this problem, the company decided to make a spot check of containers after loading—the company will test three engines from a container at random; if any of the three are nonworking, the container will not be shipped until each engine in it is checked. Suppose a given container has two nonworking engines, and find the probability that the container will not be shipped.

The container will not be shipped if the sample of three engines contains one or two defective engines. Thus, letting $P(1 \text{ defective})$ represent the probability of exactly 1 defective engine in the sample,

$$P(\text{not shipping}) = P(1 \text{ defective}) + P(2 \text{ defectives}).$$

There are $\binom{12}{3}$ ways to choose the 3 engines for testing:

$$\binom{12}{3} = \frac{12!}{3!9!} = \frac{12(11)(10)}{3(2)(1)} = 220.$$

There are $\binom{2}{1}$ ways of choosing one defective engine from the two in the container, and for each of these ways, there are $\binom{10}{2}$ ways of choosing two good engines from among the ten in the container. In total, there are

$$\binom{2}{1}\binom{10}{2} = \frac{2!}{1!1!} \cdot \frac{10!}{2!8!} = 2(45) = 90$$

ways of choosing a sample of 3 engines containing one defective. Thus,

$$P(1 \text{ defective}) = \frac{90}{220}.$$

There are $\binom{2}{2}$ ways of choosing two defective engines from the two defective engines in the container, and $\binom{10}{1}$ ways of choosing one good engine from among the ten good engines, for a total of

$$\binom{2}{2}\binom{10}{1} = \frac{2!}{2!0!} \cdot \frac{10!}{1!9!} = 1(10) = 10$$

ways of choosing a sample of three engines containing two defectives. Thus,

$$P(2 \text{ defectives}) = \frac{10}{220}$$

and

$$P(\text{not shipping}) = P(1 \text{ defective}) + P(2 \text{ defectives})$$

$$= \frac{90}{220} + \frac{10}{220} = \frac{100}{220} \approx .455.$$

The probability is $1 - .455 = .545$ that the container *will* be shipped, even though it has two defective engines. The management must decide if this probability is acceptable; if not, it may be necessary to test more than three engines from a container.   ∎

We solved this example by finding the sum $P(1 \text{ defective}) + P(2 \text{ defectives})$. We would get the same answer by finding $1 - P(\text{no defectives})$.

$$P(\text{not shipping}) = 1 - P(\text{no defectives in sample})$$

$$= 1 - \frac{\binom{2}{0}\binom{10}{3}}{\binom{12}{3}}$$

$$= 1 - \frac{1(120)}{220}$$

$$= 1 - \frac{120}{220} = \frac{100}{220} \approx .455$$

Example 3   In a common form of the card game *poker,* a hand of five cards is dealt to each player from a deck of 52 cards. There are a total of

$$\binom{52}{5} = \frac{52!}{5!47!} = 2,598,960$$

such hands possible. Find each of the following probabilities.
(a) a hand containing only hearts, called a *heart flush*

There are 13 hearts in a deck; there are

$$\binom{13}{5} = \frac{13!}{5!8!} = \frac{13(12)(11)(10)(9)}{5(4)(3)(2)(1)} = 1287$$

different hands containing only hearts. The probability of a heart flush is thus

$$P(\text{heart flush}) = \frac{1287}{2,598,960} = \frac{33}{66,640} \approx .000495.$$

(b) a flush of any suit

There are four suits in a deck, so

$$P(\text{flush}) = 4 \cdot P(\text{heart flush}) = 4 \cdot \frac{33}{66,640} \approx .00198.$$

(c) a full house of aces and eights (three aces and two eights)

There are $\binom{4}{3}$ ways to choose 3 aces from among the 4 in the deck, and $\binom{4}{2}$ ways to choose 2 eights.

$$P(\text{3 aces, 2 eights}) = \frac{\binom{4}{3} \cdot \binom{4}{2}}{2,598,960} = \frac{1}{108,290} \approx .00000923.$$

(d) any full house (3 cards of one value, 2 of another)

There are 13 values in a deck, so there are 13 choices for the first value mentioned, leaving 12 choices for the second value (order *is* important here, since a full house of aces and eights, for example, is not the same as a full house of eights and aces). Thus, since the probability of any particular full house is 1/108,290,

$$P(\text{full house}) = 13 \cdot 12 \cdot \left(\frac{1}{108,290}\right) = \frac{156}{108,290} \approx .00144. \quad \blacksquare$$

Example 4   Suppose a group of n people is in a room. Find the probability that at least two of the people have the same birthday.

Here we refer to the month and the day, not necessarily the same year. Also, we ignore leap years, and we assume that each day in the year is equally likely as a birthday. Let us first find the probability that *no two people* among five people have the same birthday. There are 365 different birthdays possible for the first of our five people, 364 for the second (so that the people have different birthdays), 363 for the third, and so on. The number of ways the five people can have different birthdays is thus the number of permutations of 365 things (days) taken 5 at a time, or

$$P(365, 5) = 365 \cdot 364 \cdot 363 \cdot 362 \cdot 361.$$

The number of ways that the five people can have the same or different birthdays is

$$365 \cdot 365 \cdot 365 \cdot 365 \cdot 365 = (365)^5.$$

Finally, the *probability* that none of the five people have the same birthday is

$$\frac{P(365, 5)}{(365)^5} = \frac{365 \cdot 364 \cdot 363 \cdot 362 \cdot 361}{365 \cdot 365 \cdot 365 \cdot 365 \cdot 365} \approx .973.$$

The probability that at least two of the five people *do* have the same birthday is $1 - .973 = .027$.

We can extend this same result for more than five people. In general, the probability that no two people among n people have the same birthday is

$$\frac{P(365, n)}{(365)^n},$$

with the probability that at least two of the n people *do* have the same birthday given by

$$1 - \frac{P(365, n)}{(365)^n}.$$

The following table shows this probability for various values of n.

Number of people, n	Probability that two have the same birthday
5	.027
10	.117
15	.253
20	.411
22	.476
23	.507
25	.569
30	.706
35	.814
40	.891
50	.970
365	1

The probability that 2 people among 23 have the same birthday is .507, a little more than half. Many people are surprised at this result—somehow it seems that a larger number of people should be required.   ▮

6.4 Exercises   *A shipment of 9 typewriters contains 2 defectives. Find the probability that a sample of the following size, drawn from the 9, will not contain a defective.*

1. 1   **2.** 2   **3.** 3   **4.** 4

Refer to Example 2. The management feels that the probability of .545 that a container will be shipped even though it contains two defectives is too high. They decide to increase the sample size chosen. Find the probability that a container will be shipped even though it contains two defectives if the sample size is increased to

5. 4;   **6.** 5.

A basket contains 6 red apples and 4 yellow apples. A sample of 3 apples is drawn. Find the probability that the sample contains

7. all red apples;   **8.** all yellow apples;

9. 2 yellow and 1 red apple;   **10.** more red than yellow apples.

Two cards are drawn at random from an ordinary deck of 52 cards.

11. How many two card hands are possible?

Find the probability that the two card hand contains

12. two aces;   **13.** at least one ace;

14. all spades;   **15.** two cards of the same suit;

16. only face cards;   **17.** no face cards;

18. no card higher than 8 (count ace as 1).

Twenty-six slips of paper are each marked with a different letter of the alphabet, and placed in a basket. A slip is pulled out, its letter recorded (in the order in which the slip was drawn), and the slip replaced. This is done five times. Find the probabilities that the "word" formed

19. is "chuck";

20. starts with p;

21. has all different letters;

22. contains no x, y, or z.

Find the probability of the following hands at poker. Assume aces are either high or low.

23. royal flush (five highest cards of a single suit)

24. straight flush (five in a row in a single suit, but not a royal flush)

25. four of a kind (four cards of the same value)

26. straight (five cards in a row, not all of the same suit) with ace either high or low

A bridge hand is made up of 13 cards from a deck of 52. Set up the probability that a hand chosen at random

27. contains only hearts;   **28.** has four aces;

29. contains exactly three aces and exactly three kings;

30. has six of one suit, five of another, and two of another.

31. Set up the probability that at least two of the 41 Presidents of the United States have had the same birthday.

32. Estimate the probability that at least two of the one hundred U.S. Senators have the same birthday.

33. Give the probability that two of the 435 members of the House of Representatives have the same birthday.

34. Show that the probability that in a group of *n* people *exactly one* pair have the same birthday is

$$\binom{n}{2} \cdot \frac{P(365, n-1)}{(365)^n}$$

35. To win a contest, a player must match four movie stars with his or her baby picture. Suppose this is done at random. Find the probability of getting no matches correct; of getting exactly two correct.

36. A contractor has hired a decorator to send three different sofas out to the contractor's model homes each week for a year. The contractor does not want exactly the same three sofas sent out twice. Find the minimum number of sofas that the decorator will need.

37. An elevator has four passengers and stops at seven floors. It is equally likely that a person will get off at any one of the seven floors. Find the probability that no two passengers leave at the same floor.

The rest of these exercises involve the idea of a circular permutation: *the number of ways of arranging distinct objects in a circle. The number of ways of arranging n distinct objects in a line is n!, but there are fewer ways for arranging the n items in a circle since the first item could be placed in any of n locations.*

38. Show that the number of ways of arranging *n* distinct items in a circle is $(n-1)!$.

Find the number of ways of arranging the following number of distinct items in a circle.

39. 4   **40.** 7   **41.** 10

42. Suppose that eight people sit at a circular table. Find the probability that two particular people are sitting next to each other.

43. A keyring contains seven keys; one black, one gold, and five silver. If the keys are arranged at random on the ring, find the probability that the black key is next to the gold key.

44. A circular table for a board of directors has ten seats for the ten attending members of the board. The chairman of the board always sits closest to the window. The vice president for sales, who is currently out of favor, will sit three positions to the chairman's left, since the chairman doesn't see so well out of his left eye. The chairman's daughter-in-law will sit opposite him. All other members take seats at random. Find the probability that a particular other member will sit next to the chairman.

6.5 Conditional Probability

The training manager for a large stockbrokerage firm has noticed that some of the firm's brokers use the firm's research advice, while other brokers tend to go with their own feelings of which stocks will go up. To see if the research department is better than just the feelings of the brokers, the manager conducted a survey of 100 brokers, with results as shown in the following table.

	Picked stocks that went up	Didn't pick stocks that went up	Totals
Used research	30	15	45
Didn't use research	30	25	55
Totals	60	40	100

Letting A represent the event "picked stocks that went up," and letting B represent the event "used research," we can find the following probabilities.

$$P(A) = \frac{60}{100} = .6 \qquad P(A') = \frac{40}{100} = .4$$

$$P(B) = \frac{45}{100} = .45 \qquad P(B') = \frac{55}{100} = .55$$

Suppose we want to find the probability that a broker using research will pick stocks that go up. From the table above, of the 45 brokers who use research, there are 30 who picked stocks that went up. Thus, the desired probability is

$$P(\text{broker who uses research picks stocks that go up}) = \frac{30}{45} = .667.$$

This is a different number than the probability that a broker picks stocks that go up, .6, since we have additional information (the broker uses research) which reduced the sample space. In other words, we found the probability that a broker picks stocks that go up, *A*, given the additional information that the broker uses research, *B*. This is called the *conditional probability* of event *A*, given that event *B* has occurred, written $P(A|B)$. In the example above,

$$P(A|B) = \frac{30}{45},$$

which can be written as

$$P(A|B) = \frac{30/100}{45/100} = \frac{P(A \cap B)}{P(B)}$$

where $P(A \cap B)$ represents, as usual, the probability that both *A* and *B* will occur.

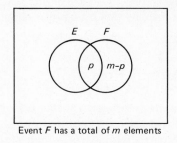

Event *F* has a total of *m* elements

Figure 7

Let us try to generalize this result. Assume that *E* and *F* are two events for a particular experiment. Assume that the sample space *S* for this experiment has *n* possible equally likely outcomes. Suppose event *F* has *m* elements, while $E \cap F$ has *p* elements ($p \leq m$). Using the fundamental principle of probability,

$$P(F) = \frac{m}{n} \quad \text{and} \quad P(E \cap F) = \frac{p}{n}.$$

We now want $P(E|F)$, the probability that *E* occurs given that *F* has occurred. Since we assume *F* has occurred, we look only at the *m* elements inside *F*. (See Figure 7). Of these *m* elements, we know there are *p* elements where *E* also occurs, since $E \cap F$ has *p* elements. Thus,

$$P(E|F) = \frac{p}{m}.$$

Divide numerator and denominator by *n* to get

$$P(E|F) = \frac{\frac{p}{n}}{\frac{m}{n}} = \frac{P(E \cap F)}{P(F)}.$$

This result is actually chosen as the definition of conditional probability.

The **conditional probability** of event E given event F, written $P(E|F)$, is defined as

$$P(E|F) = \frac{P(E \cap F)}{P(F)}, \qquad P(F) \neq 0.$$

Example 1 Use the information given in the chart at the beginning of this section to find the following probabilities.

(a) $P(B|A)$

By the definition of conditional probability,

$$P(B|A) = \frac{P(B \cap A)}{P(A)}.$$

In our example, $P(B \cap A) = 30/100$, and $P(A) = 60/100$, with

$$P(B|A) = \frac{30/100}{60/100} = \frac{1}{2}.$$

If we know that a broker picked stocks that went up, then there is a probability of 1/2 that the broker used research.

(b) $P(A'|B)$

$$P(A'|B) = \frac{P(A' \cap B)}{P(B)} = \frac{15/100}{45/100} = \frac{1}{3}.$$

(c) $P(B'|A')$

$$P(B'|A') = \frac{P(B' \cap A')}{P(A')} = \frac{25/100}{40/100} = \frac{5}{8}. \qquad \blacksquare$$

Venn diagrams can be used to illustrate problems in conditional probability. A Venn diagram for Example 1, in which the probabilities are used to indicate the number in the set defined by each region, is shown in Figure 8. In the diagram, $P(B|A)$ is found by reducing the sample space to just set A. Then $P(B|A)$ is the ratio of the number in that part of set B which is also in A to the number in set A, or $30/60 = .5$.

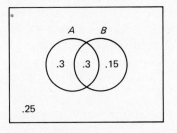

Figure 8

Example 2   Two fair coins are tossed, and it is known that at least one was heads. Find the probability that both were heads.

The sample space has four equally likely outcomes, $S = \{hh, ht, th, tt\}$. Define two events:

$$E_1 = \text{at least one head,} \quad \text{or } E_1 = \{hh, ht, th\}.$$
$$E_2 = \text{two heads,} \quad \text{or } E_2 = \{hh\}.$$

Since there are four equally likely outcomes, $P(E_1) = 3/4$. Also, $P(E_1 \cap E_2) = 1/4$. We want the probability that both were heads, given that at least one was a head; that is, we want to find $P(E_2|E_1)$. Use the definition above.

$$P(E_2|E_1) = \frac{P(E_2 \cap E_1)}{P(E_1)} = \frac{1/4}{3/4} = \frac{1}{3} \quad \blacksquare$$

In the definition of conditional probability given earlier, we can multiply both sides of the equation for $P(E|F)$ by $P(F)$ to get the following ***product rule*** for probability:

> For any events E and F,
>
> $$P(E \cap F) = P(F) \cdot P(E|F).$$

The product rule gives us a method for finding the probability that events E and F both occur, as illustrated by the next few examples.

Example 3   A class is 2/5 women and 3/5 men. Of the women, 25% are business majors. Find the probability that a student chosen at random is a woman business major.

Let B and W represent the events "business major" and "woman," respectively. We want to find $P(B \cap W)$. By the product rule,

$$P(B \cap W) = P(W) \cdot P(B|W)$$

Using the given information, $P(W) = 2/5 = .4$ and $P(B|W) = .25$. Thus

$$P(B \cap W) = .4(.25) = .10. \quad \blacksquare$$

Example 4   A company needs to hire a new director of advertising. It has decided to try to hire away either person A or person B, assistant advertising directors for its major competitor. In trying to decide between A and B, the company does research, on the campaigns managed by either A or B (none are managed by both), and finds that A is in charge of twice as many advertising campaigns as B. Also, A's campaigns have satisfactory results three out of four times, while B's campaigns have satisfactory results in only two out of five times. Suppose one advertising campaign (managed by A or B) picked at random from the competitor is selected. Find the probabilities of the following events.

(a) A is in charge of an advertising campaign that produces satisfactory results.

First construct a *tree diagram* which shows the various possible outcomes for this experiment, as shown in Figure 9. (Recall the discussion of

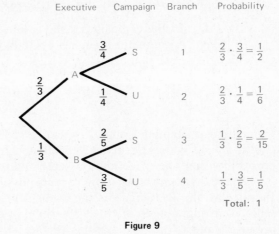

Figure 9

tree diagrams in Section 5.1.) Since A does twice as many jobs as B, the probabilities of A and B having the job are 2/3 and 1/3 respectively, as shown on the first stage of the tree. The second stage shows four different conditional probabilities. The ratings for the advertising campaigns are S (satisfactory) and U (unsatisfactory). For example, along the branch from B to S,

$$P(S|B) = 2/5$$

since B has satisfactory results in two out of five campaigns.

Each of the four composite branches in the tree is numbered, with its probability given on the right. At each point where the tree branches, the sum of the probabilities is 1. The event that A has a campaign with a satisfactory result (event A ∩ S) is associated with branch 1, so

$$P(A \cap S) = \frac{1}{2}.$$

(b) B runs the campaign and produces satisfactory results.

This event, B ∩ S, is shown on branch 3:

$$P(B \cap S) = \frac{2}{15}.$$

(c) The campaign is satisfactory.

The result S combines branches 1 and 3, so

$$P(S) = \frac{1}{2} + \frac{2}{15} = \frac{19}{30}.$$

(d) The campaign is unsatisfactory.

Event U combines branches 2 and 4, so

$$P(U) = \frac{1}{6} + \frac{1}{5} = \frac{11}{30}.$$

Alternatively, $P(U) = 1 - P(S) = 1 - 19/30 = 11/30$.

(e) Either A runs the campaign or the results are satisfactory (or both).

Event A combines branches 1 and 2, while event S combines branches 1 and 3. Thus, we use branches 1, 2, and 3.

$$P(A \cup S) = \frac{1}{2} + \frac{1}{6} + \frac{2}{15} = \frac{4}{5} \quad \blacksquare$$

Example 5   From a box containing 3 white, 2 green, and 1 red marble, two marbles are drawn one at a time without replacing the first before the second is drawn. Find the probability that one white and one green marble are drawn.

A tree diagram showing the various possible outcomes is given in Figure 10. In this diagram, W represents the event "drawing a white marble" and G represents "drawing a green marble." On the first draw, $P(W$ on the 1st$) = 3/6 = 1/2$ because three of the six marbles in the box are white. On the second draw, $P(G$ on the 2nd$|W$ on the 1st$) = 2/5$. One white marble has been removed, leaving 5, of which 2 are green.

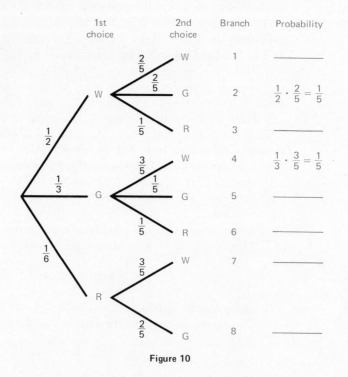

Figure 10

Now we want to find the probability of drawing one white marble and one green marble. This event can occur in two ways: drawing a white marble first and then a green one (branch 2 of the tree diagram), or drawing a green marble first and then a white one (branch 4). For branch 2, we have

$$P(W \text{ on 1st}) \cdot P(G \text{ on 2nd}|W \text{ on 1st}) = \frac{1}{2} \cdot \frac{2}{5} = \frac{1}{5}.$$

For branch 4, when the green marble is drawn first, we have ,

$$P(G \text{ on 1st}) \cdot P(W \text{ on 2nd}|G \text{ on 1st}) = \frac{1}{3} \cdot \frac{3}{5} = \frac{1}{5}.$$

Since the two events are mutually exclusive, the final probability is the sum of these two probabilities, or

$$P(\text{one } W, \text{ one } G) = P(W \text{ on 1st}) \cdot P(G \text{ on 2nd}|W \text{ on 1st})$$

$$+ P(G \text{ on 1st}) \cdot P(W \text{ on 2nd}|G \text{ on 1st}) = \frac{2}{5}. \quad \blacksquare$$

The product rule is often helpful with *stochastic processes,* where the outcome of an experiment depends on the outcomes of previous experiments. For example, the outcome of a draw of a card from a deck depends on any cards previously drawn. (We shall study stochastic processes in more detail in the next chapter.)

Example 6   Two cards are drawn without replacement from an ordinary deck. Find the probability that the first card is a heart and the second card is red.
Start with the tree diagram of Figure 11.

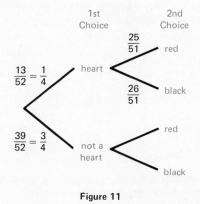

Figure 11

On the first draw, since there are 13 hearts in the 52 cards, the probability of drawing a heart is $13/52 = 1/4$. On the second draw, since a heart has been drawn already, there are 25 red cards in the remaining 51 cards. Thus, the probability of drawing a red card on the second draw, given that the first is a heart, is $25/51$. Therefore,

$$P(\text{heart on first and red on second})$$

$$= P(\text{heart on first}) \cdot P(\text{red on second}|\text{heart on first})$$

$$= \frac{1}{4} \cdot \frac{25}{51} = \frac{25}{204} \approx .1225. \quad \blacksquare$$

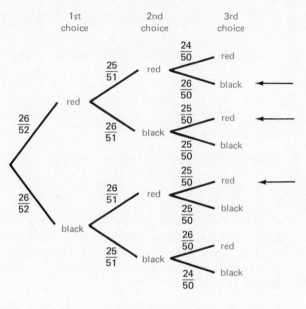

Figure 12

Example 7   Three cards are drawn without replacement from an ordinary deck. Find the probability that exactly two of the cards are red.

Here we need a tree diagram with three stages, as shown in Figure 12. The three branches indicated with arrows produce exactly two red cards from the three draws. Multiply the probabilities along each of these branches and then add.

$$P(\text{exactly two red cards}) = \frac{26}{52} \cdot \frac{25}{51} \cdot \frac{26}{50} + \frac{26}{52} \cdot \frac{26}{51} \cdot \frac{25}{50} + \frac{26}{52} \cdot \frac{26}{51} \cdot \frac{25}{50}$$

$$= \frac{50{,}700}{132{,}600} = \frac{13}{34} \approx .382. \quad \blacksquare$$

Independent Events   Suppose we toss a fair coin and get heads. The probability of heads on the next toss is still 1/2; the fact that we got heads on a given toss has no effect on the outcome of the next toss. Coin tosses are *independent events,* since knowledge of the outcome of one toss does not help decide on the outcome of the next toss. Rolls of a fair die are independent events; the fact that a 2 came up on one roll does not help increase our knowledge of the outcome of the next roll. On the other hand, the events "today is cloudy" and "today is rainy" are *dependent events;* if we know that it is cloudy, we know that there is an increased chance of rain.

If events E and F are independent, then the knowledge that E has occurred gives us no (probability) information about the occurrence or nonoccurrence of event F. That is, $P(F)$ is exactly the same as $P(F|E)$, or

$$P(F|E) = P(F).$$

This, in fact, is the formal definition of independent events:

> E and F are **independent events** if
>
> $$P(F|E) = P(F).$$

Using this definition, the product rule can be simplified for independent events.

> **Product Rule for Independent Events**   If E and F are independent events, then
>
> $$P(E \cap F) = P(E) \cdot P(F).$$

Example 8   A calculator requires a key-stroke assembly and a logic circuit. Assume that 99% of the key-stroke assemblies are satisfactory and 97% of the logic circuits are satisfactory. Find the probability that a finished calculator will be satisfactory.

If we assume that the failure of a key stroke assembly and the failure of a logic circuit are independent events, then

P(satisfactory calculator)
$= P$(satisfactory key-stroke assembly) $\cdot$ P(satisfactory logic circuit)
$= (.99)(.97) \approx .96.$

The probability of a defective calculator is thus $1 - .96 = .04.$ ∎

Example 9   When black-coated mice are crossed with brown-coated mice, a pair of genes, one from each parent, determines the coat color of the offspring. Let b represent the gene for brown and B the gene for black. If a mouse carries either one B gene and one b gene (Bb or bB) or two B genes (BB), the coat will be black. If the mouse carries two b genes (bb), the coat will be brown. Find the probability that a mouse born to a brown-coated female and a black-coated male who is known to carry the Bb combination will be brown.

To be brown-coated, the offspring must receive one b gene from each parent. The brown-coated parent carries two b genes, so that the probability of getting one b gene from the mother is 1. The probability of getting one b gene from the black-coated father is 1/2. Therefore, since these are independent events, the probability of a brown-coated offspring from these parents is $1 \cdot 1/2 = 1/2.$ ∎

6.5 Exercises   *If a single fair die is rolled, find the probability of rolling*

1. a 2, given that the number rolled was odd;

2. a 4, given that the number rolled was even;

3. an even number, given that the number rolled was 6.

If two fair dice are rolled, find the probability of rolling

4. a sum of 8, given the sum was greater than 7;

5. a sum of 6, given the roll was a "double" (two identical numbers);

6. a double, given that the sum was 9.

If two cards are drawn without replacement from an ordinary deck, find the probability that

7. the second is a heart, given that the first is a heart;

8. they are both hearts;

9. the second is black, given that the first is a spade;

10. the second is a face card, given that the first is a jack.

If five cards are drawn without replacement from an ordinary deck, find the probability that all the cards are

11. diamonds;

12. diamonds, given that the first and second were diamonds;

13. diamonds, given that the first four were diamonds;

14. clubs, given that the third was a spade;

15. the same suit.

A smooth-talking young man has a 1/3 probability of talking a policeman out of giving him a speeding ticket. The probability that he is stopped for speeding during a given weekend is 1/2. Find the probability that

16. he will receive no speeding tickets on a given weekend;

17. he will receive no speeding tickets on three consecutive weekends.

Slips of paper marked with the digits 1, 2, 3, 4, and 5 are placed in a box and mixed well. If two slips are drawn (without replacement), find the probability that

18. the first is even and the second is odd;

19. the first is a 3 and the second a number greater than 3;

20. both are even;

21. both are marked 3.

Two marbles are drawn without replacement from a jar with four black and three white marbles. Find the probability that

22. both are white;

23. both are black;

24. the second is white given that the first is black;

25. the first is black and the second is white;

26. one is black and the other is white.

The Midtown Bank has found that most customers at the tellers' windows either cash a check or make a deposit. The chart below indicates the transactions for one teller for one day.

	Cash check	No check	Totals
Make deposit	50	20	70
No deposit	30	10	40
Totals	80	30	110

Letting C represent "cashing a check" and D represent "making a deposit," express each of the following probabilities in words and find its value.

27. $P(C|D)$ **28.** $P(D'|C)$ **29.** $P(C'|D')$

30. $P(C'|D)$ **31.** $P[(C \cap D)']$

A pet shop has 10 puppies, 6 of them males. There are 3 beagles (1 male), 1 cocker spaniel (male), and 6 poodles. Construct a table similar to the one above and find the probability that one of these puppies, chosen at random, is

32. a beagle;

33. a beagle, given that it is a male;

34. a male, given that it is a beagle;

35. a cocker spaniel, given that it is a female;

36. a poodle, given that it is a male;

37. a female, given that it is a beagle.

A bicycle factory runs two assembly lines, A and B. If 95% of line A's products pass inspection, while only 90% of line B's products pass inspection, and 60% of the factory's bikes come off assembly line B (the rest off A), find the probability that one of the factory's bikes did not pass inspection and came off

38. assembly line A;   **39.** assembly line B.

40. Both of a certain pea plant's parents had a gene for red and a gene for white flowers. (See the exercises for Section 6.3.) If the offspring has red flowers, find the probability that it combined a gene for red and a gene for white (rather than two for red).

Assume that boy and girl babies are equally likely.

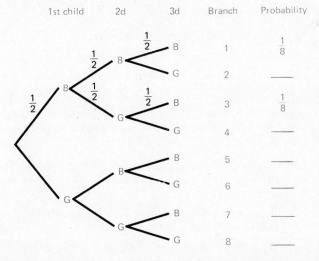

Fill in the remaining probabilities on the tree diagram above and use the information to find the probability that a family with three children has all girls, given that

41. the first is a girl;   **42.** the third is a girl;   **43.** the second is a girl;

44. at least two are girls;   **45.** at least one is a girl.

The following table shows frequencies for red-green color blindness, where M represents male and C represents color-blind.

	M	M'	Totals
C	.042	.007	.049
C'	.485	.466	.951
Totals	.527	.473	1.000

Use this table to find the following probabilities.

46. $P(M)$ **47.** $P(C)$ **48.** $P(M \cap C)$

49. $P(M \cup C)$ **50.** $P(M|C)$ **51.** $P(C|M)$

52. $P(M'|C)$

53. Are the events C and M described above dependent?

54. A scientist wishes to determine if there is any dependence between color blindness (C) and deafness (D). Given the probabilities listed in the table below, what should his findings be? (See Exercises 46–53.)

	D	D'	Totals
C	.0004	.0796	.0800
C'	.0046	.9154	.9200
Totals	.0050	.9950	1.0000

The Motor Vehicle Department has found that the probability of a person passing the test for a driver's license on the first try is .75. The probability that an individual who fails on the first test will pass on the second try is .80, and the probability that an individual who fails the first and second tests will pass the third time is .70. Find the probability than an individual

55. fails both the first and second tests;

56. will fail three times in a row;

57. will require at least two tries to pass the test.

According to a booklet put out by Frontier Airlines, 98% of all scheduled Frontier flights actually take place. (The other flights are cancelled due to weather, equipment problems, and so on.) Assume that the event that a given flight takes place is independent of the event that another flight takes place.

58. Elizabeth Thornton plans to visit her company's branch offices; her journey requires three separate flights on Frontier. What is the probability that all these flights will take place?

59. Based on the reasons we gave for a flight to be cancelled, how realistic is the assumption of independence that we made?

60. In one area, 4% of the population drives a luxury car. However, 17% of the CPAs drive a luxury car. Are the events "drive a luxury car" and "person is a CPA" independent?

61. Corporations where a computer is essential to day-to-day operations, such as banks, often have a second backup computer in case of failure by the main computer. Suppose there is a .003 chance that the main computer will fail in a given time period, and a

.005 chance that the backup computer will fail while the main computer is being repaired. Assume these failures represent independent events, and find the fraction of the time that the corporation can assume it will have computer service. How realistic is our assumption of independence?

62. A key component of a space rocket will fail with a probability of .03. How many such components must be used as backups to ensure the probability that at least one of the components will work is .999999?

In searching for a new drug with commercial possibilities, drug company researchers use the ratio

$$N_S : N_A : N_P : 1.$$

That is, if the company gives preliminary screening to N_S substances, it may find that N_A of them are worthy of further study, with N_P of these surviving into full scale development. Finally, 1 of the substances will result in a marketable drug. Typical numbers used by Smith, Kline, and French Laboratories in planning research budgets might be $2000:30:8:1$.[1] *Use this ratio in the following exercises.*

63. Suppose a compound has been chosen for preliminary screening. Find the probability that the compound will survive and become a marketable drug.

64. Find the probability that the compound will not lead to a marketable drug.

65. Suppose the number of such compounds receiving preliminary screening is a. Set up the probability that none of them produces a marketable drug. (Assume independence throughout these exercises.)

66. Use your results from Exercise 65 to set up the probability that at least one of the drugs will prove marketable.

67. Suppose now that N scientists are employed in the preliminary screening, and that each scientist can screen c compounds per year. Set up the probability that no marketable drugs will be discovered in a year.

68. Set up the probability that at least one marketable drug will be discovered.

For the following exercises, evaluate your answer in Exercise 68 for the following values of N and c. Use a calculator with an x^y key, or a computer.

69. $N = 100$, $c = 6$

70. $N = 25$, $c = 10$

Let E and F be events which are neither the empty set nor the sample space S. Identify the following as true *or* false.

71. $P(E|E) = 1$ **72.** $P(E|E') = 1$ **73.** $P(\varnothing|F) = 0$

74. $P(S|E) = P(E)$ **75.** $P(F|S) = P(F)$ **76.** $P(E|F) = P(F|E)$

77. $P(E|E \cap F) = 0$ **78.** If $P(E|F) = P(E \cap F)$, then $P(F) = 1$

79. Let E and F be mutually exclusive events such that $P(F) > 0$. Find $P(E|F)$.

80. If $E \subset F$, where $E \neq \varnothing$, find $P(F|E)$ and $P(E|F)$.

[1]Reprinted by permission of E. B. Pyle, III, B. Douglas, G. W. Ebright, W. J. Westlake, A. B. Bender, "Scientific Manpower Allocation to New Drug Screening Programs," *Management Science*, Vol. 19, No. 12, August 1973, copyright © 1973 The Institute of Management Sciences.

81. Let F_1, F_2, and F_3 be a set of pairwise mutually exclusive events (that is, $F_1 \cap F_2 = \emptyset$, $F_1 \cap F_3 = \emptyset$, and $F_2 \cap F_3 = \emptyset$), with sample space $S = F_1 \cup F_2 \cup F_3$. Let E be any event. Show that

$$P(E) = P(F_1) \cdot P(E|F_1) + P(F_2) \cdot P(E|F_2) + P(F_3) \cdot P(E|F_3).$$

82. Show that for three events E, F, and G,

$$P(E \cap F \cap G) = P(E) \cdot P(F|G) \cdot P(G|E \cap F).$$

6.6 Bayes' Formula

Suppose the probability that a person gets lung cancer, given that the person smokes a pack or more of cigarettes daily, is known. For a research project, it might be necessary to know the probability that a person smokes a pack or more of cigarettes daily, given that the person has lung cancer. More generally, if $P(E|F)$ is known for two events E and F, can $P(F|E)$ be found? It turns out that it can, using the formula to be developed in this section. To find this formula, let us start with the product rule:

$$P(E \cap F) = P(E) \cdot P(F|E),$$

which can also be written as $P(F \cap E) = P(F) \cdot P(E|F)$. From the fact that $P(E \cap F) = P(F \cap E)$, we have

$$P(E) \cdot P(F|E) = P(F) \cdot P(E|F),$$

or
$$P(F|E) = \frac{P(F) \cdot P(E|F)}{P(E)}. \qquad (1)$$

Given the two events E and F, if E occurs, then either F also occurs or F' also occurs. The probabilities of $E \cap F$ and $E \cap F'$ can be expressed as follows.

$$P(E \cap F) = P(F) \cdot P(E|F)$$
$$P(E \cap F') = P(F') \cdot P(E|F')$$

We know that $(E \cap F) \cup (E \cap F') = E$ (because F and F' together form the sample space), so that

$$P(E) = P(E \cap F) + P(E \cap F')$$

or
$$P(E) = P(F) \cdot P(E|F) + P(F') \cdot P(E|F').$$

From this, equation (1) produces the following result, a special case of Bayes' Formula, which is discussed in more generality later in this section.

$$P(F|E) = \frac{P(F) \cdot P(E|F)}{P(F) \cdot P(E|F) + P(F') \cdot P(E|F')}. \qquad (2)$$

Example 1 Experience has shown that the probability of worker error on the production line is .1, the probability that an accident will occur when there is a worker error is .3, and the probability that an accident will occur when there is no worker error is .2. Find the probability of a worker error if there is an accident.

Let A represent the event of an accident, and let E represent the event of worker error. From the information above,

$$P(E) = .1, \qquad P(A|E) = .3, \qquad \text{and } P(A|E') = .2.$$

These probabilities are shown on the tree diagram of Figure 13.

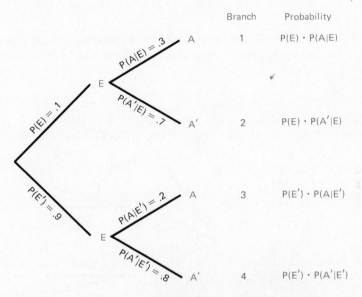

	Branch	Probability
A	1	$P(E) \cdot P(A|E)$
A'	2	$P(E) \cdot P(A'|E)$
A	3	$P(E') \cdot P(A|E')$
A'	4	$P(E') \cdot P(A'|E')$

$P(A|E) = .3$

$P(A'|E) = .7$

$P(E) = .1$

$P(E') = .9$

$P(A|E') = .2$

$P(A'|E') = .8$

Figure 13

We need to find $P(E|A)$. By equation (2) above,

$$P(E|A) = \frac{P(E) \cdot P(A|E)}{P(E) \cdot P(A|E) + P(E') \cdot P(A|E')}$$

$$= \frac{(.1)(.3)}{(.1)(.3) + (.9)(.2)} = \frac{1}{7}.$$

In a similar manner, the probability that an accident is not due to worker error is $P(E'|A)$, or

$$P(E'|A) = \frac{P(E') \cdot P(A|E')}{P(E') \cdot P(A|E') + P(E) \cdot P(A|E)}$$

$$= \frac{(.9)(.2)}{(.9)(.2) + (.1)(.3)} = \frac{6}{7}. \quad \blacksquare$$

Equation (2) above can be generalized to more than two possibilities. To do so, use the tree diagram of Figure 14. This diagram shows the paths that can produce some event E. We assume that the events $F_1, F_2, \cdots, F_n$ are pairwise mutually exclusive events (that is, events which, taken two at a time, are disjoint) whose union is the sample space, and that E is an event that has occurred.

To find the probability $P(F_i|E)$, where $1 \le i \le n$, divide the probability for the branch containing $P(E|F_i)$ by the sum of the probabilities of all the branches producing event E. That is,

$$P(F_i|E) = \frac{P(F_i) \cdot P(E|F_i)}{P(F_1) \cdot P(E|F_1) + P(F_2) \cdot P(E|F_2) + \cdots + P(F_n) \cdot P(E|F_n)}.$$

This result is known as **Bayes' Formula,** after the Reverend Thomas Bayes, whose paper on probability was published a little over two hundred years ago.

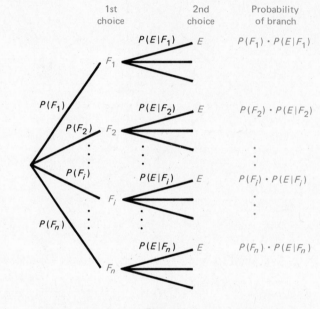

Figure 14

The basic statement of Bayes' Formula can be daunting. Actually, it is easier to remember the formula by thinking of the tree diagram that produced it. Go through the following steps.

Step 1 Start a tree diagram with branches representing events F_1, F_2, $\cdots$, F_n. Label each branch with its corresponding probability.

Step 2 From the end of each of these branches, draw a branch for event E. Label this branch with the probability of getting to it, or $P(E|F_i)$.

Step 3 You now have n different paths that result in event E. Next to each path, put its probability—the product of the probabilities that the first branch occurs, $P(F_i)$, and that the second branch occurs, $P(E|F_i)$; that is, the product $P(F_i) \cdot P(E|F_i)$.

Step 4 The desired probability is given by the probability of the branch we want, divided by the sum of the probabilities of all the branches producing event E.

Example 2   Based on past experience, a company knows that an experienced machine operator (one or more years of experience) will produce a defective item 1% of the time. People with some experience (up to one year) have a 2.5% defect rate, while new people have a 6% defect rate. At any one time, the company has 60% experienced employees, 30% with some experience, and 10% new employees. Suppose a particular item is defective; find the probability that it was produced by a new operator.

Let E represent the event "an item is defective," with F_1 representing "item was made by an experienced operator," F_2 "item was made by a person with some experience," and F_3 "item was made by a new employee." Then

$$P(F_1) = .60 \qquad P(E|F_1) = .01$$
$$P(F_2) = .30 \qquad P(E|F_2) = .025$$
$$P(F_3) = .10 \qquad P(E|F_3) = .06.$$

We need to find $P(F_3|E)$, the probability that an item was produced by a new operator, given that it is defective. First, draw a tree diagram using the given information, as in Figure 15. The steps leading to event E are shown in heavy type.

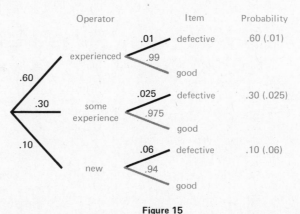

Figure 15

Since we want $P(F_3|E)$, use the bottom branch of the tree in Figure 15; divide the probability for this branch by the sum of the probabilities of all the branches leading to E, or

$$P(F_3|E) = \frac{.10(.06)}{.60(.01) + .30(.025) + .10(.06)} = \frac{.006}{.0195} = \frac{4}{13}.$$

In a similar way

$$P(F_2|E) = \frac{.30(.025)}{.60(.01) + .30(.025) + .10(.06)} = \frac{.0075}{.0195} = \frac{5}{13}.$$

Finally, $P(F_1|E) = 4/13$. Check that $P(F_1|E) + P(F_2|E) + P(F_3|E) = 1$. (That is, the defective item was made by *someone*.) ∎

Example 3 A manufacturer buys items from six different suppliers. The fraction of the total number of items that is obtained from each supplier, along with the probability that an item purchased from that supplier is defective, is shown in the following chart.

Supplier	Fraction of total supplied	Probability of defective
1	.05	.04
2	.12	.02
3	.16	.07
4	.23	:01
5	.35	.03
6	.09	.05

Find the probability that a defective item came from supplier 5.

Let F_1 be the event that an item came from supplier 1, with F_2, F_3, F_4, F_5, and F_6 defined in a similar manner. Let E be the event that an item is defective. We want to find $P(F_5|E)$. By Bayes' Formula (draw a tree),

$$P(F_5|E) = \frac{(.35)(.03)}{(.05)(.04) + (.12)(.02) + (.16)(.07) + (.23)(.01) + (.35)(.03) + (.09)(.05)}$$

$$= \frac{.0105}{.0329} \approx .319.$$

There is about a 32% chance that a defective item came from supplier 5. ∎

6.6 Exercises *For certain events, M and N, P(M) = .4, P(N|M) = .3, and P(N|M') = .4. Find each of the following.*

1. $P(M|N)$ **2.** $P(M'|N)$

For mutually exclusive events R_1, R_2, R_3, we have $P(R_1) = .05$, $P(R_2) = .6$, and $P(R_3) = .35$. Also, $P(Q|R_1) = .40$, $P(Q|R_2) = .30$, and $P(Q|R_3) = .60$. Find each of the following.

3. $P(R_1|Q)$ **4.** $P(R_2|Q)$ **5.** $P(R_3|Q)$ **6.** $P(R_1'|Q)$

Suppose you have three jars with the following contents: 2 black balls and 1 white ball in the first; 1 black ball and 2 white balls in the second; and 1 black ball and 1 white ball in the third. If the probability of selecting one of the three jars is 1/2, 1/3, and 1/6, respectively, find the probability that if a white ball is drawn, it came from

7. the second jar   **8.** the third jar.

The following table shows the fraction of the population in various income levels, as well as the probability that a person from that income level will take an airline flight within the next year.

Income level	Proportion of population	Probability of a flight during the next year
$0–$5999	12.8%	.04
$6000–$9999	14.6%	.06
$10,000–$14,999	18.5%	.07
$15,000–$19,999	17.8%	.09
$20,000–$24,999	13.9%	.12
$25,000 and over	22.4%	.13

If a person is selected at random from an airline flight, find the probability that the person has an income level of

9. $10,000–$14,999;   **10.** $25,000 and over.

The following table shows the proportion of people over 18 who are in various age categories, along with the probability that a person in a given age category will vote in a general election.

Age	Proportion of voting age population	Probability of a person of this age voting
18–21	11.0%	.48
22–24	7.6%	.53
25–44	37.6%	.68
45–64	28.3%	.64
65 or over	15.5%	.74

Suppose a voter is picked at random. Find the probability that the voter is in the following age categories.

11. 18–21   **12.** 65 or over

Of all the people applying for a certain job, 70% are qualified, and 30% are not. The personnel manager claims that she approves qualified people 85% of the time; she approves an unqualified person 20% of the time. Find each of the following probabilities.

13. A person is qualified if he or she was approved by the manager.

14. A person is unqualified if he or she was approved by the manager.

A building contractor buys 70% of his cement from supplier A, and 30% from supplier B. A total of 90% of the bags from A arrive undamaged, while 95% of the bags from B come undamaged. Give the probability that a damaged bag is from supplier

15. A;   **16.** B.

The probability that a customer of a local department store will be a "slow pay" is .02. The probability that a "slow pay" will make a large down payment when buying a refrigerator is .14. The probability that a person who is not a "slow pay" will make a large down payment when buying a refrigerator is .50. Suppose a customer makes a large down payment on a refrigerator. Find the probability that the customer is

17. a "slow pay;"   **18.** not a "slow pay."

Companies A, B, and C produce 15%, 40%, and 45% respectively of the major appliances sold in a certain area. In that area, 1% of the Company A appliances, 1 1/2% of the Company B appliances, and 2% of the Company C appliances need service within the first year. Suppose a defective appliance is chosen at random; find the probability that it was manufactured by Company

19. A;   **20.** B.

On a given weekend in the fall, a tire company can buy television advertising time for a college football game, a baseball game, or a professional football game. If the company sponsors the college game, there is a 70% chance of a high rating, a 50% chance if they sponsor a baseball game, and a 60% chance if they sponsor a professional football game. The probability of the company sponsoring these various games is .5, .2, and .3 respectively. Suppose the company does get a high rating; find the probability that it sponsored

21. a college game;   **22.** a professional football game.

According to readings in business publications, there is a 50% chance of a booming economy next summer, a 20% chance of a mediocre economy, and a 30% chance of a recession. The probabilities that a particular investment strategy will produce a huge profit under each of these possibilities are .1, .6, and .3 respectively. Suppose it turns out that the strategy does produce huge profits; find the probability that the economy was

23. booming;   **24.** in recession.

25. The probability that a person with certain symptoms has hepatitis is .8. The blood test used to confirm this diagnosis gives positive results for 90% of those who have the disease and 5% of those without the disease. What is the probability that an individual with the symptoms who reacts positively to the test has hepatitis?

A recent issue of Newsweek described a new test for toxemia, a disease that affects pregnant women. To perform the test, the woman lies on her left side and then rolls over on her back. The test is considered positive if there is a 20 mm rise in her blood pressure within one minute. The article gives the following probabilities, where T represents having toxemia at some time during the pregnancy, and N represents a negative test.

$$P(T'|N) = .90, \quad \text{and} \quad P(T|N') = .75.$$

Assume that $P(N') = .11$, and find each of the following.

26. $P(N|T)$ **27.** $P(N'|T)$

28. In a certain county, the Democrats have 53% of the registered voters, 12% of whom are under 21. The Republicans have 47% of all registered voters, of whom 10% are under 21. If Kay is a registered voter who is under 21, what is the probability that she is a Democrat?

Let F, G, and H be nonempty events with $F \cap G = \varnothing$, $F \cap H = \varnothing$, and $G \cap H = \varnothing$. Let $S = F \cup G \cup H$. Let E be any event. Prove each of the following.

29. $P(E) = P(E \cap F) + P(E \cap G) + P(E \cap H)$

30. $P(E) = P(E|F) \cdot P(F) + P(E|G) \cdot P(G) + P(E|H) \cdot P(H)$

> ## Application   *Medical Diagnosis*

When a patient is examined, information, typically incomplete, is obtained about his state of health. Probability theory provides a mathematical model appropriate for this situation, as well as a procedure for quantitatively interpreting such partial information to arrive at a reasonable diagnosis.[1]

To do this, we list the states of health that can be distinguished in such a way that the patient can be in one and only one state at the time of the examination. For each state of health H, we associate a number $P(H)$ between 0 and 1 such that the sum of all these numbers is 1. This number $P(H)$ represents the probability, before examination, that a patient is in the state of health H, and $P(H)$ may be chosen subjectively from medical experience, using any informa-

---

[1]This example is based on "Probabilistic Medical Diagnosis," Roger Wright, from *Some Mathematical Models in Biology*, Robert M. Thrall, ed., rev. ed., (The University of Michigan, 1967), by permission of Robert M. Thrall.

tion available prior to the examination. The probability may be most convenient-
ly established from clinical records, that is, a mean probability is established
for patients in general, although the number would vary from patient to patient.
Of course, the more information that is brought to bear in establishing $P(H)$, the
better the diagnosis.

For example, limiting the discussion to the condition of a patient's heart,
suppose there are exactly 3 states of health, with probabilities as follows:

	State of health H	$P(H)$
H_1	patient has a normal heart	.8
H_2	patient has minor heart irregularities	.15
H_3	patient has a severe heart condition	.05

Having selected $P(H)$, the information of the examination is processed.
First, the results of the examination must be classified. The examination itself
consists of observing the state of a number of characteristics of the patient.
Let us assume that the examination for a heart condition consists of a stethoscope examination and a cardiogram. The outcome of such an examination, C,
might be one of the following:

C_1 — stethoscope shows normal heart
and cardiogram shows normal heart;

C_2 — stethoscope shows normal heart
and cardiogram shows minor irregularities,

and so on.

It remains to assess for each state of health H the conditional probability
$P(C|H)$ of each examination outcome C using only the knowledge that a patient
is in a given state of health. (This may be based on the medical knowledge and
clinical experience of the doctor.) The conditional probabilities $P(C|H)$ will
not vary from patient to patient, so that they may be built into a diagnostic system, although they should be reviewed periodically.

Suppose the result of the examination is C_1. Let us assume the following
probabilities:

$$P(C_1|H_1) = .9$$
$$P(C_1|H_2) = .4$$
$$P(C_1|H_3) = .1.$$

Now, for a given patient, the appropriate probability associated with each state
of health H, after examination, is $P(H|C)$ where C is the outcome of the exami-
nation. This can be calculated by using Bayes' Theorem. For example, to find
$P(H_1|C_1)$ — that is, the probability that the patient has a normal heart given that
the examination showed a normal stethoscope examination and a normal cardiogram — we use Bayes' Theorem as follows:

$$P(H_1|C_1) = \frac{P(C_1|H_1)P(H_1)}{P(C_1|H_1)P(H_1) + P(C_1|H_2)P(H_2) + P(C_1|H_3)P(H_3)}$$

$$= \frac{(.9)(.8)}{(.9)(.8) + (.4)(.15) + (.1)(.05)} \approx .92.$$

Hence, the probability is about .92 that the patient has a normal heart on the basis of the examination results. This means that in 8 out of 100 patients, some abnormality will be present and not be detected by the stethoscope or the cardiogram.

Exercises   **1.** Find $P(H_2|C_1)$.

2. Assuming the following probabilities, find $P(H_1|C_2)$:

$$P(C_2|H_1) = .2, \qquad P(C_2|H_2) = .8, \qquad P(C_2|H_3) = .3.$$

3. Assuming the probabilities of Exercise 2, find $P(H_3|C_2)$.

6.7   Bernoulli Trials

Many probability problems are concerned with experiments in which an event is repeated many times. For example, we might want to find the probability of getting 7 heads in 8 tosses of a coin, or hitting a target 6 times out of 6, or finding 1 defective item in a sample of 15 items. Probability problems of this kind are called **repeated trials** problems, or **Bernoulli processes.** In each case, some outcome is designated a success, and any other outcome is considered a failure. Thus, if the probability of a success in a single trial is p, the probability of failure will be $1 - p$. Repeated trials problems, or *binomial problems* must satisfy the following conditions.

> 1. The same experiment is repeated several times.
> 2. There are only two possible outcomes, success and failure.
> 3. The repeated trials are independent.
> 4. The probability of each outcome remains the same for each trial.

Let us consider the solution of a problem of this type. Suppose we want to find the probability of getting 5 ones on 5 rolls of a die. The probability of getting a one on 1 roll is 1/6, while the probability of any other result is 5/6.

$$P(5 \text{ ones on } 5 \text{ rolls}) = P(1) \cdot P(1) \cdot P(1) \cdot P(1) \cdot P(1) = \left(\frac{1}{6}\right)^5$$

$$\approx .00013$$

Now, let us find the probability of getting a one exactly 4 times in 5 rolls of the die. The desired outcome for this experiment can occur in more than one way, as shown below, where s represents getting a success (a one), and f represents getting a failure (any other result).

$$
\begin{array}{ccccc}
s & s & s & s & f \\
s & s & s & f & s \\
s & s & f & s & s \\
s & f & s & s & s \\
f & s & s & s & s
\end{array}
$$

The probability of each of these five outcomes is

$$\left(\frac{1}{6}\right)^4\left(\frac{5}{6}\right).$$

Since the five outcomes represent mutually exclusive alternative events, we add the five probabilities.

$$P(4 \text{ ones in 5 rolls}) = 5\left(\frac{1}{6}\right)^4\left(\frac{5}{6}\right) = \frac{5^2}{6^5} \approx .0032$$

In the same way, we can compute the probability of rolling a one exactly 3 times in 5 rolls of a die. The probability of 3 successes and 2 failures will be

$$\left(\frac{1}{6}\right)^3\left(\frac{5}{6}\right)^2.$$

Again the desired outcome can occur in more than one way. Let the set $\{1, 2, 3, 4, 5\}$ represent the first, second, third, fourth, and fifth tosses. The number of 3-element subsets of this set will correspond to the number of ways in which 3 successes and 2 failures can occur. Using combinations, there are $\binom{5}{3}$ such subsets. Since $\binom{5}{3} = 5!/(3!\,2!) = 10$, we have

$$P(3 \text{ ones in 5 rolls}) = 10\left(\frac{1}{6}\right)^3\left(\frac{5}{6}\right)^2 = \frac{250}{6^5} \approx .032.$$

Suppose now that the probability of a success on one trial of a Bernoulli experiment is p, and the probability of exactly x successes in n repeated trials is needed. It is possible that the x successes could come first, followed by $n - x$ failures:

$$\underbrace{s \quad s \quad s \cdots s} \quad \underbrace{f \quad f \cdots f}. \tag{1}$$
$$x \text{ successes, then } n - x \text{ failures}$$

The probability of this result is

$$P(s\,s\,s \cdots s\,s\,f\,f \cdots f)$$
$$= \underbrace{P(s) \cdot P(s) \cdot P(s) \cdots P(s)}_{x \text{ factors}} \cdot \underbrace{P(f) \cdot P(f) \cdots P(f)}_{n - x \text{ factors}}$$
$$= \underbrace{p \cdot p \cdot p \cdots p}_{x \text{ factors}} \cdot \underbrace{(1 - p) \cdot (1 - p) \cdots (1 - p)}_{n - x \text{ factors}}$$
$$= p^x(1 - p)^{n-x}.$$

The x successes could also be obtained by rearranging the letters in (1) above. There are $\binom{n}{x}$ ways of choosing the x places where the s's occur, and the $n - x$

places where the f's occur. Thus, the probability of exactly x successes is

$$\binom{n}{x}p^x(1-p)^{n-x}.$$

In summary,

> If p is the probability of success in a single trial of a Bernoulli experiment, the probability of x successes and $n-x$ failures in n independent repeated trials of the experiment is
>
> $$\binom{n}{x} \cdot p^x \cdot (1-p)^{n-x}.$$

Example 1 The advertising agency which handles the Diet Supercola account thinks that 40% of all consumers prefer this product over its competitors. Suppose a random sample of six people is chosen. Assume that all responses are independent of each other. Find the probability of the following.

(a) Exactly 3 of the 6 people prefer Diet Supercola.
In this example, $P(\text{success}) = P(\text{prefer Diet Supercola}) = .4$. The sample is made up of six people, so $n = 6$. To find the probability that exactly 3 people prefer this drink, let $x = 3$.

$$P(\text{exactly } 3) = \binom{6}{3}(.4)^3(1 - .4)^{6-3}$$

$$= 20(.4)^3(.6)^3$$

$$= 20(.064)(.216)$$

$$= .27648$$

(b) None of the 6 people prefer Diet Supercola.
Let $x = 0$.

$$P(\text{exactly } 0) = \binom{6}{0}(.4)^0(1 - .4)^6 = 1(1)(.6)^6 \approx .0467 \quad \blacksquare$$

Example 2 At a certain school in northern Michigan, 80% of the students ski. If five students at this school are selected, and their responses are independent, then the probability that exactly one of the five students skis is

$$P(\text{exactly } 1) = \binom{5}{1}(.8)^1(.2)^4 = .0064, \quad \blacksquare$$

while the probability that exactly four of the five students ski is

$$P(\text{exactly } 4) = \binom{5}{4}(.8)^4(.2)^1 = .4096.$$

Example 3 Find each of the following probabilities.

(a) the probability of getting exactly seven heads in eight tosses of a fair coin
The probability of success, getting a head in a single toss, is 1/2. The probability of a failure, getting a tail, is $1 - 1/2 = 1/2$. Thus,

$$P(\text{7 heads in 8 tosses}) = \binom{8}{7}\left(\frac{1}{2}\right)^7\left(\frac{1}{2}\right)^1 = 8\left(\frac{1}{2}\right)^8 = .03125.$$

(b) the probability of 2 fours in 8 rolls of a die
The probability of success, a 4, is 1/6, while the probability of failure (a number other than 4), is 5/6.

$$P(2 \text{ fours in 8 tosses}) = \binom{8}{2}\left(\frac{1}{6}\right)^2\left(\frac{5}{6}\right)^6 \approx .2605 \quad \blacksquare$$

Example 4 Assuming that selection of items for a sample can be treated as independent trials, find the probability of the occurrence of one defective item in a random sample of 15 items from a production line, if the probability that any one item is defective is .01.

The probability of success (a defective item), is .01, while the probability of failure (an acceptable item) is .99. Thus,

$$P(1 \text{ defective in 15 items}) = \binom{15}{1}(.01)^1(.99)^{14}$$

$$= 15(.01)(.99)^{14}$$

$$\approx .130. \quad \blacksquare$$

Example 5 A new style of shoe is sweeping the country. In one area, 30% of all the shoes are of this type. Assume that these sales are independent events, and find the following probabilities.

(a) Out of 10 customers in a shoe store, at least 8 buy the new shoe style.
Let success be "buy the new style", so that $P(\text{success}) = .3$. For at least 8 people out of 10 to buy the shoe, it must be sold to 8, 9, or 10 people. Thus,

$$P(\text{at least } 8) = P(8) + P(9) + P(10)$$

$$= \binom{10}{8}(.3)^8(.7)^2 + \binom{10}{9}(.3)^9(.7)^1 + \binom{10}{10}(.3)^{10}(.7)^0$$

$$\approx .0014467 + .0001378 + .0000059$$

$$= .0015904.$$

(b) Out of 10 customers in a shoe store, no more than 7 buy the new shoe style.
"No more than 7" means 0, 1, 2, 3, 4, 5, 6, or 7 people buy the shoe. We could add $P(0)$, $P(1)$, and so on, but it is easier to use the formula $P(E) = 1 - P(E')$. The complement of "no more than 7" is "8 or more." Thus,

$$P(\text{no more than } 7) = 1 - P(8 \text{ or more})$$

$$= 1 - .0015904 \quad \text{(answer from part (a))}$$

$$= .9984096. \quad \blacksquare$$

The Probability of k Trials for m Successes   In the rest of this section, we will find the probability that k trials will be needed to guarantee m successes in a Bernoulli experiment. Suppose that a salesperson in a very competitive business makes a sale in one client visit out of five, so $P(\text{sale}) = .2$. The probability of a sale on the first call is .2. The probability that the *first* sale will be on the *second* call is

$$P(\text{no sale on first}) \cdot P(\text{sale on second}) = .8(.2) = .16.$$

The probability that the *first* sale will be on the *third* call is

$$(.8)^2(.2) = .128.$$

In general, the probability that the first sale will be on the k-th call is

$$(.8)^{k-1}(.2).$$

Example 6   How many calls must this salesperson make to have an 80% chance of making a sale?

There is a .2 chance of making a sale on the first call, a .16 chance of making the first sale on the second call, a .128 chance of making the first sale on the third call, and so on. The probability of a sale by the k-th call is the sum of all the probabilities of sales on calls 1, 2, 3, $\cdots$, k. A calculator gives the results of the following table.

Call number	Probability that first sale is on that call	Total of all probabilities up to and including this call
1	$(.8)^0(.2) = .2$	.2
2	$(.8)^1(.2) = .16$	.36
3	$(.8)^2(.2) = .128$	.488
4	$(.8)^3(.2) = .1024$	.5904
5	$(.8)^4(.2) \approx .082$	$\approx .672$
6	$(.8)^5(.2) \approx .066$	$\approx .738$
7	$(.8)^6(.2) \approx .052$	$\approx .790$
8	$(.8)^7(.2) \approx .042$	$\approx .832$

The salesperson must make eight calls to have an 80% chance of making one sale. ∎

We can generalize this result: let p be the probability of success on one trial in a Bernoulli experiment. Then to find the probability that k trials will be needed to guarantee m successes, we must assume the k-th trial was a success, and that $m - 1$ successes were distributed in some order among the other $k - 1$ trials. The desired probability is thus

$$\left[\binom{k-1}{m-1} p^{(m-1)} \cdot (1-p)^{(k-1)-(m-1)} \right] \cdot p$$

or

$$\binom{k-1}{m-1} p^m \cdot (1-p)^{k-m}.$$

Example 7   Find the probability that the salesperson of Example 6 will require 9 calls to make 3 sales.

Let $k = 9$ and $m = 3$. We know that $p = .2$. The desired probability is

$$\binom{9-1}{3-1}(.2)^3(1 - .2)^{9-3} = \binom{8}{2}(.2)^3(.8)^6 \approx .0587. \quad ∎$$

6.7 Exercises   *Suppose that a family has 5 children. Also, suppose that the probability of having a girl is 1/2. Find the probability that the family will have*

1. exactly 2 girls;
2. exactly 3 girls;
3. no girls;
4. no boys;
5. at least 4 girls;
6. at least 3 boys;
7. no more than 3 boys;
8. no more than 4 girls.

A die is rolled 12 times. Find the probability of rolling

9. exactly 12 ones;
10. exactly 6 ones;
11. exactly 1 one;
12. exactly 2 ones;
13. no more than 3 ones;
14. no more than 1 one.

A coin is tossed 5 times. Find the probability of getting

15. all heads;
16. exactly 3 heads;
17. no more than 3 heads;
18. at least 3 heads.

A factory tests a random sample of 20 transistors for defectives. The probability that a particular transistor will be defective has been established by past experience to be .05.

19. What is the probability that there are no defectives in the sample?
20. What is the probability that the number of defectives in the sample is at most 2?

A company gives prospective employees a 6-question multiple-choice test. Each question has 5 possible answers, so that there is a 1/5 or 20% chance of answering a question correctly just by guessing. Find the probability of answering, by chance,

21. exactly 2 questions correctly;
22. no questions correctly;
23. at least 4 correctly;
24. no more than 3 correctly.
25. Five out of the fifty clients of a certain stockbroker will lose their life savings as a result of his advice. Find the probability that in a sample of 3 clients, exactly 1 loses all his money. (Assume independence).

According to a recent article in a business publication, only 20% of the population of the United States has never had a Big Mac hamburger at McDonalds. Assume independence and find the probability that in a random sample of 10 people

26. exactly 2 never had a Big Mac;
27. exactly 5 never had a Big Mac;
28. 3 or fewer never had a Big Mac;
29. 4 or more *have* had a Big Mac.

A new drug cures 70% of the people taking it. Suppose 20 people take the drug; find the probability that

30. exactly 18 are cured;
31. exactly 17 are cured;
32. at least 17 are cured;
33. at least 18 are cured.

In a 10-question multiple-choice biology test with 5 choices for each question, a student who did not prepare guesses on each item. Find the probability that he answers

34. exactly 6 questions correctly;
35. exactly 7 correctly;
36. at least 8 correctly;
37. less than 8 correctly.

Assume that the probability that a person will die within a month after a certain operation is 20%. Find the probability that in 3 such operations

38. all 3 people survive;

39. exactly 1 person survives;

40. at least 2 people survive;

41. no more than 1 person survives.

Six mice from the same litter, all suffering from a vitamin A deficiency are fed a certain dose of carrots. If the probability of recovery under such treatment is .70, find the probability that

42. none recover;

43. exactly 3 of the 6 recover;

44. all recover;

45. no more than 3 recover.

46. In an experiment on the effects of a radiation dose on cells, a beam of radioactive particles is aimed at a group of 10 cells. Find the probability that 8 of the cells will be hit by the beam, if the probability that any single cell will be hit is .6. (Assume independence.)

47. The probability of a mutation of a given gene under a dose of 1 roentgen of radiation is approximately 2.5×10^{-7}. What is the probability that in 10,000 genes, at least 1 mutation occurs?

48. A new drug being tested causes a serious side effect in 5 out of 100 patients. What is the probability that no side effects occur in a sample of 10 patients taking the drug?

An economist feels that the probability that a person at a certain income level will buy a new car this year is .2. Find the probability that among 12 such people,

49. exactly 4 buy a new car;

50. exactly 6 buy a new car;

51. no more than 3 buy a new car;

52. at least 3 buy a new car.

Find the probability that the following numbers of tosses of a fair coin will be required to obtain three heads.

53. 5   **54.** 6   **55.** 8   **56.** 10

Find the probability that the following numbers of rolls of a fair die will be required to get 4 fives.

57. 6   **58.** 10   **59.** 12   **60.** 16

The probability that a given exploration team sent out by a mining company will find commercial quantities of iron ore is .15. How many such teams must the company send out to have the following probabilities of finding ore?

61. 60%   **62.** 75%   **63.** 80%

64. Suppose we find the probability of r successes out of n trials for a Bernoulli experiment having probability p. Show that the result is the same as for the probability of $n - r$ successes out of n trials for a Bernoulli experiment having probability $1 - p$.

Key Words

deterministic phenomena	event
random phenomena	simple event
experiment	certain event

trial	impossible event
outcome	mutually exclusive events
sample space	probability of an event
probability distribution	stochastic processes
addition principle	independent events
complement of an event	dependent events
odds	Bayes' formula
subjective probability	repeated trials
conditional probability	Bernoulli experiments
tree diagram	binomial problems
product rule	

Chapter 6   Review Exercises

Write sample spaces for the following.

1. a die is rolled

2. a card is drawn from a deck containing only the thirteen spades

3. the weight of a person is measured to the nearest half pound; the scale will not measure more than 300 pounds

4. a coin is tossed four times

An urn contains five balls labeled 3, 5, 7, 9, and 11, while a second urn contains four red and two green balls. An experiment consists of pulling one ball from each urn, in turn. Write each of the following.

5. the sample space

6. event *E,* the first ball is greater than 5

7. event *F,* the second ball is green

8. Are the outcomes in the sample space equally likely?

A company sells typewriters and copiers. Let E be the event "a customer buys a typewriter," and let F be the event "a customer buys a copier." Write each of the following using ∩, ∪, or ', as necessary.

9. A customer buys neither

10. A customer buys at least one

When a single card is drawn from an ordinary deck, find the probability that it will be

11. a heart;

12. a red queen;

13. a face card;

14. black or a face card;

15. red, given it is a queen;

16. a jack, given it is a face card;

17. a face card, given it is a king.

Find the odds in favor of a card drawn from an ordinary deck being

18. a club   **19.** a black jack   **20.** a red face card or a queen

A sample shipment of five swimming pool filters is chosen at random. The probability of exactly 0, 1, 2, 3, 4, or 5 filters being defective is given in the following table.

Number defective	0	1	2	3	4	5
Probability	.31	.25	.18	.12	.08	.06

Find the probability that the following number of filters is defective.

21. no more than 3   **22.** at least 3

The square shows the four possible (equally likely) combinations when both parents are carriers of the sickle cell anemia trait. Each carrier parent has normal cells (N) and trait cells (T).

	2nd parent	
	N_2	T_2
1st parent N_1		N_1T_2
T_1		

23. Complete the table.

24. If the disease occurs only when two trait cells combine, find the probability that a child born to these parents will have sickle cell anemia.

25. The child will carry the trait but not have the disease if a normal cell combines with a trait cell. Find this probability.

26. Find the probability that the child is neither a carrier nor has the disease.

Find the probability for the following sums when two fair dice are rolled.

27. 8   **28.** 0   **29.** at least 10   **30.** no more than 5

31. odd and greater than 8   **32.** 12, given it is greater than 10

33. 7, given that one die is 4   **34.** at least 9, given that one die is 5

Suppose $P(E) = .51$, $P(F) = .37$, and $P(E \cap F) = .22$. Find each of the following probabilities.

35. $P(E \cup F)$ **36.** $P(E \cap F')$ **37.** $P(E' \cup F)$ **38.** $P(E' \cap F')$

A basket contains 4 black, 2 blue, and 5 green balls. A sample of 3 balls is drawn. Find the probability that the sample contains

39. all black balls;   **40.** all blue balls;

41. 2 black balls and 1 green ball;   **42.** 2 black balls;

43. 2 green and 1 blue ball;   **44.** 1 blue ball.

Suppose two cards are drawn without replacement from an ordinary deck of 52. Find the probability that

45. both cards are red;   **46.** both cards are spades;

47. at least one card is a spade;

48. the second card is red given that the first card was a diamond;

49. the second card is a face card, given that the first card was not;

50. the second card is a five, given that the first card was the five of diamonds.

The table below shows the results of a survey of 1000 *new or used car buyers of a certain model car.*

	Satisfied	Not satisfied	Totals
New	300	100	400
Used	450	150	600
Totals	750	250	1000

Let S represent the event "satisfied", and N the event "bought a new car." Find each of the following.

51. $P(N \cap S)$ **52.** $P(N \cup S')$ **53.** $P(N|S)$

54. $P(N'|S)$ **55.** $P(S|N')$ **56.** $P(S'|N')$

Of the appliance repair shops listed in the phone book, 80% are competent and 20% are not. A competent shop can repair an appliance correctly 95% of the time; an incompetent shop can repair an appliance correctly 60% of the time. Suppose an appliance was repaired correctly. Find the probability that it was repaired by

57. a competent shop;   **58.** an incompetent shop.

Suppose an appliance was repaired incorrectly. Find the probability that it was repaired by

59. a competent shop;   **60.** an incompetent shop.

61. Box A contains 5 red balls and 1 black ball; box B contains 2 red and 3 black balls. A box is chosen at random, and a ball is selected from it. The probability of choosing box A is 3/8. If the selected ball is black, what is the probability that it came from box A?

62. Find the probability that the ball in Exercise 61 came from box B, given that it is red.

Suppose a family plans six children, and the probability that a particular child is a girl is 1/2. Find the probability that the family will have

63. exactly 3 girls;   **64.** all girls;

65. at least 4 girls;   **66.** no more than 2 boys.

A certain machine used to manufacture screws produces a defective rate of .01. A random sample of 20 screws is selected. Find the probability that the sample contains

67. exactly 4 defective screws;   **68.** exactly 3 defective screws;

69. no more than 4 defective screws.

70. *Set up* the probability that the sample has 12 or more defective screws. (Do not evaluate.)

71. An oil company finds oil with 14% of the wells that it drills. How many wells must the company drill to have the following probabilities of finding oil?

   (a) 2/3   **(b)** 3/4

72. *Randomized Response Method for Getting Honest Answers to Sensitive Questions.*[1] Basically, this is a method to guarantee an individual that answers to sensitive questions will be anonymous, thus encouraging a truthful response. This method is, in effect, an application of the formula for finding the probability of an intersection and operates as follows. Two questions A and B are posed, one of which is sensitive and the other not. The probability of receiving a "yes" to the nonsensitive question must be known. For example, one could ask

A: Does your Social Security number end in an odd digit? (Nonsensitive)
B: Have you ever intentionally cheated on your income taxes? (Sensitive)

We know that P (answer yes | answer A) = 1/2. We wish to approximate P(answer yes | answer B). The subject is asked to flip a coin and answer A if the coin comes up heads and otherwise to answer B. In this way, the interviewer does not know which question the subject is answering. Thus, a "yes" answer is not incriminating. There is no way for the interviewer to know whether the subject is saying "Yes, my Social Security number ends in an odd digit" or "Yes, I have intentionally cheated on my income taxes." The percentage of subjects in the group answering "yes" is used to approximate P (answer yes).

(a) Use the fact that the event "answer yes" is the union of the event "answer yes and answer A" with the event "answer yes and answer B" to prove that

P(answer yes | answer B)

$$= \frac{P(\text{answer yes}) - P(\text{answer yes} \mid \text{answer A}) \cdot P(\text{answer A})}{P(\text{answer B})}$$

(b) If this technique is tried on 100 subjects and 60 answered "yes," what is the approximate probability that a person randomly selected from the group has intentionally cheated on income taxes?

From *Applied Statistics With Probability* by J. S. Milton and J. J. Corbet. Copyright © 1979 by Litton Educational Publishing, Inc. Reprinted by permission of D. Van Nostrand Company.

Markov Chains

In Chapter 6 we briefly studied *stochastic processes,* where the outcome of an experiment depends on the outcome of previous experiments. In this chapter we study a special type of stochastic process called a **Markov chain;** here the outcome of an experiment depends only on the outcome of the previous experiment. That is, given the present state of the system, future states are independent of past states. It turns out that such experiments are common enough in applications to make their study worthwhile. Markov chains are named after the Russian mathematician A. A. Markov, 1856–1922, who started the theory of stochastic processes.

7.1 Basic Properties of Markov Chains

Transition Matrix   In sociology, it is convenient to classify people by income as *lower class, middle class,* and *upper class.* The strongest determinant of the income class of an individual turns out to be the income class of the individual's parents. For example, if we say that an individual in the lower income class is in *state 1,* an individual in the middle income class is in *state 2,* and an individual in the upper income class is in *state 3,* then we might have the following probabilities of change in income class from one generation to the next.

		Next generation		
	State	1	2	3
Current	1	.65	.28	.07
generation	2	.15	.67	.18
	3	.12	.36	.52

This chart shows that if an individual is in state 1 (lower income class) then there is a probability of .65 that any offspring will be in the lower class, a probability of .28 that offspring will be in the middle class, and a probability of .07 that offspring will be in the upper class.

The symbol p_{ij} will be used for the probability of transition from state i to state j, in one generation. For example, p_{23} represents the probability that a person in state 2 will have offspring in state 3; from the table above,

$$p_{23} = .18.$$

Also from the table, $p_{31} = .12$, $p_{22} = .67$, and so on.

The table above can be written as a matrix, with the states indicated at the side and top; this matrix is called a **transition matrix.** If P represents the transition matrix for the table above, then

$$P = \begin{array}{c} \\ 1 \\ 2 \\ 3 \end{array} \begin{array}{ccc} 1 & 2 & 3 \\ \left[\begin{array}{ccc} .65 & .28 & .07 \\ .15 & .67 & .18 \\ .12 & .36 & .52 \end{array}\right] \end{array}.$$

A transition matrix has several features:
  (a) it is square, since all possible states must be used both as rows and as columns;
  (b) all entries are between 0 and 1, inclusive; this is because all entries represent probabilities;
  (c) the sum of the entries in any row must be 1, since the numbers in the row give the probability of changing from the state at the left to one of the states indicated across the top.

Markov Chains   A transition matrix, such as matrix P above, also shows two of the key features of a *Markov chain:*
  (a) the outcome of each experiment is one of a set of discrete states;
  (b) the outcome of an experiment depends only on the present state, and not on any past states.

For example, in the transition matrix above, a person is assumed to be in one of three discrete states (lower, middle, or upper class) with any offspring in one of these same three discrete states. We assumed that the probability that an offspring is in states 1, 2, or 3 depends only on the social class of the parents.

Example 1   A small town has only two drycleaners, Johnson and NorthClean. Johnson's manager desires to increase the firm's market share by an extensive advertising campaign. After the campaign, a market research firm finds that there is a probability of .8 that a customer of Johnson's will bring their next batch of dirty items to Johnson, and a .35 chance that a NorthClean customer will switch to Johnson for their next batch. Write a transition matrix showing this information.

Here we must assume that the probability that a customer comes to a given cleaners depends only on where the last load of clothes was taken. If there is an .8 chance that a Johnson customer will return to Johnson, then there must be a

$1 - .8 = .2$ chance that the customer will switch to NorthClean. In the same way, there is a $1 - .35 = .65$ chance that a NorthClean customer will return to North-Clean. These probabilities give the following transition matrix.

$$
\begin{array}{c}
 & \textit{Second load} \\
 & \text{Johnson} \quad \text{NorthClean} \\
\textit{First load} \quad \begin{array}{c} \text{Johnson} \\ \text{NorthClean} \end{array} \left[\begin{array}{cc} .8 & .2 \\ .35 & .65 \end{array} \right]
\end{array}
$$

We shall come back to this transition matrix later in this section (See Example 4). ∎

Look again at the transition matrix for social class changes,

$$
P = \begin{array}{c} 1 \\ 2 \\ 3 \end{array} \overset{\begin{array}{ccc} 1 & 2 & 3 \end{array}}{\left[\begin{array}{ccc} .65 & .28 & .07 \\ .15 & .67 & .18 \\ .12 & .36 & .52 \end{array} \right]}.
$$

This matrix shows the probability of change in social class from one generation to the next. Now let us investigate the probabilities for changes in social class over *two* generations. For example, if a parent is upper class (state 3), what is the probability that a grandchild will be in state 2?

To find out, start with a tree diagram as shown in Figure 1; the various probabilities come from the transition matrix above.

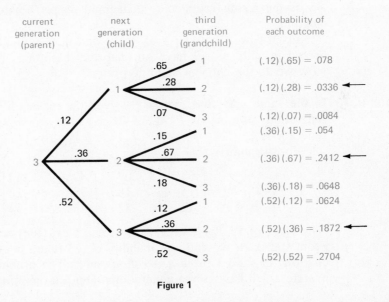

Figure 1

The arrows point to the outcomes "grandchild in state 2"; the grandchild can get to state 2 after having had parents in either state 1, state 2, or state 3. The probability that a parent in state 3 will have a grandchild in state 2 is given by the sum of the probabilities indicated with arrows, or

$$.0336 + .2412 + .1872 = .4620.$$

We used p_{ij} to represent the probability of changing from state i to state j in one generation. We can use this notation to write the probability that a parent in state 3 will have a grandchild in state 2:

$$p_{31} \cdot p_{12} + p_{32} \cdot p_{22} + p_{33} \cdot p_{32}.$$

This sum of products of probabilities should remind you of matrix multiplication—it is nothing more than one step in the process of multiplying matrix P by itself. In particular, it is row 3 of P times column 2 of P. Using P^2 to represent the matrix product $P \cdot P$, then P^2 gives the probabilities of a transition from one state to another in *two* repetitions of an experiment. Generalizing,

> P^k gives the probabilities of a transition from one state to another in k repetitions of an experiment.

Example 2   For our transition matrix P of social class changes,

$$P^2 = \begin{bmatrix} .65 & .28 & .07 \\ .15 & .67 & .18 \\ .12 & .36 & .52 \end{bmatrix} \cdot \begin{bmatrix} .65 & .28 & .07 \\ .15 & .67 & .18 \\ .12 & .36 & .52 \end{bmatrix} \approx \begin{bmatrix} .47 & .39 & .13 \\ .22 & .56 & .22 \\ .19 & .46 & .34 \end{bmatrix}$$

(Here we rounded the numbers in the product to match the number of decimals in matrix P.) The entry in row 3, column 2 of P^2 gives the probability that a person in state 3 will have an offspring in state 2 two generations later. This number, .46, is the result (rounded to 2 decimal places) found through use of the tree diagram.

From row 1, column 3 of P^2 we find the number .13, the probability that a person in state 1 will have an offspring in state 3, but two generations later. How would the entry .47 be interpreted?   ∎

Example 3   In the same way that matrix P^2 gives the probability of transitions after *two* generations, the matrix $P^3 = P \cdot P^2$ gives the probabilities of change after *three* generations.

For our matrix P,

$$P^3 = P \cdot P^2 = \begin{bmatrix} .65 & .28 & .07 \\ .15 & .67 & .18 \\ .12 & .36 & .52 \end{bmatrix} \cdot \begin{bmatrix} .47 & .39 & .13 \\ .22 & .56 & .22 \\ .19 & .46 & .34 \end{bmatrix} \approx \begin{bmatrix} .38 & .44 & .17 \\ .25 & .52 & .23 \\ .23 & .49 & .27 \end{bmatrix}$$

(The rows of P^3 don't necessarily total 1 exactly because of rounding errors.) From matrix P^3 we find a probability of .25 that a person in state 2 will have an offspring in state 1 *three generations* later. The probability is .52 that a person in state 2 will have, three generations later, an offspring in state 2.   ∎

Example 4   Let us return to the transition matrix for the cleaners,

		Second load	
		Johnson	NorthClean
First load	Johnson	.8	.2
	NorthClean	.35	.65

As this matrix shows, there is a .8 chance that persons bringing their first load to Johnson will also bring their second load to Johnson, and so on. To find the probabilities for the third load, the second stage of this Markov chain, we need to find the square of the transition matrix. If C represents the transition matrix, then

$$C^2 = C \cdot C = \begin{bmatrix} .8 & .2 \\ .35 & .65 \end{bmatrix} \cdot \begin{bmatrix} .8 & .2 \\ .35 & .65 \end{bmatrix} = \begin{bmatrix} .71 & .29 \\ .51 & .49 \end{bmatrix}.$$

From C^2, the probability that a person bringing their first load of clothes to Johnson will also bring their third load to Johnson is .71; the probability that a person bringing their first load to NorthClean will bring their third load to North-Clean is .49.

The cube of matrix C gives the probabilities for the fourth load, the third step in our experiment.

$$C^3 = C \cdot C^2 = \begin{bmatrix} .67 & .33 \\ .58 & .42 \end{bmatrix}$$

The probability is .58, for example, that persons bringing their first load to North-Clean will bring their fourth load to Johnson.   ∎

Distribution of States   Look again at the transition matrix for social class changes:

$$P = \begin{bmatrix} .65 & .28 & .07 \\ .15 & .67 & .18 \\ .12 & .36 & .52 \end{bmatrix}.$$

Suppose we have the following initial distribution of people in the three social classes.

class	state	proportion
lower	1	21%
middle	2	68%
upper	3	11%

To find how these proportions would change after one generation, we could use the tree diagram of Figure 2.

To find the proportion of people in state 2 after one generation, add the numbers indicated with arrows.

$$.0588 + .4556 + .0396 = .5540$$

In a similar way, we can find the proportion of people in state 1 after one generation.

$$.1365 + .1020 + .0132 = .2517$$

For state 3, we have

$$.0147 + .1224 + .0572 = .1943$$

The initial distribution of states, 21%, 68%, and 11%, becomes, after one generation, 25.17% in state 1, 55.4% in state 2, and 19.43% in state 3. These

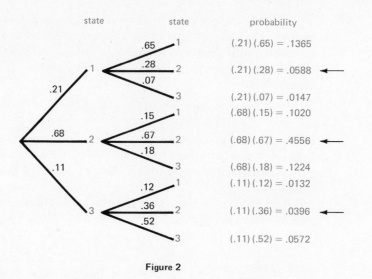

state   state   probability

.65   1   (.21)(.65) = .1365

1   .28   2   (.21)(.28) = .0588 ←

.07

.21   3   (.21)(.07) = .0147

.15   1   (.68)(.15) = .1020

.68   2   .67   2   (.68)(.67) = .4556 ←

.18

.11   3   (.68)(.18) = .1224

.12   1   (.11)(.12) = .0132

3   .36   2   (.11)(.36) = .0396 ←

.52

3   (.11)(.52) = .0572

Figure 2

distributions can be written as *probability vectors* (where the percents have been changed to decimals rounded to the nearest hundredth):

$$[.21 \quad .68 \quad .11] \quad \text{and} \quad [.25 \quad .55 \quad .19]$$

respectively. A ***probability vector*** is a matrix of only one row, having nonnegative entries, with the sum of the entries 1.

    The work we did with the tree diagram in finding the distribution of states after one generation is exactly the work that would have been required in multiplying the initial probability vector, $[.21 \quad .68 \quad .11]$, and the transition matrix P:

$$[.21 \quad .68 \quad .11] \cdot \begin{bmatrix} .65 & .28 & .07 \\ .15 & .67 & .18 \\ .12 & .36 & .52 \end{bmatrix} \approx [.25 \quad .55 \quad .19].$$

In a similar way, the distribution of social classes after two generations can be found by multiplying the initial probability vector and the square of P, the matrix P^2. Using P^2 from above,

$$[.21 \quad .68 \quad .11] \cdot \begin{bmatrix} .47 & .39 & .13 \\ .22 & .56 & .22 \\ .19 & .46 & .34 \end{bmatrix} \approx [.27 \quad .51 \quad .21]$$

    In the next section we shall find a long-range prediction for the proportions of the population in each social class. Our work in this section is summarized below.

> **Suppose a Markov chain has initial probability vector $I = [i_1 \quad i_2 \quad i_3 \quad \cdots \quad i_n]$, and transition matrix P. The probability vector after n repetitions of the experiment is**
>
> $$I \cdot P^n.$$

7.1 Exercises   *Which of the following could be a probability vector?*

1. $[\frac{2}{3} \quad \frac{1}{2}]$

2. $[\frac{1}{2} \quad 1]$

3. $[0 \quad 1]$

4. $[.1 \quad .1]$

5. $[.4 \quad .2 \quad 0]$

6. $[\frac{1}{4} \quad \frac{1}{8} \quad \frac{5}{8}]$

7. $[.07 \quad .04 \quad .37 \quad .52]$

8. $[.3 \quad -.1 \quad .8]$

9. $[0 \quad -.2 \quad .6 \quad .6]$

Which of the following could be transition matrices, by definition?

10. $\begin{bmatrix} .5 & 0 \\ 0 & .5 \end{bmatrix}$

11. $\begin{bmatrix} \frac{2}{3} & \frac{1}{3} \\ 1 & 0 \end{bmatrix}$

12. $\begin{bmatrix} \frac{1}{4} & \frac{3}{4} \\ \frac{1}{2} & \frac{1}{2} \end{bmatrix}$

13. $\begin{bmatrix} \frac{1}{4} & \frac{3}{4} & 0 \\ 2 & 0 & 1 \\ 1 & \frac{2}{3} & 3 \end{bmatrix}$

14. $\begin{bmatrix} \frac{1}{3} & \frac{1}{3} & \frac{1}{3} \\ 0 & 1 & 0 \\ \frac{1}{2} & 0 & \frac{1}{2} \end{bmatrix}$

15. $\begin{bmatrix} \frac{1}{3} & \frac{1}{2} & 1 \\ 0 & 1 & 0 \\ \frac{1}{2} & \frac{1}{2} & 1 \end{bmatrix}$

16. $\begin{bmatrix} \frac{1}{3} & \frac{1}{2} & 1 \\ \frac{1}{3} & 0 & 0 \\ \frac{1}{3} & \frac{1}{2} & 0 \end{bmatrix}$

17. $\begin{bmatrix} .9 & .1 & 0 \\ .1 & .6 & .3 \\ 0 & .3 & .7 \end{bmatrix}$

18. $\begin{bmatrix} .6 & .2 & .2 \\ .9 & .02 & .08 \\ 0 & 0 & .6 \end{bmatrix}$

Find the first three powers of each of the following transition matrices, for example, A, A^2 and A^3. For each transition matrix, find the probability that state 1 changes to state 2 after three repetitions of the experiment.

19. $A = \begin{bmatrix} 1 & 0 \\ .8 & .2 \end{bmatrix}$

20. $B = \begin{bmatrix} .7 & .3 \\ 0 & 1 \end{bmatrix}$

21. $C = \begin{bmatrix} .5 & .5 \\ .72 & .28 \end{bmatrix}$

22. $D = \begin{bmatrix} .3 & .2 & .5 \\ 0 & 0 & 1 \\ .6 & .1 & .3 \end{bmatrix}$

23. $E = \begin{bmatrix} .8 & .1 & .1 \\ .3 & .6 & .1 \\ 0 & 1 & 0 \end{bmatrix}$

24. $F = \begin{bmatrix} .01 & .9 & .09 \\ .72 & .1 & .18 \\ .34 & 0 & .66 \end{bmatrix}$

25. Years ago, about 10% of all cars sold were small, while 90% were large. This has changed drastically; now of the people buying a car in a given year, 20% of small car owners will switch to a large car, while 60% of large car owners will switch to a small car.
(a) Write a transition matrix using this information.
(b) Write a probability vector for the initial distribution of cars.
(c) Square the transition matrix and find the distribution of cars after 2 years.
Find the distribution of cars after
(d) 3 years;   (e) 4 years;   (f) 5 years.

26. In the example in the text, we used the transition matrix

$$\begin{array}{c} \\ \text{Johnson} \\ \text{NorthClean} \end{array} \begin{array}{cc} \text{Johnson} & \text{NorthClean} \\ \begin{bmatrix} .8 & .2 \\ .35 & .65 \end{bmatrix} \end{array}$$

Suppose now that we assume that each customer brings in one load of clothes each week. Use various powers of the transition matrix to find the probability that a customer bringing a load of clothes to Johnson initially also brings a load to Johnson after
(a) 1 week;   (b) 2 weeks;   (c) 3 weeks;   (d) 4 weeks.

27. Suppose Johnson has a 40% market share initially, with NorthClean having a 60% share. Use this information to write a probability vector; use this vector along with the transition matrix above to find the share of the market for each firm after
(a) 1 week;   (b) 2 weeks;   (c) 3 weeks;   (d) 4 weeks.

28. An insurance company classifies its drivers into three groups: G_0 (no accidents), G_1 (one accident), and G_2 (more than one accident). The probability that a driver in G_0 will stay in G_0 after one year is .85, that the driver will become a G_1 is .10, and that the driver will become a G_2 is .05. A driver in G_1 cannot move to G_0 (this company has a long memory). There is a .8 probability that a G_1 driver will stay in G_1. A G_2 driver must stay in G_2. Write a transition matrix using this information.

29. Suppose that the company of Exercise 28 accepts 50,000 new policyholders, all of whom are in G_0. Find the number in each group after
    (a) 1 year   (b) 2 years   (c) 3 years   (d) 4 years

30. The difficulty with the mathematical model of Exercises 28 and 29 is that no "grace period" is provided; there should be a certain positive probability of moving from G_1 or G_2 back to G_0 (say, after four years with no accidents). A new system with this feature might produce the following transition matrix.

$$\begin{bmatrix} .85 & .10 & .05 \\ .15 & .75 & .10 \\ .10 & .30 & .60 \end{bmatrix}$$

Suppose that when this new policy is adopted, the company has 50,000 policyholders, all in G_0. Find the number in each group after

(a) 1 year;   (b) 2 years;   (c) 3 years.

31. Research done by the Gulf Oil Corporation[1] produced the following transition matrix for the probability that during a given year a person with one form of home heating would switch to another.

| | Will switch to | | |
	Oil	Gas	Electric
Oil	.825	.175	0
Now has Gas	.060	.919	.021
Electric	.049	0	.951

The current share of the market held by these three types of heat is given by the vector [.26   .60   .14]. Find the share of the market held by each type of heat after
(a) 1 year;   (b) 2 years;   (c) 3 years.

32. In one state, a land use survey showed that 35% of all land was used for agricultural purposes, while 10% was urban. Ten years later, of the agricultural land, 15% had become urban and 80% had remained agricultural. (The remainder lay idle.) Of the idle land, 20% had become urbanized and 10% had been converted for agricultural use. Of the urban land, 90% remained urban and 10% was idle. Assume that these trends continue.
(a) Write a transition matrix using this information.
(b) Write a probability vector for the initial distribution of land.

Find the land use pattern after
(c) ten years;   (d) twenty years.

33. In a survey investigating change in housing patterns in one urban area, it was found

Reprinted by permission of Ali Ezzati, "Forcasting Market Shares of Alternative Home Heating Units," *Management Science*, Vol. 21, No. 4, December 1974, copyright © 1974 The Institute of Management Sciences.

that 75% of the population lived in single-family dwellings and 25% in multiple housing of some kind. Five years later, in a follow-up survey, of those who had been living in single family dwellings, 90% still did so, but 10% had moved to multiple family dwellings. Of those in multiple family housing, 95% were still living in that type of housing, while 5% had moved to single-family dwellings. Assume that these trends continue.

(a) Write a transition matrix for this information.

(b) Write a probability vector for the initial distribution of housing.

What percent of the population can be expected in each category

(c) five years later;   (d) ten years later?

34. At the end of June in a Presidential election year, 40% of the voters were registered as liberal, 45% as conservative, and 15% as independent. Over a one month period, the liberals retained 80% of their constituency, while 15% switched to conservative and 5% to independent. The conservatives retained 70%, and lost 20% to the liberals. The independents retained 60% and lost 20% each to the conservatives and liberals. Assume that these trends continue.

(a) Write a transition matrix using this information.

(b) Write a probability vector for the initial distribution.

Find the percent of each type of voter at the end of

(c) July;   (d) August;   (e) September;   (f) October.

7.2   Regular Markov Chains

We have seen that by starting with a transition matrix P and an initial probability vector, the n-th power of P permits us to find the probability vector for n repetitions of an experiment. In this section we try to decide what happens to our initial probability vector "in the long run," that is, as n gets larger and larger.

For example, let us use the transition matrix associated with the dry cleaners example of the previous section,

$$\begin{bmatrix} .8 & .2 \\ .35 & .65 \end{bmatrix}.$$

We assumed that the initial probability vector, the market share for each firm at the beginning of our experiment, was [.4   .6]. By finding powers of the transition matrix, we get the market shares shown in the following table. (See Exercise 27 of Section 7.1.)

Week	Johnson	NorthClean
Start	[.4	.6]
1	[.53	.47]
2	[.59	.41]
3	[.62	.38]
4	[.63	.37]
5	[.63	.37]
12	[.64	.36]

The results seem to approach the probability vector [.64   .36].

What happens if we try an initial probability vector other than [.4   .6]? Suppose we try [.75   .25]; using the same powers of the transition matrix as above, we get the following results.

Week	Johnson	NorthClean
Start	[.75	.25]
1	[.69	.31]
2	[.66	.34]
3	[.65	.35]
4	[.64	.36]
5	[.64	.36]
6	[.64	.36]

Here the results also seem to be approaching the probability vector [.64   .36], the same vector approached with the initial probability vector [.4   .6]. In either case, the long-range trend is for a market share of about 64% for Johnson and 36% for NorthClean. Based on our example, this long-range trend does not depend on the initial distribution of market shares.

Regular Transition Matrices   One of the many applications of Markov chains is in finding these long-range predictions. It is not possible to make long-range predictions with all transition matrices, but there is a large set of transition matrices with which long-range predictions *are* possible. Such predictions are always possible with **regular transition matrices**. A transition matrix is **regular** if some power of the matrix contains all positive entries. A Markov chain is a **regular Markov chain** if its transition matrix is regular.

Example 1   Decide if the following transition matrices are regular.

(a) $A = \begin{bmatrix} .75 & .25 & 0 \\ 0 & .5 & .5 \\ .6 & .4 & 0 \end{bmatrix}$

Square A.

$A^2 = \begin{bmatrix} .5625 & .3125 & .125 \\ .3 & .45 & .25 \\ .45 & .35 & .2 \end{bmatrix}$

All entries in A^2 are positive, so that matrix A is regular.

(b) $B = \begin{bmatrix} .5 & 0 & .5 \\ 0 & 1 & 0 \\ 0 & 0 & 1 \end{bmatrix}$

Find various powers of B.

$B^2 = \begin{bmatrix} .25 & 0 & .75 \\ 0 & 1 & 0 \\ 0 & 0 & 1 \end{bmatrix}; \quad B^3 = \begin{bmatrix} .125 & 0 & .875 \\ 0 & 1 & 0 \\ 0 & 0 & 1 \end{bmatrix}; \quad B^4 = \begin{bmatrix} .0625 & 0 & .9375 \\ 0 & 1 & 0 \\ 0 & 0 & 1 \end{bmatrix}$

Further powers of B will still give the same zero entries, so that no power of matrix B contains all positive entries. For this reason, B is not regular. ∎

Suppose that v is a probability vector. It turns out that for a regular Markov chain with a transition matrix P, there exists a single vector V such that $v \cdot P^n$ approaches closer and closer to V as n gets larger and larger.

> Suppose a Markov chain is regular, and has a transition matrix P. Then for any probability vector v, there is a unique vector V such that
>
> $$v \cdot P^n \approx V$$
>
> for large values of n.
>   Vector V is called the **equilibrium vector** or the **fixed vector** of the Markov chain.

In the example with Johnson Cleaners, we found that the equilibrium vector V is approximately $[.64 \quad .36]$.

We could find vector V by finding P^n for larger and larger values of n, and then looking for a vector that the product $v \cdot P^n$ approaches. However, such an approach can be very tedious and prone to error. To find a better way, start with the fact that for a large value of n,

$$v \cdot P^n \approx V,$$

as mentioned above. From this result, $v \cdot P^n \cdot P \approx V \cdot P$, so that

$$v \cdot P^n \cdot P = v \cdot P^{n+1} \approx VP.$$

Since $v \cdot P^n \approx V$ for large values of n, we also have $v \cdot P^{n+1} \approx V$ for large values of n (the product $v \cdot P^n$ approaches V, so that $v \cdot P^{n+1}$ must also approach V.) Thus, $v \cdot P^{n+1} \approx V$ and $v \cdot P^{n+1} \approx VP$, which suggests that

$$VP = V.$$

> If a Markov chain is regular and has transition matrix P, then there exists a probability vector V such that
>
> $$VP = V.$$

This vector V gives the long range trend of the Markov chain. Vector V is found by solving a system of linear equations, as shown in the next examples.

Example 2   Find the long range trend for the Markov chain in our drycleaning example with transition matrix

$$\begin{bmatrix} .8 & .2 \\ .35 & .65 \end{bmatrix}.$$

This matrix is regular since all entries are positive. Let P represent this transition matrix, and let V be the probability vector $[v_1 \quad v_2]$. We want to find V such that

$$VP = V,$$

or

$$[v_1 \quad v_2] \begin{bmatrix} .8 & .2 \\ .35 & .65 \end{bmatrix} = [v_1 \quad v_2].$$

Use matrix multiplication on the left.

$$[.8v_1 + .35v_2 \quad .2v_1 + .65v_2] = [v_1 \quad v_2]$$

Set corresponding entries from the two matrices equal to get

$$.8v_1 + .35v_2 = v_1 \quad \text{and} \quad .2v_1 + .65v_2 = v_2.$$

Simplify each of these equations.

$$-.2v_1 + .35v_2 = 0 \qquad .2v_1 - .35v_2 = 0$$

These last two equations are really the same. (The system of equations obtained from $VP = V$ is always dependent.) To find the values of v_1 and v_2, recall that $V = [v_1 \quad v_2]$ was a probability vector, so that

$$v_1 + v_2 = 1.$$

To find v_1 and v_2 we must solve the system

$$-.2v_1 + .35v_2 = 0$$
$$v_1 + v_2 = 1.$$

From the second equation, $v_1 = 1 - v_2$. Substitute $1 - v_2$ for v_1 in the first equation.

$$-.2(1 - v_2) + .35v_2 = 0$$
$$-.2 + .2v_2 + .35v_2 = 0$$
$$.55v_2 = .2$$
$$v_2 = \frac{4}{11} \approx .364$$

Since $v_1 = 1 - v_2$, we have $v_1 = 7/11 \approx .636$, with equilibrium vector $V = [7/11 \quad 4/11] \approx [.636 \quad .364]$. ∎

If we begin taking powers of the transition matrix P of Example 1 we get (rounded to two decimals).

$$P^2 = \begin{bmatrix} .71 & .29 \\ .51 & .49 \end{bmatrix} \qquad P^3 = \begin{bmatrix} .67 & .33 \\ .58 & .42 \end{bmatrix} \qquad P^4 = \begin{bmatrix} .65 & .35 \\ .62 & .38 \end{bmatrix}$$

$$P^5 = \begin{bmatrix} .65 & .35 \\ .63 & .37 \end{bmatrix} \qquad P^6 = \begin{bmatrix} .64 & .36 \\ .63 & .37 \end{bmatrix} \qquad P^{10} = \begin{bmatrix} .64 & .36 \\ .64 & .36 \end{bmatrix}$$

As these results suggest, higher and higher powers of the transition matrix P approach a matrix having all rows identical; these identical rows have as entries the entries of the equilibrium vector V. This agrees with what we said above: the initial state doesn't matter. Irrespective of the initial probability vector, the system will still approach a fixed vector V.

Let us summarize the results of this section.

Suppose a regular Markov chain has a transition matrix P.

(1) For any initial probability vector v, the product $v \cdot P^n$ approaches a unique vector V as n gets larger and larger. Vector V is called the *equilibrium* or *fixed vector*.

(2) Vector V has the property that $VP = V$.

(3) To find V, solve a system of equations obtained from the matrix equation $VP = V$, and from the fact that the sum of the entries of V is 1.

(4) The powers P^n approach closer and closer to a matrix whose rows are made up of the entries of the equilibrium vector V.

Example 3   Find the equilibrium vector for the transition matrix

$$K = \begin{bmatrix} .2 & .6 & .2 \\ .1 & .1 & .8 \\ .3 & .3 & .4 \end{bmatrix}.$$

Matrix K has all positive entries and thus is regular. For this reason, an equilibrium matrix V must exist such that $VK = V$. Let $V = [v_1 \quad v_2 \quad v_3]$. Then

$$[v_1 \quad v_2 \quad v_3] \begin{bmatrix} .2 & .6 & .2 \\ .1 & .1 & .8 \\ .3 & .3 & .4 \end{bmatrix} = [v_1 \quad v_2 \quad v_3].$$

Use matrix multiplication on the left.

$$[.2v_1 + .1v_2 + .3v_3 \quad .6v_1 + .1v_2 + .3v_3 \quad .2v_1 + .8v_2 + .4v_3]$$
$$= [v_1 \quad v_2 \quad v_3]$$

Putting corresponding entries equal gives three equations.

$$.2v_1 + .1v_2 + .3v_3 = v_1$$
$$.6v_1 + .1v_2 + .3v_3 = v_2$$
$$.2v_1 + .8v_2 + .4v_3 = v_3$$

Simplifying these equations gives

$$-.8v_1 + .1v_2 + .3v_3 = 0$$
$$.6v_1 - .9v_2 + .3v_3 = 0$$
$$.2v_1 + .8v_2 - .6v_3 = 0.$$

Since V is a probability vector,

$$v_1 + v_2 + v_3 = 1.$$

We now have a system of four equations in three unknowns.

$$-.8v_1 + .1v_2 + .3v_3 = 0$$
$$.6v_1 - .9v_2 + .3v_3 = 0$$
$$.2v_1 + .8v_2 - .6v_3 = 0$$
$$v_1 + v_2 + v_3 = 1.$$

This system can be solved with the Gaussian method of Chapter 2. Start with the augmented matrix

$$\begin{bmatrix} -.8 & .1 & .3 & | & 0 \\ .6 & -.9 & .3 & | & 0 \\ .2 & .8 & -.6 & | & 0 \\ 1 & 1 & 1 & | & 1 \end{bmatrix}.$$

The solution of this system is $v_1 = 5/23$, $v_2 = 7/23$, and $v_3 = 11/23$, and

$$V = \begin{bmatrix} \dfrac{5}{23} & \dfrac{7}{23} & \dfrac{11}{23} \end{bmatrix} \approx \begin{bmatrix} .22 & .30 & .48 \end{bmatrix}. \quad\blacksquare$$

7.2 Exercises

Which of the following matrices are regular?

1. $\begin{bmatrix} .2 & .8 \\ .9 & .1 \end{bmatrix}$
2. $\begin{bmatrix} .22 & .78 \\ .43 & .57 \end{bmatrix}$
3. $\begin{bmatrix} 1 & 0 \\ .6 & .4 \end{bmatrix}$
4. $\begin{bmatrix} .55 & .45 \\ 0 & 1 \end{bmatrix}$

5. $\begin{bmatrix} 0 & 1 & 0 \\ .4 & .2 & .4 \\ 1 & 0 & 0 \end{bmatrix}$
6. $\begin{bmatrix} .3 & .5 & .2 \\ 1 & 0 & 0 \\ .5 & .1 & .4 \end{bmatrix}$

Find the equilibrium vector for each of the following transition matrices.

7. $\begin{bmatrix} \frac{1}{4} & \frac{3}{4} \\ \frac{1}{2} & \frac{1}{2} \end{bmatrix}$
8. $\begin{bmatrix} \frac{2}{3} & \frac{1}{3} \\ \frac{1}{8} & \frac{7}{8} \end{bmatrix}$
9. $\begin{bmatrix} .3 & .7 \\ .4 & .6 \end{bmatrix}$
10. $\begin{bmatrix} .8 & .2 \\ .1 & .9 \end{bmatrix}$

11. $\begin{bmatrix} .1 & .1 & .8 \\ .4 & .4 & .2 \\ .1 & .2 & .7 \end{bmatrix}$
12. $\begin{bmatrix} .5 & .2 & .3 \\ .1 & .4 & .5 \\ .2 & .2 & .6 \end{bmatrix}$

13. $\begin{bmatrix} .25 & .35 & .4 \\ .1 & .3 & .6 \\ .55 & .4 & .05 \end{bmatrix}$
14. $\begin{bmatrix} .16 & .28 & .56 \\ .43 & .12 & .45 \\ .86 & .05 & .09 \end{bmatrix}$

Find the equilibrium vector for each of the following transition matrices. These matrices were first used in the Exercises of Section 7.1.

15. car sizes, Exercise 25,
$$\begin{bmatrix} .8 & .2 \\ .6 & .4 \end{bmatrix}$$

16. housing patterns, Exercise 33,
$$\begin{bmatrix} .90 & .10 \\ .05 & .95 \end{bmatrix}$$

17. insurance categories, Exercise 28,
$$\begin{bmatrix} .85 & .10 & .05 \\ 0 & .80 & .20 \\ 0 & 0 & 1 \end{bmatrix}$$

18. "modified" insurance categories, Exercise 30,
$$\begin{bmatrix} .85 & .10 & .05 \\ .15 & .75 & .10 \\ .10 & .30 & .60 \end{bmatrix}$$

19. land use, Exercise 32,

$$\begin{bmatrix} .80 & .15 & .05 \\ 0 & .90 & .10 \\ .10 & .20 & .70 \end{bmatrix}$$

20. voting registration, Exercise 34,

$$\begin{bmatrix} .80 & .15 & .05 \\ .20 & .70 & .10 \\ .20 & .20 & .60 \end{bmatrix}$$

21. home heating systems, Exercise 31,

$$\begin{bmatrix} .825 & .175 & 0 \\ .060 & .919 & .021 \\ .049 & 0 & .951 \end{bmatrix}$$

22. The probability that a complex assembly line works correctly depends on whether or not the line worked correctly the last time it was used. There is a .9 chance that the line will work correctly if it worked correctly the time before, and a .7 chance that it will work correctly if it did *not* work correctly the time before. Set up a transition matrix with this information and find the long run probability that the line will work correctly.

23. Suppose improvements are made in the assembly line of Exercise 22, so that the transition matrix becomes

$$\begin{array}{cc} & \text{Works} \quad \text{Doesn't} \\ \begin{array}{c} \text{Works} \\ \text{Doesn't} \end{array} & \begin{bmatrix} .95 & .05 \\ .80 & .20 \end{bmatrix} \end{array}$$

Find the long run probability now that the line will work properly.

24. A certain genetic defect is carried only by males. Suppose the probability is .95 that a male offspring will have the defect if his father did, with the probability .10 that a male offspring will have the defect if his father did not have it. Find the long range prediction for the fraction of the males in the population who will have the defect.

25. Each month, a sales manager classifies her salespeople as low, medium, or high producers. There is a .4 chance that a low producer one month will become a medium producer the following month, and a .1 chance that a low producer will become a high producer. A medium producer will become a low or high producer, respectively, with probabilities .25 and .3. A high producer will become a low or medium producer, respectively, with probabilities .05 and .4. Find the long range trend for the proportion of low, medium, and high producers.

26. The weather in a certain spot is classified as fair, cloudy without rain, or rainy. A fair day is followed by a fair day 60% of the time, and by a cloudy day 25% of the time. A cloudy day is followed by a cloudy day 35% of the time, and by a rainy day 25% of the time. A rainy day is followed by a cloudy day 40% of the time, and by another rainy day 25% of the time. Find the long range prediction for the proportion of fair, cloudy, and rainy days.

27. At one liberal arts college, students are classified as humanities majors, science majors, or undecideds. There is a 20% chance that a humanities major will change to a science major from one year to the next, and a 45% chance that a humanities major will change to undecided. A science major will change to humanities with probability .15, and to undecided with probability .35. An undecided will switch to humanities or science with probabilities of .5 and .3 respectively. Find the long range prediction for the fraction of students in each of these three majors.

28. A large group of mice is kept in a cage having connected compartments A, B, and C. Mice in compartment A move to B with probability .3 and to C with probability .4.

Mice in B move to A or C with probabilities of .15 and .55, respectively. Mice in C move to A and B with probabilities of .3 and .6 respectively. Find the long range prediction for the fraction of mice in each of the compartments.

29. The manager of the slot machines at a major casino makes a decision about whether or not to "loosen up" the slots so that the customers get a larger playback. The manager tells only one other person, a person whose word cannot be trusted. In fact, there is only a probability p, where $0 < p < 1$, that this person will tell the truth. Suppose this person tells several other people, each of whom tell several people, what the manager's decision is. Suppose there is always a probability p that the decision is passed on as heard. Find the long range prediction for the fraction of the people who will hear the decision correctly. (Hint: use a transition matrix; let the first row be $[p \quad 1-p]$, with second row $[1-p \quad p]$.)

30. Find the equilibrium vector for the transition matrix

$$\begin{bmatrix} p & 1-p \\ 1-q & q \end{bmatrix}$$

where $0 < p < 1$ and $0 < q < 1$. Under what conditions is this matrix regular?

31. Show that the transition matrix

$$K = \begin{bmatrix} \frac{1}{4} & 0 & \frac{3}{4} \\ 0 & 1 & 0 \\ 0 & 0 & 1 \end{bmatrix}$$

has more than one vector V such that $VK = V$. Why does this not violate the statements of this section?

32. Let

$$P = \begin{bmatrix} a_{11} & a_{12} \\ a_{21} & a_{22} \end{bmatrix}$$

be a regular transition matrix having *column* sums of 1. Show that the equilibrium vector for P is $[1/2 \quad 1/2]$.

7.3   Absorbing Markov Chains

Suppose a Markov chain has transition matrix

$$P = \begin{array}{c} \\ 1 \\ 2 \\ 3 \end{array} \begin{array}{ccc} 1 & 2 & 3 \\ \begin{bmatrix} .3 & .6 & .1 \\ 0 & 1 & 0 \\ .6 & .2 & .2 \end{bmatrix} \end{array}$$

The matrix shows that p_{12}, the probability of going from state 1 to state 2, is .6, while p_{22}, the probability of staying in state 2 is 1. Thus, once state 2 is entered, it is impossible to leave. For this reason, state 2 is called an *absorbing state*.
  Generalizing from this example,

State i of a Markov chain is an **absorbing state** if $p_{ii} = 1$.

Using the idea of an absorbing state, we can define an absorbing Markov chain.

> A Markov chain is an **absorbing chain** if and only if the following two conditions are satisfied:
>
> (1) The chain has at least one absorbing state.
> (2) It is possible to go from any nonabsorbing state to an absorbing state (perhaps in more than one step).

Example 1   Identify all absorbing states in the Markov chains having the following matrices. Decide if the Markov chain is absorbing.

(a)
$$\begin{array}{c} \\ 1 \\ 2 \\ 3 \end{array} \begin{array}{ccc} 1 & 2 & 3 \\ \left[\begin{array}{ccc} 1 & 0 & 0 \\ .3 & .5 & .2 \\ 0 & 0 & 1 \end{array}\right] \end{array}$$

Since $p_{11} = 1$ and since $p_{33} = 1$, both state 1 and state 3 are absorbing states. (Once these states are reached, they cannot be left.) The only nonabsorbing state is state 2. There is a .3 probability of going from state 2 to the absorbing state 1, so that it is possible to go from the nonabsorbing state to an absorbing state. This Markov chain is thus absorbing.

(b)
$$\begin{bmatrix} .6 & 0 & .4 & 0 \\ 0 & 1 & 0 & 0 \\ .9 & 0 & .1 & 0 \\ 0 & 0 & 0 & 1 \end{bmatrix}$$

States 2 and 4 are absorbing, with states 1 and 3 nonabsorbing. From state 1, it is possible to go only to states 1 or 3; from state 3 it is possible to go only to states 1 or 3. Thus, neither nonabsorbing state leads to an absorbing state, so that this Markov chain is nonabsorbing.   ∎

Example 2   *(Gambler's Ruin)* Suppose players A and B have a coin tossing game going on — a fair coin is tossed and the player predicting the toss correctly wins $1 from the other player. Suppose the players have a total of $6 between them, and that the game goes on until one player has no money (is ruined).

Let us agree that the states of this system are the amounts of money held by player A. There are seven possible states: A can have 0, 1, 2, 3, 4, 5, or 6 dollars. When either state 0 or state 6 is reached, the game is over. In any other state, the amount of money held by player A will increase by $1, or decrease by $1, with each of these events having probability 1/2 (since we assume a fair coin). For example, in state 3 (A has $3), there is 1/2 chance of changing to state 2, and a

1/2 chance of changing to state 4. Thus, $p_{32} = 1/2$ and $p_{34} = 1/2$. The probability of changing from state 3 to any other state is 0. Using this information gives the following 7×7 transition matrix.

$$
G =
\begin{array}{c c}
& \begin{array}{c c c c c c c} 0 & 1 & 2 & 3 & 4 & 5 & 6 \end{array} \\
\begin{array}{c} 0 \\ 1 \\ 2 \\ 3 \\ 4 \\ 5 \\ 6 \end{array} &
\left[\begin{array}{c c c c c c c}
1 & 0 & 0 & 0 & 0 & 0 & 0 \\
\frac{1}{2} & 0 & \frac{1}{2} & 0 & 0 & 0 & 0 \\
0 & \frac{1}{2} & 0 & \frac{1}{2} & 0 & 0 & 0 \\
0 & 0 & \frac{1}{2} & 0 & \frac{1}{2} & 0 & 0 \\
0 & 0 & 0 & \frac{1}{2} & 0 & \frac{1}{2} & 0 \\
0 & 0 & 0 & 0 & \frac{1}{2} & 0 & \frac{1}{2} \\
0 & 0 & 0 & 0 & 0 & 0 & 1
\end{array} \right]
\end{array}
$$

Based on our definition, states 0 and 6 are absorbing—once these states are reached, they can never be left, and the game is over. It is possible to get from one of the nonabsorbing states, 1, 2, 3, 4, or 5, to one of the absorbing states, so the Markov chain is absorbing.

To find the long term trend of the game, we can find various powers of the transition matrix. Use a computer or a programmable calculator to verify our results.

$$
G^6 =
\left[\begin{array}{c c c c c c c}
1.0000 & .0000 & .0000 & .0000 & .0000 & .0000 & .0000 \\
.6875 & .0781 & .0000 & .1406 & .0000 & .0625 & .0313 \\
.4531 & .0000 & .2188 & .0000 & .2031 & .0000 & .1250 \\
.2188 & .1406 & .0000 & .2813 & .0000 & .1406 & .2188 \\
.1250 & .0000 & .2031 & .0000 & .2188 & .0000 & .4531 \\
.0313 & .0625 & .0000 & .1406 & .0000 & .0781 & .6875 \\
.0000 & .0000 & .0000 & .0000 & .0000 & .0000 & 1.0000
\end{array} \right]
$$

$$
G^{10} =
\left[\begin{array}{c c c c c c c}
1.0000 & .0000 & .0000 & .0000 & .0000 & .0000 & .0000 \\
.7539 & .0400 & .0000 & .0791 & .0000 & .0391 & .0879 \\
.5479 & .0000 & .1191 & .0000 & .1182 & .0000 & .2148 \\
.3418 & .0791 & .0000 & .1582 & .0000 & .0791 & .3418 \\
.2148 & .0000 & .1182 & .0000 & .1191 & .0000 & .5479 \\
.0879 & .0391 & .0000 & .0791 & .0000 & .0400 & .7539 \\
.0000 & .0000 & .0000 & .0000 & .0000 & .0000 & 1.0000
\end{array} \right]
$$

As these results suggest, the system tends to one of the absorbing states, so that the probability is 1 that one of the two gamblers will eventually be wiped out. ∎

In fact, the following can be shown.

Regardless of the original state of an absorbing Markov chain, in a finite number of steps the chain will enter an absorbing state and then stay in that state.

Example 3 Estimate the long term trend for the transition matrix

$$P = \begin{bmatrix} .3 & .2 & .5 \\ 0 & 1 & 0 \\ 0 & 0 & 1 \end{bmatrix}.$$

Both states 2 and 3 are absorbing, and since it is possible to go from non-absorbing state 1 to an absorbing state, the chain will eventually enter either state 2 or state 3. To find the long term trend, let us find various powers of P.

$$P^2 = \begin{bmatrix} .09 & .26 & .65 \\ 0 & 1 & 0 \\ 0 & 0 & 1 \end{bmatrix} \qquad P^4 = \begin{bmatrix} .0081 & .2834 & .7085 \\ 0 & 1 & 0 \\ 0 & 0 & 1 \end{bmatrix}$$

$$P^8 = \begin{bmatrix} .0001 & .2857 & .7142 \\ 0 & 1 & 0 \\ 0 & 0 & 1 \end{bmatrix} \qquad P^{16} = \begin{bmatrix} .0000 & .2857 & .7142 \\ 0 & 1 & 0 \\ 0 & 0 & 1 \end{bmatrix}$$

Based on these powers, it appears that the transition matrix is approaching closer and closer to the matrix

$$\begin{bmatrix} 0 & .29 & .71 \\ 0 & 1 & 0 \\ 0 & 0 & 1 \end{bmatrix}.$$

If the system is originally in state 1, there is no chance it will end up in state 1, but a .29 chance that it will end up in state 2 and a .71 chance it will end up in state 3. If the system was originally in state 2 it will end up in state 2; a similar statement can be made for state 3. ▌

This example suggests two further properties of absorbing systems:

> (1) **The powers of the transition matrix get closer and closer to some particular matrix.**
> (2) **The long-term trend depends on the initial state — changing the initial state can change the final result.**

This second fact is different from regular Markov chains, where the final result is independent of the initial state.

We would prefer a method for finding the final probabilities of entering an absorbing state without finding all the powers of the transition matrix, as we did in Example 3. We don't really need to worry about the absorbing states (to enter an absorbing state is to stay there). Thus, it is necessary only to work with the non-absorbing states. To see how this is done, let us use as an example the transition matrix from the gambler's ruin problem of Example 2. Rewrite the matrix so that the rows and columns corresponding to the absorbing states come first.

		absorbing		nonabsorbing				
		0	6	1	2	3	4	5
0		1	0	0	0	0	0	0
6		0	1	0	0	0	0	0
1		$\frac{1}{2}$	0	0	$\frac{1}{2}$	0	0	0
2		0	0	$\frac{1}{2}$	0	$\frac{1}{2}$	0	0
3		0	0	0	$\frac{1}{2}$	0	$\frac{1}{2}$	0
4		0	0	0	0	$\frac{1}{2}$	0	$\frac{1}{2}$
5		0	$\frac{1}{2}$	0	0	0	$\frac{1}{2}$	0

Let I_2 represent the 2×2 identity matrix in the upper left hand corner, let θ (the Greek letter *theta*) represent the matrix of zeros in the upper right, let R represent the matrix in the lower left, and let Q represent the matrix in the lower right. Using these symbols, G can be written as

$$G = \left[\begin{array}{c|c} I_2 & \theta \\ \hline R & Q \end{array}\right].$$

Now define the **fundamental matrix** for an absorbing Markov chain as matrix F, where

$$F = [I_5 - Q]^{-1}.$$

Here I_5 is the 5×5 identity matrix corresponding in size to matrix Q, so that the difference $I_5 - Q$ exists.

For the gambler's ruin problem,

$$F = \left(\begin{bmatrix} 1 & 0 & 0 & 0 & 0 \\ 0 & 1 & 0 & 0 & 0 \\ 0 & 0 & 1 & 0 & 0 \\ 0 & 0 & 0 & 1 & 0 \\ 0 & 0 & 0 & 0 & 1 \end{bmatrix} - \begin{bmatrix} 0 & \frac{1}{2} & 0 & 0 & 0 \\ \frac{1}{2} & 0 & \frac{1}{2} & 0 & 0 \\ 0 & \frac{1}{2} & 0 & \frac{1}{2} & 0 \\ 0 & 0 & \frac{1}{2} & 0 & \frac{1}{2} \\ 0 & 0 & 0 & \frac{1}{2} & 0 \end{bmatrix}\right)^{-1}$$

$$= \begin{bmatrix} 1 & -\frac{1}{2} & 0 & 0 & 0 \\ -\frac{1}{2} & 1 & -\frac{1}{2} & 0 & 0 \\ 0 & -\frac{1}{2} & 1 & -\frac{1}{2} & 0 \\ 0 & 0 & -\frac{1}{2} & 1 & -\frac{1}{2} \\ 0 & 0 & 0 & -\frac{1}{2} & 1 \end{bmatrix}^{-1}$$

	1	2	3	4	5
1	$\frac{5}{3}$	$\frac{4}{3}$	1	$\frac{2}{3}$	$\frac{1}{3}$
2	$\frac{4}{3}$	$\frac{8}{3}$	2	$\frac{4}{3}$	$\frac{2}{3}$
= 3	1	2	3	2	1
4	$\frac{2}{3}$	$\frac{4}{3}$	2	$\frac{8}{3}$	$\frac{4}{3}$
5	$\frac{1}{3}$	$\frac{2}{3}$	1	$\frac{4}{3}$	$\frac{5}{3}$

The inverse was found using techniques of Chapter 2.

Finally, use the fundamental matrix F along with matrix R found above to get the product FR.

$$FR = \begin{bmatrix} \frac{5}{3} & \frac{4}{3} & 1 & \frac{2}{3} & \frac{1}{3} \\ \frac{4}{3} & \frac{8}{3} & 2 & \frac{4}{3} & \frac{2}{3} \\ 1 & 2 & 3 & 2 & 1 \\ \frac{2}{3} & \frac{4}{3} & 2 & \frac{8}{3} & \frac{4}{3} \\ \frac{1}{3} & \frac{2}{3} & 1 & \frac{4}{3} & \frac{5}{3} \end{bmatrix} \begin{bmatrix} \frac{1}{2} & 0 \\ 0 & 0 \\ 0 & 0 \\ 0 & 0 \\ 0 & \frac{1}{2} \end{bmatrix} = \begin{matrix} & \begin{matrix} 0 & \ \ 6 \end{matrix} \\ \begin{matrix} 1 \\ 2 \\ 3 \\ 4 \\ 5 \end{matrix} & \begin{bmatrix} \frac{5}{6} & \frac{1}{6} \\ \frac{2}{3} & \frac{1}{3} \\ \frac{1}{2} & \frac{1}{2} \\ \frac{1}{3} & \frac{2}{3} \\ \frac{1}{6} & \frac{5}{6} \end{bmatrix} \end{matrix}$$

The product matrix FR gives the probability that if the system was originally in a nonabsorbing state, it ended up in either of the two absorbing states. For example, the probability is 2/3 that if the system was originally in state 2, it ended up in state 0; the probability is 5/6 that if the system was originally in state 5 it ended up in state 6, and so on.

Based on our original statement of the gambler's ruin problem, if player A starts with \$2 (state 2), there is a 2/3 chance of ending in state 0 (player A is ruined); if player A starts with \$5 (state 5) there is a 1/6 chance of player A being ruined, and so on.

Let us summarize what we have learned about absorbing Markov chains.

(1) Regardless of the initial state, in a finite number of steps the chain will enter an absorbing state and then stay in that state.

(2) The powers of the transition matrix get closer and closer to some particular matrix.

(3) The long term trend depends on the initial state.

(4) Let G be the transition matrix for an absorbing Markov chain. Rearrange the rows and columns of G so that the absorbing states come first. Matrix G will have the form

$$G = \left[\begin{array}{c|c} I_n & \theta \\ \hline R & Q \end{array} \right]$$

where I_n is an identity matrix, and θ is a matrix of all zeros. The fundamental matrix is defined as

$$F = [I_m - Q]^{-1}$$

where I_m has the same order as Q.

(5) The product FR gives the matrix of probabilities that a particular initial nonabsorbing state will lead to a particular absorbing state.

Example 4 Find the long term trend for the transition matrix,

$$P = \begin{matrix} & \begin{matrix} 1 & \ 2 & \ 3 \end{matrix} \\ \begin{matrix} 1 \\ 2 \\ 3 \end{matrix} & \begin{bmatrix} .3 & .2 & .5 \\ 0 & 1 & 0 \\ 0 & 0 & 1 \end{bmatrix} \end{matrix},$$

of Example 3.

Rewrite the matrix so that absorbing states 2 and 3 come first.

$$
\begin{array}{c} \\ 2 \\ 3 \\ 1 \end{array}
\begin{array}{c} \begin{array}{ccc} 2 & 3 & 1 \end{array} \\
\left[\begin{array}{cc|c} 1 & 0 & 0 \\ 0 & 1 & 0 \\ .2 & .5 & .3 \end{array}\right] \end{array}
$$

Here $R = [.2 \quad .5]$ and $Q = [.3]$. Find the fundamental matrix F.

$$F = [I_1 - Q]^{-1} = [1 - .3]^{-1} = [.7]^{-1} = [1/.7]$$

The product FR is

$$FR = [1/.7][.2 \quad .5] = [2/7 \quad 5/7] \approx [.286 \quad .714].$$

If the system starts in the nonabsorbing state 1, there is a 2/7 chance of ending up in the absorbing state 2, and a 5/7 chance of ending in the absorbing state 3. ■

7.3 Exercises

Find all absorbing states for the following transition matrices. Which are transition matrices for an absorbing Markov chain?

1. $\begin{bmatrix} .15 & .05 & .8 \\ 0 & 1 & 0 \\ .4 & .6 & 0 \end{bmatrix}$
2. $\begin{bmatrix} .1 & .5 & .4 \\ .2 & .2 & .6 \\ 0 & 0 & 1 \end{bmatrix}$
3. $\begin{bmatrix} .4 & 0 & .6 \\ 0 & 1 & 0 \\ .9 & 0 & .1 \end{bmatrix}$
4. $\begin{bmatrix} .5 & .5 & 0 \\ .8 & .2 & 0 \\ 0 & 0 & 1 \end{bmatrix}$

5. $\begin{bmatrix} .2 & .5 & .1 & .2 \\ 0 & 1 & 0 & 0 \\ .9 & .02 & .04 & .04 \\ 0 & 0 & 0 & 1 \end{bmatrix}$
6. $\begin{bmatrix} 1 & 0 & 0 & 0 \\ .9 & .1 & 0 & 0 \\ 0 & 0 & 1 & 0 \\ .6 & 0 & .4 & 0 \end{bmatrix}$

7. $\begin{bmatrix} .1 & .8 & 0 & .1 \\ 0 & 1 & 0 & 0 \\ 1 & 0 & 0 & 0 \\ 0 & 0 & 0 & 1 \end{bmatrix}$
8. $\begin{bmatrix} .32 & .41 & .16 & .11 \\ .42 & .30 & 0 & .28 \\ 0 & 0 & 0 & 1 \\ 1 & 0 & 0 & 0 \end{bmatrix}$

Find the fundamental matrix F for each of the following absorbing matrices. Also find the product matrix FR.

9. $\begin{bmatrix} 1 & 0 & 0 \\ 0 & 1 & 0 \\ .2 & .3 & .5 \end{bmatrix}$
10. $\begin{bmatrix} 1 & 0 & 0 \\ .6 & .1 & .3 \\ 0 & 0 & 1 \end{bmatrix}$

11. $\begin{bmatrix} .8 & .15 & .05 \\ 0 & 1 & 0 \\ 0 & 0 & 1 \end{bmatrix}$
12. $\begin{bmatrix} .42 & .37 & .21 \\ 0 & 1 & 0 \\ 0 & 0 & 1 \end{bmatrix}$

13. $\begin{bmatrix} 1 & 0 & 0 \\ 0 & 1 & 0 \\ \frac{1}{3} & \frac{1}{3} & \frac{1}{3} \end{bmatrix}$
14. $\begin{bmatrix} 1 & 0 & 0 \\ \frac{3}{8} & \frac{1}{8} & \frac{1}{2} \\ 0 & 0 & 1 \end{bmatrix}$

15. $\begin{bmatrix} 1 & 0 & 0 & 0 \\ \frac{1}{3} & 0 & \frac{2}{3} & 0 \\ 0 & 0 & 1 & 0 \\ \frac{1}{4} & \frac{1}{4} & \frac{1}{4} & \frac{1}{4} \end{bmatrix}$
16. $\begin{bmatrix} \frac{1}{4} & \frac{1}{2} & 0 & \frac{1}{4} \\ 0 & 1 & 0 & 0 \\ 0 & 0 & 1 & 0 \\ \frac{1}{2} & 0 & 0 & \frac{1}{2} \end{bmatrix}$

17.
$$\begin{bmatrix} 1 & 0 & 0 & 0 & 0 \\ 0 & 1 & 0 & 0 & 0 \\ .1 & .2 & .3 & .2 & .2 \\ .3 & .5 & .1 & 0 & .1 \\ 0 & 0 & 0 & 0 & 1 \end{bmatrix}$$

18.
$$\begin{bmatrix} .4 & .2 & .3 & 0 & .1 \\ 0 & 1 & 0 & 0 & 0 \\ 0 & 0 & 1 & 0 & 0 \\ .1 & .5 & .1 & .1 & .2 \\ 0 & 0 & 0 & 0 & 1 \end{bmatrix}$$

19. Write a transition matrix for a gambler's ruin problem when player A and player B start with a total of $4.

(a) Find matrix F for this transition matrix, and find the product matrix FR.

(b) Suppose player A starts with $1. What is the probability of ruin for A?

(c) Suppose player A starts with $3. What is the probability of ruin for A?

20. Suppose player B (Exercise 19) slips in a coin that is slightly "loaded"—such that the probability that B wins a particular toss changes from 1/2 to 3/5. Suppose that A and B start the game with a total of $5.

(a) If B starts with $3, find the probability that A will be ruined.

(b) If B starts with $1, find the probability that A will be ruined.

21. At a particular two-year college, a student has a probability of .25 of flunking out during a given year, a .15 probability of having to repeat the year, and a .6 probability of finishing the year. Use the following states.

State	Meaning
1	freshman
2	sophomore
3	has flunked out
4	has graduated

(a) Write a transition matrix. Find F and FR.

(b) Find the probability that a freshman will graduate.

22. A rat is placed at random in one of the compartments of the maze pictured below. The probability that a rat in compartment 1 will move to compartment 2 is .3; to compartment 3 is .2; and to compartment 4 is .1. A rat in compartment 2 will move to compartments 1, 4, or 5 with probabilities .2, .6, and .1 respectively. A rat in compartment 3 cannot leave that compartment. A rat in compartment 4 will move to 1, 2, 3, or 5 with probabilities of .1, .1, .4, and .3, respectively. A rat in compartment 5 cannot leave that compartment.

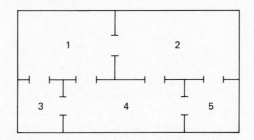

(a) Set up a transition matrix using this information. Find matrices F and FR. Find the probability that a rat ends up in compartment 5 if it was originally in compartment

(b) 1;   (c) 2;   (d) 3;   (e) 4.

It can be shown that the probability of ruin for player A in a game such as the one described in this section is

$$x_a = \frac{b}{a+b} \quad \text{if } r = 1, \quad \text{and} \quad x_a = \frac{r^a - r^{a+b}}{1 - r^{a+b}} \quad \text{if } r \neq 1$$

where a is the initial amount of money that player A has, b is the initial amount that player B has, r = (1 − p)/p, and p is the probability that player A will win on a given play.

23. Find the probability that A will be ruined if $a = 10$, $b = 30$, and $p = .49$.

24. Find the probability in Exercise 23 if p changes to .50.

25. Complete the following chart, assuming $a = 10$ and $b = 10$.

p	.1	.2	.3	.4	.5	.6	.7	.8	.9
x_a									

Key Words

state	equilibrium vector
transition matrix	fixed vector
Markov chain	absorbing state
probability vector	absorbing chain
regular transition matrix	Gambler's ruin
regular Markov chain	fundamental matrix

Chapter 7   Review Exercises

Which of the following could be transition matrices?

1. $\begin{bmatrix} .4 & .6 \\ 1 & 0 \end{bmatrix}$ **2.** $\begin{bmatrix} -.2 & 1.2 \\ .8 & .2 \end{bmatrix}$

3. $\begin{bmatrix} .8 & .2 & 0 \\ 0 & 1 & 0 \\ .1 & .4 & .5 \end{bmatrix}$ **4.** $\begin{bmatrix} .6 & .2 & .3 \\ .1 & .5 & .4 \\ .3 & .3 & .4 \end{bmatrix}$

For each of the following transition matrices, (a) find the first three powers; (b) find the probability that state 2 changes to state 1 after three repetitions of the experiment.

5. $C = \begin{bmatrix} .6 & .4 \\ 1 & 0 \end{bmatrix}$ **6.** $D = \begin{bmatrix} .3 & .7 \\ .5 & .5 \end{bmatrix}$

7. $E = \begin{bmatrix} .2 & .5 & .3 \\ .1 & .8 & .1 \\ 0 & 1 & 0 \end{bmatrix}$ **8.** $F = \begin{bmatrix} .14 & .12 & .74 \\ .35 & .28 & .37 \\ .71 & .24 & .05 \end{bmatrix}$

Use the following transition matrices T, along with the given initial distribution D, to find the distribution after two repetitions of the experiment. Also predict the long range distribution.

9. $D = [.3 \quad .7]$; $T = \begin{bmatrix} .4 & .6 \\ .5 & .5 \end{bmatrix}$ **10.** $D = [.8 \quad .2]$; $T = \begin{bmatrix} .7 & .3 \\ .2 & .8 \end{bmatrix}$

11. $D = [.2 \quad .4 \quad .4]$; $T = \begin{bmatrix} .6 & .2 & .2 \\ .3 & .3 & .4 \\ .5 & .4 & .1 \end{bmatrix}$

12. $D = [.1 \quad .1 \quad .8]$; $T = \begin{bmatrix} .2 & .3 & .5 \\ .1 & .1 & .8 \\ .7 & .1 & .2 \end{bmatrix}$

Currently, 35% of all hot dogs sold in one area are made by Dogkins, while 65% are made by Long Dog. Suppose that Dogkins starts a heavy advertising campaign, with the campaign producing the following transition matrix.

		After campaign	
		Dogkins	Long Dog
Before	Dogkins	.8	.2
campaign	Long Dog	.4	.6

13. Find the share of the market for each company after
   (a) the campaign;   **(b)** three such campaigns.

14. Predict the long-range market share for Dogkins.

A credit card company classifies its customers in three groups, nonusers in a given month, light users, and heavy users. The transition matrix for these states is

	nonuser	light	heavy
nonuser	.8	.15	.05
light	.25	.55	.2
heavy	.04	.21	.75

Suppose the initial distribution for the three states is $[.4 \quad .4 \quad .2]$. *Find the distribution after*

15. 1 month;

16. 2 months;

17. 3 months.

18. What is the long range prediction for the distribution of users?

A medical researcher is studying the risk of heart attack in men. She first divides men into three weight categories, thin, normal, and overweight. By studying the ancestors, children, and grandchildren of these men, the researcher comes up with the transition matrix

	thin	normal	overweight
thin	.3	.5	.2
normal	.2	.6	.2
overweight	.1	.5	.4

Find the probability that a person of normal weight has a thin

19. child;   **20.** grandchild;   **21.** great-grandchild.

Find the probability that an overweight man will have an overweight

22. child;   **23.** grandchild;   **24.** great-grandchild.

Suppose that the distribution of men by weight is initially given by [.2   .55   .25]. *Find the distribution after*

25. 1 generation;   **26.** 2 generations;   **27.** 3 generations.

28. Find the long range prediction for the distribution of weights.

Which of the following transition matrices are regular?

29. $\begin{bmatrix} 0 & 1 \\ .2 & .8 \end{bmatrix}$ **30.** $\begin{bmatrix} .4 & .2 & .4 \\ 0 & 1 & 0 \\ .6 & .3 & .1 \end{bmatrix}$ **31.** $\begin{bmatrix} 1 & 0 & 0 \\ 0 & 1 & 0 \\ .3 & .5 & .2 \end{bmatrix}$

Which of the following are transition matrices for an absorbing Markov chain?

32. $\begin{bmatrix} 1 & 0 & 0 \\ .5 & .1 & .4 \\ 0 & 1 & 0 \end{bmatrix}$ **33.** $\begin{bmatrix} .2 & 0 & .8 \\ 0 & 1 & 0 \\ .7 & 0 & .3 \end{bmatrix}$ **34.** $\begin{bmatrix} .5 & .1 & .1 & .3 \\ 0 & 0 & 1 & 0 \\ 1 & 0 & 0 & 0 \\ .1 & .8 & .05 & .05 \end{bmatrix}$

Find the fundamental matrix F for each of the following absorbing matrices. Also find the matrix FR.

35. $\begin{bmatrix} .2 & .5 & .3 \\ 0 & 1 & 0 \\ 0 & 0 & 1 \end{bmatrix}$ **36.** $\begin{bmatrix} 1 & 0 & 0 \\ 0 & 1 & 0 \\ .3 & .1 & .6 \end{bmatrix}$

37. $\begin{bmatrix} \frac{1}{5} & \frac{1}{5} & \frac{2}{5} & \frac{1}{5} \\ 0 & 1 & 0 & 0 \\ \frac{1}{2} & \frac{1}{4} & \frac{1}{8} & \frac{1}{8} \\ 0 & 0 & 0 & 1 \end{bmatrix}$ **38.** $\begin{bmatrix} .3 & .5 & .1 & .1 \\ .4 & .1 & .3 & .2 \\ 0 & 0 & 1 & 0 \\ 0 & 0 & 0 & 1 \end{bmatrix}$

People in genetics sometimes study the problem of mating offspring from the same two parents; two of these offspring are then mated, and so on. Let A be a dominant gene for some trait, and a the recessive gene. The original offspring can carry genes AA, Aa, or aa. There are six possible ways that these offspring can mate.

State	Mating
1	AA and AA
2	AA and Aa
3	AA and aa
4	Aa and Aa
5	Aa and aa
6	aa and aa

Using these states gives the following transition matrix.

$$
\begin{array}{c}
 \\
1 \\
2 \\
3 \\
4 \\
5 \\
6
\end{array}
\begin{array}{cccccc}
1 & 2 & 3 & 4 & 5 & 6 \\
\left[\begin{array}{cccccc}
1 & 0 & 0 & 0 & 0 & 0 \\
\frac{1}{4} & \frac{1}{2} & 0 & \frac{1}{4} & 0 & 0 \\
0 & 0 & 1 & 0 & 0 & 0 \\
\frac{1}{16} & \frac{1}{4} & \frac{1}{8} & \frac{1}{4} & \frac{1}{4} & \frac{1}{16} \\
0 & 0 & 0 & \frac{1}{4} & \frac{1}{2} & \frac{1}{4} \\
0 & 0 & 0 & 0 & 0 & 1
\end{array}\right]
\end{array}
$$

39. Identify the absorbing states.

40. Find matrix Q.

41. Find F, and the product FR.

42. If the system starts in state 4, find the probability it will end in state 3.

8

Statistics and Probability Distributions

The study of statistics deals with the collection and summarization of data, and methods of drawing conclusions about a population based on data from a sample of the population. Such methods have become increasingly useful in a variety of fields—for example, manufacturing, government, agriculture, medicine, the social sciences, and in all types of research. In this chapter we give a brief introduction to some of the key topics from statistical theory.

8.1 Basic Properties of Probability Distributions

Random Variables   A bank is interested in improving its services to the public. The manager decides to begin by finding out how much time the tellers spend on each transaction. She decides to time the transactions to the nearest minute. To each transaction, then, will be assigned one of the numbers 0, 1, 2, 3, 4, · · ·. That is, if T represents the experiment of timing a transaction, then T may take on any of the values from the list 0, 1, 2, 3, 4, · · ·. Since the value that T takes on for a particular transaction is random, T is called a *random variable*.

> A **random variable** is a function that assigns a real number to each outcome of an experiment.

It is common to use upper case letters, such as X, or Y, for random variables. Lower case letters, such as x, or y, are then used for a particular value of the random variable.

Frequency Distributions   Suppose that the bank manager finds the times for 75 different transactions, with results as shown in Table 1. As the table shows, the shortest transaction time was $t = 1$ minute, and there were 3 transactions of 1-minute duration. The longest time was $t = 10$ minutes. Only one transaction took that long.

Table 1

Time	Frequency
1	3
2	5
3	9
4	12
5	15
6	11
7	10
8	6
9	3
10	1
Total:	75

In Table 1, the ten values which the random variable T assumed in the experiment are listed in the first column and the number of occurrences corresponding to each of these values, the **frequency** of that value, is given in the second column. Table 1 is an example of a **frequency distribution,** which is simply a table listing the frequencies for each value a random variable may assume.

Now suppose after several weeks of instituting new procedures to speed up transactions, the manager takes another survey, this time of 57 transactions, with the times as shown in the frequency distribution of Table 2.

Table 2

Time	Frequency
1	4
2	5
3	8
4	10
5	12
6	17
7	0
8	1
9	0
10	0
Total:	57

Do the results in Table 2 indicate an improvement? It is hard to compare the two tables, since one is based on 75 transactions and the other on 57. To make the data more comparable, we can add a column to each table which will give the relative frequency of each transaction time. These results are shown in Tables 3 and 4. Where necessary, decimals are rounded to the nearest hundredth. To find a **relative frequency,** divide each frequency by the total of the frequencies. Here the individual frequencies are divided by 75 or 57 respectively.

Table 3

Time	Frequency	Relative frequency
1	3	$\frac{3}{75} = .04$
2	5	$\frac{5}{75} \approx .07$
3	9	$\frac{9}{75} = .12$
4	12	$\frac{12}{75} = .16$
5	15	$\frac{15}{75} = .20$
6	11	$\frac{11}{75} \approx .15$
7	10	$\frac{10}{75} \approx .13$
8	6	$\frac{6}{75} = .08$
9	3	$\frac{3}{75} = .04$
10	1	$\frac{1}{75} \approx .01$

Table 4

Time	Frequency	Relative frequency
1	4	$\frac{4}{57} \approx .07$
2	5	$\frac{5}{57} \approx .09$
3	8	$\frac{8}{57} \approx .14$
4	10	$\frac{10}{57} \approx .18$
5	12	$\frac{12}{57} \approx .21$
6	17	$\frac{17}{57} \approx .30$
7	0	$\frac{0}{57} = 0$
8	1	$\frac{1}{57} \approx .02$
9	0	$\frac{0}{57} = 0$
10	0	$\frac{0}{57} = 0$

Whether the differences in relative frequency between the distributions in Tables 3 and 4 are interpreted as desirable or undesirable depends on management goals. If the manager wanted to eliminate the most time-consuming transactions, the results appear to be desirable. However, before the new procedures were followed, the largest relative frequency of transactions, .20 of all transactions, was for a transaction of 5 minutes. After the new procedures, the largest relative frequency, .30, corresponds to a transaction of 6 minutes. At any rate, the results shown in the two tables are easier to compare using relative frequencies.

The relative frequencies of Tables 3 and 4 can be considered as probabilities. A table, such as Table 3 or Table 4, which gives a set of values that a random variable may take, along with the corresponding probabilities, is called a **probability distribution.** The sum of the probabilities shown in a probability distribution must always be 1.

Example 1   Many plants have seed pods with a variable number of seeds. One variety of green beans has no more than 6 seeds per pod. Suppose that examination of thirty such bean pods gave the results shown in Table 5. Here the random variable X tells the number of seeds per pod. Give a probability distribution for these results.

Table 5

X	Frequency
0	3
1	4
2	6
3	8
4	5
5	3
6	1
Total:	30

The probabilities are found by computing the relative frequencies. A total of 30 bean pods were examined, so divide each frequency by 30 to get the probabilities shown in the distribution of Table 6. Some of the results have been rounded to the nearest hundredth.

Table 6

X	Frequency	Probability
0	3	$\frac{3}{30} = .10$
1	4	$\frac{4}{30} \approx .13$
2	6	$\frac{6}{30} = .20$
3	8	$\frac{8}{30} \approx .27$
4	5	$\frac{5}{30} \approx .17$
5	3	$\frac{3}{30} = .10$
6	1	$\frac{1}{30} \approx .03$
Total:	30	

As shown in Table 6, the probability that the random variable X takes on the value 2 is 6/30, or .20. This is often written as

$$P(X = 2) = .20.$$

Also, $P(X = 5) = .10$, and $P(X = 6) \approx .03$. ∎

Instead of writing the probability distribution of the number of seeds as a table, we could write the same information as a set of ordered pairs:

$$\{(0, .10), (1, .13), (2, .20), (3, .27), (4, .17), (5, .10), (6, .03)\}.$$

There is just one probability for each value of the random variable. Thus, a probability distribution defines a function, called a **probability distribution function,** or, simply, a **probability function.** We shall use the terms "probability distribution" and "probability function" interchangeably. The function described in Example 1 is a **discrete function,** since it has a finite number of ordered pairs. A **continuous** probability distribution function has an infinite number of values of the random variable, corresponding to an interval on the number line. We discuss continuous probability distribution functions in Sections 8.4 and 8.5.

The information in a probability distribution is often displayed graphically in a special kind of bar graph called a **histogram.** The bars all have the same width. The heights of the bars are determined by the frequencies. A histogram

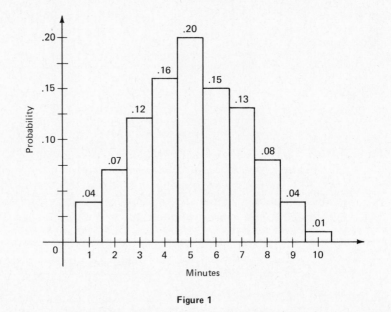

Figure 1

for the data of Table 3 is shown in Figure 1. A histogram shows clearly important characteristics of a distribution which may not be evident in tabular form. For example, a histogram shows at a glance the relative sizes of the probabilities and any symmetry in the distribution.

Note that the area of the bar above $T = 1$ in Figure 1 is given by the product of 1 and .04, or $.04 \times 1 = .04$. Since each bar has a width of 1, its area is equal to the probability which corresponds to that value of T. The probability that a particular value will occur is thus given by the area of the appropriate bar of the graph. Thus, the probability of a transaction time less than four minutes, for example, is given by the sum of the areas for $T = 1$, $T = 2$, and $T = 3$. This area, which is shaded in Figure 2, corresponds to 23% of the total area, since

$$P(T < 4) = P(T = 1) + P(T = 2) + P(T = 3)$$
$$= .04 + .07 + .12 = .23.$$

Example 2   Construct a histogram for the probability distribution of Example 1. Then shade the area which gives the probability that the number of seeds will be more than 4.

A histogram for this distribution is shown in Figure 3. Since

$$P(X \text{ is more than } 4) = P(X = 5) + P(X = 6),$$

the last two bars are shaded. This shaded portion of the histogram represents

$$P(X = 5) + P(X = 6) = .10 + .03 = .13,$$

or 13%, of the total area of the histogram.   ■

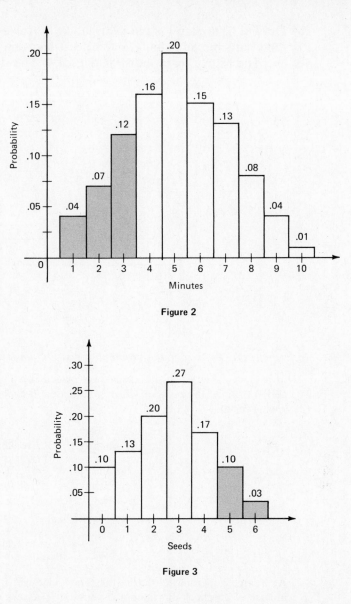

Figure 2

Figure 3

Example 3   **(a)** Give the probability distribution for the number of heads showing when two coins are tossed.

Let X represent the random variable, number of heads. Then X can take on the values 0, 1, or 2. Now find the probability of each outcome. The results are shown in Table 7.

Table 7

X	$P(X)$
0	$\frac{1}{4}$
1	$\frac{1}{2}$
2	$\frac{1}{4}$

(b) Draw a histogram for the distribution of Table 7 and shade the part which represents the probability that at least one coin comes up heads.

The histogram is shown in Figure 4. The shaded portion represents

$$P(X \geq 1) = P(X = 1) + P(X = 2) = \frac{3}{4}. \quad \blacksquare$$

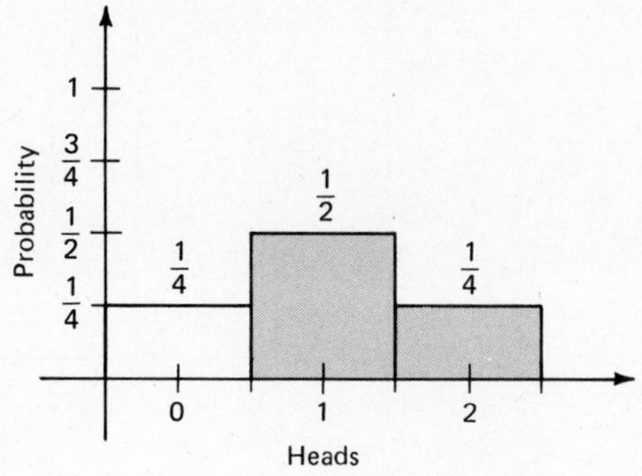

Figure 4

8.1 Exercises   *In Exercises 1–6, (a) give the probability distribution, and (b) sketch its histogram.*

1. In a seed-viability test 50 seeds were placed in 10 rows of five seeds each. After a period of time, the number which germinated in each row were counted with the following results.

Number germinated	Frequency
0	0
1	0
2	1
3	3
4	4
5	2
Total:	10

2. At a large supermarket during the 5-o'clock rush, the number of customers waiting in each of 10 check-out lines was counted. The results are shown below.

Number waiting	Frequency
2	1
3	2
4	4
5	2
6	0
7	1
Total:	10

3. At a training program for police officers, a class of 25 each took 6 shots at a target. The total number of bullseyes are shown in the table below.

Number of bullseyes	Frequency
0	0
1	1
2	0
3	4
4	10
5	8
6	2
Total:	25

4. A class of 42 students took a 10-point quiz. The frequency of scores is given below.

Number of points	Frequency
5	2
6	5
7	10
8	15
9	7
10	3
Total:	42

5. Five mice are innoculated for a disease. After an incubation period, the number who contract the disease is noted. The experiment is repeated twenty times with the following results.

Number with the disease	Frequency
0	3
1	5
2	6
3	3
4	2
5	1
Total:	20

6. The telephone company kept track of the calls for the correct time during a 24-hour period for two weeks. The results are shown below.

Number of calls	Frequency
28	1
29	1
30	2
31	3
32	2
33	2
34	2
35	1
Total:	14

For each of the following experiments, let X determine a random variable, and use your knowledge of probability to prepare a probability distribution.

7. Four coins are tossed and the number of heads is observed each time.

8. Two dice are rolled and the total number of points is noted.

9. Three cards are drawn from a deck. The number of aces are counted.

10. Two balls are drawn from a bag in which there are 4 white balls and two black balls. The number of black balls is counted.

11. A ballplayer with a batting average of .290 comes to bat four times in a game. The number of hits is counted.

12. Five cards are drawn from a deck. The number of black threes is counted.

For each of the following draw a histogram and shade the region which gives the indicated probability.

13. The experiment of Exercise 7; $P(X \leq 2)$

14. The experiment of Exercise 8; $P(X \geq 11)$

15. The experiment of Exercise 9; P(at least one ace)

16. The experiment of Exercise 10; P(at least one black ball)

17. The experiment of Exercise 11; $P(X = 2$ or $X = 3)$

18. The experiment of Exercise 12; $P(1 \leq X \leq 2)$

19. The frequency with which letters occur in a large sample of any written language does not vary much. Therefore, determining the frequency of each letter in a coded message is usually the first step in deciphering it. The percent frequencies of the letters in the English language are as follows.

Letter	%	Letter	%	Letters	%
E	13	S, H	6	W, G, B	1.5
T	9	D	4	V	1
A, O	8	L	3.5	K, X, J	0.5
N	7	C, U, M	3	Q, Z	0.2
I, R	6.5	F, P, Y	2		

Use the introductory paragraph of this exercise as a sample of the English language. Find the percent frequency for each letter in the sample. Compare your results with the frequencies given above.

20. The following message is written in a code in which the frequency of the symbols is the main key to the solution.

)?— —8))y* + 8506*3 × 6;4?*7*& ×*—6.48 () 985)?

(8 + 2: ;48)81&?(;46*3)y*;48&(+8(*509 + &8 () 8 = 8

(5* — 8 — 5(81?098;4&+)&15*50:)6)6*; ?6;6&*0? — 7

(a) Find the frequency of each symbol.

(b) By comparing the high-frequency symbols with the high-frequency letters in English, and the low-frequency symbols with the low-frequency letters, try to decipher the message. (Hint: Look for repeated two-symbol combinations and double letters for added clues. Try to identify vowels first.)

8.2 Expected Value

In working with experimental data, it is often useful to have a single number, a typical, or "average" number, which is representative of the entire set of data. Just about everyone is familiar with some type of average. For example, we compare our heights and weights to those of the typical or "average" person on weight charts. Students are familiar with the "class average" and their own "average" at any time in a given course.

Let's take a look at some statistics. In a recent year, a citizen of the United States could expect to complete about 12.5 years of school, to be a member of a household earning $20,091 per year, and to live in a household of 2.78 people. What do we mean here by "expect"? We know of many people who have completed less than 12.5 years of school; many others have completed more. Many households have less income than $20,091 per year; many others have more. The idea of a household of 2.78 people is a little hard to swallow. The numbers all refer to *averages*. When the term "expect" is used in this way, it refers to *mathematical expectation,* which we shall see is a kind of average.

The **arithmetic mean,** or **average,** of a set of numbers is the sum of the numbers, divided by the total number of numbers. To write the sum of the n numbers $x_1, x_2, x_3, \cdots, x_n$ in a compact way, it is customary to use **summation notation:** using the Greek letter Σ (sigma), the sum $x_1 + x_2 + x_3 + \cdots + x_n$ is written

$$x_1 + x_2 + x_3 + \cdots + x_n = \sum_{i=1}^{n} x_i.$$

The symbol $\bar{x}$ (read x-bar) is used to represent the mean, so that the mean of the n numbers $x_1, x_2, x_3, \cdots, x_n$ is

$$\bar{x} = \frac{\sum_{i=1}^{n} x_i}{n}.$$

For example, the mean of the set of numbers 2, 3, 5, 6, 8 is

$$\frac{2 + 3 + 5 + 6 + 8}{5} = \frac{24}{5} = 4.8.$$

What about an average value for a random variable? Can we use the mean to find it? As an example, let us find the average number of offspring for a certain species of pheasant, given the probability distribution below.

Number of offspring	Frequency	Probability
0	8	.08
1	14	.14
2	29	.29
3	32	.32
4	17	.17
Total:	100	

We might be tempted to find the typical number of offspring by averaging the numbers 0, 1, 2, 3, and 4, which represent the numbers of offspring possible. This won't work, however, since the various numbers of offspring do not occur with equal probability: for example, 3 offspring are much more common than 0 or 1 offspring. We can take the differing probabilities of occurrence into account by finding a **weighted average.** This is done by multiplying each of the possible numbers of offspring by its corresponding probability, as follows:

$$\text{typical number of offspring} = 0(.08) + 1(.14) + 2(.29) + 3(.32) + 4(.17)$$
$$= 0 + .14 + .58 + .96 + .68$$
$$= 2.36.$$

Based on the data above, the typical family of pheasants has 2.36 offspring.

It is certainly not possible for a pair of pheasants to produce 2.36 offspring. However, if the number of offspring produced by many different pairs of pheasants are found, then the average, or the mean, of these numbers will be about 2.36.

We can use the idea of this example to define the mean, or expected value, of a probability distribution. This is done as follows.

Suppose the random variable X can take on the n values $x_1, x_2, x_3, \cdots, x_n$. Also, suppose the probabilities that each of these values occurs are respectively $p_1, p_2, p_3, \cdots, p_n$. Then the **expected value** of the random variable is

$$E(X) = x_1 p_1 + x_2 p_2 + x_3 p_3 + \cdots + x_n p_n.$$

The symbol μ (the Greek letter mu) is used for the expected value of the random variable X. Note that the expected value of a random variable may be a number which can never occur on any one trial of the experiment (as with the pheasant example above). The expected value gives an approximation of the result of averaging the outcomes of a great many trials.

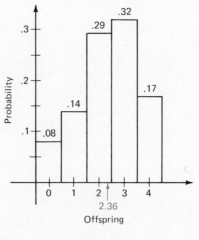

Figure 5

Physically, the expected value of a probability distribution represents a balance point. Figure 5 shows a histogram for the distribution of the pheasant offspring. If we imagine the histogram to be a series of weights with magnitudes represented by the heights of the bars, then the system would balance if supported at the point corresponding to the expected value.

Example 1 The local church decides to raise money by raffling a microwave oven worth $400. A total of 2000 tickets are sold at $1 each. Find the expected value of winning for a person who buys one ticket in the raffle.

Here the random variable represents the possible amounts of net winnings, where net winnings = amount of winning − cost of ticket. The net winnings of the person winning the oven are $400 (amount of winning) − $1 (cost of ticket) = $399. The net winnings for each losing ticket are $0 − $1 = −$1.

The probability of winning is 1 in 2000, or 1/2000, while the probability of losing is 1999/2000. In summary, we have the information shown in the chart.

Outcome (net winning)	Probability
$399	$\dfrac{1}{2000}$
−$1	$\dfrac{1999}{2000}$

The expected winnings for a person buying one ticket are thus

$$399\left(\frac{1}{2000}\right) + (-1)\left(\frac{1999}{2000}\right) = \frac{399}{2000} - \frac{1999}{2000} = -\frac{1600}{2000} = -.80.$$

On the average, a person buying one ticket in the raffle will lose $.80, or 80¢.

It is not possible to lose 80¢ in this raffle—you either lose $1, or you win a $400 prize. However, if you bought tickets in many such raffles over a long period of time, you would find that you lose 80¢ per ticket, on the average. ∎

Example 2 What is the expected number of girls in a family having three children?

Some families with three children will have 0 girls, others will have 1 girl, and so on. We need to find the probabilities associated with 0, 1, 2, or 3 girls in a family of three children. To find these probabilities we first write the sample space S of all possible three-child families: $S = \{$ggg, ggb, bgg, gbb, bgb, bbg, bbb, gbg$\}$. From this sample space, we get the probabilities shown in the following chart.

Outcome (number of girls)	Probability
0	$\frac{1}{8}$
1	$\frac{3}{8}$
2	$\frac{3}{8}$
3	$\frac{1}{8}$

The expected number of girls can now be found by multiplying each outcome (number of girls) by its corresponding probability and finding the sum of these values. Thus,

$$\text{expected number of girls} = 0 \cdot \frac{1}{8} + 1 \cdot \frac{3}{8} + 2 \cdot \frac{3}{8} + 3 \cdot \frac{1}{8}$$

$$= \frac{3}{8} + \frac{6}{8} + \frac{3}{8}$$

$$= \frac{12}{8} = \frac{3}{2} = 1\frac{1}{2}.$$

On the average, a three-child family will have 1 1/2 girls. ▐

Example 3   Each day Donna and Mary toss a coin to see who buys the coffee (40¢ a cup). One tosses and the other calls the outcome. If the person who calls the outcome is correct, the other buys the coffee; otherwise the caller pays. Find Donna's expected winnings.

Let's assume that an honest coin is used, that Mary tosses the coin, and that Donna calls the outcome. The possible results and corresponding probabilities are shown below.

	Possible results			
Result of toss	H	H	T	T
Call	H	T	H	T
Caller wins?	Yes	No	No	Yes
Probability	$\frac{1}{4}$	$\frac{1}{4}$	$\frac{1}{4}$	$\frac{1}{4}$

Donna wins a 40¢ cup of coffee whenever the results and calls match, and loses a 40¢ cup when there is no match. Her expected winnings are

$$(.40)\left(\frac{1}{4}\right) + (-.40)\left(\frac{1}{4}\right) + (-.40)\left(\frac{1}{4}\right) + (.40)\left(\frac{1}{4}\right) = 0.$$

On the average, over the long run, Donna neither wins nor loses. ▐

A game with an expected value of 0 (such as the one of Example 3) is called a **fair game.** Casinos do not offer fair games. If they did, they would win on the average $0, and have a hard time paying the help! Casino games have expected winnings for the house that vary from 1 1/2¢ per dollar to 60¢ per dollar. Exercises 18–21 below ask you to find the expected winnings for certain games of chance.

The idea of expected value can be very useful in decision making, as shown by the next example.

Example 4   At age 50, you receive a letter from the Mutual of Mauritania Insurance Company. According to the letter, you must tell the company immediately which of the following two options you will choose: take $20,000 at age 60 (if you are alive, $0 otherwise) or $30,000 at age 70 (again, if you are alive, $0 otherwise). Based on the idea of expected value, which should you choose?

Life insurance companies have constructed elaborate tables showing the probability of a person living a given number of years into the future. By consulting a recent such table, we find that the probability of living from age 50 to age 60 is .88, while the probability of living from age 50 to 70 is .64. The expected values of the two options are given below.

First option: $(20,000)(.88) + (0)(.12) = 17,600.$

Second option: $(30,000)(.64) + (0)(.36) = 19,200.$

Based strictly on expected values, we would choose the second option.   ■

8.2 Exercises   *Find the expected value for each of the following random variables.*

1.

X	2	3	4	5
$P(X = x)$	.1	.4	.3	.2

2.

Y	4	6	8	10
$P(Y = y)$	.4	.4	.05	.15

3.

Z	9	12	15	18	21
$P(Z = z)$	.14	.22	.36	.18	.10

4.

X	30	32	36	38	44
$P(X = x)$	.31	.30	.29	.06	.04

Find the expected value for the random variable X having probability functions graphed as follows.

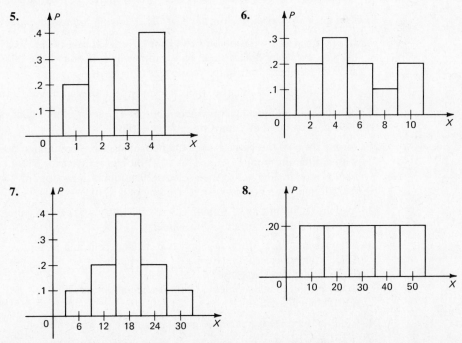

5.

6.

7.

8.

9. A raffle offers a first prize of $100, and two second prizes of $40 each. One ticket costs $1, and 500 tickets are sold. Find the expected winnings for a person who buys one ticket. Is this a fair game?

10. A raffle offers a first prize of $1000, two second prizes of $300 each, and twenty prizes of $10 each. If 10,000 tickets are sold at 50¢ each, find the expected winnings for a person buying one ticket. Is this a fair game?

Many of the following exercises use the idea of combinations, which we discussed in Chapter 5.

11. If 3 marbles are drawn from a bag containing 3 yellow and 4 white marbles, what is the expected number of yellow marbles in the sample?

12. If 5 apples in a barrel of 25 apples are known to be rotten, what is the expected number of rotten apples in a sample of 2 apples?

13. A delegation of 3 is selected from a city council made up of 5 liberals and 4 conservatives.

    (a) What is the expected number of liberals on the committee?
    (b) What is the expected number of conservatives?

14. From a group of 2 women and 5 men, a delegation of 2 is selected. Find the expected number of women in the delegation.

15. In a club with 20 senior and 10 junior members, what is the expected number of junior members on a 3-member committee?

16. If 2 cards are drawn at one time from a deck of 52 cards, what is the expected number of diamonds?

17. Suppose someone offers to pay you $5 if you draw 2 diamonds in the game of Exercise 16. He says that you should pay 50¢ for the chance to play. Is this a fair game?

Find the expected winnings for each of the following games of chance.

18. In one form of roulette, you bet $1 on "even." If one of the 18 even numbers comes up, you get your dollar back, plus another one. If one of the 20 noneven (18 odd, 0, and 00) numbers comes up, you lose.

19. In another form of roulette, there are only 19 noneven numbers (no 00).

20. Numbers is an illegal game where you bet $1 on any three digit number from 000 to 999. If your number comes up, you get $500.

21. In one form of the game Keno, the house has a pot containing 80 balls, each marked with a different number from 1 to 80. You buy a ticket for $1 and mark one of the 80 numbers on it. The house then selects 20 numbers at random. If your number is among the 20, you get $3.20 (for a net winning of $2.20).

22. Use the assumptions of Example 3 to find Mary's expected winnings. If Mary tosses and Donna calls, is it still a fair game?

23. Suppose one day Mary brings a two-headed coin and uses it to toss for the coffee. Since Mary tosses, Donna calls.
    (a) Is this still a fair game?
    (b) What is Donna's expected gain if she calls heads?
    (c) If she calls tails?

24. Find the expected number of girls in a family of four children.

25. Find the expected number of boys in a family of five children.

26. Jack must choose at age 40 to inherit either $25,000 at age 50 (if he is still alive) or $30,000 at age 55 (if he is still alive). If the probabilities for a person of age 40 to live to be 50 and 55 are .90 and .85, respectively, which choice gives him the larger expected inheritance?

27. An insurance company has written 100 policies of $10,000, 500 of $5000, and 1000 policies of $1000 on people of age 20. If experience shows that the probability of dying during the twentieth year of life is .001, how much can the company expect to pay out during the year the policies were written?

28. A builder is considering a job which promises a profit of $30,000 with a probability of .7 or a loss (due to bad weather, strikes, and such) of $10,000 with a probability of .3. What is the expected profit?

29. Experience has shown that a ski lodge will be full (160 guests) during the Christmas holidays if there is a heavy snow pack in December, while a light snowfall in December means that they will have only 90 guests. What is the expected number of guests if the probability for a heavy snow in December is .40? (Assume that there must either be a light snowfall or a heavy snowfall.)

30. A magazine distributor offers a first prize of $100,000, two second prizes of $40,000 each, and two third prizes of $10,000 each. A total of 2,000,000 entries are received in the contest. Find the expected winnings if you submit one entry to the contest. If it would cost you 25¢ in time, paper, and stamps to enter, would it be worth it?

31. The local Saab dealer gets complaints about his cars, as shown in the following table.

Number of complaints per day	0	1	2	3	4	5	6
Probability	.01	.05	.15	.26	.33	.14	.06

Find the expected number of complaints per day.

32. I can take one of two jobs. With job A, there is a 50% chance that I will make $60,000 per year after 5 years, and a 50% chance of making $30,000. With job B, there is a 30% chance that I will make $90,000 per year after 5 years and a 70% chance that I will make $20,000. Based strictly on expected value, which job should I take?

33. Levi Strauss and Company[1] uses expected value to help its salespeople rate their accounts. For each account, a salesperson estimates potential additional volume and the probability of getting it. The product of these gives the expected value of the potential, which is added to the existing volume. The totals are then classified as A, B, or C as follows: below $40,000, class C; between $40,000 and $55,000, class B; above $55,000, class A. Complete the following chart for one of its salespeople.

Account no.	Existing volume	Potential add'l vol.	Probability of getting it	Expected value of potent'l	Existing vol. + expected value of potent'l	Class
1	$15,000	$10,000	.25	$2,500	$17,500	C
2	40,000	0	—	—	40,000	C
3	20,000	10,000	.20			
4	50,000	10,000	.10			
5	5,000	50,000	.50			
6	0	100,000	.60			
7	30,000	20,000	.80			

[1]This example was supplied by James McDonald, Levi Strauss and Company, San Francisco.

34. At the end of play in a major golf tournament, two players, an "old pro" and a "new kid" are tied. Suppose first prize is $80,000 and second prize is $20,000. Find the expected winnings for the old pro if
    (a) both players are of equal ability,
    (b) the new kid will freeze up, giving the old pro a 3/4 chance of winning.

35. In a certain animal species, the probability that a healthy adult female will have no offspring in a given year is .31, while the probability of 1, 2, 3, or 4 offspring are respectively .21, .19, .17, and .12. Find the expected number of offspring.

36. According to an article in a magazine not known for its accuracy, a male decreases his life expectancy by one year, on the average, for every point that his blood pressure is above 120. The average life expectancy for a male is 76 years. Find the life expectancy for a male whose blood pressure is
    (a) 135;   (b) 150;   (c) 115;   (d) 100.
    (e) Suppose a certain male has a blood pressure of 145. Find his life expectancy. How would you interpret the result to him?

37. One of the few methods that can be used in an attempt to cut the severity of a hurricane is to *seed* the storm. In this process, silver iodide crystals are dropped into the storm. Unfortunately, silver iodide crystals sometimes cause the storm to *increase* its speed. Wind speeds may also increase or decrease even with no seeding. The probabilities and amounts of property damage in the following tree diagram are from an article by R. A. Howard, J. E. Matheson, and D. W. North, "The Decision to Seed Hurricanes."[1]
    (a) Find the expected amount of damage under each option, "seed" and "do not seed."
    (b) To minimize total expected damage, what option should be chosen?

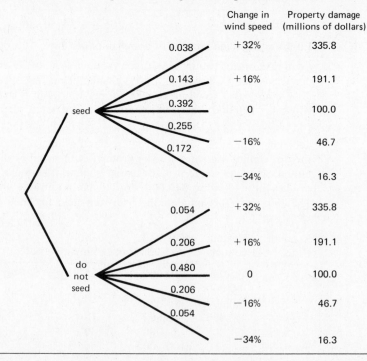

[1]"The Decision to Seed Hurricanes," Howard, R. A. et al., *Science*, Vol. 176, pp. 1191–1202, Figure 9, 16 June 1972. Copyright 1972 by the American Association for the Advancement of Science.

38. A recent McDonald's contest offered the following cash prizes and probabilities of winning on one visit.

Prize	Probability
$100,000	$\dfrac{1}{176,402,500}$
$25,000	$\dfrac{1}{39,200,556}$
$5000	$\dfrac{1}{17,640,250}$
$1000	$\dfrac{1}{1,568,022}$
$100	$\dfrac{1}{282,244}$
$5	$\dfrac{1}{7056}$
$1	$\dfrac{1}{588}$

Suppose you spend $1 to buy a bus pass that lets you go to 25 different McDonald's and pick up entry forms. Find your expected value.

Application   Optimal Inventory for a Service Truck

For many different items it is difficult or impossible to take the item to a central repair facility when service for the item is required. Washing machines, large television sets, office copiers, and computers are only a few examples of such items. Service for items of this type is commonly performed by sending a repair person to the item, with the person driving to the item in a truck containing various parts that might be required in repairing the item. Ideally, the truck should contain all the parts that might be required in repairing the item. However, most parts would be needed only infrequently, so that inventory costs for the parts would be high.

   An optimum policy for deciding on the parts to stock on a truck would require that the probability of not being able to repair an item without a trip back to the warehouse for needed parts be as low as possible, consistent with minimum inventory costs. An analysis similar to the one below was developed at the Xerox Corporation.[1]

   To set up a mathematical model for deciding on the optimum truck stocking policy, let us assume that a broken machine might require one of 5 different

---

[1]Reprinted by permission of Stephen Smith, John Chambers, and Eli Shlifer, "Optimal Inventories Based on Job Completion Rate for Repairs Requiring Multiple Items," *Management Science*, Vol. 26, No. 8, August 1980, copyright © 1980 The Institute of Management Sciences.

parts (we could assume any number of different parts—we use 5 to simplify the notation). Suppose also that the probability that a particular machine requires part 1 is p_1, that it requires part 2 is p_2, and so on. Assume also that the failure of different part types are independent, and that at most one part of each type is used on a given job.

Suppose that, on the average, a repair person makes N service calls per time period. If the repair person is unable to make a repair because at least one of the parts is unavailable, there is a penalty cost, L, corresponding to wasted time for the repair person, an extra trip to the parts depot, customer unhappiness, and so on. For each of the parts carried on the truck, an average inventory cost is incurred. Let H_i be the average inventory cost for part i, where $1 \le i \le 5$.

Let M_1 represent a policy of carrying only part 1 on the repair truck, M_{24} represent a policy of carrying only parts 2 and 4, with M_{12345} and M_0 representing policies of carrying all parts and no parts, respectively.

For policy M_{35}, carrying parts 3 and 5 only, the expected cost per time period per repair person, written $C(M_{35})$, is

$$C(M_{35}) = (H_3 + H_5) + NL[1 - (1 - p_1)(1 - p_2)(1 - p_4)]$$

(The expression in brackets represents the probability of needing at least one of the parts not carried, 1, 2, or 4 here.) As a further example, $C(M_{125})$ is

$$C(M_{125}) = (H_1 + H_2 + H_5) + NL[1 - (1 - p_3)(1 - p_4)],$$

while

$$C(M_{12345}) = (H_1 + H_2 + H_3 + H_4 + H_5) + NL[1 - 1]$$
$$= H_1 + H_2 + H_3 + H_4 + H_5,$$

and

$$C(M_0) = NL[1 - (1 - p_1)(1 - p_2)(1 - p_3)(1 - p_4)(1 - p_5)].$$

To find the best policy, evaluate $C(M_0)$, $C(M_1)$, $\cdots$, $C(M_{12345})$ and choose the smallest result. (A general method of solution is given in the Management Science paper.)

Example   Suppose that for a particular item, only 3 possible parts might need to be replaced. By studying past records of failures of the item, and finding necessary inventory costs, suppose that the following values have been found.

p_1	p_2	p_3	H_1	H_2	H_3
.09	.24	.17	$15	$40	$9

Suppose $N = 3$ and L is $54. Then, as an example,

$$C(M_1) = H_1 + NL[1 - (1 - p_2)(1 - p_3)]$$
$$= 15 + 3(54)[1 - (1 - .24)(1 - .17)]$$
$$= 15 + 3(54)[1 - (.76)(.83)]$$
$$\approx 15 + 59.81$$
$$= 74.81.$$

Thus, if policy M_1 is followed (carrying only part 1 on the truck), the expected cost per repair person per time period is $74.81. Also,

$$C(M_{23}) = H_2 + H_3 + NL[1 - (1 - p_1)]$$
$$= 40 + 9 + 3(54)[.09]$$
$$= 63.58,$$

so that M_{23} is a better policy than M_1. By finding the expected values for all other possible policies (see the exercises below), the optimum policy may be chosen. ∎

Exercises

1. Refer to the example above and find each of the following.
   (a) $C(M_0)$ (b) $C(M_2)$ (c) $C(M_3)$ (d) $C(M_{12})$
   (e) $C(M_{13})$ (f) $C(M_{123})$.

2. Which policy leads to lowest expected cost?

3. In the example above, $p_1 + p_2 + p_3 = .09 + .24 + .17 = .50$. Why is it not necessary that the probabilities add to 1?

4. Suppose an item to be repaired might need one of n different parts. How many different policies would then need to be evaluated?

Application   *Bidding on a Potential Oil Field—Signal Oil*   ❮

Signal Oil, with headquarters in Los Angeles, is a major petroleum company, controlling, along with other companies, Mack Truck. In this example we use probability and expected values to help determine the best bid price for a new off-shore oil field. The company has used all the modern methods of oil exploration to help interpret the economic potential of each tract.[1]

Two uncontrollable (and therefore uncertain) variables dominate a problem of this type: (a) the amount of commercial oil reserves that might be found in a tract, and (b) the length of time that would be required to develop and begin commercial production using these reserves. Another important variable is the amount to be bid for the right to develop the tract. Although the bid is a variable, it is not subject to uncertainty, but is under the control of the company. The company must analyze the effects of bids of various sizes along with the variables involving uncertainty so that the proper bid can be made.

The following chart shows the probabilities of various events. Commercial production includes events B_2, B_3, and B_4. Note that commercial production is

---

[1]This example was supplied by Kenneth P. King, Senior Planning Analyst, Signal Oil Company.

given a 20% chance of occurring, with an 80% chance of the occurrence of less than a commercially profitable level of oil reserves.

Event B_j	Oil Reserves Millions of barrels	Chance of occurrence
B_1	0.0	.80
B_2	19.0	.06
B_3	25.5	.10
B_4	30.6	.04

Any delay in beginning the commercial development of the field adversely affects the overall profitability of the project. This delay can be caused by seasonal weather variation in the offshore area, together with its relative isolation and the uncertainty of drilling rig availability. Beginning development in a shorter-than-normal time would require a concerted speedup effort that would incur cost increases over the normal period of development. This additional cost, however, is somewhat offset by the fact that the income from the field would be received sooner. The chart below shows the probabilities of various lengths of time required for commercial development to begin.

Event A_i	Years From Bid to Start of Drilling Years	Chance of occurrence
A_1	2	.75
A_2	1	.13
A_3	3	.12

The time required for drilling to begin is independent of the quantity of reserves in the field. Hence, for each possible value of i and j, we have

$$P(A_i \text{ and } B_j) = P(A_i) \cdot P(B_j).$$

For example, $P(2 \text{ years' delay and } 25.5 \text{ million barrels}) = P(A_1 \text{ and } B_3) = P(A_1) \cdot P(B_3) = (.75)(.10) = .075$. The chart below shows the probabilities for all possible cases, along with the payoffs to the company for different bid levels.

Case	Event	Probability	Payoff (in millions of dollars) $0 Bid	$2	$5	$10
1	A_1 and B_1	.600	−1.1	−2.3	−4.0	−6.9
2	A_1 and B_2	.045	10.4	8.4	5.4	0.4
3	A_1 and B_3	.075	17.0	15.0	12.0	7.0
4	A_1 and B_4	.030	22.2	20.2	17.2	12.2
5	A_2 and B_1	.104	−1.1	−2.3	−4.0	−6.9
6	A_2 and B_2	.008	13.2	11.2	8.2	3.2
7	A_2 and B_3	.013	21.5	19.5	16.5	11.5
8	A_2 and B_4	.005	27.7	25.7	22.7	17.7
9	A_3 and B_1	.096	−1.1	−2.3	−4.0	−6.9
10	A_3 and B_2	.007	7.5	5.5	2.5	−2.5
11	A_3 and B_3	.012	13.0	11.0	8.0	3.0
12	A_3 and B_4	.005	17.5	15.5	12.5	7.5
	Total:	1.000				

Now the company must calculate the expected value for each different bid level. For example, the expected value at a bid level of $2 million is given by

$$E(\text{bid of \$2 million}) = (-2.3)(.600) + (8.4)(.045)$$
$$+ (15.0)(.075) + \cdots + (15.5)(.005).$$

If the expected values for various possible bid levels are found in the same way and plotted, we get the graph in Figure 1. Using techniques from mathematics of finance (the payoffs above are actually present values), the company knows that the expected value of a profitable bid must be $0 or more. As shown in Figure 1, this means that $3.5 million is the most the company can bid for this particular tract.

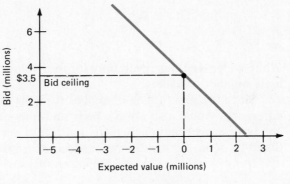

Figure 1

Exercises

1. Calculate $E(\text{bid of \$2 million})$.

2. Calculate $E(\text{bid of \$5 million})$.

8.3 Variance and Standard Deviation

The mean of a distribution gives us an average value of the distribution, but the mean tells us nothing about the spread of the numbers in the distribution. For example, suppose seven measurements of the thickness (in cm) of a copper wire produced by one machine were

.010, .010, .009, .008, .007, .009, .010,

and seven measurements of the same type of wire produced by another machine were

.014, .004, .013, .005, .009, .004, .014.

The mean of both samples is .009, yet the two samples are quite different; the amount of dispersion or variation within the samples is different. In addition to the mean, we need another kind of measure, which describes how much the numbers in a distribution vary.

Since the mean represents the center of the distribution, one way to measure the variation within a set of numbers might be to find the average of their distances from the mean. That is, if the numbers are $x_1, x_2, \cdots, x_n$ and the mean is $\bar{x}$, find the differences $x_1 - \bar{x}, x_2 - \bar{x}, \cdots, x_n - \bar{x}$, and then find the mean of the differences. However, it turns out that the sum of these differences is always 0, so that the mean would also be 0. To see why, let us look at the four numbers x_1, x_2, x_3, x_4 having mean $\bar{x}$. The sum of the four differences is

$$\sum_{i=1}^{4} (x_i - \bar{x}) = (x_1 - \bar{x}) + (x_2 - \bar{x}) + (x_3 - \bar{x}) + (x_4 - \bar{x}) = x_1 + x_2 + x_3 + x_4 - 4(\bar{x})$$

By definition, $\bar{x} = (x_1 + x_2 + x_3 + x_4)/4$, giving

$$\bar{x} = x_1 + x_2 + x_3 + x_4 - 4\left(\frac{x_1 + x_2 + x_3 + x_4}{4}\right)$$

$$= 0.$$

While we proved this result only for four values, the proof can be extended to any finite number of values.

Since the sum of the differences from the mean is always 0, the mean of these differences would also always be 0, and thus not be a good measure of the variability of a distribution. It turns out that a very useful measure of variability is found by *squaring* the differences from the mean.

For example, let us use the seven measurements given above,

$$.010, \quad .010, \quad .009, \quad .008, \quad .007, \quad .009, \quad .010.$$

The mean of these numbers is .009. By subtracting the mean from each of the seven values, we get the differences

$$.001, \quad .001, \quad 0, \quad -.001, \quad -.002, \quad 0, \quad .001.$$

(Check that the sum of these differences is 0.) Now square each difference, getting

$$.000001, \quad .000001, \quad 0, \quad .000001, \quad .000004, \quad 0, \quad .000001.$$

Next, find the mean of these squares, which is .00000114 (rounded).

This number, the mean of the squares of the differences, is called the **variance** of the distribution. If X is the random variable for the distribution, then the variance is written Var(X). The variance gives a measure of the variation of the numbers in the distribution, but, since we used the squared differences to get it, the size of the variance does not reflect the actual amount of variation. To correct this problem, another measure of variation is used, the **standard deviation,** which is the square root of the variance. The symbol σ (the Greek lower case sigma) is used for standard deviation. The standard deviation of the distribution discussed above is

$$\sigma = \sqrt{.00000114} \approx .001.$$

Example 1   Find the standard deviation of the seven measurements of copper wire produced by the second machine in the example above.

It is best to arrange the work in columns.

x	$x - \bar{x}$	$(x - \bar{x})^2$
.014	.005	.000025
.004	−.005	.000025
.013	.004	.000016
.005	−.004	.000016
.009	0	0
.004	−.005	.000025
.014	.005	.000025
	Total:	.000132

As we have seen, the column $x - \bar{x}$ should always have a sum of 0. This is a good way to check your work at that point. Now, to get the variance, divide the sum of the $(x - \bar{x})^2$ column by the number of values in the set, seven in this case. Then take the square root to get the standard deviation.

$$\text{variance} = \frac{.000132}{7} \approx .0000189$$

$$\sigma = \sqrt{.0000189} \approx .004$$

Note that both measures of variation, the variance and the standard deviation, are larger for this sample than for the first sample. Thus, the first machine produces copper wire with less variation than the second. ∎

The concept of variance is generalized as follows:

The **variance** of a set of n numbers $x_1, x_2, x_3, \cdots x_n$, with a mean $\bar{x}$, is

$$\text{Var}(x) = \frac{\Sigma(x - \bar{x})^2}{n}.$$

The **standard deviation** of the set is

$$\sigma = \sqrt{\frac{\Sigma(x - \bar{x})^2}{n}}.$$

Variation for a probability distribution is measured in a similar way.

If a random variable X takes on the n values $x_1, x_2, x_3, \cdots, x_n$ with respective probabilities $p_1, p_2, p_3, \cdots, p_n$, and if its expected value is $E(X) = \mu$, then the **variance** of X is

$$\text{Var}(X) = p_1(x_1 - \mu)^2 + p_2(x_2 - \mu)^2 + \cdots + p_n(x_n - \mu)^2.$$

The **standard deviation** of X is

$$\sigma = \sqrt{\text{Var}(X)}.$$

Example 2   Find the variance and the standard deviation of the number of pheasant offspring given the following probability distribution.

Table 8

X	p_i
0	.08
1	.14
2	.29
3	.32
4	.17

We need the expected value of the distribution. In Section 8.2, we found $\mu = 2.36$ for this distribution.

To use the formula in the box above it is easiest to work in columns as we did in Example 1.

Table 9

X	p_i	$x_i - \mu$	$(x_i - \mu)^2$	$p_i(x_i - \mu)^2$
0	.08	−2.36	5.57	.45
1	.14	−1.36	1.85	.26
2	.29	−.36	.13	.04
3	.32	.64	.41	.13
4	.17	1.64	2.69	.46
				1.34

The total of the last column gives the variance, 1.34. To find the standard deviation, take the square root of the variance.

$$\sigma = \sqrt{1.34} \approx 1.16 \quad \blacksquare$$

Chebyshev's Theorem   Suppose we know only the mean, or expected value, μ of a distribution, along with the standard deviation σ. What then can be said about the values of the distribution? For example, if σ is very small, we would expect most of the values of the distribution to be close to μ, while a larger value of σ would suggest more spread in the values. One estimate of the fraction of values that lie within a specified distance of the mean is given by **Chebyshev's Theorem,** named after the Russian mathematician P. L. Chebyshev, 1821–94.

> *Chebyshev's Theorem*   For any distribution of numbers with mean μ and standard deviation σ, the probability that a number will lie within k standard deviations of the mean is at least
>
> $$1 - \frac{1}{k^2}.$$
>
> That is,
>
> $$P(\mu - k\sigma \leq X \leq \mu + k\sigma) \geq 1 - \frac{1}{k^2}.$$

Example 3   By Chebyshev's Theorem, at least

$$1 - \frac{1}{3^2} = 1 - \frac{1}{9} = \frac{8}{9},$$

or about 89%, of the numbers in any distribution lie within 3 standard deviations of the mean. The graph in Figure 6 shows a geometric interpretation of this result.   ▊

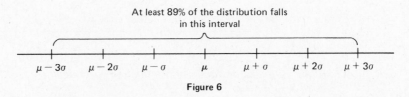

At least 89% of the distribution falls in this interval

$$\mu - 3\sigma \quad \mu - 2\sigma \quad \mu - \sigma \quad \mu \quad \mu + \sigma \quad \mu + 2\sigma \quad \mu + 3\sigma$$

Figure 6

Suppose a distribution has mean 52 and standard deviation 3.5. Then "3 standard deviations" is $3 \times 3.5 = 10.5$, and "three standard deviations from the mean" is

$$52 - 10.5 \quad \text{to} \quad 52 + 10.5,$$

or

$$41.5 \quad \text{to} \quad 62.5.$$

By Example 3, at least 89% of the values in this distribution will lie between 41.5 and 62.5.

Chebyshev's Theorem gets much of its importance from the fact that it applies to *any* distribution—we need know only the mean and the standard deviation. Other results that we give later produce more accurate estimates, but only after we have additional information about the distribution.

Example 4   The Forever Power Company claims that their batteries have a mean life of 26.2 hours with a standard deviation of 4.1 hours. In a shipment of 100 batteries, about how many will have a life within 2 standard deviations of the mean—that is, between $26.2 - (4.1 \times 2) = 18$ and $26.2 + (4.1 \times 2) = 34.4$ hours?

Use Chebyshev's Theorem with $k = 2$. At least

$$1 - \frac{1}{2^2} = 1 - \frac{1}{4} = \frac{3}{4},$$

or 75%, of the batteries will have a life within 2 standard deviations of the mean. Thus, at least $75\% \times 100 = 75$ of the batteries will last between 18 and 34.4 hours.   ▊

8.3 Exercises   *Find the standard deviation for each of the following sets of numbers.*

  1. 42; 38; 29; 74; 82; 71; 35

  2. 122; 132; 141; 158; 162; 169; 180

  3. 241; 248; 251; 257; 252; 287

4. 51; 58; 62; 64; 67; 71; 74; 78; 82; 93

5. 3; 7; 4; 12; 15; 18; 19; 27; 24; 11

6. 15; 42; 53; 7; 9; 12; 28; 47; 63; 14

Find the variance and standard deviation for each of the following probability distributions.

7.

x_i	2	3	4	5
p_i	.1	.3	.4	.2

8.

x_i	10	20	30	40
p_i	.1	.5	.3	.1

9.

x_i	.01	.02	.03	.04	.05
p_i	.1	.5	.2	.1	.1

10.

x_i	100	105	110	115	120
p_i	.01	.08	.20	.50	.21

Find the standard deviation of the random variable in each of the following. All exercises refer to Section 8.2.

11. the number of yellow marbles in Exercise 11

12. the number of rotten apples in Exercise 12

13. the number of liberals on the committee in Exercise 13

14. the number of women in the delegation in Exercise 14

Each of the following gives histograms of two probability distributions. Decide from the graphs without any calculations which distribution has the greatest variance.

15.

16.

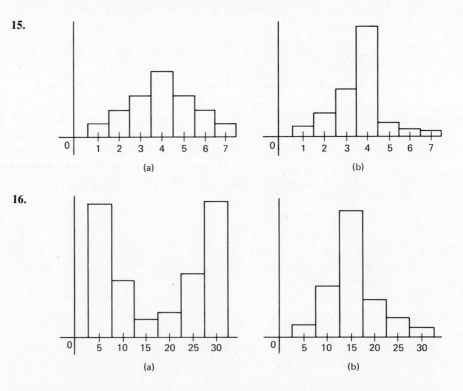

(a)   (b)

(a)   (b)

For each of the following, the histogram of a probability distribution is given. Calculate the variance.

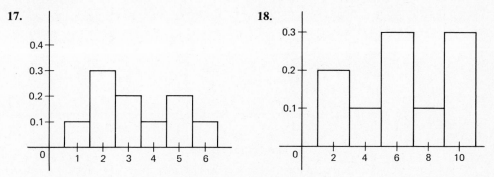

17.

18.

19. Use Chebyshev's Theorem to find the fraction of a distribution that lies within the following numbers of standard deviations from the mean.

(a) 2   (b) 4   (c) 5

20. A probability distribution has an expected value of 50 and a standard deviation of 6. Use Chebyshev's Theorem to tell what percent of the numbers lie between each of the following values.

(a) 38 and 62   (b) 32 and 68   (c) 26 and 74   (d) 20 and 80
(e) less than 38 or more than 62   (f) less than 32 or more than 68

21. The weekly wages of the seven employees of Harold's Hardware Store are $180, $190, $240, $256, $300, $360, and $714.

(a) Find the mean and standard deviation of this distribution.
(b) How many of the seven employees earn within one standard deviation of the mean?
(c) How many earn within two standard deviations of the mean?
(d) What does Chebyshev's Theorem give as the number earning within two standard deviations of the mean?

22. The Forever Power Company conducted tests on the life of its batteries and those of a competitor (Brand X). They found that their batteries had a mean life in hours of 26.2 with a standard deviation of 4.1 (see Example 4). Their results for a sample of 10 Brand X batteries were as follows: 15, 18, 19, 23, 25, 25, 28, 30, 34, 38.

(a) Find the mean and standard deviation for Brand X batteries.
(b) Which batteries have a more uniform life in hours?
(c) Which batteries have the highest average life in hours?

23. The Quaker Oats Company conducted a survey to determine if a proposed premium, to be included in their cereal, was appealing enough to generate new sales. Four cities were used as test markets, where the cereal was distributed with the premium, and four cities as control markets, where the cereal was distributed without the premium.

---

[1]This example was supplied by Jeffery S. Berman, Senior Analyst, Marketing Information, Quaker Oats Company.

The eight cities were chosen on the basis of their similarity in terms of population, per capita income, and total cereal purchase volume. The results were as follows.

		Percent change in average market shares per month
Test cities	1	+18
	2	+15
	3	+7
	4	+10
Control cities	1	+1
	2	−8
	3	−5
	4	0

(a) Find the mean of the change in market share for the four test cities.

(b) Find the mean of the change in market share for the four control cities.

(c) Find the standard deviation of the change in market share for the test cities.

(d) Find the standard deviation of the change in market share for the control cities.

(e) Find the difference between the means of (a) and (b). This difference represents the estimate of the percent change in sales due to the premium.

(f) The two standard deviations from (c) and (d) of Exercise 23 were used to calculate an "error" of ± 7.95 for the estimate in (e). With this amount of error, what is the smallest and largest estimate of the increase in sales?

On the basis of the interval estimate of part (f) the company decided to mass produce the premium and distribute it nationally.

24. Show that the formula for variance given in the text can be rewritten as

$$\frac{1}{n^2} \left[n \cdot \Sigma(x^2) - (\Sigma x)^2 \right].$$

8.4 The Normal Distribution

The bank transaction times, in the example discussed in Section 8.1, were timed to the nearest minute. Theoretically at least, they could have been timed to the nearest tenth of a minute, or hundredth of a minute, or even more accurately. Actually it is possible for the transaction times to take on any real number value greater than 0. As noted in Section 8.1, a distribution in which the random variable can take any real number value within some interval is a **continuous distribution.**

The distribution of heights (in inches) of college freshmen women is another example of a continuous distribution, since their heights include infinitely many possible measurements, such as 53 in, 58.5 in, 66.3 in, 72.666 . . . in, and so on. Figure 7 shows the continuous distribution of heights of college freshmen women. Here the most frequent heights occur near the center of the interval shown.

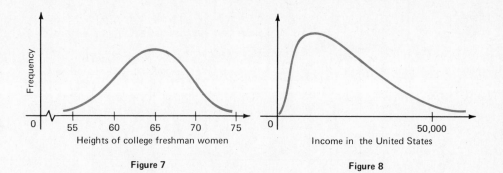

Figure 7

Figure 8

Another continuous curve, which approximates the distribution of yearly incomes in the United States, is shown in Figure 8. From the graph, it can be seen that the most frequent incomes are grouped near the low end of the interval. This kind of distribution, where the peak is not at the center, is called **skewed.**

Many different experiments produce probability distributions which at least approximately come from a very important class of continuous distributions called **normal probability distributions.** The distribution shown in Figure 7 is approximately a normal distribution, while the one of Figure 8 is not. The distribution of the lengths of the leaves of a certain tree would approximate a normal distribution, as should the distribution of the actual weights of cereal boxes that have an average weight of 16 ounces.

Each normal probability distribution has associated with it a bell-shaped curve, called a **normal curve,** such as the one in Figure 9. This curve is symmetric about a vertical line drawn through the mean, μ. Vertical lines drawn at points $+1\sigma$ and -1σ from the mean show where the direction of "curvature" of the graph changes. (For those who have studied calculus, these points are the inflection points of the graph.)

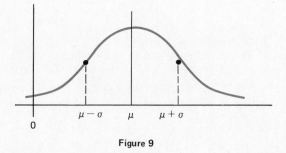

Figure 9

A normal curve never touches the x-axis—it extends indefinitely in both directions. The area under a normal curve is always the same: 1. Many different normal curves have the same mean. For example, a larger value of σ produces a "flatter" normal curve, while smaller values of σ produce more values near the mean, which results in a "taller" normal curve. See Figure 10.

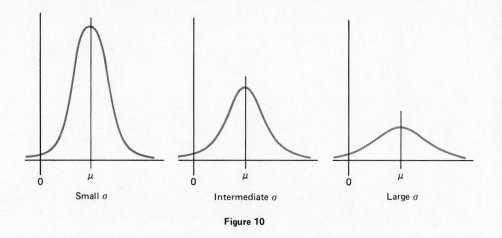

Small σ   Intermediate σ   Large σ

Figure 10

An experiment which has normally distributed outcomes and its associated normal curve are connected by the fact that the probability that an experiment produces a result between *a* and *b* is equal to the area under the normal curve from *a* to *b*. That is, the shaded area in Figure 11 gives the probability that the experimental outcome is between *a* and *b*. (Notice the connection to our work with histograms in Section 8.1.)

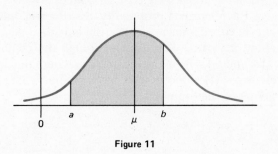

Figure 11

Since a normal curve is symmetric about the mean, and since the total area under a normal curve is 1, the probability that a particular outcome is below the mean is 1/2. A normal curve comes from a continuous distribution, with an infinite number of possible values, so that the probability of the occurrence of any particular value is 0.

Let X be a random variable with a normal probability distribution. Then

(a) $P(a \leq X \leq b)$ is the area under the associated normal curve between a and b;
(b) $P(X < \mu) = 1/2$;
(c) $P(X > \mu) = 1/2$;
(d) $P(X = x) = 0$ for any real number x;
(e) $P(X < x) = P(X \leq x)$ for any real number x.

Part (e) follows from part (d).

The equation of the normal curve having mean μ and standard deviation σ is given by

$$y = \frac{1}{\sigma\sqrt{2\pi}}\, e^{-[(x-\mu)/\sigma]^2/2},$$

where $e \approx 2.7182818$. To find probabilities from normal curves, we would need to use this equation, along with calculus. Doing so would produce a separate table for each pair of values of μ and σ—an infinite number of different tables. We solve this problem by using just one table, the table for the normal curve where $\mu = 0$ and $\sigma = 1$, to find values for any normal curve.

The normal curve having $\mu = 0$ and $\sigma = 1$ is called the **standard normal curve.** Table 2 in the Appendix gives the areas under the standard normal curve, along with a sketch of the curve. The values in this table include the total area under the standard normal curve to the left of the number z.

Example 1   Find the following areas from the table for the standard normal curve.

(a) to the left of $z = 1.25$.

Look up 1.25 in Table 2. The corresponding area is .8944. Thus the shaded area shown in Figure 12 is .8944. This area represents 89.44% of the total area under the normal curve.

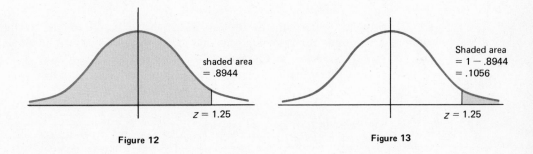

shaded area = .8944

$z = 1.25$

Shaded area = $1 - .8944$ = .1056

$z = 1.25$

Figure 12   Figure 13

(b) to the right of $z = 1.25$.

In part (a) of this example we found that the area to the left of $z = 1.25$ is .8944. The total area under the normal curve is 1, so that the area to the right of $z = 1.25$ is

$$1 - .8944 = 1.0000 - .8944 = .1056.$$

See Figure 13, where the shaded area represents 10.56% of the total area under the normal curve.

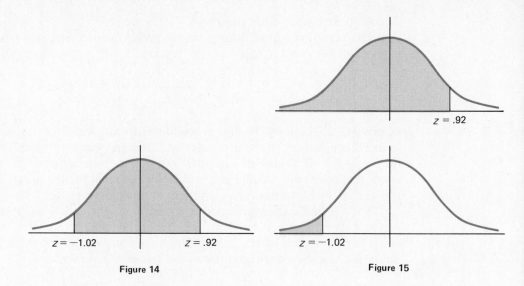

Figure 14   Figure 15

(c) between $z = -1.02$ and $z = .92$

To find this area, shaded in Figure 14, start with the area to the left of $z = .92$ and subtract the area to the left of $z = -1.02$. See Figure 15. We have $.8212 - .1539$, giving the result $.6673$. ∎

What do we do with a normal distribution that does not have $\mu = 0$ and $\sigma = 1$? In this case, we use the following theorem, which is stated without proof.

Suppose a normal distribution has mean μ and standard deviation σ. The area under the associated normal curve that is to the left of the value x is exactly the same as the area to the left of

$$z = \frac{x - \mu}{\sigma}$$

for the standard normal curve.

Using this result, Table 2 can be used for *any* normal curve, no matter the values of μ and σ. The number z in the theorem is called a **z-score**.

Example 2   A normal distribution has mean 46 and standard deviation 7.2. Find the following areas under the associated normal curve.

(a) to the left of 50

Find the appropriate z-score using $x = 50$, $\mu = 46$, and $\sigma = 7.2$. Round to the nearest hundredth.

$$z = \frac{50 - 46}{7.2} = \frac{4}{7.2} \approx .56$$

From Table 2, the desired area is $.7123$.

(b) to the right of 39

$$z = \frac{39 - 46}{7.2} = \frac{-7}{7.2} \approx -.97$$

The area to the *left* of $z = -.97$ is .1660, so that the area to the *right* is

$$1 - .1660 = .8340.$$

(c) between 32 and 43

Find z-scores for both values.

$$z = \frac{32 - 46}{7.2} = \frac{-14}{7.2} \approx -1.94 \quad \text{and} \quad z = \frac{43 - 46}{7.2} = \frac{-3}{7.2} \approx -.42$$

Start with the area to the left of $z = -.42$ and subtract the area to the left of $z = -1.94$, which gives

$$.3372 - .0262 = .3110. \quad \blacksquare$$

The z-scores are actually standard deviation multiples—that is, a z-score of 2.5 corresponds to a value 2.5 standard deviations above the mean. For example, looking up $z = 1.00$ and $z = -1.00$ in Table 2 shows that

$$.8413 - .1587 = .6826,$$

or 68.26%, of the area under a normal curve lies within one standard deviation of the mean. Also,

$$.9772 - .0228 = .9544,$$

or 95.44% of the area lies within two standard deviations of the mean. These results, summarized in Figure 16, can be used to get a quick estimate of results when working with normal curves.

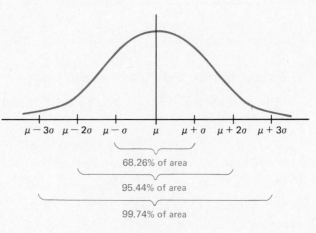

Figure 16

Example 3   Suppose that the average salesperson for Dixie Office Supplies drives $\mu = 1200$ miles per month in a company car, with standard deviation $\sigma = 150$ miles. Assume that the number of miles driven is closely approximated by a normal curve. Find the percent of all drivers traveling

(a) between 1200 and 1600 miles per month.

We first need to find the number of standard deviations above the mean that corresponds to 1600 miles. This is done by finding the z-score for 1600.

$$z = \frac{x - \mu}{\sigma}$$

$$= \frac{1600 - 1200}{150} \qquad \text{Let } x = 1600, \mu = 1200, \sigma = 150$$

$$= \frac{400}{150}$$

$$z \approx 2.67$$

From Table 2, the area to the left of $z = 2.67$ is .9962. Since $\mu = 1200$, the value 1200 corresponds to $z = 0$, the area to the left of $z = 0$ is .5000. Thus,

$$.9962 - .5000 = .4962,$$

or 49.62% of the drivers travel between 1200 and 1600 miles per month. See Figure 17.

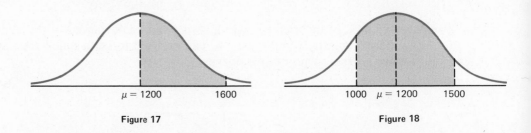

Figure 17   Figure 18

(b) between 1000 and 1500 miles per month.

As shown in Figure 18, we need to find the z-score for both $x = 1000$ and $x = 1500$.

For $x = 1000$,   For $x = 1500$,

$$z = \frac{1000 - 1200}{150} \qquad\qquad z = \frac{1500 - 1200}{150}$$

$$= \frac{-200}{150} \qquad\qquad\qquad\quad = \frac{300}{150}$$

$$z \approx -1.33 \qquad\qquad\qquad\quad z = 2.00$$

From the table, $z = -1.33$ leads to an area of .0918, while $z = 2.00$ corresponds to .9772. Thus, a total of $.9772 - .0918 = .8854$, or 88.54%, of all drivers travel between 1000 and 1500 miles per month.   ■

Suppose a normal distribution has $\mu = 1000$ and $\sigma = 150$. Then the techniques of this section can be used to show that 95.44% of all values lie within 2 standard deviations of the mean; that is, between

$$1000 - (2 \times 150) = 700 \quad \text{and} \quad 1000 + (2 \times 150) = 1300.$$

Chebyshev's Theorem of Section 8.3 says that *at least*

$$1 - \frac{1}{2^2} = 1 - \frac{1}{4} = \frac{3}{4},$$

or 75% of the values lie between 700 and 1300. The difference here, 95.44%, and at least 75%, comes from the fact that Chebyshev's Theorem has the important property of applying to *any* distribution, while the methods of this section apply only to *normal* distributions. Thus it should not be surprising that having more information (a normal distribution) should produce more accurate results (95.44% instead of "at least 75%.")

8.4 Exercises   *Find the percent of the area under a normal curve between the mean and the following number of standard deviations from the mean.*

1. 2.50   2. 1.68   3. 0.45   4. 0.81

5. −1.71   6. −2.04   7. 3.11   8. 2.80

Find the percent of the total area under the normal curve between the following z-scores.

9. $z = 1.41$ and $z = 2.83$ 10. $z = 0.64$ and $z = 2.11$

11. $z = -2.48$ and $z = -0.05$ 12. $z = -1.74$ and $z = -1.02$

13. $z = -3.11$ and $z = 1.44$ 14. $z = -2.94$ and $z = -0.43$

15. $z = -0.42$ and $z = 0.42$ 16. $z = -1.98$ and $z = 1.98$

Find a z-score satisfying the following conditions. (Hint: use Table 2 backwards.)

17. 5% of the total area is to the left of z

18. 1% of the total area is to the left of z

19. 15% of the total area is to the right of z

20. 25% of the total area is to the right of z

A certain type of light bulb has an average life of 500 hours, with a standard deviation of 100 hours. The length of life of the bulb can be closely approximated by a normal curve. An amusement park buys and installs 10,000 such bulbs. Find the total number that can be expected to last

21. at least 500 hours;   22. less than 500 hours;

23. between 500 and 650 hours;   24. between 300 and 500 hours;

25. between 650 and 780 hours;   26. between 290 and 540 hours;

27. less than 740 hours;   **28.** more than 300 hours;

29. more than 790 hours;   **30.** less than 410 hours.

A box of oatmeal must contain 16 ounces. The machine that fills the oatmeal boxes is set so that, on the average, a box contains 16.5 ounces. The boxes filled by the machine have weights that can be closely approximated by a normal curve. What fraction of the boxes filled by the machine are underweight if the standard deviation is

31. .5 ounce;   **32.** .3 ounce;   **33.** .2 ounce;   **34.** .1 ounce?

The chickens at Colonel Thompson's Ranch have a mean weight of 1850 grams with a standard deviation of 150 grams. The weights of the chickens are closely approximated by a normal curve. Find the percent of all chickens having weights

35. more than 1700 grams;

36. less than 1800 grams;

37. between 1750 grams and 1900 grams;

38. between 1600 grams and 2000 grams;

39. less than 1550 grams;

40. more than 2100 grams.

In nutrition, the Recommended Daily Allowance of vitamins is a number set by the government as a guide to an individual's daily vitamin intake. Actually, vitamin needs vary drastically from person to person, but the needs are very closely approximated by a normal curve. To calculate the Recommended Daily Allowance, the government first finds the average need for vitamins among people in the population, and the standard deviation. The Recommended Daily Allowance is then defined as the mean plus 2.5 times the standard deviation.

41. What percentage of the population will receive adequate amounts of vitamins under this plan?

Find the recommended daily allowance for the following vitamins.

42. mean = 1800 units, standard deviation = 140 units

43. mean = 159 units, standard deviation = 12 units

44. mean = 1200 units, standard deviation = 92 units

Assume the following distributions are all normal, and use the areas under the normal curve given in Table 2 to answer the questions.

45. A machine produces bolts with an average diameter of .25 inches and a standard deviation of .02 inches. What is the probability that a bolt will be produced with a diameter greater than .3 inches?

46. The mean monthly income of the trainees of an engineering firm is $600 with a standard deviation of $100. Find the probability that an individual trainee earns less than $500 per month.

47. A machine which fills quart milk cartons is set up to average 32.2 ounces per carton, with a standard deviation of 1.2 ounces. What is the probability that a filled carton will contain less than 32 ounces of milk?

48. The average contribution to the campaign of Polly Potter, a candidate for city council, was $50 with a standard deviation of $15. How many of the 200 people who contributed to Polly's campaign gave between $30 and $100?

49. At the Discount Market, the average weekly grocery bill is $32.25 with a standard deviation of $9.50. What are the largest and smallest amounts spent by the middle 50% of this market's customers?

50. The mean clotting time of blood is 7.45 seconds with a standard deviation of 3.6 seconds. What is the probability that an individual's blood clotting time will be less than 7 seconds or greater than 8 seconds?

51. The average size of the fish in Lake Amotan is 12.3 inches with a standard deviation of 4.1 inches. Find the probability of catching a fish there which is longer than 18 inches.

52. To be graded extra large, an egg must weigh at least 2.2 ounces. If the average weight for an egg is 1.5 ounces with a standard deviation of .4 ounces, how many of five dozen eggs would you expect to grade extra large?

One professor uses the following grading system for assigning letter grades in a course.

Grade	Score in class is
A	greater than $\mu + \frac{3}{2}\sigma$
B	$\mu + \frac{1}{2}\sigma$ to $\mu + \frac{3}{2}\sigma$
C	$\mu - \frac{1}{2}\sigma$ to $\mu + \frac{1}{2}\sigma$
D	$\mu - \frac{3}{2}\sigma$ to $\mu - \frac{1}{2}\sigma$
F	below $\mu - \frac{3}{2}\sigma$

What percent of the students receive the following grades?

53. A   54. B   55. C

56. Do you think this system would be more likely to be fair in a large freshman class in psychology or in a graduate seminar of five students? Why?

A teacher gives a test to a large group of students. The results are closely approximated by a normal curve. The mean is 74, with a standard deviation of 6. The teacher wishes to give A's to the top 8% of the students and F's to the bottom 8%. A grade of B is given to the next 15%, with D's given similarly. All other students get C's. Find the bottom cutoff (rounded to the nearest whole number) for the following grades. (Hint: use the table in the Appendix backwards.)

57. A   58. B   59. C   60. D

Application   *Inventory Control*

A department store must control its inventory carefully.[1] It should not reorder too often, because it then builds up a large warehouse full of merchandise, which is expensive to hold. On the other hand, it must reorder sufficiently often to be sure of having sufficient stock to meet customer demand. The company desires a simple chart that can be used by its employees to determine the best possible time to reorder merchandise. The merchandise level on hand will be checked periodically. At the end of each period, if the level on hand is less than some

---

[1]Example supplied by Leonard W. Cooper, Operations Research Project Director, Federated Department Stores.

predetermined level given in the chart, which considers sales rate and waiting time for orders, the item will be reordered. The example uses the following variables.

$F =$ frequency of stock review (in weeks)

$r =$ acceptable risk of being out of stock (in percent)

$P =$ level of inventory at which reordering should occur

$L =$ waiting time for order to arrive (in weeks)

$S =$ sales rate in units per week

$M =$ minimum level of merchandise to guarantee that the probability of being out of stock is no higher than r

$z =$ z-score (from Table 2 at back of book) corresponding to r

Goods should not be reordered until inventory on hand has declined to a level less than the rate of sales per week, S, times the sum of the number of weeks until the next stock review, F, and the expected waiting time in weeks, L, plus a minimum level of merchandise, M, necessary to guarantee that the probability of being out of stock is no higher than r. That is, $P = S(F + L) + M$.

To find M, which depends on S, F, and L, we shall assume that both sales and waiting time are normally distributed. With this assumption, a formula from more advanced statistics courses permits us to write $M = z\sqrt{2S(F + L)}$, where z is the z-score (from Table 2) corresponding to r, and $\sqrt{2S(F + L)}$ is the standard deviation of normally distributed deviations in sales rate and waiting time.

Combining these two formulas, we have

$$P = S(F + L) + z\sqrt{2S(F + L)}.$$

Suppose the firm wishes to be 95% sure of having goods to sell, so that $r = 5\%$. From Table 2, we find $z = 1.64$. Hence, for $r = 5\%$,

$$P = S(F + L) + 1.64\sqrt{2S(F + L)}.$$

Based on this formula, the chart on page 319 was prepared.

To use the chart, for example, if an item is reviewed every 4 weeks ($F = 4$), the waiting time for a reorder is 3 weeks ($L = 3$), and the rate of sales is 2 per week ($S = 2$), then the item should be reordered when the number on hand falls at or below 23.

Exercises

1. Suppose an item is reviewed weekly, and the waiting time for a reorder is 4 weeks. If the average sales per week of the item is 12 units, and the current inventory level is 85 units, should it be reordered? What if the inventory level is 50 units?

2. Suppose an item is reviewed every four weeks. If orders require a 5-week waiting time, and if sales average 3 units per week, should the item be reordered if current inventory is 50 units? What if current inventory is 30?

REORDER LEVELS (P)
Reorder merchandise when inventory on hand falls below the levels given in the chart.
$r = 5\%$, $F = 1$ week

Rate of Sales (S)	Waiting Time in Weeks for Order (L)					
(Units/Week)	**1**	**2**	**3**	**4**	**5**	**6**
9	28	39	50	61	71	81
10	30	43	55	66	78	89
11	33	46	59	72	85	97
12	35	50	64	78	92	105
13	38	54	69	84	99	113
14	40	57	83	89	105	121
15	43	61	78	95	112	129

$r = 5\%$, $F = 4$ weeks

Rate of Sales (S)	Waiting Time in Weeks for Order (L)					
(Units/Week)	**1**	**2**	**3**	**4**	**5**	**6**
1	10	12	13	15	16	17
2	17	20	23	25	28	30
3	24	28	32	35	39	43
4	30	35	40	45	50	55

8.5   The Normal Curve Approximation to the Binomial Distribution

A **binomial distribution** is an example of a discrete distribution where the experiment is a series of n independent trials in which we observe the number of successes and failures. The probability of a success cannot change from one trial to the next. (Recall from Chapter 6 that an experiment of this kind is called a *Bernoulli experiment*.) Examples of binomial distributions include tossing a coin, rolling a 5 on a die, or selecting a defective radio from a large batch of radios produced by a factory.

Let us consider an experiment in which a die is tossed 5 times. Each time that the toss results in a 1 or a 2, the toss is considered a success; otherwise it is a failure. How many successes are possible in such an experiment? Since each trial can result in either success or failure, it is possible to have from 0 to 5 successes. Consider the probability of each of these six outcomes. In Section 6.7, we saw that for independent repeated trials, the required probabilities are given by

$$P(X = x) = \binom{n}{x} p^x (1 - p)^{n-x},$$

where n is the number of trials, x is the number of successes, p is the probability of a single trial, and $P(X = x)$ gives the probability that x of the n trials result in successes. In this example, $n = 5$ and $p = 1/3$, since either a 1 or a 2 results in a success. The results for this experiment are tabulated in Table 10.

Table 10

x	$P(X = x)$
0	$\binom{5}{0}\left(\frac{1}{3}\right)^0\left(\frac{2}{3}\right)^5 = \frac{32}{243}$
1	$\binom{5}{1}\left(\frac{1}{3}\right)^1\left(\frac{2}{3}\right)^4 = \frac{80}{243}$
2	$\binom{5}{2}\left(\frac{1}{3}\right)^2\left(\frac{2}{3}\right)^3 = \frac{80}{243}$
3	$\binom{5}{3}\left(\frac{1}{3}\right)^3\left(\frac{2}{3}\right)^2 = \frac{40}{243}$
4	$\binom{5}{4}\left(\frac{1}{3}\right)^4\left(\frac{2}{3}\right)^1 = \frac{10}{243}$
5	$\binom{5}{5}\left(\frac{1}{3}\right)^5\left(\frac{2}{3}\right)^0 = \frac{1}{243}$

By definition, the mean μ of a probability distribution is given by the expected value of X. Expected value is found by finding the products of outcomes and probabilities. For the distribution of Table 10,

$$\mu = 0\left(\frac{32}{243}\right) + 1\left(\frac{80}{243}\right) + 2\left(\frac{80}{243}\right) + 3\left(\frac{40}{243}\right) + 4\left(\frac{10}{243}\right) + 5\left(\frac{1}{243}\right)$$

$$= \frac{405}{243} = 1\frac{2}{3}.$$

For a binomial distribution, which is a special kind of probability distribution, it can be shown that the method for finding the mean reduces to the formula

$$\mu = np,$$

where n is the number of trials and p is the probability of success for a single trial. Using this simplified formula, the computation of the mean in the example above is

$$\mu = np = 5\left(\frac{1}{3}\right) = 1\frac{2}{3},$$

which agrees with the result we obtained using the expected value.

Like the mean, the variance, $\text{Var}(X)$, of a probability distribution is an expected value—the expected value of the squared deviations from the mean, $(x - \mu)^2$. To find the variance for the example given above, first find the quantities $(x - \mu)^2$. (Recall, the mean μ is 5/3.)

Table 11

x	$P(X = x)$	$x - \mu$	$(x - \mu)^2$
0	$\dfrac{32}{243}$	$\dfrac{-5}{3}$	$\dfrac{25}{9}$
1	$\dfrac{80}{243}$	$\dfrac{-2}{3}$	$\dfrac{4}{9}$
2	$\dfrac{80}{243}$	$\dfrac{1}{3}$	$\dfrac{1}{9}$
3	$\dfrac{40}{243}$	$\dfrac{4}{3}$	$\dfrac{16}{9}$
4	$\dfrac{10}{243}$	$\dfrac{7}{3}$	$\dfrac{49}{9}$
5	$\dfrac{1}{243}$	$\dfrac{10}{3}$	$\dfrac{100}{9}$

We now find $\text{Var}(X)$ by finding the products $[(x - \mu)^2][P(X = x)]$.

$$\text{Var}(X) = \frac{25}{9}\left(\frac{32}{243}\right) + \frac{4}{9}\left(\frac{80}{243}\right) + \frac{1}{9}\left(\frac{80}{243}\right) + \frac{16}{9}\left(\frac{40}{243}\right) + \frac{49}{9}\left(\frac{10}{243}\right) + \frac{100}{9}\left(\frac{1}{243}\right)$$

$$= \frac{10}{9} = 1\frac{1}{9}$$

To find the standard deviation σ, find $\sqrt{10/9}$ to get $\sqrt{10}/3$, or approximately 1.05.

Just as with the mean, the variance of a binomial distribution can be found with a relatively simple formula. Again, it can be shown that

$$\text{Var}(X) = np(1 - p) \quad \text{or} \quad \sigma = \sqrt{np(1 - p)}.$$

By substituting the appropriate values for n and p from the example into this new formula,

$$\text{Var}(X) = 5\left(\frac{1}{3}\right)\left(\frac{2}{3}\right) = 10/9 = 1\frac{1}{9},$$

which agrees with our previous result.

Summarizing, we start with an experiment which is a series of n independent repeated trials, where the probability of a success in a single trial is always p. We observe the number of successes, x, in the n trials.

The binomial distribution is the probability distribution where the probability that exactly x successes will occur in n trials is found by the formula

$$\binom{n}{x} p^x (1 - p)^{n-x}.$$

The mean and variance of a binomial distribution are respectively

$$\mu = np \quad \text{and} \quad \text{Var}(X) = np(1 - p).$$

The standard deviation is

$$\sigma = \sqrt{np(1 - p)}.$$

Example 1   The probability that a plate picked at random from the assembly line in a china factory will be defective is .01. A sample of three is to be selected. Write the distribution for the number of defective plates in the sample, and give its mean and standard deviation.

Since three plates will be selected, the possible number of defective plates ranges from 0 to 3. Here, n (the number of trials) is 3, and p (the probability of selecting a defective on a single trial) is .01. The distribution and the probability of each outcome are shown in Table 12.

Table 12

x	$p(X = x)$
0	$\binom{3}{0}(.01)^0(.99)^3 = .970$
1	$\binom{3}{1}(.01)(.99)^2 = .029$
2	$\binom{3}{2}(.01)^2(.99) = .0003$
3	$\binom{3}{3}(.01)^3(.99)^0 = .000001$

The mean of the distribution is

$$\mu = np = 3(.01) = .03.$$

The standard deviation is

$$\sigma = \sqrt{np(1 - p)} = \sqrt{3(.01)(.99)} = \sqrt{.0297} = .17. \quad \blacksquare$$

The binomial distribution is extremely useful, but its use can lead to complicated calculations if n is large. However, the normal curve of the previous

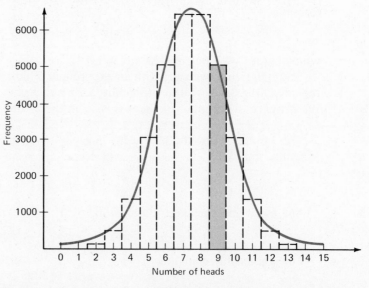

Figure 19

section can be used to get a good approximation to the binomial distribution. This approximation was first discovered by Abraham DeMoivre in 1718 for the case $p = 1/2$. The result was generalized and given great visibility among mathematicians by the French mathematician Laplace in a book published in 1812.

To see how the normal curve is used, look at the bar graph and normal curve in Figure 19. This histogram shows the expected number of heads if one coin is tossed 15 times, with the experiment repeated 32,768 times. Since the probability of heads on one toss is 1/2 and $n = 15$, the mean of this distribution is

$$\mu = np = 15\left(\frac{1}{2}\right) = 7.5.$$

The standard deviation is

$$\sigma = \sqrt{15\left(\frac{1}{2}\right)\left(1 - \frac{1}{2}\right)}$$
$$= \sqrt{15\left(\frac{1}{2}\right)\left(\frac{1}{2}\right)}$$
$$= \sqrt{3.75}$$
$$\sigma = 1.94.$$

In Figure 19 we have superimposed the normal curve with $\mu = 7.5$ and $\sigma = 1.94$ over the bar graph of the distribution.

Suppose we need to know the fraction of the time that we would get exactly 9 heads in the 15 tosses. We could work this out using the methods above. After performing a huge amount of arithmetic, we would get .153. This answer is the same fraction that would be found by dividing the area of the shaded bar in Figure 19 by the total area of all 16 bars in the graph. (Note: Some of the bars at the extreme left and right ends of the graph are too short to show up.)

As the graph suggests, the area of the shaded bar is also approximately equal to the area under the normal curve from $x = 8.5$ to $x = 9.5$. The normal curve runs higher than the top of the bar in the left half, but lower in the right half.

To find the area under the normal curve from $x = 8.5$ to $x = 9.5$, we first need to find z-scores, as we did in the last section. To do so, we use the mean and the standard deviation for the distribution, which we have already calculated, to get z-scores for $x = 8.5$ and $x = 9.5$.

For $x = 8.5$ For $x = 9.5$

$$z = \frac{8.5 - 7.5}{1.94} \qquad z = \frac{9.5 - 7.5}{1.94}$$

$$= \frac{1.00}{1.94} \qquad = \frac{2.00}{1.94}$$

$$z \approx .52 \qquad\qquad z \approx 1.03$$

From Table 2, $z = .52$ gives an area of .6985, while $z = 1.03$ gives .8485. To find the desired result, subtract these two numbers.

$$.8485 - .6985 = .1500.$$

This answer (.1500) is not far from the exact answer, .153, that we found above.

Example 2 About 6% of the bolts produced by a certain machine are defective.

(a) Find the probability that in a sample of 100 bolts, 3 or fewer are defective.

Notice that this problem satisfies the conditions of the definition of a binomial distribution. For this reason, we can use the normal curve approximation, first finding the mean and the standard deviation. Here $n = 100$ and $p = 6\% = .06$. Thus

$$\mu = 100(.06) \qquad \sigma = \sqrt{100(.06)(1 - .06)}$$

$$\mu = 6. \qquad\qquad = \sqrt{100(.06)(.94)}$$

$$= \sqrt{5.64}$$

$$\sigma \approx 2.37$$

As the graph of Figure 20 shows, we need to find the area to the left of $x = 3.5$ (since we want 3 or fewer defective bolts). The z-score corresponding to $x = 3.5$ is

$$z = \frac{3.5 - 6}{2.37} = \frac{-2.5}{2.37} \approx -1.05.$$

From the Table, $z = -1.05$ leads to an area of .1469, so that the probability of getting 3 or fewer defective bolts in a set of 100 bolts is .1469, or 14.69%.

(b) Find the probability of getting exactly 11 defective bolts in a sample of 100 bolts.

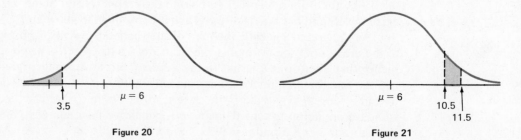

Figure 20   Figure 21

As Figure 21 shows, we need the area between $x = 10.5$ and $x = 11.5$.

$$\text{If } x = 10.5, \text{ then } z = \frac{10.5 - 6}{2.37} \approx 1.90.$$

$$\text{If } x = 11.5, \text{ then } z = \frac{11.5 - 6}{2.37} \approx 2.32.$$

Look in Table 2; $z = 1.90$ gives an area of .9713, while $z = 2.32$ yields .9898. The final answer is the difference of these numbers, or

$$.9898 - .9713 = .0185.$$

There is about a 1.85% chance of having exactly 11 defective bolts. ∎

The normal curve approximation to the binomial distribution is usually quite accurate, especially for practical problems. For *n* up to say, 15 or 20, it is usually not too difficult to actually calculate the binomial probabilities directly. For larger values of *n,* a rule of thumb is that the normal curve approximation can be used as long as both *np* and $n(1 - p)$ are at least 5.

8.5 Exercises   *In Exercises 1–6, several binomial experiments are described. For each one, give (a) the distribution; (b) the mean; (c) the standard deviation.*

1. A die is rolled six times and the number of 1's that come up is tallied. Write the distribution of 1's that can be expected to occur.

2. A 6-item multiple choice test has four possible answers for each item. A student selects all his answers randomly. Give the distribution of correct answers.

3. To maintain quality control on the production line, the Bright Lite Company randomly selects three light bulbs each day for testing. Experience has shown a defective rate of .02. Write the distribution for the number of defectives in the daily samples.

4. In a taste test, each member of a panel of four is given two glasses of Supercola, one made using the old formula and one with the new formula, and asked to identify the new formula. Assuming the panelists operate independently, write the distribution of the number of successful identifications, if each judge actually guesses.

5. The probability that a radish seed will germinate is .7. Joe's mother gives him four seeds to plant. Write the distribution for the number of seeds which germinate.

6. Five patients in Ward 8 of Memorial Hospital have a disease with a known mortality rate of .1. Write the distribution of the number who survive.

Work the following problems involving binomial experiments.

7. The probability that an infant will die in the first year of life is about .025. In a group of 500 babies, what are the mean and standard deviation of the number of babies who can be expected to die in their first year of life?

8. The probability that a particular kind of mouse will have a brown coat is 1/4. In a litter of 8, assuming independence, how many could be expected to have a brown coat? With what standard deviation?

9. A certain drug is effective 80% of the time. Give the mean and standard deviation of the number of patients using the drug who recover, out of a group of 64 patients.

10. The probability that a newborn infant will be a girl is .49. If 50 infants are born on Susan B. Anthony's birthday, how many can be expected to be girls? With what standard deviation?

For the remaining exercises, use the normal curve approximation to the binomial distribution.
Suppose 16 coins are tossed. Find the probability of getting exactly

11. 8 heads;   12. 7 heads;   13 10 tails;   14. 12 tails.

Suppose 1000 coins are tossed. Find the probability of getting each of the following.
(Hint: $\sqrt{250} = 15.8$)

15. exactly 500 heads   16. exactly 510 heads   17. 480 heads or more

18. less than 470 tails   19. less than 518 heads   20. more than 550 tails

A die is tossed 120 *times. Find the probability of getting each of the following. (Hint:* $\sigma = 4.08$)

21. exactly 20 fives

22. exactly 24 sixes

23. exactly 17 threes

24. exactly 22 twos

25. more than 18 threes

26. fewer than 22 sixes

Two percent of the quartz heaters produced in a certain plant are defective. Suppose the plant produced 10,000 *such heaters last month. Find the probability that among these heaters*

27. fewer than 170 were defective;

28. more than 222 were defective.

A new drug cures 80% *of the patients to whom it is administered. It is given to* 25 *patients. Find the probability that among these patients*

29. exactly 20 are cured;

30. exactly 23 are cured;

31. all are cured;

32. no one is cured;

33. 12 or fewer are cured;

34. between 17 and 23 are cured.

35. An experimental drug causes a rash in 15% of all people taking it. If the drug is given to 12,000 people, find the probability that more than 1700 people will get the rash.

36. In one state, 55% of the voters expect to vote for Jones. Suppose 1400 people are asked for the name of the person they expect to vote for. Find the probability that at least 700 people will say that they expect to vote for Jones.

8.6 Curve Fitting—The Least Squares Method

To produce a mathematical model, we frequently must work with numerical data; we want to obtain an equation which describes the data. The simplest kind of equation is a linear equation, so we usually start by trying to fit a straight line through the data points. Figure 22 shows several data points, and a straight line which fits through the points fairly well.

How do we decide on the "best" possible line? The line that has proven to be "best" in many different applications is the one in which the sum of the squares of the vertical distances from the data points to the line is as small as possible.

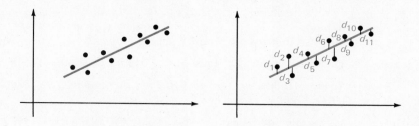

Figure 22   Figure 23

We call such a line the **least squares line**. In Figure 23 the distances from the data points to the line are represented by d_1, d_2, d_3, and so on. Then the least squares line is found by minimizing the sum $(d_1)^2 + (d_2)^2 + (d_3)^2 + \cdots + (d_n)^2$.

The method of finding the equation of the least squares line requires calculus, and will not be given here; we give only the results. We use y' instead of y to distinguish these predicted values from the y-values of the given pairs of data points. The **least squares line** $y' = mx + b$ which provides the best fit to the data points (x_1, y_1), (x_2, y_2), . . . , (x_n, y_n) has

slope
$$m = \frac{n(\Sigma xy) - (\Sigma x)(\Sigma y)}{n(\Sigma x^2) - (\Sigma x)^2}$$

and y'-intercept
$$b = \frac{\Sigma y - m(\Sigma x)}{n}.$$

Example 1   A college registrar wants to see if college grade-point averages and high school grade-point averages can be closely approximated by a linear function. To begin, the registrar chooses 10 students at random and finds their college and high school grade-point averages, with the results shown in Table 13. Find the least squares line which relates these averages.

Table 13

Student	1	2	3	4	5	6	7	8	9	10
High school GPA. x	2.5	2.8	2.9	3.0	3.2	3.3	3.3	3.4	3.6	3.9
College GPA. y	2.0	2.5	2.5	2.8	3.0	3.0	3.5	3.3	3.4	3.6

A graph of the ten pairs of averages is shown in Figure 24. This graph is called a **scatter diagram.** In order to use the formulas above to get m and b, we need Σx, Σy, Σxy, and Σx^2. Find these sums as shown in Table 14. (The column headed y^2 will be used later.)

Table 14

x	y	xy	x^2	y^2
2.5	2.0	5.00	6.25	4.00
2.8	2.5	7.00	7.84	6.25
2.9	2.5	7.25	8.41	6.25
3.0	2.8	8.40	9.00	7.84
3.2	3.0	9.60	10.24	9.00
3.3	3.0	9.90	10.89	9.00
3.3	3.5	11.55	10.89	12.25
3.4	3.3	11.22	11.56	10.89
3.6	3.4	12.24	12.96	11.56
3.9	3.6	14.04	15.21	12.96
31.9	29.6	96.20	103.25	90.00

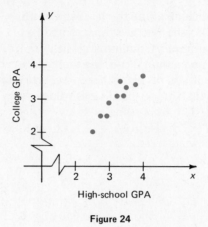

Figure 24

We now compute m and b as shown below.

$$m = \frac{10(96.2) - (31.9)(29.6)}{10(103.25) - (31.9)^2} \approx 1.19$$

$$b = \frac{29.6 - 1.19(31.9)}{10} \approx -.84$$

The least squares line is $y' = mx + b$, or

$$y' = 1.19x - .84.$$

This equation can be used to predict y from a given value of x. See Example 2.   ▮

Example 2   In Example 1, we ended up with the least squares line $y' = 1.19x - .84$, where x represents high school grade-point averages and y' represents predicted college averages. Use this equation to answer the following question.

Lupe Renoso had a grade-point average of 3.5 in high school. Predict her college average to the nearest tenth.

Let $x = 3.5$ in our equation $y' = 1.19x - .84$.

$$y' = 1.19(3.5) - .84$$

$$= 4.165 - .84$$

$$y' = 3.325$$

$$\approx 3.3$$

We would predict that Renoso will have a college grade-point average of about 3.3. (In more advanced courses, you will see methods for deciding how much faith to place in this prediction.)   ▮

Correlation   Once we find an equation for the line of best fit (the least squares line), we might well ask, "Just how good is this line for prediction purposes?" If the points already observed fit the line quite closely, then we can expect future pairs of scores to do so. If the points are widely scattered about even the "best fitting" line, then predictions are not likely to be accurate.

In order to have a quantitative basis for confidence in our predictions, we need a measure of the "goodness of fit" of our original data to the prediction line. One such measure is called the **coefficient of correlation,** denoted r. We can calculate r by using the following formula.

$$r = \frac{n(\Sigma xy) - (\Sigma x)(\Sigma y)}{\sqrt{n(\Sigma x^2) - (\Sigma x)^2} \cdot \sqrt{n(\Sigma y^2) - (\Sigma y)^2}}$$

The coefficient of correlation, r, is always equal to or between 1 and -1. Values of exactly 1 or -1 indicate that the data points lie *exactly* on the least squares line. If $r = 1$, the least squares line has a positive slope; $r = -1$ gives a negative slope. If $r = 0$, there is no linear correlation between the data points. (However, some other nonlinear function might provide an excellent fit for the data.) Scatter diagrams which correspond to these values of r are shown in Figure 25.

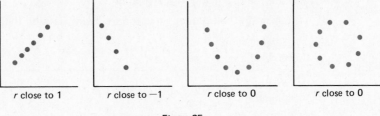

| r close to 1 | r close to -1 | r close to 0 | r close to 0 |

Figure 25

Example 3   Find r for the data of Example 1.

In Example 1, we found that $\Sigma x = 31.9, \Sigma y = 29.6, \Sigma xy = 96.20, \Sigma x^2 = 103.25$, and $\Sigma y^2 = 90.00$. To find the coefficient of correlation between the college grade-point averages and the high school grade-point averages given above, substitute these values into the formula for r.

$$r = \frac{n(\Sigma xy) - (\Sigma x)(\Sigma y)}{\sqrt{n(\Sigma x^2) - (\Sigma x)^2} \cdot \sqrt{n(\Sigma y^2) - (\Sigma y)^2}}$$

$$= \frac{10(96.2) - (31.9)(29.6)}{\sqrt{10(103.25) - (31.9)^2} \cdot \sqrt{10(90) - (29.6)^2}}$$

$$r = \frac{17.76}{\sqrt{14.89} \cdot \sqrt{23.84}} \approx .94$$

As discussed above, the fact that the coefficient of correlation is close to 1 indicates that there is a close relationship between the two grade-point averages. ∎

8.6 Exercises

1. In a study to determine the linear relationship between the size (in decimeters) of an ear of corn (y) and the amount (in tons per acre) of fertilizer used (x), the following data were collected.

$$n = 10 \qquad \Sigma xy = 75$$
$$\Sigma x = 30 \qquad \Sigma x^2 = 100$$
$$\Sigma y = 24 \qquad \Sigma y^2 = 80$$

(a) Find an equation for the least squares line.
(b) Find the coefficient of correlation.
(c) If 3 tons per acre of fertilizer are used, what length (in decimeters) would the equation in (a) predict for an ear of corn?

2. In an experiment to determine the linear relationship between temperatures on the Celsius scale (y) and on the Fahrenheit scale (x), a student got the following results.

$$n = 5 \qquad \Sigma xy = 28{,}050$$
$$\Sigma x = 376 \qquad \Sigma x^2 = 62{,}522$$
$$\Sigma y = 120 \qquad \Sigma y^2 = 13{,}450$$

(a) Find an equation for the least squares line.
(b) Find the reading on the Celsius scale that corresponds to a reading of 120° Fahrenheit, using the equation of part (a).
(c) Find the coefficient of correlation.

3. A sample of 10 adult men gave the following data on their heights and weights.

Height (inches)	(x)	62	62	63	65	66	67	68	68	70	72
Weight (pounds)	(y)	120	140	130	150	142	130	135	175	149	168

(a) Find the equation of the least squares line.
Using the results of (a), predict the weight of a man whose height is
(b) 60 inches;
(c) 70 inches.
(d) Compute the coefficient of correlation.

4. (This problem is appropriate only for those who have studied common logarithms.) Sometimes the scatter diagram of the data does not have a linear pattern. This is particularly true in biological applications. In these applications, however, often the scatter diagram of the *logarithms* of the data has a linear pattern. A least squares line then can be used to predict the logarithm of any desired value from which the value itself can be found. Suppose that a certain kind of bacterium grows in number as shown in Table A. The actual number of bacteria present at each time period is replaced with the common logarithm of that number (Table B).

Table A		*Table B*	
Time in hours	Number of bacteria	Time x	Log y
0	1000	0	3.0000
1	1649	1	3.2172
2	2718	2	3.4343
3	4482	3	3.6515
4	7389	4	3.8686
5	12182	5	4.0857

We can now find a least squares line which will predict y, given x.
(a) Plot the original pairs of numbers. The pattern should be nonlinear.
(b) Plot the log values against the time values. The pattern should be almost linear.
(c) Find the equation of the least squares line. (First round off the log values to the nearest hundredth.)
(d) Predict the log value for a time of 7 hours. Find the number whose logarithm is your answer. This will be the predicted number of bacteria.

It is sometimes possible to get a better prediction for a variable by considering its relationship with more than one other variable. For example, one should be able to predict college GPAs more precisely if both high school GPAs and scores on the ACT are considered. To do this, we alter the equation used to find a least squares line by adding a term for the new variable as follows. If y represents college GPAs, x_1 high school GPAs, and x_2 ACT scores, then y', the predicted GPAs, is given by

$$y' = ax_1 + bx_2 + c.$$

*This equation represents a **least squares plane**. The equations for the constants a, b, and c are more complicated than those given in the text for m and b, so that calculating a least squares equation for three variables is more likely to require the aid of a computer.*

5. Alcoa[1] used a least squares line with two independent variables, x_1 and x_2, to predict the effect on revenue of the price of aluminum forged truck wheels, as follows.

$x_1 =$ the average price per wheel

$x_2 =$ large truck production in thousands

$y =$ sales of aluminum forged truck wheels in thousands

Using data for the past eleven years, the company found the equation of the least squares line to be

$$y' = 49.2755 - 1.1924x_1 + 0.1631x_2,$$

for which the correlation coefficient was .902. The following figures were then forecast for truck production.

1982	1983	1984	1985	1986	1987
160.0	165.0	170.0	175.0	180.0	185.0

Three possible price levels per wheel were considered: $42, $45, and $48.
(a) Use the least squares plane equation given above to find the estimated sales of wheels (y') for 1984 at each of the three price levels.
(b) Repeat part (a) for 1987.
(c) For which price level, on the basis of the 1984 and 1987 figures, are total estimated sales greatest?
(By comparing total estimated sales for the years 1982 through 1987 at each of the three price levels, the company found that the selling price of $42 per wheel would generate the greatest sales volume over the six-year period.)

---

[1]This example supplied by John H. Van Denender, Public Relations Department, Aluminum Company of America.

6. Records show that the annual sales of the EZ Life Company in 5-year periods for the last 20 years were as follows.

Year (x)	Sales (in millions) (y)
1960	1.0
1965	1.3
1970	1.7
1975	1.9
1980	2.1

The company wishes to estimate sales from these records for the next few years. Code the years so that $1960 = 0$, $1961 = 1$, and so on.

(a) Plot the 5 points on a graph.

(b) Find the equation of the least squares line, and graph it on the graph of part (a).

(c) Predict the company's sales for 1984 and 1985.

(d) Compute the correlation coefficient.

7. Sales, in thousands of dollars, of a certain company are shown here.

Year (x)	0	1	2	3	4	5
Sales (y)	48	59	66	75	80	90

Find the equation of the least squares line. Find the coefficient of correlation.

8. The admission test scores of 8 students were compared with their grade-point averages after one year of college. The results are shown below.

Admission test score (x)	19	20	22	24	25	26	27	29
Grade-point average (y)	2.2	2.4	2.7	2.6	3.0	3.5	3.4	3.8

(a) Plot the 8 points on a graph.

(b) Find the equation of the least squares line and graph it on the graph of (a).

(c) Using the results of part (b), predict the grade-point average of a student with an admission test score of 28.

(d) Compute the coefficient of correlation.

9. The following data, furnished by a major brewery, which asked that its name not be given, were used to determine if there is a relationship between repair costs and barrels of beer produced. The data in thousands are given for a 10-month period.

Month	Barrels of beer X	Repairs Y
Jan	369	299
Feb	379	280
Mar	482	393
April	493	388
May	496	385
June	567	423
July	521	374
Aug	482	357
Sept	391	106
Oct	417	332

(a) Find the equation of the least squares line.

(b) Find the coefficient of correlation.

(c) If 500,000 barrels of beer are produced, what will the equation from part (a) give as the predicted repair costs?

Key Words

random variable	variance
frequency distribution	standard deviation
probability distribution	Chebyshev's Theorem
probability distribution function	skewed
discrete distribution function	normal distribution
continuous distribution function	normal curve
histogram	standard normal distribution
mathematical expectation	z-score
arithmetic average	binomial distribution
mean	least squares line
expected value	scatter diagram
fair game	coefficient of correlation

Chapter 8   Review Exercises

For each of the following (a) give a probability distribution and (b) sketch its histogram.

1.

x	1	2	3	4	5
frequency	3	7	9	3	2

2.

x	8	9	10	11	12	13	14
frequency	1	0	2	5	8	4	3

3. A coin is tossed three times and the number of heads are recorded.

4. A pair of dice are rolled and the number of points showing are recorded.

5. Patients in groups of five were given a new treatment for a fatal disease. The experiment was repeated ten times with the following results.

Number who survived	frequency
0	1
1	1
2	2
3	3
4	3
5	0
	Total: 10

For each of the following give the probability which corresponds to the shaded region of the figure.

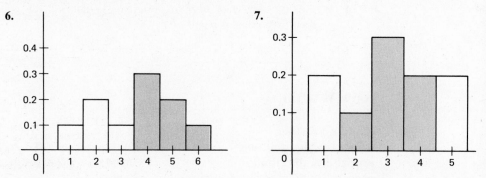

6.

7.

Solve the following word problems.

8. You pay $6 to play in a game where you will roll a die, with payoffs as follows: $8 for a 6, $7 for a 5, $4 otherwise. What is your mathematical expectation? Is the game fair?

9. A lottery has a first prize of $5000, two second prizes of $1000 each, and two $100 third prizes. A total of ten thousand tickets is sold, at $1 each. Find the expected winnings of a person buying one ticket.

10. Find the expected number of girls in a family of five children.

11. A developer can buy a piece of property that will produce a profit of $16,000 with probability .7, or a loss of $9000 with probability .3. What is the expected profit?

12. Game boards for a recent United Airlines contest could be obtained by sending a self addressed stamped envelope to a certain address. The prize was a ticket for any city to which United flies. Assume that the value of the ticket was $1000 (we might as well go first class), and that the probability that a particular game board would win was 1/4000. If the stamps to enter the contest cost 30¢, and envelopes are 1¢ each, find the expected winnings for a person ordering one game board.

Find the expected value of the random variable in each of the following.

13. the data in Exercise 2

14. the experiment of Exercise 3

15. the experiment of Exercise 5

16. 3 cards are drawn from a standard deck of 52 cards.
(a) What is the expected number of aces?
(b) What is the expected number of clubs?

17. Suppose someone offers to pay you $100 if you draw three cards from a standard deck of 52 cards and all the cards are clubs. What should you pay for the chance to win if it is a fair game?

Find the variance and standard deviation of the random variable in the following.

18. the data in Exercise 3

19. the experiment of Exercise 5

20. The annual returns of two stocks for three years are given below.

	1980	1981	1982
stock I	11%	−1%	14%
stock II	9%	5%	10%

(a) Find the mean and standard deviation for each stock over the three-year period.

(b) If you are looking for security with an 8% return, which of these two stocks would you choose?

21. The weight gains of 2 groups of 10 rats fed on two different experimental diets were as follows:

weight gains										
diet A	1	0	3	7	1	1	5	4	1	4
diet B	2	1	1	2	3	2	1	0	1	0

Compute the mean and standard deviation for each group.

(a) Which diet produced the greatest mean gain?

(b) Which diet produced the most consistent gain?

22. A probability distribution has an expected value of 28 and a standard deviation of 4. Use Chebyshev's Theorem to decide what percent of the distribution is

(a) between 20 and 36;   (b) less than 23.2 or greater than 32.8.

23. (a) Find the percent of the area under a normal curve within 2.5 standard deviations of the mean.

(b) Compare your answer to part (a) with the result using Chebyshev's Theorem.

24. Find the percent of the total area under the normal curve which corresponds to

(a) $z \geq 2.2$,   (b) between $z = -1.3$ and $z = .2$.

25. Find a z-score such that 8% of the area under the curve is to the right of z.

26. A machine which fills quart milk cartons is set to fill them with 32.1 oz. If the actual contents of the cartons vary normally with a standard deviation of .1 oz., what percent of the cartons contain less than a quart (32 oz.)?

27. The probability that a can of beer from a certain brewery is defective is .005. A sample of 4 cans is selected at random. Write a distribution for the number of defective cans in the sample, and give its mean and standard deviation.

28. Find the probability of getting the following number of heads in 15 tosses of a coin.

(a) exactly 7

(b) between 7 and 10

(c) at least 10

29. The probability that a small business will go bankrupt in its first year is .21. For 50 such small businesses, find the following probabilities.

(a) exactly 8 go bankrupt.

(b) no more than 2 go bankrupt.

30. On standard IQ tests, the mean is 100, with a standard deviation of 15. The results are very close to fitting a normal curve. Suppose an IQ test is given to a very large group of people. Find the percentage of those people whose IQ score is
    (a) more than 130;
    (b) less than 85;
    (c) between 85 and 115.

31. The residents of a certain Eastern suburb average 42 minutes a day commuting to work, with a standard deviation of 12 minutes. Assume that commuting times are closely approximated by a normal curve and find the percent of the residents of this suburb who commute
    (a) at least 50 minutes per day;   (b) no more than 35 minutes per day;
    (c) between 32 and 40 minutes per day;   (d) between 38 and 60 minutes per day.

32. About 6% of the frankfurters produced by a certain machine are overstuffed, and thus defective. Find the probability that in a sample of 500 frankfurters,
    (a) 25 or fewer are overstuffed (Hint: $\sigma = 5.3$);
    (b) exactly 30 are overstuffed;
    (c) more than 40 are overstuffed.

33. Find the least squares line for the following data.

x	3	5	7	8
y	4	11	20	23

34. Use your equation from Exercise 33 to predict y when x is 6.

35. Find the coefficient of correlation for the data in Exercise 33.

36. The following data show the connection between blood sugar levels, x, and cholesterol levels, y, for 8 different patients.

Patient	1	2	3	4	5	6	7	8
Blood sugar level, x	130	138	142	159	165	200	210	250
Cholesterol level, y	170	160	173	181	201	192	240	290

    For this data, $\Sigma x = 1394, \Sigma y = 1607, \Sigma xy = 291,990, \Sigma x^2 = 255,214$, and $\Sigma y^2 = 336,155$.

    (a) Find the equation of the least squares line, $y' = mx + b$.
    (b) Predict the cholesterol level for a person whose blood sugar level is 190.
    (c) Find r.

9

Decision Theory

John F. Kennedy once remarked that he had assumed that as President it would be difficult to choose between distinct, opposite alternatives when a decision needed to be made. Actually, however, he said that he found such decisions easy to make; the hard decisions came when he was faced with choices that were not as clear-cut. Most decisions that we are faced with fall in this last category — decisions which must be made under conditions of uncertainty. In this chapter we look at *decision theory,* which provides a systematic way to attack problems of decision making when not all alternatives are clear and unambiguous.

9.1 Decision Making

In Chapter 8 we saw how to use expected values to help make a decision. We extend these ideas in this section, where we consider decision making in the face of uncertainty. Let us begin with an example.

Freezing temperatures are endangering the orange crop in central California. A farmer can protect his crop by burning smudge pots — the heat from the pots keeps the oranges from freezing. However, burning the pots is expensive; the cost is $2000. The farmer knows that if he burns smudge pots he will be able to sell his crop for a net profit (after smudge pot costs are deducted) of $5000, provided that the freeze does develop and wipes out many of the other orange growers in California. If he does nothing he will either lose $1000 in planting costs if it does freeze, or make a profit of $4800 if it does not freeze. (If it does not freeze, there will be a large supply of oranges, and thus his profit will be lower than if there was a small supply.)

What should the farmer do? He should begin by carefully defining the problem. First he must decide on the **states of nature,** the possible alternatives over which he has no control. Here there are two: freezing temperatures, or no freezing temperatures. Next, the farmer should list the things he can control—his actions or **strategies.** The farmer has two possible strategies: use smudge pots or not use smudge pots. The consequences of each action under each state of nature, called **payoffs,** can be summarized in a **payoff matrix,** as shown below. The payoffs in this case represent the profit for each possible combination of events.

States of nature

		Freeze	No freeze
Strategies of farmer	Use smudge pots	$5000	$2800
	Do not use pots	−$1000	$4800

To get the $2800 entry in the payoff matrix, we took the profit if there is no freeze, $4800, and subtracted the $2000 cost of using the smudge pots.

Once the farmer makes the payoff matrix, what then? The farmer might be an optimist (some might call him a gambler); in this case he might assume that the best will happen and go for the biggest number on the matrix ($5000). To get this profit, he must adopt the strategy "use smudge pots."

On the other hand, the farmer might be a pessimist. As a pessimist, he would want to minimize the worst thing that could happen. If he uses smudge pots, the worst thing that could happen to him would be a profit of $2800, which will result if there is no freeze. If he does not use smudge pots, he might well face a loss of $1000. To minimize the worst thing that could happen to him, he once again should adopt the strategy "use smudge pots."

Suppose the farmer decides that he is neither an optimist nor a pessimist, but would like further information before choosing a strategy. For example, he might call the weather forecaster and ask for the probability of a freeze. Suppose the forecaster says that this probability is only .1. What should the farmer do? He should recall our discussion of expected value from the previous chapter and work out the expected profit for each of his two possible strategies. If the probability of a freeze is .1, then the probability that there will be no freeze is .9. Using this information, we get the following expected values.

If smudge pots are used: $5000(.1) + 2800(.9) = 3020$

If no smudge pots are used: $-1000(.1) + 4800(.9) = 4220$

Here the maximum expected profit, $4220, is obtained if smudge pots are *not* used.

As the example and problem have shown, as the farmer's beliefs about the probabilities of a freeze change, so might his choice of strategies.

Example 1   A small Christmas card manufacturer must decide in February about the type of cards she should emphasize in her fall line of cards. She has three possible strategies: emphasize modern cards, emphasize old-fashioned cards, or emphasize a mixture of the two. Her success is dependent on the state of the economy in December. If the economy is strong, she will do well with her modern cards, while in a weak economy people long for the old days and buy old-fashioned cards. In an in-between economy, her mixture of lines would do the best. She first prepares a payoff matrix for all three possibilities. The numbers in the matrix represent her profits in thousands of dollars.

States of nature

		Weak economy	In-between	Strong economy
	Modern	$40	$85	$120
Strategies	Old-fashioned	$106	$46	$83
	Mixture	$72	$90	$68

(a) If the manufacturer is an optimist, she should aim for the biggest number on the matrix, the $120 (representing $120,000 in profit). Her strategy in this case would be to produce modern cards.

(b) A pessimistic manufacturer wants to avoid the worst of all bad things that can happen. If she produces modern cards, the worst that can happen is a profit of $40,000. For old-fashioned cards, the worst is a profit of $46,000, while the worst that can happen from a mixture is a profit of $68,000. Her strategy here is to use a mixture.

(c) Suppose the manufacturer reads an article in *The Wall Street Journal* that claims that leading experts feel there is a 50% chance of a weak economy at Christmas, a 20% chance of an in-between economy, and a 30% chance of a strong economy. The manufacturer should now use this information to find her expected profit for each possible strategy.

Modern: $40(.50) + 85(.20) + (120)(.30) = 73$

Old-fashioned: $106(.50) + 46(.20) + 83(.30) = 87.1$

Mixture: $72(.50) + 90(.20) + 68(.30) = 74.4$

Here the best strategy is old-fashioned cards; the expected profit is 87.1, or $87,100. ∎

9.1 Exercises

1. An investor has $20,000 to invest in stocks. She has two possible strategies: buy conservative blue-chip stocks or buy highly speculative stocks. There are two states of nature: the market goes up or the market goes down. The following payoff matrix shows the net amounts she will have under the various circumstances.

	Market up	Market down
Buy blue-chip	$25,000	$18,000
Buy speculative	$30,000	$11,000

What should the investor do if she is
(a) an optimist;
(b) a pessimist?
(c) Suppose there is a .7 probability of the market going up. What is the best strategy? What is the expected profit?
(d) What is the best strategy if the probability of a market rise is .2?

2. A developer has $100,000 to invest in land. He has a choice of two parcels (at the same price), one on the highway and one on the coast. With both parcels, his ultimate profit depends on whether he faces light opposition from environmental groups or heavy opposition. He estimates that the payoff matrix is as follows (the numbers represent his profit).

$$
\begin{array}{cc}
 & \textit{Opposition} \\
 & \begin{array}{cc} \text{Light} & \text{Heavy} \end{array}
\end{array}
$$

$$
\textit{Parcels} \begin{array}{c} \text{Highway} \\ \text{Coast} \end{array} \begin{bmatrix} \$70{,}000 & \$30{,}000 \\ \$150{,}000 & -\$40{,}000 \end{bmatrix}
$$

What should the developer do if he is
(a) an optimist;　(b) a pessimist?
(c) Suppose the probability of heavy opposition is .8. What is his best strategy? What is the expected profit?
(d) What is the best strategy if the probability of heavy opposition is only .4?

3. Hillsdale College has sold out all tickets for a jazz concert to be held in the stadium. If it rains, the show will have to be moved to the gym, which has a much smaller capacity. The dean must decide in advance whether to set up the seats and the stage in the gym or in the stadium, or both, just in case. The payoff matrix below shows the net profit in each case.

$$
\begin{array}{cc}
 & \textit{States of nature} \\
 & \begin{array}{cc} \text{Rain} & \text{No rain} \end{array}
\end{array}
$$

$$
\textit{Strategies} \begin{array}{c} \text{Set up in stadium} \\ \text{Set up in gym} \\ \text{Set up both} \end{array} \begin{bmatrix} -\$1550 & \$1500 \\ \$1000 & \$1000 \\ \$750 & \$1400 \end{bmatrix}
$$

What strategy should the dean choose if she is
(a) an optimist;　(b) a pessimist?
(c) If the weather forecaster predicts rain with a probability of .6, what strategy should she choose to maximize expected profit? What is the maximum expected profit?

4. An analyst must decide what fraction of the items produced by a certain machine are defective. He has already decided that there are three possibilities for the fraction of defective items: .01, .10, and .20. He may recommend two courses of action: repair the machine or make no repairs. The payoff matrix below represents the *costs* to the company in each case.

$$
\begin{array}{cc}
 & \textit{States of nature} \\
 & \begin{array}{ccc} .01 & .10 & .20 \end{array}
\end{array}
$$

$$
\textit{Strategies} \begin{array}{c} \text{Repair} \\ \text{No repair} \end{array} \begin{bmatrix} \$130 & \$130 & \$130 \\ \$25 & \$200 & \$500 \end{bmatrix}
$$

What strategy should the analyst recommend if he is
(a) an optimist;　(b) a pessimist?
(c) Suppose the analyst is able to estimate probabilities for the three states of nature as follows.

Fraction of defectives	Probability
.01	.70
.10	.20
.20	.10

Which strategy should he recommend? Find the expected cost to the company if this strategy is chosen.

5. The research department of the Allied Manufacturing Company has developed a new process which it believes will result in an improved product. Management must decide whether or not to go ahead and market the new product. The new product may be better than the old or it may not be better. If the new product is better, and the company decides to market it, sales should increase by $50,000. If it is not better and they replace the old product with the new product on the market, they will lose $25,000 to competitors. If they decide not to market the new product they will lose $40,000 if it is better, and research costs of $10,000 if it is not.
   (a) Prepare a payoff matrix.
   (b) If management believes the probability that the new product is better to be .4, find the expected profits under each strategy and determine the best action.

6. A businessman is planning to ship a used machine to his plant in Nigeria. He would like to use it there for the next four years. He must decide whether or not to overhaul the machine before sending it. The cost of overhaul is $2600. If the machine fails when in operation in Nigeria, it will cost him $6000 in lost production and repairs. He estimates the probability that it will fail at .3 if he does not overhaul it, and .1 if he does overhaul it. Neglect the possibility that the machine might fail more than once in the four years.
   (a) Prepare a payoff matrix.
   (b) What should the businesman do to minimize his expected costs?

7. A contractor prepares to bid on a job. If all goes well, his bid should be $30,000, which will cover his costs plus his usual profit margin of $4500. However, if a threatened labor strike actually occurs, his bid should be $40,000 to give him the same profit. If there is a strike and he bids $30,000, he will lose $5500. If his bid is too high, he may lose the job entirely, while if it is too low, he may lose money.
   (a) Prepare a payoff matrix.
   (b) If the contractor believes that the probability of a strike is .6, how much should he bid?

8. A community is considering an anti-smoking campaign.[1] The city council will choose one of three possible strategies: a campaign for everyone over age 10 in the community, a campaign for youths only, or no campaign at all. The two states of nature are a true cause-effect relationship between smoking and cancer and no cause-effect relationship. The costs to the community (including loss of life and productivity) in each case are as shown below.

| | | *States of nature* | |
		Cause-effect relationship	No cause-effect relationship
Strategies	Campaign for all	$100,000	$800,000
	Campaign for youth	$2,820,000	$20,000
	No campaign	$3,100,100	$0

What action should the city council choose if it is
   (a) optimistic;   (b) pessimistic?
   (c) If the Director of Public Health estimates that the probability of a true cause-effect relationship is .8, which strategy should the city council choose?

---

[1]This problem is based on an article by B. G. Greenberg in the September 1969 issue of the *Journal of the American Statistical Association.*

*Sometimes the numbers (or payoffs) in a payoff matrix do not represent money (profits or costs, for example), but utility. A **utility** is a number which measures the satisfaction (or lack of it) that results from a certain action. The numbers must be assigned by each individual, depending on how he or she feels about a situation. For example, one person might assign a utility of +20 for a week's vacation in San Francisco, with −6 being assigned if the vacation were moved to Sacramento. Work the following problems in the same way as those above.*

9. A politician must plan her reelection strategy. She can emphasize jobs or she can emphasize the environment. The voters can be concerned about jobs or about the environment. A payoff matrix showing the utility of each possible outcome is shown below.

$$\begin{array}{c} & & \textit{Voters} \\ & & \text{Jobs} \quad \text{Environment} \\ \textit{Candidate} \begin{array}{c} \text{Jobs} \\ \text{Environment} \end{array} & \left[\begin{array}{cc} +25 & -10 \\ -15 & +30 \end{array} \right] \end{array}$$

The political analysts feel that there is a .35 chance that the voters will emphasize jobs. What strategy should the candidate adopt? What is its expected utility?

10. In an accounting class, the instructor permits the students to bring a calculator or a reference book (but not both) to an examination. The examination itself can emphasize either numerical problems or definitions. In trying to decide which aid to take to an examination, a student first decides on the utilities shown in the following payoff matrix.

$$\begin{array}{c} & & \textit{Exam emphasizes} \\ & & \text{Numbers} \quad \text{Definitions} \\ \textit{Student chooses} \begin{array}{c} \text{Calculator} \\ \text{Book} \end{array} & \left[\begin{array}{cc} +50 & 0 \\ +10 & +40 \end{array} \right] \end{array}$$

(a) What strategy should the student choose if the probability that the examination will emphasize numbers is .6? What is the expected utility in this case?

(b) Suppose the probability that the examination emphasizes numbers is .4. What strategy should be chosen by the student?

> *Application* **Decision Making in Life Insurance**

When a life insurance company receives an application from an agent requesting insurance on the life of an individual, it knows from experience that the applicant will be in one of three possible states of risk, with proportions as shown.[1]

States of risk	Proportions
s_1 = Standard risk	.90
s_2 = Substandard risk (greater risk)	.07
s_3 = Sub-substandard risk (greatest risk)	.03

[1]This example was supplied by Donald J. vanKeuren, actuary of Metropolitan Life Insurance Company, and Dave Halmstad, senior actuarial assistant. It is based on a paper by Donald Jones.

A particular applicant could be correctly placed if all possible information about the applicant were known. This is not realistic in a practical situation; the company's problem is to obtain the maximum information at the lowest possible cost.

The company can take any of three possible strategies when it receives the application.

Strategies

a_1 = Offer a standard policy
a_2 = Offer a substandard policy (higher rates)
a_3 = Offer a sub-substandard policy (highest rates)

The payoff matrix in Table I below shows the payoffs associated with the possible strategies of the company and the states of the applicant. Here M represents the face value of the policy in thousands of dollars (for a \$30,000 policy we have $M = 30$). For example, if the applicant is substandard (s_2) and the company offers him or her a standard policy (a_1), the company makes a profit of 13M (13 times the face value of the policy in thousands). Strategy a_2 would result in a larger profit of 20M. However, if the prospective customer is a standard risk (s_1) but the company offers a substandard policy (a_2), the company loses \$50 (the cost of preparing a policy) since the customer would reject the policy because it has higher rates than he or she could obtain elsewhere.

Table 1

		States of nature		
		s_1	s_2	s_3
Strategies of company	a_1	20M	13M	3M
	a_2	−50	20M	10M
	a_3	−50	−50	20M

Before deciding on the policy to be offered, the company can perform any of three experiments to help it decide.

e_0 = No inspection report (no cost)

e_1 = Regular inspection report (cost: \$5)

e_2 = Special life report (cost: \$20)

On the basis of this report, the company can classify the applicant as follows.

T_1 = Applicant seems to be a standard risk

T_2 = Applicant seems to be a substandard risk

T_3 = Applicant seems to be a sub-substandard risk

We let $P(s|T)$ represent the probability that an applicant is in state s when the report indicates that he or she is in state T. For example, $P(s_1|T_2)$ represents the probability that an applicant is a standard risk (s_1) when the report indicates that he or she is a substandard risk (T_2). These probabilities, shown in Table 2, are based on Bayes' formula.

Table 2

True state	Regular report			Special report		
	$P(s_i\vert T_1)$	$P(s_i\vert T_2)$	$P(s_i\vert T_3)$	$P(s_i\vert T_1)$	$P(s_i\vert T_2)$	$P(s_i\vert T_3)$
s_1	.9695	.8411	.7377	.9984	.2081	.2299
s_2	.0251	.1309	.1148	.0012	.7850	.0268
s_3	.0054	.0280	.1475	.0004	.0069	.7433

From Table 2 we see that $P(s_2\vert T_2)$, the probability that an applicant actually is substandard (s_2) if the regular report indicates substandard (T_2) is only .1309, while $P(s_2\vert T_2)$, using the special report, is .7850.

We now have probabilities and payoffs, and we can use them to find expected values for each possible strategy the company might adopt. There are many possibilities here: the company can use one of three experiments, the experiments can indicate one of three states, the company can offer one of three policies, and the applicant can actually be in one of three states. Figure 1 shows some of these possibilities in a *decision tree*.

In order to find an optimum strategy for the company, let us consider an example. Suppose the company decides to perform experiment e_2 (special life report) with the report indicating a substandard risk, T_2. Then the expected values E_1, E_2, E_3 for the three possible actions a_1, a_2, a_3, respectively, are as shown below. (Recall: M is a variable, representing the face amount of the policy in thousands.)

For action a_1 (offer standard policy):
$$E_1 = [P(s_1\vert T_2)](20M) + [P(s_2\vert T_2)](13M) + [P(s_3\vert T_2)](3M)$$
$$= (.2081)(20M) + (.7850)(13M) + (.0069)(3M)$$
$$= 4.162M + 10.205M + .0207M$$
$$\approx 14.388M.$$

For action a_2 (offer a substandard policy):
$$E_2 = [P(s_1\vert T_2)](-50) + [P(s_2\vert T_2)](20M) + [P(s_3\vert T_2)](10M)$$
$$= (.2081)(-50) + (.7850)(20M) + (.0069)(10M)$$
$$= 15.769M - 10.405.$$

For action a_3 (offer a sub-substandard policy):
$$E_3 = [P(s_1\vert T_2)](-50) + [P(s_2\vert T_2)](-50) + [P(s_3\vert T_2)](20M)$$
$$= .138M - 49.655.$$

Strategy a_2 is better than a_3 (for any positive M, $15.769M - 10.405 > .138M - 49.655$). Hence, we are reduced to a choice between strategies a_1 and a_2. Strategy a_2 is superior if it leads to a higher expected value than a_1. This happens for all values of M such that
$$15.769M - 10.405 > 14.388M$$
$$1.381M > 10.405$$
$$M > 7.535.$$

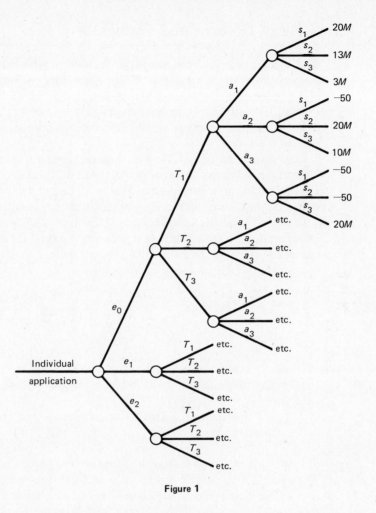

Figure 1

If the applicant applies for more than $7535 of insurance, the company should use strategy a_2; otherwise it should use a_1.

Similar analyses can be performed for all possible strategies from the decision tree above to find the best strategy. It turns out that the company will maximize its expected profits if it offers a standard policy to all people applying for less than $50,000 in life insurance, with a special report form required for all others.

Exercises

1. Find the expected values for each strategy a_1, a_2, and a_3 if the insurance company performs experiment e_1 (a regular report) with the report indicating that the applicant is a substandard risk.

2. Find the expected values for each action if the company performs e_1 with the report indicating that the applicant is a standard risk.

3. Find the expected values for each strategy if e_0 (no report) is selected. (Hint: use the proportions given for the three states s_1, s_2, and s_3 as the probabilities.)

9.2 Strictly Determined Games

The word *game* in the title of this section may have led you to think of checkers or perhaps some card game. While **game theory** does have some application to these recreational games, it was developed in the 1940's as a means of analyzing competitive situations in business, warfare, and social situations. Game theory deals with how to make decisions when in competition with an aggressive opponent.

A game can be set up with a payoff matrix, such as the one shown below. This game involves the two players A and B, and is called a **two-person game.** Player A can choose either of the two rows, 1 or 2, while player B can choose either column 1 or column 2. A player's choice is called a **strategy,** just as before. The payoff is at the intersection of the row and column selected. As a general agreement, a positive number represents a payoff from B to A; a negative number represents a payoff from A to B. For example, if A chooses row 2 and B chooses column 2, then B pays $4 to A.

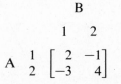

Example 1 In the payoff matrix shown above, suppose A chooses row 1 and B chooses column 2. Who gets what?

Row 1 and column 2 lead to the number −1. This number represents a payoff of $1 from A to B. ▌

We assumed above that the numbers in the payoff matrix represented money. They could just as well represent goods or other property.

In the game above, no money enters the game from the outside; whenever one player wins, the other loses. Such a game is called a **zero-sum game.** The stock market is not a zero-sum game. Stocks can go up or down according to outside forces. Therefore, it is possible that all investors can make or lose money.

We only discuss two-person zero-sum games in the rest of this chapter. Each player can have many different options. In particular, an $m \times n$ matrix game is one in which player A has m strategies (rows) and player B has n strategies (columns).

Dominant Strategies In the rest of this section we look for the best possible strategy for each player. Let us begin with the 3×3 game defined by the following matrix.

$$
\begin{array}{c c}
 & \begin{array}{c c c} 1 & \;\;2 & \;\;3 \end{array} \\
\begin{array}{c} 1 \\ 2 \\ 3 \end{array} &
\left[\begin{array}{r r r}
-3 & -6 & 10 \\
3 & 0 & -9 \\
5 & -4 & -8
\end{array} \right]
\end{array}
$$

From B's viewpoint, strategy 2 is better than strategy 1 no matter which strategy A selects. This can be seen by comparing the two columns. If A chooses row 1, receiving $6 from A is better than receiving $3; in row 2 breaking even is better than paying $3, and in row 3, getting $4 from A is better than paying $5. Therefore, B should never select strategy 1. Strategy 2 is said to *dominate* strategy 1, and strategy 1 (the dominated strategy) can be removed from consideration, producing the following reduced matrix.

$$
\begin{array}{c} \\ 1 \\ 2 \\ 3 \end{array}
\begin{array}{c} 2 \quad3 \\ \begin{bmatrix} -6 & 10 \\ 0 & -9 \\ -4 & -8 \end{bmatrix} \end{array}
$$

Either player may have dominated strategies. In fact, after a dominated strategy for one player is removed, the other player may then have a dominated strategy where there was none before.

A row for A **dominates** another row if every entry in the first row is *larger* than the corresponding entry in the second row. For a column for B to dominate another, each entry must be *smaller*.

In the 3 × 2 matrix above, neither player now has a dominated strategy. From A's viewpoint strategy 1 is best if B chooses strategy 3, while strategy 2 is best if B chooses strategy 1. Verify that there are no dominated strategies for either player.

Example 2   Find any dominated strategies in the games having the following payoff matrices.

(a)

$$
\begin{array}{c} \\ 1 \\ 2 \end{array}
\begin{array}{c} 1 \quad2 \quad3 \quad4 \\ \begin{bmatrix} -8 & -4 & -6 & -9 \\ -3 & 0 & -9 & 12 \end{bmatrix} \end{array}
$$

Here every entry in column 3 is smaller than the corresponding entry in column 2. Thus, column 3 dominates column 2. By removing the dominated column 2, the final game is as follows.

$$
\begin{array}{c} \\ 1 \\ 2 \end{array}
\begin{array}{c} 1 \quad3 \quad4 \\ \begin{bmatrix} -8 & -6 & -9 \\ -3 & -9 & 12 \end{bmatrix} \end{array}
$$

(b)

$$
\begin{array}{c} \\ 1 \\ 2 \\ 3 \end{array}
\begin{array}{c} 1 \quad2 \\ \begin{bmatrix} 3 & -2 \\ 0 & 8 \\ 6 & 4 \end{bmatrix} \end{array}
$$

Each entry in row 3 is greater than the corresponding entry in row 1, so that row 3 dominates row 1. Removing row 1 gives the following game.

$$
\begin{array}{c} \\ 2 \\ 3 \end{array}
\begin{array}{c} 1 \quad 2 \\ \begin{bmatrix} 0 & 8 \\ 6 & 4 \end{bmatrix} \end{array}
$$
∎

Strictly Determined Games   Consider the following game.

$$
\begin{array}{c}
 & & \text{B} \\
 & & \begin{array}{ccc} 1 & 2 & 3 \end{array} \\
\text{A} \begin{array}{c} 1 \\ 2 \\ 3 \end{array} &
\left[\begin{array}{ccc} -9 & 11 & -4 \\ 2 & 3 & 5 \\ -1 & -9 & 6 \end{array} \right]
\end{array}
$$

Which strategies should the players choose? The goal of game theory is to find **optimum strategies,** those which are the most profitable to the respective players. The payoff which results from each player's choosing the optimum strategy is called the **value** of the game.

The simplest strategy for a player is to consistently choose a certain row (or column). Such a strategy is called a **pure strategy,** in contrast to strategies requiring the random choice of a row (or column); these alternate strategies are discussed in the next section.[1]

To choose a pure strategy in the game above, player A could choose row 1, in hopes of getting the payoff of $11. However, B would quickly discover this, and start playing column 1. By playing column 1, B would receive $9 from A. If A chooses row 2 consistently, then B would again minimize outgo by choosing column 1 (a payoff of $2 by B to A is better than paying $3 or $5, respectively, to A). By choosing row 3 consistently, A would cause B to choose column 2. In summary, by choosing a given row consistently, A would cause the following actions by B.

A chooses pure strategy	then B would choose	with payoff
row 1	column 1	$9 to B
row 2	column 1	$2 to A
row 3	column 2	$9 to B

Based on these results, A's optimum strategy is to choose row 2; in this way A will guarantee a payoff of $2 per play of the game, no matter what B does.

The optimum pure strategy in this game for A (the *row* player), is found by identifying the *smallest* number in each row of the payoff matrix; the row giving the *largest* such number gives the optimum strategy.

By going through a similar analysis for player B, we find that B should choose that column which will minimize the amount that A can win. In the game above, B will pay $2 to A if B consistently chooses column 1. By choosing column 2 consistently, B will pay $11 to A, and by choosing column 3 player B will pay $6 to A. The optimum strategy for B is thus to choose column 1 — with each play of the game B will pay $2 to A.

The optimum pure strategy in this game for B (the column player) is to identify the *largest* number in each column of the payoff matrix, and then choose the column producing the *smallest* such number.

---

[1]In this section we solve (find the optimum strategies for) only games which have optimum *pure* strategies.

   In the game above, the entry 2 is both the *smallest* entry in its *row* and the *largest* entry in its *column*. Such an entry is called a **saddle point.** (See Figure 1.) As example 3(c) shows, there may be more than one such entry, though then the entries will have the same value.

> In a game with a saddle point, the optimum pure strategy for player A is to choose the row containing the saddle point, while the optimum strategy for B is to choose the column containing the saddle point.

A game with a saddle point is called a **strictly determined game.** By using these optimum strategies, A and B will ensure that the same amount always changes hands with each play of the game; this amount, given by the saddle point, is the value of the game. The value of the game above is $2. A game having a value of 0 is a **fair game;** our game is not fair.

The name *saddle point* comes from a saddle. The seat of the saddle is the maximum from one direction and the minimum from another direction.

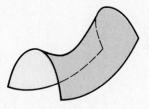

Figure 1

Example 3   Find the saddle points in the following games.

(a)   1   2

$$\begin{array}{c} 1 \\ 2 \\ 3 \\ 4 \end{array} \begin{bmatrix} 2 & 2 \\ 0 & 4 \\ 1 & 6 \\ 3 & 7 \end{bmatrix}$$

   The number that is both the smallest number in its row and the largest number in its column is 3. Thus, 3 is the saddle point, and the game has value 3. The strategies producing the saddle point can be written (4, 1). (A's strategy is written first.)

(b)   1   2

$$\begin{array}{c} 1 \\ 2 \end{array} \begin{bmatrix} 6 & 5 \\ 2 & 3 \end{bmatrix}$$

   The saddle point is 5, at strategies (1, 2).

(c)   1   2   3   4

$$\begin{array}{c} 1 \\ 2 \end{array} \begin{bmatrix} 4 & 6 & 4 & 12 \\ -8 & -9 & 3 & 2 \end{bmatrix}$$

   The saddle point, 4, occurs with either of two strategies, (1, 1), or (1, 3).

The value of the game is 4. (Note that none of the games in parts (a), (b), or (c) of this example are fair games: none had a value of 0.)

(d)   1   2   3

$$\begin{array}{c} 1 \\ 2 \end{array} \begin{bmatrix} 3 & 6 & -2 \\ 8 & -3 & 5 \end{bmatrix}$$

There is no number which is both the smallest number in its row and the largest number in its column, so that the game has no saddle point. Since the game has no saddle point, it is not strictly determined. In the next section we shall look at methods for finding optimum strategies for such games. ∎

9.2 Exercises   *In the following game, decide on the payoff when the indicated strategies are used.*

$$\begin{array}{cc} & B \\ & \begin{array}{ccc} 1 & 2 & 3 \end{array} \\ A \begin{array}{c} 1 \\ 2 \\ 3 \end{array} & \begin{bmatrix} 6 & -4 & 0 \\ 3 & -2 & 6 \\ -1 & 5 & 11 \end{bmatrix} \end{array}$$

1. (1, 1)   **2.** (1, 2)   **3.** (2, 2)   **4.** (2, 3)   **5.** (3, 1)   **6.** (3, 2)

7. Does the game have any dominated strategies?

8. Does it have a saddle point?

Remove any dominated strategies in the following games. (From now on, we will save space by deleting the names of the strategies.)

9. $\begin{bmatrix} 0 & -2 & 8 \\ 3 & -1 & -9 \end{bmatrix}$
   10. $\begin{bmatrix} 6 & 5 \\ 3 & 8 \\ -1 & -4 \end{bmatrix}$
   11. $\begin{bmatrix} 1 & 4 \\ 4 & -1 \\ 3 & 5 \\ -4 & 0 \end{bmatrix}$

12. $\begin{bmatrix} 2 & 3 & 1 & -5 \\ -1 & 5 & 4 & 1 \\ 1 & 0 & 2 & -3 \end{bmatrix}$
   13. $\begin{bmatrix} 8 & 12 & -7 \\ -2 & 1 & 4 \end{bmatrix}$
   14. $\begin{bmatrix} 6 & 2 \\ -1 & 10 \\ 3 & 5 \end{bmatrix}$

When it exists, find the saddle point and the value of the game for each of the following. Identify any games that are strictly determined.

15. $\begin{bmatrix} 3 & 5 \\ 2 & -5 \end{bmatrix}$
   16. $\begin{bmatrix} 7 & 8 \\ -2 & 15 \end{bmatrix}$
   17. $\begin{bmatrix} 3 & -4 & 1 \\ 5 & 3 & -2 \end{bmatrix}$

18. $\begin{bmatrix} -4 & 2 & -3 & -7 \\ 4 & 3 & 5 & -9 \end{bmatrix}$
   19. $\begin{bmatrix} -6 & 2 \\ -1 & -10 \\ 3 & 5 \end{bmatrix}$
   20. $\begin{bmatrix} 1 & 4 & -3 & 1 & -1 \\ 2 & 5 & 0 & 4 & 10 \\ 1 & -3 & 2 & 5 & 2 \end{bmatrix}$

21. $\begin{bmatrix} 2 & 3 & 1 \\ -1 & 4 & -7 \\ 5 & 2 & 0 \\ 8 & -4 & -1 \end{bmatrix}$
   22. $\begin{bmatrix} 3 & 8 & -4 & -9 \\ -1 & -2 & -3 & 0 \\ -2 & 6 & -4 & 5 \end{bmatrix}$

23. $\begin{bmatrix} -6 & 1 & 4 & 2 \\ 9 & 3 & -8 & -7 \end{bmatrix}$
   24. $\begin{bmatrix} 6 & -1 \\ 0 & 3 \\ 4 & 0 \end{bmatrix}$

25. When a football team has the ball and is planning its next play, it can choose one of several plays or strategies. The success of the chosen play depends largely on how well the other team "reads" the chosen play. Suppose a team with the ball (team A) can choose from three plays, while the opposition (team B) has four possible strategies. The numbers shown in the following payoff matrix represent yards of gain to team A.

$$\begin{bmatrix} 9 & -3 & -4 & 16 \\ 12 & 9 & 6 & 8 \\ -5 & -2 & 3 & 18 \end{bmatrix}$$

Find the saddle point. Find the value of the game.

26. Two armies, A and B, are involved in a war game. Each army has available three different strategies, with payoffs as shown below. These payoffs represent square kilometers of land with positive numbers representing gains by A.

$$\begin{bmatrix} 3 & -8 & -9 \\ 0 & 6 & -12 \\ -8 & 4 & -10 \end{bmatrix}$$

Find the saddle point and the value of the game.

27. Write a payoff matrix for the child's game *stone, scissors, paper*. Each of two children writes down one of these three words, *stone, scissors,* or *paper*. If the words are the same, the game is a tie. Otherwise, *stone* beats *scissors* (since stone can break scissors), *scissors* beats *paper* (since scissors can cut paper), and *paper* beats *stone* (since paper can hide stone). The winner receives $1 from the loser; no money changes hands in case of a tie. Is the game strictly determined?

28. John and Joann play a finger matching game—each shows one or two fingers at the same time. If the sum of the number of fingers showing is even, Joann pays John that number of dollars; for an odd sum, John pays Joann. Find the payoff matrix for this game. Is the game strictly determined?

29. Two merchants are planning competing stores to serve an area of three small cities. The fraction of the total population that live in each city is shown in the figure. If both merchants locate in the same city, merchant A will get 65% of the total business. If the merchants locate in different cities, each will get 80% of the business in the city it is in, and A will get 60% of the business from the city not containing B. Payoffs are measured by the number of percentage points above or below 50%. Write a payoff matrix for this game. Is this game strictly determined?

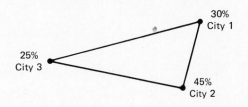

30. Suppose the payoff matrix for a game has at least three rows. Also, suppose that row 1 dominates row 2, and row 2 dominates row 3. Show that row 1 must dominate row 3.

9.3   Mixed Strategies

As we saw earlier, not every game has a saddle point. However, two-person zero-sum games still have optimum strategies, even if the strategy is not as simple as the ones we saw earlier. In a game with a saddle point, the optimum strategy for player A is to pick the row containing the saddle point. Such a strategy is called a *pure strategy,* since the same row is always chosen.

If there is no saddle point, then it will be necessary for both players to mix their strategies. For example, A will sometimes play row 1, sometimes row 2, and so on. If this were done in some specific pattern, the competitor would soon guess it and play accordingly.

For this reason, it is best to mix strategies according to previously determined probabilities. For example, if a player has only two strategies and has decided to play them with equal probability, the random choice could be made by tossing a fair coin, letting heads represent one strategy and tails the other. This would result in each strategy being used about equally over the long run. However, on a particular play it would not be possible to predetermine the strategy to be used. Some other device, such as a spinner, is necessary for more than two strategies, or when the probabilities are not 1/2.

Example 1   Suppose a game has payoff matrix

$$\begin{bmatrix} -1 & 2 \\ 1 & 0 \end{bmatrix},$$

where the entries represent dollar winnings. Suppose player A chooses row 1 with probability 1/3 and row 2 with probability 2/3, and player B chooses each column with probability 1/2. Find the expected value of the game.

We assume that rows and columns are chosen independently, so that, for example,

$$P(\text{row } 1, \text{column } 1) = P(\text{row } 1) \cdot P(\text{column } 1) = \frac{1}{3} \cdot \frac{1}{2} = \frac{1}{6}$$

$$P(\text{row } 1, \text{column } 2) = P(\text{row } 1) \cdot P(\text{column } 2) = \frac{1}{3} \cdot \frac{1}{2} = \frac{1}{6}$$

$$P(\text{row } 2, \text{column } 1) = P(\text{row } 2) \cdot P(\text{column } 1) = \frac{2}{3} \cdot \frac{1}{2} = \frac{1}{3}$$

$$P(\text{row } 2, \text{column } 2) = P(\text{row } 2) \cdot P(\text{column } 2) = \frac{2}{3} \cdot \frac{1}{2} = \frac{1}{3}.$$

The table below lists the probability of each possible outcome, along with the payoff to player A.

Outcome	Probability of outcome	Payoff for A
row 1, column 1	1/6	−1
row 1, column 2	1/6	2
row 2, column 1	1/3	1
row 2, column 2	1/3	0

The expected value of the game is given by the sum of the products of the probabilities and the payoffs, or

$$\text{expected value} = \frac{1}{6}(-1) + \frac{1}{6}(2) + \frac{1}{3}(1) + \frac{1}{3}(0) = \frac{1}{2}.$$

In the long run, for a great many plays of the game, the playoff to A will average 1/2 dollar per play of the game. It is important to note that as the mixed strategies used by A and B are changed, the expected value of the game may well change. (See Example 2 below.) ∎

Let us generalize the work of Example 1. Let the payoff matrix for a 2 × 2 game be

$$M = \begin{bmatrix} a_{11} & a_{12} \\ a_{21} & a_{22} \end{bmatrix}.$$

Let player A choose row 1 with probability p_1 and row 2 with probability p_2, where $p_1 + p_2 = 1$. Write these probabilities as the row matrix

$$A = [p_1 \quad p_2].$$

Let player B choose column 1 with probability q_1 and column 2 with probability q_2, where $q_1 + q_2 = 1$. Write this as the column matrix

$$B = \begin{bmatrix} q_1 \\ q_2 \end{bmatrix}.$$

The probability of choosing row 1 and column 1 is

$$P(\text{row 1, column 1}) = P(\text{row 1}) \cdot P(\text{column 1}) = p_1 \cdot q_1.$$

In the same way, the probabilities of each possible outcome are shown in the table below, along with the payoff matrix for each outcome.

Outcome	Probability of outcome	Payoff for A
row 1, column 1	$p_1 \cdot q_1$	a_{11}
row 1, column 2	$p_1 \cdot q_2$	a_{12}
row 2, column 1	$p_2 \cdot q_1$	a_{21}
row 2, column 2	$p_2 \cdot q_2$	a_{22}

The expected value for this game is

$$(p_1 \cdot q_1) \cdot a_{11} + (p_1 \cdot q_2) \cdot a_{12} + (p_2 \cdot q_1) \cdot a_{21} + (p_2 \cdot q_2) \cdot a_{22}.$$

This same result can be written as the matrix product

$$\text{expected value} = [p_1 \quad p_2] \begin{bmatrix} a_{11} & a_{12} \\ a_{21} & a_{22} \end{bmatrix} \begin{bmatrix} q_1 \\ q_2 \end{bmatrix} = AMB.$$

The same method works for games larger than 2 × 2: let the payoff matrix for a game have dimension $m \times n$; call this matrix $M = [a_{ij}]$. Let the mixed strategy for player A be given by the row matrix

$$A = [p_1 \quad p_2 \quad p_3 \cdots p_m],$$

and the mixed strategy for player B be given by the column matrix

$$B = \begin{bmatrix} q_1 \\ q_2 \\ \vdots \\ q_n \end{bmatrix}.$$

The expected value for this game is the product

$$AMB = [p_1 \quad p_2 \cdots p_m] \begin{bmatrix} a_{11} & a_{12} & \cdots & a_{1n} \\ a_{21} & a_{22} & \cdots & a_{2n} \\ \vdots & & & \vdots \\ a_{m1} & a_{m2} & \cdots & a_{mn} \end{bmatrix} \begin{bmatrix} q_1 \\ q_2 \\ \vdots \\ q_n \end{bmatrix}.$$

Example 2 In the game of Example 1, having payoff matrix

$$M = \begin{bmatrix} -1 & 2 \\ 1 & 0 \end{bmatrix},$$

suppose player A chooses row 1 with probability .2, and player B chooses column 1 with the probability .6. Find the expected value of the game.

If A chooses row 1 with probability .2, then row 2 is chosen with probability $1 - .2 = .8$, giving

$$A = [.2 \quad .8].$$

In the same way,

$$B = \begin{bmatrix} .6 \\ .4 \end{bmatrix}.$$

The expected value of this game is given by the product AMB, or

$$AMB = [.2 \quad .8] \begin{bmatrix} -1 & 2 \\ 1 & 0 \end{bmatrix} \begin{bmatrix} .6 \\ .4 \end{bmatrix}$$

$$= [.6 \quad .4] \begin{bmatrix} .6 \\ .4 \end{bmatrix}$$

$$= [.52].$$

On the average, these two strategies will produce a payoff of $.52, or 52¢, for A for each play of the game. This is a little better payoff than the 50¢ found in Example 1. ∎

It turns out, however, that B could cause this payoff to decline by a change of strategy. (Check this by choosing different matrices for B.) For this reason, player A needs to develop an *optimum strategy*—a strategy that will produce the best possible payoff no matter what B does. Just as in the previous section, this is done by finding the largest of the smallest possible amounts that can be won.

We need to find values of p_1 and p_2 so that the probability vector $[p_1 \quad p_2]$ produces an optimum strategy. Start with our payoff matrix

$$M = \begin{bmatrix} -1 & 2 \\ 1 & 0 \end{bmatrix}$$

and assume that A chooses row 1 with probability p_1. If player B chooses column 1, then player A's expectation is given by E_1, where

$$E_1 = -1 \cdot p_1 + 1 \cdot p_2 = -p_1 + p_2.$$

Since $p_1 + p_2 = 1$, we have $p_2 = 1 - p_1$, and

$$E_1 = -p_1 + 1 - p_1$$
$$E_1 = 1 - 2p_1.$$

If B chooses column 2, then A's expected value is given by E_2, where

$$E_2 = 2 \cdot p_1 + 0 \cdot p_2$$
$$E_2 = 2p_1.$$

Draw graphs of $E_1 = 1 - 2p_1$ and $E_2 = 2p_1$; see Figure 2.

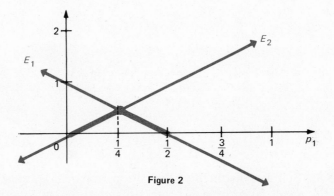

Figure 2

As we said, A needs to maximize the smallest amounts that can be won. On the graph, the smallest amounts that can be won are represented by the points of E_2 up to the intersection point. To the right of the intersection point, the smallest amounts that can be won are represented by the points of the line E_1. Player A can maximize the smallest amounts that can be won by choosing the point of intersection itself, the peak of the heavily shaded line in Figure 2.

To find this point of intersection, we need to find the simultaneous solution of the two equations. At the point of intersection, we have $E_1 = E_2$. Substitute $1 - 2p_1$ for E_1 and $2p_1$ for E_2.

$$E_1 = E_2$$
$$1 - 2p_1 = 2p_1$$
$$1 = 4p_1$$
$$1/4 = p_1$$

Thus, player A should choose strategy 1 with probability 1/4, and strategy 2 with probability $1 - 1/4 = 3/4$. By doing so, expected winnings will be maximized. To find the maximum winnings (which is also the value of the game), we can substitute 1/4 for p_1 in either E_1 or E_2. If we choose E_2,

$$E_2 = 2p_1 = 2\left(\frac{1}{4}\right) = \frac{1}{2},$$

that is, 1/2 dollar, or 50¢. By going through a similar argument for player B, we can find that the optimum strategy for player B is to choose each column with probability 1/2; in this case the value also turns out to be 50¢. In Example 2, A's winnings were 52¢; however, that was because B was not using his optimum strategy.

In the game above, we found that player A can maximize expected winnings by playing row 1 with probability 1/4 and row 2 with probability 3/4. Such a strategy is called a **mixed strategy.** To actually decide which row to use on a given game, player A could use a spinner, such as the one in Figure 3.

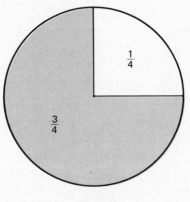

Figure 3

Example 3 Boll weevils threaten the cotton crop near Hattiesburg. Charles Dawkins owns a small farm; he can protect his crop by spraying with a potent (and expensive) insecticide. In trying to decide what to do, Dawkins first sets up a payoff matrix. The numbers in the matrix represent his profits.

		States of nature	
		Boll weevil attack	No attack
Strategies	Spray	\$14,000	\$7000
	Don't spray	−\$3000	\$8000

Let p_1 represent the probability with which Dawkins chooses strategy 1, so that $1 - p_1$ is the probability with which he chooses strategy 2. If nature chooses strategy 1 (an attack), then Dawkins' expected value is

$$E_1 = 14{,}000p_1 - 3000(1 - p_1)$$
$$= 14{,}000p_1 - 3000 + 3000p_1$$
$$E_1 = 17{,}000p_1 - 3000.$$

For nature's strategy 2 (no attack), Dawkins has an expected value of

$$E_2 = 7000p_1 + 8000(1 - p_1)$$
$$= 7000p_1 + 8000 - 8000p_1$$
$$E_2 = 8000 - 1000p_1.$$

As suggested by our work above, to maximize his expected profit, Dawkins should find the value of p_1 for which $E_1 = E_2$.

$$E_1 = E_2$$
$$17,000p_1 - 3000 = 8000 - 1000p_1$$
$$18,000p_1 = 11,000$$
$$p_1 = 11/18$$

Thus, $p_2 = 1 - p_1 = 1 - 11/18 = 7/18$.

Dawkins will maximize his expected profit if he chooses strategy 1 with probability 11/18 and strategy 2 with probability 7/18. His expected profit from this mixed strategy, $[11/18 \quad 7/18]$, can be found by substituting 11/18 for p_1 in either E_1 or E_2. If we choose E_1,

$$\text{expected profit} = 17,000\left(\frac{11}{18}\right) - 3000 = \frac{133,000}{18} \approx \$7400. \quad \blacksquare$$

Let us now obtain a formula for the optimum strategy in a game that is not strictly determined. Start with the matrix

$$M = \begin{bmatrix} a_{11} & a_{12} \\ a_{21} & a_{22} \end{bmatrix},$$

the payoff matrix of a game that is not strictly determined. Assume that A chooses row 1 with probability p_1. The expected value for A, assuming that B plays column 1, is E_1, where

$$E_1 = a_{11} \cdot p_1 + a_{21} \cdot (1 - p_1).$$

The expected value for A if B chooses column 2 is E_2, where

$$E_2 = a_{12} \cdot p_1 + a_{22} \cdot (1 - p_1).$$

As above, the optimum strategy for player A is found by letting $E_1 = E_2$.

$$a_{11} \cdot p_1 + a_{21} \cdot (1 - p_1) = a_{12} \cdot p_1 + a_{22} \cdot (1 - p_1)$$

Solve this equation for p_1.

$$a_{11} \cdot p_1 + a_{21} - a_{21} \cdot p_1 = a_{12} \cdot p_1 + a_{22} - a_{22} \cdot p_1$$
$$a_{11} \cdot p_1 - a_{21} \cdot p_1 - a_{12} \cdot p_1 + a_{22} \cdot p_1 = a_{22} - a_{21}$$
$$p_1(a_{11} - a_{21} - a_{12} + a_{22}) = a_{22} - a_{21}$$
$$p_1 = \frac{a_{22} - a_{21}}{a_{11} - a_{21} - a_{12} + a_{22}}$$

Since $p_2 = 1 - p_1$,

$$p_2 = 1 - \frac{a_{22} - a_{21}}{a_{11} - a_{21} - a_{12} + a_{22}}$$

$$= \frac{a_{11} - a_{21} - a_{12} + a_{22} - (a_{22} - a_{21})}{a_{11} - a_{21} - a_{12} + a_{22}}$$

$$= \frac{a_{11} - a_{12}}{a_{11} - a_{21} - a_{12} + a_{22}}.$$

This result is valid only if $a_{11} - a_{21} - a_{12} + a_{22} \neq 0$; it turns out that this condition is satisfied if the game is not strictly determined.

There is a similar result for player B, which is included in the following summary.

Let a non-strictly determined game have payoff matrix

$$\begin{bmatrix} a_{11} & a_{12} \\ a_{21} & a_{22} \end{bmatrix}.$$

The optimum strategy for player A is $[p_1 \quad p_2]$, where

$$p_1 = \frac{a_{22} - a_{21}}{a_{11} - a_{21} - a_{12} + a_{22}} \quad \text{and} \quad p_2 = \frac{a_{11} - a_{12}}{a_{11} - a_{21} - a_{12} + a_{22}}.$$

The optimum strategy for player B is $\begin{bmatrix} q_1 \\ q_2 \end{bmatrix}$, where

$$q_1 = \frac{a_{22} - a_{12}}{a_{11} - a_{21} - a_{12} + a_{22}} \quad \text{and} \quad q_2 = \frac{a_{11} - a_{21}}{a_{11} - a_{21} - a_{12} + a_{22}}.$$

The value of the game is

$$\frac{a_{11}a_{22} - a_{12}a_{21}}{a_{11} - a_{21} - a_{12} + a_{22}}.$$

Example 4 Suppose a game has payoff matrix

$$\begin{bmatrix} 5 & -2 \\ -3 & -1 \end{bmatrix}.$$

Here $a_{11} = 5$, $a_{12} = -2$, $a_{21} = -3$, and $a_{22} = -1$. To find the optimum strategy for player A, first find p_1.

$$p_1 = \frac{-1 - (-3)}{5 - (-3) - (-2) + (-1)} = \frac{2}{9}$$

Player A should play row 1 with probability 2/9 and row 2 with probability $1 - 2/9 = 7/9$.

For player B,

$$q_1 = \frac{-1 - (-2)}{5 - (-3) - (-2) + (-1)} = \frac{1}{9}.$$

Player B should choose column 1 with probability 1/9, and column 2 with probability 8/9. The value of the game is

$$\frac{5(-1) - (-2)(-3)}{5 - (-3) - (-2) + (-1)} = \frac{-11}{9}.$$

On the average, B will receive 11/9 dollar from A per play of the game. ■

9.3 Exercises

1. Suppose a game has payoff matrix

$$\begin{bmatrix} 3 & -4 \\ -5 & 2 \end{bmatrix}.$$

Suppose that player B uses the strategy $\begin{bmatrix} .3 \\ .7 \end{bmatrix}$. Find the expected value of the game if player A uses the strategy

(a) [.5   .5];   (b) [.1   .9];   (c) [.8   .2];   (d) [.2   .8].

2. Suppose a game has payoff matrix

$$\begin{bmatrix} 0 & -4 & 1 \\ 3 & 2 & -4 \\ 1 & -1 & 0 \end{bmatrix}.$$

Find the expected value of the game for the following strategies for players A and B.

(a) $A = [.1 \quad .4 \quad .5]; B = \begin{bmatrix} .2 \\ .4 \\ .4 \end{bmatrix}$

(b) $A = [.3 \quad .4 \quad .3]; B = \begin{bmatrix} .8 \\ .1 \\ .1 \end{bmatrix}$

Find the optimum strategy for both player A and player B in each of the following games. Find the value of the game. Be sure to look for a saddle point first.

3. $\begin{bmatrix} 5 & 1 \\ 3 & 4 \end{bmatrix}$ 4. $\begin{bmatrix} -4 & 5 \\ 3 & -4 \end{bmatrix}$ 5. $\begin{bmatrix} -2 & 0 \\ 3 & -4 \end{bmatrix}$ 6. $\begin{bmatrix} 6 & 2 \\ -1 & 10 \end{bmatrix}$

7. $\begin{bmatrix} 4 & -3 \\ -1 & 7 \end{bmatrix}$ 8. $\begin{bmatrix} 0 & 6 \\ 4 & 0 \end{bmatrix}$ 9. $\begin{bmatrix} -2 & \frac{1}{2} \\ 0 & -3 \end{bmatrix}$ 10. $\begin{bmatrix} 6 & \frac{3}{4} \\ \frac{2}{3} & -1 \end{bmatrix}$

11. $\begin{bmatrix} \frac{8}{3} & -\frac{1}{2} \\ \frac{3}{4} & -\frac{5}{12} \end{bmatrix}$ 12. $\begin{bmatrix} -\frac{1}{2} & \frac{2}{3} \\ \frac{7}{8} & -\frac{3}{4} \end{bmatrix}$ 13. $\begin{bmatrix} -1 & 2 \\ 3 & 1 \end{bmatrix}$ 14. $\begin{bmatrix} 8 & 18 \\ -4 & 2 \end{bmatrix}$

Remove any dominated strategies and then find the optimum strategies for each player and the value of the game.

15. $\begin{bmatrix} -4 & 9 \\ 3 & -5 \\ 8 & 7 \end{bmatrix}$ 16. $\begin{bmatrix} 3 & 4 & -1 \\ -2 & 1 & 0 \end{bmatrix}$ 17. $\begin{bmatrix} 8 & 6 & 3 \\ -1 & -2 & 4 \end{bmatrix}$

18. $\begin{bmatrix} -1 & 6 \\ 8 & 3 \\ -2 & 5 \end{bmatrix}$ 19. $\begin{bmatrix} 9 & -1 & 6 \\ 13 & 11 & 8 \\ 6 & 0 & 9 \end{bmatrix}$ 20. $\begin{bmatrix} 4 & 8 & -3 \\ 2 & -1 & 1 \\ 7 & 9 & 0 \end{bmatrix}$

21. Suppose Allied Manufacturing Company (see Exercise 5, Section 9.1) decides to put its new product on the market with a big television and radio advertising campaign. At the same time, the company finds out that its major competitor, Bates Manufacturing, has also decided to launch a big advertising campaign for a similar product. The payoff matrix below shows the increased sales (in millions) for Allied, as well as the decreased sales for Bates.

$$\begin{array}{cc} & \textit{Bates} \\ & \text{TV} \quad \text{Radio} \end{array}$$

$$\textit{Allied} \begin{array}{c} \text{TV} \\ \text{Radio} \end{array} \begin{bmatrix} 1.0 & -.7 \\ -.5 & .5 \end{bmatrix}$$

Find the optimum strategy for Allied Manufacturing and the value of the game.

22. The payoffs in the table below represent the differences between Boeing Aircraft Company's profit and its competitor's profit for two prices (in millions) on commercial jet transports, with positive payoffs being in Boeing's favor. What should Boeing's price strategy be?[1]

$$\begin{array}{c} \textit{Competitor's} \\ \textit{price strategy} \\ 4.75 \quad 4.9 \end{array}$$

$$\textit{Boeing's strategy} \begin{array}{c} 4.9 \\ 4.75 \end{array} \begin{bmatrix} -4 & 2 \\ 2 & 0 \end{bmatrix}$$

23. The number of cases of African flu has reached epidemic levels. The disease is known to have two strains with similar symptoms. Doctor De Luca has two medicines available: the first is 60% effective against the first strain and 40% effective against the second. The second medicine is completely effective against the second strain but ineffective against the first. Use the matrix below to decide which medicine she should use and the results she can expect.

$$\begin{array}{c} \textit{Strain} \\ 1 \quad 2 \end{array}$$

$$\textit{Medicine} \begin{array}{c} 1 \\ 2 \end{array} \begin{bmatrix} .6 & .4 \\ 0 & 1 \end{bmatrix}$$

24. Players A and B play a game in which each show either one or two fingers at the same time. If there is a match, A wins the amount equal to the total number of fingers shown. If there is no match, B wins the amount of dollars equal to the number of fingers shown.
    (a) Write the payoff matrix
    (b) Find optimum strategies for A and B and the value of the game.

25. Repeat Exercise 24 if each player may show either 0 or 2 fingers with the same sort of payoffs.

---

[1]From "Pricing, Investment, and Games of Strategy," by Georges Brigham in *Management Sciences Models and Techniques*, Vol. 1. Copyright © 1960 Pergamon Press, Ltd. Reprinted with permission.

26. In the game of matching coins, two players each flip a coin. If both coins match (both show heads or both show tails), player A wins \$1. If there is no match, player B wins \$1, as in the payoff matrix. Find the optimum strategies for the two players and the value of the game.

$$\begin{bmatrix} 1 & -1 \\ -1 & 1 \end{bmatrix}$$

27. The Huckster[1] Merrill has a concession at Yankee Stadium for the sale of sunglasses and umbrellas. The business places quite a strain on him, the weather being what it is. He has observed that he can sell about 500 umbrellas when it rains, and about 100 when it is sunny; in the latter case he can also sell 1000 sunglasses. Umbrellas cost him 50 cents and sell for \$1; glasses cost 20 cents and sell for 50 cents. He is willing to invest \$250 in the project. Everything that is not sold is considered a total loss.

He assembles the facts regarding profit in a table.

$$\begin{array}{c} & \textit{Selling during} \\ & \text{Rain} \quad \text{Shine} \end{array}$$

$$\textit{Buying for} \begin{array}{c} \text{Rain} \\ \text{Shine} \end{array} \begin{bmatrix} 250 & -150 \\ -150 & 350 \end{bmatrix}$$

He immediately takes heart, for this is a mixed-strategy game, and he should be able to find a stabilizing strategy which will save him from the vagaries of the weather. Find the best mixed strategy for Merrill.

28. The Squad Car[1] This is a somewhat more harrowing example. A police dispatcher was conveying information and opinion, as fast as she could speak, to Patrol Car 2, cruising on the U.S. Highway: ". . . in a Cadillac; just left Hitch's Tavern on the old Country Road. Direction of flight unknown. Suspect Plesset is seriously wounded but may have an even chance if he finds a good doctor, like Doctor Haydon, soon—even Veterinary Paxson might save him, but his chances would be halved. Plesset shot Officer Flood, who has a large family."

Deputy Henderson finally untangled the microphone from the riot gun and his size 14 shoes. He replied: "Roger. We can cut him off if he heads for Haydon's and we have a fifty-fifty chance of cutting him off at the State Highway if he heads for the vet's. We must cut him off because we can't chase him—Deputy Root got this thing stuck in reverse a while ago, and our cruising has been a disgrace to the department ever since."

The headquarter's carrier-wave again hummed in the speaker, but the dispatcher's musical voice was now replaced by the grating tones of Sheriff Lipp. "If you know anything else, don't tell it. He has a hi-fi radio in that Cad. Get him."

Root suddenly was seized by an idea and stopped struggling with the gearshift. "Henderson, we may not need a gun tonight, but we need a pencil: this is just a two-by-two game. The dispatcher gave us all the dope we need." "You gonna use *her* estimates?" "You got better ones? She's got intuition; besides, that's information from headquarters. Now let's see . . . . Suppose we head for Haydon's. And suppose Plesset does too; then we rack up one good bandit, if you don't trip on that gun again. But if he heads for Paxson, the chances are three out of four that old doc will kill him."

"I don't get it." "Well, it didn't come easy. Remember, Haydon would have an even chance—one-half—of saving him. He'd have half as good a chance with Paxson;

---

[1]From *The Compleat Strategyst* by J. D. Williams. Published 1966, by McGraw-Hill Book Company. Reprinted by permission of The Rand Corporation. This is an excellent nontechnical book on game theory.

and half of one-half is one-quarter. So the chance he dies must be three-quarters—subtracting from one, you know."

"Yeah, it's obvious." "Huh. Now if we head for Paxson's it's tougher to figure. First of all, *he* may go to Haydon's, in which case we have to rely on the doc to kill him, of which the chance is only one-half."

"You ought to subtract that from one." "I did. Now suppose he too heads for Paxson's. Either of two things can happen. One is, we catch him, and the chance is one-half. The other is, we don't catch him—and again the chance is one-half—but there is a three-fourths chance that the doc will have a lethal touch. So the overall probability that he will get by us, but not by the doc, is one-half times three-fourths, or three-eighths. Add to that the one-half chance that he doesn't get by us, and we have seven-eighths."

"I don't like this stuff. He's probably getting away while we're doodling." "Relax. He has to figure it out too, doesn't he? And he's in worse shape than we are. Now let's see what we have."

$$
\begin{array}{c}
\textit{Cad goes to} \\
\quad\text{Haydon} \quad \text{Paxson}^{\bullet}
\end{array}
$$

$$
\textit{Patrol car goes to} \quad
\begin{array}{c}
\text{Haydon} \\
\text{Paxson}
\end{array}
\left[
\begin{array}{cc}
1 & \frac{3}{4} \\
\frac{1}{2} & \frac{7}{8}
\end{array}
\right]
$$

"Fractions aren't so good in this light," Root continues. "Let's multiply everything by eight to clean it up. I hear it doesn't hurt anything."

$$
\begin{array}{c}
\textit{Cad} \\
\text{Haydon} \quad \text{Paxson}
\end{array}
$$

$$
\textit{Patrol car} \quad
\begin{array}{c}
\text{Haydon} \\
\text{Paxson}
\end{array}
\left[
\begin{array}{cc}
8 & 6 \\
4 & 7
\end{array}
\right]
$$

"It is now clear that this is a very messy business . . ." "I know." "There is no single strategy which we can safely adopt. I shall therefore compute the best mixed strategy." What mixed strategy should deputies Root and Henderson pursue?

> ## Application   *Decision Making in the Military*

This example has been reproduced with only minor change from the *Journal of the Operations Research Society of America,* November 1954, pages 365–369.[1] The article is titled "Military Decision and Game Theory," by O. G. Haywood, Jr. This case is presented unedited so that you can get an idea of the type of articles published in the journals.

A military commander may approach decision with either of two philosophies. He may select his course of action on the basis of his estimate of what his enemy *is able to do* to oppose him. Or, he may make his selection on the

---

[1]Reprinted by permission from *Operations Research,* Volume 3, Issue 6, 1954. Copyright 1954 Operations Research Society of America. No further reproduction permitted without the consent of the copyright owner.

basis of his estimate of what his enemy *is going to do*. The former is a doctrine of decision based on enemy capabilities; the latter, on enemy intentions.

The doctrine of decision of the armed forces of the United States is a doctrine based on enemy capabilities. A commander is enjoined to select the course of action which offers the greatest promise of success in view of the enemy capabilities. The process of decision, as approved by the Joint Chiefs of Staff and taught in all service schools, is formalized in a five-step analysis called the *Estimate of the Situation*. These steps are illustrated in the following analysis of an actual World War II battle situation.

General Kenney was Commander of the Allied Air Forces in the Southwest Pacific Area. The struggle for New Guinea reached a critical stage in February 1943. Intelligence reports indicated a Japanese troop and supply convoy was assembling at Rabaul (see Figure 1). Lae was expected to be the unloading point. With this general background Kenney proceeded to make his five-step Estimate of the Situation.

Figure 1 *The Rabaul-Lae Convoy Situation*. The problem is the distribution of reconnaissance to locate a convoy which may sail by either one of two routes.

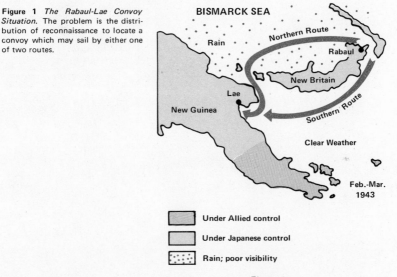

Figure 1

Step 1.   The Mission
General MacArthur as Supreme Commander had ordered Kenney to intercept and inflict maximum destruction on the convoy. This then was Kenney's mission.

Step 2.   Situation and Courses of Action
The situation as outlined above was generally known. One new critical factor was pointed out by Kenney's staff. Rain and poor visibility were predicted for the area north of New Britain. Visibility south of the island would be good.

The Japanese commander had two choices for routing his convoy from Rabaul to Lae. He could sail north of New Britain, or he could go south of that island. Either route required three days.

Kenney considered two courses of action, as he discusses in his memoirs. He could concentrate most of his reconnaissance aircraft either along the northern route where visibility would be poor, or along the southern route where clear weather was predicted. Mobility being one of the great advantages of air power, his bombing force could strike the convoy on either route once it was spotted.

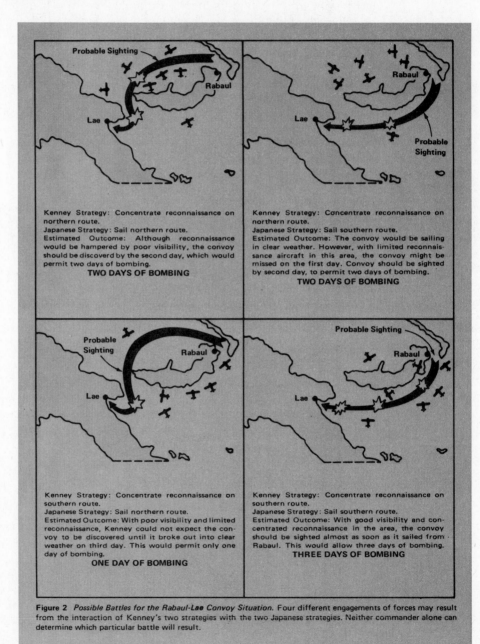

Kenney Strategy: Concentrate reconnaissance on northern route.
Japanese Strategy: Sail northern route.
Estimated Outcome: Although reconnaissance would be hampered by poor visibility, the convoy should be discovered by the second day, which would permit two days of bombing.
TWO DAYS OF BOMBING

Kenney Strategy: Concentrate reconnaissance on northern route.
Japanese Strategy: Sail southern route.
Estimated Outcome: The convoy would be sailing in clear weather. However, with limited reconnaissance aircraft in this area, the convoy might be missed on the first day. Convoy should be sighted by second day, to permit two days of bombing.
TWO DAYS OF BOMBING

Kenney Strategy: Concentrate reconnaissance on southern route.
Japanese Strategy: Sail northern route.
Estimated Outcome: With poor visibility and limited reconnaissance, Kenney could not expect the convoy to be discovered until it broke out into clear weather on third day. This would permit only one day of bombing.
ONE DAY OF BOMBING

Kenney Strategy: Concentrate reconnaissance on southern route.
Japanese Strategy: Sail southern route.
Estimated Outcome: With good visibility and concentrated reconnaissance in the area, the convoy should be sighted almost as soon as it sailed from Rabaul. This would allow three days of bombing.
THREE DAYS OF BOMBING

Figure 2 *Possible Battles for the Rabaul-Lae Convoy Situation.* Four different engagements of forces may result from the interaction of Kenney's two strategies with the two Japanese strategies. Neither commander alone can determine which particular battle will result.

Step 3.   Analysis of the Opposing Courses of Action
With each commander having two alternative courses of action, four possible conflicts could ensue. These conflicts are pictured in Figure 2.

Step 4.   Comparison of Available Courses of Action
If Kenney concentrated on the northern route, he ensured one of the two battles of the top row of sketches. However, he alone could not determine which one of these two battles in the top row would result from his decision. Similarly, if Kenney concentrated on the southern route, he ensured one of the battles of the lower row. In the same manner, the Japanese commander could not select a particular battle, but could by his decision assure that the battle would be one of those pictured in the left column or one of those in the right column.

Kenney sought a battle which would provide the maximum opportunity for bombing the convoy. The Japanese commander desired the minimum exposure to bombing. But neither commander could determine the battle which would result from his own decision. Each commander had full and independent freedom to select either one of his alternative strategies. He had to do so with full realization of his opponent's freedom of choice. The particular battle which resulted would be determined by the two independent decisions.

The U.S. doctrine of decision—the doctrine that a commander base his action on his estimate of what the enemy is capable of doing to oppose him—dictated that Kenney select the course of action which offered the greatest promise of success in view of all of the enemy capabilities. If Kenney concentrated his reconnaissance on the northern route, he could expect two days of bombing regardless of his enemy's decision. If Kenney selected his other strategy, he must accept the possibility of a less favorable outcome.

Step 5.   The Decision
Kenney concentrated his reconnaissance aircraft on the northern route.

Discussion
Let us assume that the Japanese commander used a similar philosophy of decision, basing his decision on his enemy's capabilities. Considering the four battles as sketched, the Japanese commander could select either the left or the right column, but could not select the row. If he sailed the northern route, he exposed the convoy to a maximum of two days of bombing. If he sailed the southern route, the convoy might be subjected to three days of bombing. Since he sought minimum exposure to bombing, he should select the northern route.

These two independent choices were the actual decisions which led to the conflict known in history as the Battle of the Bismarck Sea. Kenney concentrated his reconnaissance on the northern route; the Japanese convoy sailed the northern route; the convoy was sighted approximately one day after it sailed; and Allied bombing started shortly thereafter. Although the Battle of the Bismarck Sea ended in a disastrous defeat for the Japanese, we cannot say the Japanese commander erred in his decision. A similar convoy had reached Lae with minor losses two months earlier. The need was critical, and the Japanese were prepared to pay a high price. They did not know that Kenney had modified a number of his aircraft for low-level bombing and had perfected a deadly technique. The U.S. victory was the result of careful planning, thorough training, resolute execution, and tactical surprise of a new weapon—not of error in the Japanese decision.

Exercises

1. Use the results of Figure 2 to make a 2 × 2 game. Let the payoffs represent the number of days of bombing.

2. Find any saddle point and optimum strategy for the game.

3. Read the rest of the article used as the source for this example and prepare a discussion of the game theory aspects of the Avranches-Gap situation.

Key Words

states of nature	dominated strategy
strategies	optimum strategy
payoff matrix	value of the game
utility	pure strategy
decision tree	strictly determined game
two-person game	saddle point
zero-sum game	mixed strategy

Chapter 9   Review Exercises

In labor-management relations, both labor and management can adopt either a friendly or a hostile attitude. The results are shown in the following payoff matrix. The numbers give the wage gains made by an average worker.

$$
\begin{array}{cc}
 & \text{Management} \\
 & \begin{array}{cc} \text{Friendly} & \text{Hostile} \end{array} \\
\text{Labor} \begin{array}{c} \text{Friendly} \\ \text{Hostile} \end{array} & \left[\begin{array}{cc} \$600 & \$800 \\ \$400 & \$950 \end{array} \right]
\end{array}
$$

1. Suppose the chief negotiator for labor is an optimist. What strategy should he choose?

2. What strategy should he choose if he is a pessimist?

3. The chief negotiator for labor feels that there is a 70% chance that the company will be hostile. What strategy should he adopt? What is the expected payoff?

4. Just before negotiations begin, a new management is installed in the company. There is only a 40% chance that the new management will be hostile. What strategy should be adopted by labor?

A candidate for city council can come out in favor of a new factory, be opposed to it, or waffle on the issue. The change in votes for the candidate depends on what her opponent does, with payoffs as shown.

$$\begin{array}{c} & & \textit{Opponent} \\ & & \text{favors} \quad \text{waffles} \quad \text{opposes} \\ \textit{Candidate} \begin{array}{c} \text{favors} \\ \text{waffles} \\ \text{opposes} \end{array} & \left[\begin{array}{ccc} 0 & -1000 & -4000 \\ 1000 & 0 & -500 \\ 5000 & 2000 & 0 \end{array} \right] \end{array}$$

5. What should the candidate do if she is an optimist?

6. What should she do if she is a pessimist?

7. Suppose the candidate's campaign manager feels there is a 40% chance that the opponent will favor the plant, and a 35% chance that he will waffle. What strategy should the candidate adopt? What is the expected change in the number of votes?

8. The opponent conducts a new poll which shows strong opposition to the new factory. This changes the probability he will favor the factory to 0 and the probability he will waffle to .7. What strategy should our candidate adopt? What is the expected change in the number of votes now?

Use the following payoff matrix and decide on the payoff if the given strategies are used.

$$\begin{bmatrix} -2 & 5 & -6 & 3 \\ 0 & -1 & 7 & 5 \\ 2 & 6 & -4 & 4 \end{bmatrix}$$

9. (1, 1)   **10.** (1, 4)   **11.** (2, 3)   **12.** (3, 4)

13. Are there any dominated strategies in this game?

14. Is there a saddle point?

Remove any dominated strategies in the following games.

15. $\begin{bmatrix} -11 & 6 & 8 & 9 \\ -10 & -12 & 3 & 2 \end{bmatrix}$ **16.** $\begin{bmatrix} -1 & 9 & 0 \\ 4 & -10 & 6 \\ 8 & -6 & 7 \end{bmatrix}$ **17.** $\begin{bmatrix} -2 & 4 & 1 \\ 3 & 2 & 7 \\ -8 & 1 & 6 \\ 0 & 3 & 9 \end{bmatrix}$ **18.** $\begin{bmatrix} 3 & -1 & 4 \\ 0 & 4 & -1 \\ 1 & 2 & -3 \\ 0 & 0 & 2 \end{bmatrix}$

Find any saddle points for the following games. Give the value of the game. Identify any fair games.

19. $\begin{bmatrix} -2 & 3 \\ -4 & 5 \end{bmatrix}$ **20.** $\begin{bmatrix} -4 & 0 & 2 & -5 \\ 6 & 9 & 3 & 8 \end{bmatrix}$ **21.** $\begin{bmatrix} -4 & -1 \\ 6 & 0 \\ 8 & -3 \end{bmatrix}$

22. $\begin{bmatrix} 4 & -1 & 6 \\ -3 & -2 & 0 \\ -1 & -4 & 3 \end{bmatrix}$ **23.** $\begin{bmatrix} 8 & 1 & -7 & 2 \\ -1 & 4 & -3 & 3 \end{bmatrix}$ **24.** $\begin{bmatrix} 2 & -9 \\ 7 & 1 \\ 4 & 2 \end{bmatrix}$

Find the optimum strategies for each of the following games. Find the value of the game.

25. $\begin{bmatrix} 1 & 0 \\ -2 & 3 \end{bmatrix}$ **26.** $\begin{bmatrix} 2 & -3 \\ -3 & 5 \end{bmatrix}$ **27.** $\begin{bmatrix} -3 & 5 \\ 1 & 0 \end{bmatrix}$ **28.** $\begin{bmatrix} 8 & -3 \\ -6 & 2 \end{bmatrix}$

For each of the following games, remove any dominated strategies, then solve the game. Find the value of the game.

29. $\begin{bmatrix} -4 & 8 & 0 \\ -2 & 9 & -3 \end{bmatrix}$ **30.** $\begin{bmatrix} 1 & 0 & 3 & -3 \\ 4 & -2 & 4 & -1 \end{bmatrix}$ **31.** $\begin{bmatrix} 2 & -1 \\ -4 & 5 \\ -1 & -2 \end{bmatrix}$ **32.** $\begin{bmatrix} 8 & -6 \\ 4 & -8 \\ -9 & 9 \end{bmatrix}$

10

Mathematics of Finance

Not too many years ago, money could be borrowed by the largest and most secure corporations for 3% (and home mortgages could be had for 4 1/2%). Today, however, even the largest corporations must pay at least 12% or 13% for their money, and over 20% at times. Thus it is important that the management of corporations and consumers (who pay 21% or more to Sears or Wards) alike have a good understanding of the cost of borrowing money. The cost of borrowing money is called **interest.** The formulas for interest are developed in this chapter.

10.1 Simple Interest and Discount

Interest on loans of a year or less is usually calculated as **simple interest;** simple interest is interest that is charged only on the amount borrowed and not on past interest. The amount borrowed is the *principal, P.* The *rate* of interest, *r*, is given as a percent per year, and *t* is the *time,* measured in years. Simple interest, *I*, is the product of the principal, rate, and time.

Simple Interest
I = interest $\qquad r$ = rate per year
P = principal $\qquad t$ = time in years
$I = Prt$

Example 1 Find the total amount that must be repaid in each of the following.

(a) a loan of $2500, made on June 5; to be repaid on February 5 with interest of 14%.

There are 8 months from June 5 to February 5, so that the loan is for $8/12 = 2/3$ of a year. The interest owed is

$$I = Prt$$

$$I = 2500(.14)\left(\frac{2}{3}\right)$$

$$I = \$233.33.$$

(Here we have rounded to the nearest cent, as is customary in financial problems.)

The total amount that must be repaid is given by the sum of the amount borrowed and the interest, or

$$\$2500 + \$233.33 = \$2733.33.$$

(b) a loan of $11,280 for 85 days at 11% interest

It is common to assume 360 days in a year when working problems with simple interest. We shall make such an assumption in this book. (Interest found assuming 360 days is called **ordinary interest**; interest found by assuming 365 days is called **exact interest**.)

The interest in our example is

$$\$11,280(.11)\left(\frac{85}{360}\right) = \$292.97.$$

The total amount to be repaid is $11,280 + $292.97 = $11,572.97.   ∎

Present Value   A sum of money that can be deposited today to yield some larger amount in the future is called the **present value** of that future amount. Let P be the present value of some amount A at some time t (in years) in the future. Assume a rate of interest r. The simple interest on today's P dollars is

$$I = Prt;$$

since we now have P dollars we will have a total of

$$P + Prt = P(1 + rt)$$

dollars in the future. We said that P is the present value of A dollars, so that

$$P(1 + rt) = A$$

or

$$P = \frac{A}{1 + rt}.$$

Present Value of an Amount

P = present value r = interest rate per year
A = future amount t = time in years

$$P = \frac{A}{1 + rt}$$

Example 2   Find the present value of the following future amounts.

(a) $10,000 in one year, if interest is 13%

Here $A = 10{,}000$, $t = 1$, and $r = .13$. Thus,

$$P = \frac{10{,}000}{1 + (.13)(1)} = \frac{10{,}000}{1.13} = 8849.56.$$

If $8849.56 were deposited today, at 13% interest, a total of $10,000 would be in the account in one year. These two sums, $8849.56 today, and $10,000 in a year, are equivalent (at 13%); one becomes the other in a year.

(b) $32,000 in four months at 9% interest

$$P = \frac{32{,}000}{1 + (.09)(4/12)} = \frac{32{,}000}{1.03} = 31{,}067.96. \quad \blacksquare$$

Example 3   Because of a court settlement, John Walker owes $5000 to Arnold Parker. The money must be paid in ten months, with no interest. Suppose Walker wishes to pay the money today. What amount should Parker be willing to accept? Assume an interest rate of 11%.

The amount that Parker should be willing to accept is given by the present value:

$$P = \frac{5000}{1 + (.11)(10/12)} = \frac{5000}{1.09167} = 4580.14.$$

Parker should be willing to accept $4580.14 in settlement. $\blacksquare$

Simple Discount   It is not an uncommon practice to have interest deducted from the amount of a loan before giving the balance to the borrower. The money that is deducted is called the **discount,** with the money actually received by the borrower called the **proceeds.**

Example 4   Elizabeth Thornton agrees to pay $8500 to her banker in nine months. The banker subtracts a discount of 15% and gives the balance to Thornton. Find the amount of the discount and the proceeds.

The discount is found in the same way that simple interest is found.

$$\text{discount} = 8500(.15)\left(\frac{9}{12}\right) = 956.25$$

The proceeds are found by subtracting the discount from the original amount.

$$\text{proceeds} = \$8500 - \$956.25 = \$7543.75 \quad \blacksquare$$

Example 5   Find the actual rate of interest paid by Thornton in Example 4.

The rate of 15% stated in Example 4 is not the actual rate of interest since it applies to the total amount of $8500 and not to the amount actually borrowed, or $7543.75. To find the actual rate of interest, use the formula for simple interest, $I = Prt$, with r the unknown. From Example 4, $I = \$956.25$, $P = \$7543.75$, and $t = 9/12$. Substitute these values into $I = Prt$.

$$I = Prt$$
$$956.25 = 7543.75(r)\left(\frac{9}{12}\right)$$
$$\frac{956.25}{7543.75(9/12)} = r$$
$$.169 \approx r$$

The actual interest rate (called the *effective rate*) is about 16.9%. ▮

Let D represent the amount of discount on a loan. Then $D = Art$, where A is the total amount of the loan (amount borrowed plus interest), and r is the stated rate of interest. The amount actually received, the proceeds, can be written as $p = A - D$, or $p = A - Art$, from which $p = A(1 - rt)$.
Let us summarize the formulas for discount interest.

Discount Interest

p = proceeds r = rate of interest per year
D = discount t = time in years
A = total amount

$$p = A - D$$
$$p = A - Art$$
$$p = A(1 - rt)$$

One common use of discount interest is in **discounting a note,** a process by which a promissory note due at some time in the future can be converted to cash now.

Example 6   Jim Levy owes $4250 to Jenny Toms. The loan is payable in one year, at 12% interest. Toms needs cash to buy a new car, so three months before the loan is payable she goes to her bank to have the loan discounted. The bank charges a 16% discount fee. Find the amount of cash she will receive from the bank.
First find the amount that Levy must pay to Toms. The interest is

$$I = Prt$$
$$I = 4250(.12)(1) = 510.$$

Levy must repay a total of $4250 + $510 = $4760.
The bank applies its discount rate to this total.

$$\text{amount of discount} = 4760(.16)\left(\frac{3}{12}\right) = 190.40$$

(Remember that the loan was discounted three months before it was due.) Toms actually receives

$$\$4760 - \$190.40 = \$4569.60$$

in cash from the bank. Three months later, the bank would get $4760 from Levy. ▮

10.1 Exercises   *Find the simple interest in each of the following.*

1. $1000 at 12% for one year
2. $4500 at 10% for one year
3. $25,000 at 13% for nine months
4. $3850 at 9% for eight months
5. $1974 at 11.2% for seven months
6. $3724 at 14.1% for eleven months

In the following exercises, assume a 360 day year. Also, assume 30 days in each month.

7. $12,000 at 14% for 72 days
8. $38,000 at 10% for 216 days
9. $5147.18 at 11.6% for 58 days
10. $2930.42 at 13.9% for 123 days
11. $7980 at 15%; the loan was made May 7 and is due on September 19
12. $5408 at 12%; the loan was made August 16 and is due on December 30

In the following exercises, assume 365 days in a year, and use the exact number of days in a month. (Assume 28 days for February.)

13. $7800 at 14%; made on July 7 and due October 25
14. $11,906 at 15%; made on February 19 and due May 31
15. $2579 at 9.5%; made on October 4 and due March 15
16. $37,098 at 12.2%; made on September 12 and due July 30

Find the present value of the following future amounts. Assume 360 days in a year.

17. $15,000 for 8 months, money earns 10%
18. $48,000 for 9 months, money earns 14%
19. $5200 for 3 months, money earns 12.4%
20. $6892 for 7 months, money earns 9.2%
21. $15,402 for 125 days, money earns 10.7%
22. $29,764 for 310 days, money earns 13.2%

Find the proceeds for Exercises 23 – 26. Assume 360 days in a year.

23. $7150, discount rate 12%, length of loan 11 months
24. $9450, discount rate 14%, length of loan 7 months
25. $358, discount rate 15.1%, length of loan 183 days
26. $509, discount rate 16.4%, length of loan 238 days
27. Donna Sharp borrowed $25,900 from her father to start a flower shop. She repaid him in eleven months, with interest of 13.4%. Find the total amount she repaid.
28. A corporation accountant forgot to pay the firm's income tax of $725,896.15 on time. The government charged a penalty of 13.7% interest for the 34 days the money was late. Find the total amount, tax and penalty, that was paid. (Use a 365 day year.)

29. Tuition of $1769 will be due when the spring term begins, in four months. What amount should a student deposit today, at 6.25%, to have enough to pay the tuition?

30. A firm of attorneys has ordered seven new IBM typewriters, at a cost of $2104 each. The machines will not be delivered for seven months. What amount could the firm deposit in an account paying 10.42% to have enough to pay for the machines?

31. Roy Gerard needs $5196 to pay for remodeling work on his house. He plans to repay the loan in 10 months. His bank loans money at a discount rate of 17%. Find the amount of his loan.

32. Mary Collins decides to go back to college. To get to school she buys a small car for $6100. She decides to borrow the money at the bank, where they charge a 15.9% discount rate. If she will repay the loan in 7 months, find the amount of the loan.

33. Marge Prullage signs a $4200 note at the bank. The bank charges a 17.2% discount rate. Find the net proceeds if the note is for ten months. Find the effective interest rate charged by the bank.

34. A bank charges a 14.6% discount rate on money borrowed for 90 days. Find the effective rate.

35. Helen Spence owes $7000 to the Eastside Music Shop. She has agreed to pay the money in 7 months, at an interest rate of 11%. Two months before the loan is due to be paid, the store discounts it at the bank. The bank charges a 16.3% discount rate. How much money does the store receive?

36. A building contractor gives a $13,500 note to a plumber. The note is due in nine months, with interest of 13%. Three months after the note is signed, the plumber discounts it at the bank. The bank charges a 17.5% discount rate. How much money does the plumber actually receive?

10.2 Compound Interest

Simple interest is normally used for loans of a year or less; for longer periods **compound interest** is used. With compound interest, interest is charged on interest, as well as principal. To find a formula for compound interest, first suppose that P dollars is deposited at a rate of interest i per year. (While we used r with simple interest, it is common to use i for compound interest.) The interest earned during the first year is found by the formula for simple interest:

$$\text{first year interest} = P \cdot i \cdot 1 = Pi.$$

At the end of one year, the amount on deposit will be the sum of the original principal and the interest earned, or

$$P + Pi = P(1 + i). \tag{1}$$

If the deposit earns compound interest, the interest earned during the second year is found from the total amount on deposit at the end of the first year. Thus, the interest earned during the second year (again found by the formula for simple interest), is given by

$$P(1 + i)(i)(1) = P(1 + i)i, \tag{2}$$

so that the total amount on deposit at the end of the second year is given by the sum of the amounts from (1) and (2) above, or

$$P(1 + i) + P(1 + i)i = P(1 + i) \cdot (1 + i)$$
$$= P(1 + i)^2.$$

In the same way, the total amount on deposit at the end of three years is

$$P(1 + i)^3.$$

Generalizing, in j years, the total amount on deposit will be $P(1 + i)^j$, called the **compound amount.**

Interest can be compounded more than once a year. Suppose interest is compounded m times per year (m *periods* per year), at a rate i per year, so that i/m is the rate for each period. Suppose that interest is compounded for n periods. Then the following formula for the compound amount can be derived in the same way as was the previous formula.

Compound Amount

A = compound amount i = interest rate per year

P = principal m = number of periods per year

n = number of periods

$$A = P\left(1 + \frac{i}{m}\right)^n$$

Example 1   Suppose $1000 is deposited for 6 years in an account paying 5% per year compounded annually.

(a) Find the compound amount.

In the formula above, $P = 1000$, $i = 5\% = .05$, $m = 1$, and $n = 6$. The compound amount is

$$A = P\left(1 + \frac{i}{m}\right)^n$$

$$A = 1000\left(1 + \frac{.05}{1}\right)^6$$

$$A = 1000(1.05)^6.$$

We could find $(1.05)^6$ by using a calculator, or by using special compound interest tables. Table 3 at the back of this book is such a table. To find $(1.05)^6$, look for 5% across the top and 6 (for 6 periods) down the side. You should find 1.34010; thus $(1.05)^6 \approx 1.34010$, and

$$A = 1000(1.34010) = 1340.10$$

or $1340.10, which represents the final amount on deposit.

(b) Find the actual amount of interest earned.

Take the compound amount and subtract the initial deposit.

$$\text{amount of interest} = \$1340.10 - \$1000 = \$340.10 \quad \blacksquare$$

Example 2   Find the amount of interest earned by a deposit of $1000 for 6 years at 8% compounded quarterly.

Interest compounded quarterly is compounded four times a year. In 6 years, there are $4 \cdot 6 = 24$ quarters, or 24 periods. Interest of 8% per year is 8%/4, or 2%, per quarter. The compound amount is thus

$$1000(1 + .02)^{24} = 1000(1.02)^{24}.$$

The value of $(1.02)^{24}$ can be found with a calculator or in Table 3; locate 2% across the top and 24 periods at the left. You should find the number 1.60844. Thus,

$$A = 1000(1.60844) = 1608.44,$$

or $1608.44. The compound amount is $1608.44 and the interest earned is $1608.44 − $1000 = $608.44.   ■

Example 3   Find the compound amount if $900 is deposited at 6% compounded semiannually for 8 years.

In 8 years there are $8 \cdot 2 = 16$ semiannual periods. If interest is 6% per year, then 6%/2 = 3% is earned per semiannual period. Look in Table 3 for 3% and 16 periods, finding the number 1.60471. The compound amount is

$$A = 900(1.03)^{16}$$
$$A = 900(1.60471) = 1444.24,$$

or $1444.24.   ■

The more often interest is compounded within a given time period, the more interest will be earned. Using a calculator with an x^y key, or a compound interest table more complete than the one in this text, and using the formula above, we can get the results shown in the chart below.

Interest on $1000 at 6%
per year for 10 Years

Compounded	Interest
Not at all (simple interest)	$600
Yearly	790.85
Semiannually	806.11
Quarterly	814.02
Monthly	819.40
Daily	822.03
Hourly	822.11

As suggested by the chart, it makes a big difference whether interest is compounded or not. Interest differs by $190.85 when simple interest is compared to interest compounded annually. However, increasing the frequency of compounding makes smaller and smaller differences in the amount of interest earned. In fact, it can be shown that even if interest is compounded at intervals of time as small as one chooses (such as each hour, each minute, or each second), the total amount of interest earned will be only slightly more than for daily compounding. This is true even for a process called **continuous compounding** which can be loosely described as compounding at every instant. (See Examples 4 and 5 below.)

It turns out, although we shall not prove it here, that the formula for continuous compounding involves the number e. The number e, like the number π, is irrational: these two numbers cannot be written as quotients of integers. (See the Appendix to this book.) We can, however, give a decimal approximation for e:

$$e \approx 2.718281828459.$$

Using this number e, we have the following result.

Continuous Compounding

A = compound amount i = interest rate per year
P = principal n = number of years

$$A = Pe^{ni}$$

Example 4   Suppose \$5000 is deposited in an account paying 8% compounded continuously for five years. Find the compound amount.

Let $P = 5000$, $n = 5$, and $i = .08$. Then

$$A = 5000e^{5(.08)} = 5000e^{.4}.$$

From Table 8, or a calculator, $e^{.4} \approx 1.49182$, and

$$A = 5000(1.49182) = 7459.10$$

or \$7459.10. Check that daily compounding would have produced a compound amount about 30¢ less.   ▌

Example 5   Suppose \$24,000 is deposited at 12% for 9 years. Find the interest earned by (a) daily, (b) hourly, and (c) continuous compounding.

(a) The compound amount with daily compounding is

$$24,000\left(1 + \frac{.12}{360}\right)^{9 \cdot 360} = 70,659.59,$$

which includes interest of

$$\$70,659.59 - \$24,000 = \$46,659.59.$$

(b) In one year there are $360 \times 24 = 8640$ hours, while in 9 years of 360 days each, there are $9 \cdot 360 \cdot 24 = 77,760$ hours. The compound amount is

$$24,000\left(1 + \frac{.12}{8640}\right)^{77,760} = 70,671.77$$

which includes interest of $\$70,671.77 - \$24,000 = \$46,671.77$, only \$12.18 more than when interest is compounded daily.

(c) The compound amount for continuous compounding is

$$24{,}000e^{9(.12)} = 24{,}000e^{1.08}$$

$$= 24{,}000(2.9446795) \qquad use\ a\ calculator$$

$$= 70{,}672.31.$$

The interest earned, $70,672.31 − $24,000 = $46,672.21, is only 54¢ more than when interest is compounded hourly. ∎

If we deposit $1 at 4% compounded quarterly, we can use a calculator or Table 3 to find that at the end of one year, the compound amount is $1.0406, an increase of 4.06% over the original $1. The actual increase of 4.06% in our money is somewhat higher than the stated increase of 4%. To differentiate between these two numbers, 4% is called the **nominal** or **stated** rate of interest, while 4.06% is called the **effective** rate.

Example 6 Find the effective rate corresponding to a nominal rate of 6% compounded semi-annually.

Look in Table 3 for 3% and 2 periods. You should find the number 1.06090. Thus, $1 will increase to $1.06090, an actual increase of 6.09%. The effective rate is 6.09%. ∎

In general, the effective rate of interest is given by the following formula.

Effective Rate

i = interest rate per year

m = number of times per year of compounding

$$\left(1 + \frac{i}{m}\right)^m - 1$$

Present Value with Compound Interest   The formula for interest compounded annually, $A = P(1 + i)^n$, has four variables, $A, P, i,$ and n. If we know the values of any three of these variables, we can then find the value of the fourth. In particular, if we know A, the amount of money we wish to end up with, and also know i and n, then we can find P. Here P is the amount that we should deposit today to produce A dollars in n years. The next example shows how this works.

Example 7 Joan must pay a lump sum of $6000 in 5 years. What amount deposited today at 8% compounded annually will amount to $6000 in 5 years?

Here $A = 6000$, $i = .08$, $n = 5$, and P is unknown. Substituting these values into the formula for the compound amount gives

$$6000 = P(1 + .08)^5$$

or $$6000 = P(1.08)^5.$$

To solve for P, multiply both sides of this last equation by $(1.08)^{-5}$. Recall: $(1.08)^5 \cdot (1.08)^{-5} = (1.08)^{5+(-5)} = (1.08)^0 = 1$. Doing so gives

$$6000\,(1.08)^{-5} = P.$$

To evaluate $(1.08)^{-5}$, use a calculator with an x^y key, or use Table 4, constructed to find present value. Look for 8% across the top and 5 periods at the side. You should find .68058. Therefore,

$$P = 6000\,(.68058) = 4083.48,$$

or $4083.48. If Joan deposits $4083.48 for 5 years in an account paying 8% compounded annually, she will have $6000 when she needs it. (Note that corresponding entries in Table 3 and 4 are *reciprocals* of each other: that is, their product is 1.) ∎

As this last example shows, $6000 in 5 years is the same as $4083.48 today (if money can be deposited at 8% compounded annually). Recall from Section 10.1 that an amount that can be deposited today to yield a given sum in the future is called the **present value** of this future sum.

Example 8 Find the present value of $16,000 in 9 years if money can be deposited at 12% compounded semiannually.

In 9 years there are $2 \cdot 9 = 18$ semiannual periods. A rate of 12% per year is 6% each semiannual period. Look in Table 4 for 6% across the top and 18 periods down the side. You should find .35034. The present value is

$$16,000\,(.35034) = 5605.44.$$

A deposit of $5605.44 today, at 12% compounded semiannually, will produce a total deposit of $16,000 in 9 years. ∎

We can also solve $A = P(1 + i)^n$ for n, as the following example shows.

Example 9 Suppose the general level of inflation in the economy averages 8% per year. Find the number of years it would take for the general level of prices to double.

We want to find the number of years it will take for $1 worth of goods and services to cost $2. That is, we want to find n in the equation

$$2 = 1(1 + .08)^n,$$

where $A = 2$, $P = 1$, and $i = .08$. This equation simplifies to

$$2 = (1.08)^n.$$

We could find n by using logarithms or certain calculators, but we can find a reasonable approximation by reading down the 8% column of Table 3. Read down this column until you come to the number closest to 2. The number closest to 2 is 1.99900; this number corresponds to 9 periods. Thus, the general level of prices will double in about 9 years. ∎

10.2 Exercises *Find the compound amount when the following deposits are made.*

1. $1000 at 4% compounded annually for 8 years
2. $1000 at 8% compounded annually for 10 years
3. $4500 at 6% compounded annually for 20 years

4. $810 at 8% compounded annually for 12 years

5. $470 at 6% compounded semiannually for 12 years

6. $15,000 at 8% compounded semiannually for 11 years

7. $46,000 at 12% compounded semiannually for 5 years

8. $1050 at 10% compounded semiannually for 13 years

9. $7500 at 8% compounded quarterly for 9 years

10. $8000 at 8% compounded quarterly for 4 years

11. $6500 at 12% compounded quarterly for 6 years

12. $9100 at 12% compounded quarterly for 4 years

Find the amount of interest earned by the following deposits.

13. $6000 at 8% compounded annually for 8 years

14. $21,000 at 6% compounded annually for 5 years

15. $43,000 at 10% compounded semiannually for 9 years

16. $7500 at 8% compounded semiannually for 5 years

17. $2196.58 at 8% compounded quarterly for 4 years

18. $4915.73 at 12% compounded quarterly for 3 years

Find the present value of the sums in Exercises 19–26.

19. $4500 at 6% compounded annually for 9 years

20. $11,500 at 8% compounded annually for 12 years

21. $3800 at 8% compounded annually for 7 years

22. $10,750 at 6% compounded annually for 11 years

23. $2000 at 10% compounded semiannually for 8 years

24. $2000 at 12% compounded semiannually for 8 years

25. $8800 at 12% compounded quarterly for 5 years

26. $7500 at 8% compounded quarterly for 9 years

27. If money can be invested at 8% compounded quarterly, which is larger, $1000 now or $1210 in 5 years?

28. If money can be invested at 6% compounded annually, which is larger, $10,000 now or $15,000 in 10 years?

Find the compound amount if $20,000 *is invested at 8% compounded continuously for the following number of years. (Use a calculator with an e^x or an x^y key for Exercises* 33 *and* 34.)

29. 1   30. 5   31. 10   32. 15   33. 3   34. 7

Find the present value of $17,200 *for the following number of years, if money can be deposited at 11.4% compounded continuously. (Use a calculator with an e^x or an x^y key.)*

35. 2   36. 4   37. 7   38. 10

Find the effective rate corresponding to each of the following nominal rates.

39. 4% compounded semiannually   40. 8% compounded quarterly

41. 8% compounded semiannually

42. 10% compounded semiannually

43. 12% compounded semiannually

44. 12% compounded quarterly

Use the ideas of Example 9 in the text to answer questions 45 – 50. Find the time it would take for the general level of prices in the economy to double if the average annual inflation rate is

45. 4%;   **46.** 5%;   **47.** 6%;   **48.** 10%.

49. The consumption of electricity has increased historically at 6% per year. If it continued to increase at this rate indefinitely, find the number of years before the electric utilities would need to double the amount of generating capacity.

50. Suppose a conservation campaign coupled with higher rates caused the demand for electricity to increase at only 2% per year, as it has recently. (See Exercise 49.) Find the number of years before the utilities would need to double generating capacity.

*Under certain conditions, Swiss banks pay **negative interest**—they charge you. (You didn't think all that secrecy was free?) Suppose a bank "pays" —2.4% interest compounded annually. Use a calculator and find the compound amount for a deposit of $150,000 after*

51. 2 years;   **52.** 4 years;   **53.** 8 years;   **54.** 12 years.

10.3 Annuities

If you deposit $1500 at the end of each year for the next six years, in an account paying 6% per year, compounded annually, how much will you have altogether?

A sequence of payments such as this is called an **annuity.** If the payments are made at the end of the time period, and if the frequency of payments is the same as the frequency of compounding, the annuity is called an **ordinary annuity.** The time between payments is the **payment period,** with the time from the beginning of the first payment period to the end of the last called the **term** of the annuity. The **amount** of the annuity, the final sum on deposit, is defined as the sum of the compound amounts of all the payments, compounded to the end of the term.

Let us return to the problem of depositing $1500 at the end of each year for six years. The first payment will produce a compound amount of

$$1500(1 + .06)^5 = 1500(1.06)^5.$$

Use 5 as the exponent instead of 6 since the money is deposited at the *end* of the first year, and thus earns interest for only five years. The second payment of $1500 will produce a compound amount of $1500(1.06)^4$. As shown in Figure 1, the amount of the annuity is

$$1500(1.06)^5 + 1500(1.06)^4 + 1500(1.06)^3 + 1500(1.06)^2 + 1500(1.06)^1 + 1500.$$

(The last payment earns no interest at all.) From Table 3, this sum is

$1500(1.33823) + 1500(1.26248) + 1500(1.19102) + 1500(1.12360)$
$$+ 1500(1.06) + 1500$$

$$= 2007.35 + 1893.72 + 1786.53 + 1685.40 + 1590 + 1500$$

$$= 10,463.00,$$

or \$10,463.00.

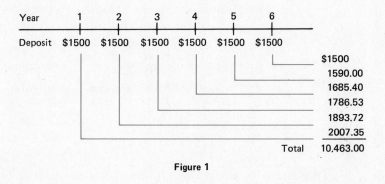

Figure 1

To generalize this result, suppose that a payment of R dollars is paid into an account at the end of each period for n periods, at a rate of interest i per period. The first payment of R dollars will produce a compound amount of $R(1 + i)^{n-1}$ dollars, the second payment produces $R(1 + i)^{n-2}$ dollars, and so on; the final payment earns no interest and contributes just R dollars to the total. If A represents the amount of the annuity, then

$$A = R(1 + i)^{n-1} + R(1 + i)^{n-2} + R(1 + i)^{n-3} + \cdots + R(1 + i) + R. \qquad (1)$$

To simplify this result, first multiply both sides by $1 + i$, to get

$$(1 + i)A = R(1 + i)^n + R(1 + i)^{n-1} + R(1 + i)^{n-2} +$$
$$\cdots + R(1 + i)^2 + R(1 + i). \qquad (2)$$

Subtract equation (1) from equation (2); a great many terms on the right disappear, leaving

$$(1 + i)A - A = R(1 + i)^n - R. \qquad (3)$$

On the left, $(1 + i)A - A = A + iA - A = iA$, while factoring on the right gives $R[(1 + i)^n - 1]$. Thus, equation (3) becomes

$$iA = R[(1 + i)^n - 1],$$

and, finally,

$$A = R\left[\frac{(1 + i)^n - 1}{i}\right].$$

The quantity in brackets is commonly written $s_{\overline{n}|i}$ (read "s-angle-n-at-i"), so that

$$A = R \cdot s_{\overline{n}|i}.$$

Values of $s_{\overline{n}|i}$ can be found by a calculator or from Table 5.

To check our result, go back to the annuity of $1500 at the end of each year for 6 years; interest was 6% per year compounded annually. Look at Table 5, finding the number 6.97532. Multiply this number and 1500:

$$\text{amount} = 1500(6.97532) = 10{,}462.98,$$

or $10,462.98. This result differs by 2¢ from the result we found above due to rounding errors.

The general formulas for the amount of an annuity are the following.

Amount of an Annuity

A = amount of annuity i = interest per period

R = amount of each payment n = number of periods

$$A = R\left[\frac{(1+i)^n - 1}{i}\right], \qquad \text{or} \qquad A = R \cdot s_{\overline{n}|i}$$

Example 1   Tom Bleser is an athlete who feels that his playing career will last 7 years. To prepare for his future, he deposits $22,000 at the end of each year for 7 years in an account paying 6% compounded annually. How much will he have on deposit after 7 years?

His payments form an ordinary annuity with $R = 22{,}000$, $n = 7$, and $i = .06$. The amount of this annuity is (by the formula above)

$$A = 22{,}000\left[\frac{(1.06)^7 - 1}{.06}\right].$$

From Table 5, the number in brackets, $s_{\overline{7}|.06}$, is 8.39384, so that

$$A = 22{,}000(8.39384) = 184{,}664.48,$$

or $184,664.48. ▮

Example 2   Suppose $1000 is deposited at the end of each six month period for 8 years in an account paying 6% compounded semiannually. Find the amount of the annuity.

Interest of $6\%/2 = 3\%$ is earned semiannually. In 8 years there are $8 \cdot 2 = 16$ semiannual periods. We need to find $s_{\overline{16}|.03}$ from Table 5. By looking in the 3% column and down 16 periods, we find $s_{\overline{16}|.03} = 20.15688$. Thus, the $1000 deposits will lead to an amount of

$$A = 1000(20.15688) = 20{,}156.88,$$

or $20,156.88. ▮

Just as we can solve the formula $A = P(1 + i)^n$ for its various variables, we can also use the formula for the amount of an annuity to find the values of variables other than A. In Example 3 below we are given A, the amount of money wanted at the end, and we need to find R, the amount of each payment.

Example 3 Betsy Martens wants to buy an expensive movie camera three years from now. She wants to deposit an equal amount at the end of each quarter for three years in order to accumulate enough money to pay for the camera. The camera costs $2400, and the bank pays 8% interest compounded quarterly. Find the amount of each of the twelve deposits she will make.

This example describes an ordinary annuity with $A = 2400$, $i = .02$ ($8\%/4 = 2\%$) and $n = 3 \cdot 4 = 12$ periods. The unknown here is the amount of each payment, R. If we use the formula for the amount of an annuity from above, then

$$2400 = R \cdot s_{\overline{12}|.02}.$$

From Table 5,

$$2400 = R \cdot (13.41209)$$

$$R = 178.94,$$

or $178.94 (dividing both sides by 13.41209). ∎

In Example 3 we had to divide 2400 by 13.41209, a difficult task if no calculator is available. Dividing by 13.41209 gives the same result as multiplying by the *reciprocal* of 13.41209. (Recall: the reciprocal of a nonzero real number x is $1/x$.) Table 7 in the back of this book gives values of $1/s_{\overline{n}|i}$. Look at Table 7 for 2%, 12 periods, and find that $1/s_{\overline{12}|.02} = .07456$. Using the numbers from Example 3,

$$R = 2400 \cdot \frac{1}{s_{\overline{12}|.02}} = 2400(.07456) = 178.94,$$

the same payment as found above.

With ordinary annuities, payments are made at the *end* of each time period. With **annuities due**, the payments are made at the *beginning* of the time period. To find the amount of an annuity due, treat each payment as if it were made at the *end* of the *preceding* period. Thus, use Table 5 to find $s_{\overline{n}|i}$ for *one additional* period; to compensate for this, subtract the amount of one payment.

Example 4 Find the amount of an annuity due if payments of $500 are made at the beginning of each quarter for 7 years, in an account paying 12% compounded quarterly.

In 7 years, there are 28 quarterly periods. Look in row 29 (28 + 1) of the table. Use the $12\%/4 = 3\%$ column of the table. You should find the number 45.21885. Multiply this number by 500, the amount of each payment.

$$500(45.21885) = 22,609.43,$$

or $22,609.43. Subtract the amount of one payment from this result.

$$\$22,609.43 - \$500 = \$22,109.43$$

The account will contain a total of $22,109.43 after 7 years. ∎

10.3 Exercises *Find each of the following values. Use Table 5 or Table 7.*

1. $s_{\overline{12}|.05}$ **2.** $s_{\overline{20}|.06}$ **3.** $s_{\overline{16}|.04}$ **4.** $s_{\overline{40}|.02}$ **5.** $s_{\overline{20}|.01}$

6. $s_{\overline{18}|.015}$ **7.** $1/s_{\overline{15}|.04}$ **8.** $1/s_{\overline{30}|.015}$ **9.** $1/s_{\overline{19}|.06}$ **10.** $1/s_{\overline{24}|.05}$

Find the value of the following ordinary annuities. Interest is compounded annually.

11. $R = 100$, $i = .06$, $n = 10$

12. $R = 1000$, $i = .06$, $n = 12$

13. $R = 10,000$, $i = .05$, $n = 19$

14. $R = 100,000$, $i = .08$, $n = 23$

15. $R = 8500$, $i = .06$, $n = 30$

16. $R = 11,200$, $i = .08$, $n = 25$

17. $R = 46,000$, $i = .06$, $n = 32$

18. $R = 29,500$, $i = .05$, $n = 15$

Find the value of each of the following ordinary annuities. Payments are made and interest is compounded as given.

19. $R = 800$, 10% interest compounded semiannually for 12 years

20. $R = 4600$, 8% interest compounded quarterly for 9 years

21. $R = 15,000$, 12% interest compounded quarterly for 6 years

22. $R = 42,000$, 6% interest compounded semiannually for 12 years

Find the amount of each of the following annuities due. Assume that interest is compounded annually.

23. $R = 600$, $i = .06$, $n = 8$

24. $R = 1400$, $i = .08$, $n = 10$

25. $R = 20,000$, $i = .08$, $n = 6$

26. $R = 4000$, $i = .06$, $n = 11$

Find the amounts of each of the following annuities due.

27. payments of $1000 made at the beginning of each year for 9 years at 8% compounded annually

28. $750 deposited at the beginning of each year for 15 years at 6% compounded annually

29. $100 deposited at the beginning of each quarter for 9 years at 8% compounded quarterly

30. $1500 deposited at the beginning of each semiannual period for 11 years at 10% compounded semiannually

Find the periodic payment that will amount to the following sums under the given conditions.

31. $A = \$10,000$, interest is 8% compounded annually, payments made at the end of each year for 12 years

32. $A = \$100,000$, interest is 10% compounded semiannually, payments made at the end of each semiannual period for 9 years

33. $A = \$50,000$, interest is 12% compounded quarterly, payments made at the end of each quarter for 8 years

34. Pat Dillon deposits $12,000 at the end of each year for 9 years in an account paying 8% interest compounded annually. Find the final amount she will have on deposit.

35. Pat's brother-in-law works in a bank which pays 6% compounded annually. If she deposits her money in this bank, instead of the one of Exercise 34, how much would she have in her account?

36. How much would Pat lose over 9 years by using her brother-in-law's bank? (See Exercises 34 and 35.)

37. Pam Parker deposits $2435 at the beginning of each year for 8 years in an account paying 6% compounded annually. She then leaves that money alone, with no further deposits, for an additional 5 years. Find the final amount on deposit after the entire 13 year period.

38. Chuck deposits $10,000 at the beginning of each year for 12 years in an account paying 6% compounded annually. He then puts the total amount on deposit in another account paying 6% compounded semiannually for another 9 years. Find the final amount on deposit after the entire 21-year period.

39. Ray Berkowitz needs $10,000 in 8 years. What amount can he deposit at the end of each quarter at 8% compounded quarterly so that he will have his $10,000?

40. Find Berkowitz's quarterly deposit (see Exercise 39) if the money is deposited at 12% compounded quarterly.

41. Barb Silverman wants to buy an $18,000 car in 6 years. How much money must she deposit at the end of each quarter in an account paying 12% compounded quarterly, so that she will have enough to pay for her car?

42. Harv's Meats knows that it must buy a new deboner machine in 4 years. The machine costs $12,000. In order to accumulate enough money to pay for the machine, Harv decides to deposit a sum of money at the end of each six months in an account paying 10% compounded semiannually. How much should each payment be?

10.4 Present Value of an Annuity; Amortization

As we saw in the previous section, if deposits of R dollars are made at the end of each period for n periods, at a rate of interest i per period, then the account will contain

$$A = R \cdot s_{\overline{n}|i} = R\left[\frac{(1 + i)^n - 1}{i}\right]$$

dollars after n periods. Let us now find the *lump sum P* that can be deposited today at a rate of interest i per period which will amount to the same A dollars in n periods.

First recall that P dollars deposited today will amount to $P(1 + i)^n$ dollars after n periods at a rate of interest i per period. We want this amount, $P(1 + i)^n$, to be the same as A, the amount of the annuity. Substituting $P(1 + i)^n$ for A in the formula above gives

$$P(1 + i)^n = R\left[\frac{(1 + i)^n - 1}{i}\right].$$

We want to solve this equation for P. To do so, multiply both sides of the equation by $(1 + i)^{-n}$.

$$P = R(1 + i)^{-n}\left[\frac{(1 + i)^n - 1}{i}\right]$$

Use the distributive property and the fact that $(1 + i)^{-n} \cdot (1 + i)^n = 1$.

$$P = R\left[\frac{(1 + i)^{-n}(1 + i)^n - (1 + i)^{-n}}{i}\right]$$

$$P = R\left[\frac{1 - (1 + i)^{-n}}{i}\right]$$

The amount P is called the **present value of the annuity.** The quantity in brackets is abbreviated as $a_{\overline{n}|i}$, so

$$a_{\overline{n}|i} = \frac{1 - (1 + i)^{-n}}{i}.$$

Values of $a_{\overline{n}|i}$ are given in Table 6.

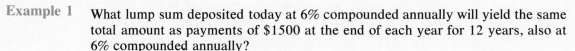

Present Value of an Annuity	
$P =$ present value	$i =$ interest rate per period
$R =$ amount of each payment	$n =$ number of periods
$P = R\left[\dfrac{1 - (1 + i)^{-n}}{i}\right]$ or $P = R \cdot a_{\overline{n}|i}$	

Example 1   What lump sum deposited today at 6% compounded annually will yield the same total amount as payments of $1500 at the end of each year for 12 years, also at 6% compounded annually?

We want to find the present value of an annuity of $1500 per year for 12 years at 6% compounded annually.

Using Table 6, we have $a_{\overline{12}|.06} = 8.38384$, so

$$A = 1500(8.38384) = 12{,}575.76,$$

or $12,575.76. A lump sum deposit of $12,575.76 today at 6% compounded annually will yield the same total after 12 years as deposits of $1500 at the end of each year for 12 years at 6% compounded annually.

Let's check this result. The compound amount in 12 years of a deposit of $12,575.76 today at 6% compounded annually can be found by the formula $A = P(1 + i)^n$. From Table 3, $12,575.76 will produce a total of

$$(12{,}575.76)(2.01220) = 25{,}304.94$$

or $25,304.94. On the other hand, from Table 5, deposits of $1500 at the end of each year for 12 years, at 6% compounded annually, give an amount of

$$1500(16.86994) = 25{,}304.91$$

or $25,304.91. (The difference of 3¢ is due to rounding errors.)

In summary, there are two ways that we can have $25,304.91 in 12 years at 6% compounded annually—a single deposit of $12,575.76 today, or payments of $1500 at the end of each year for 12 years. ◼

Example 2   Mr. Jones and Ms. Gonsalez are both graduates of the Forestvire Institute of Technology. They both agree to contribute to the endowment fund of FIT. Mr. Jones says that he will give $500 at the end of each year for 9 years. Ms. Gonsalez would rather give a lump sum today. What lump sum can she give that will be equivalent to Mr. Jones' annual gifts, if the endowment fund earns 8% compounded annually?

Here $R = 500$, $n = 9$, and $i = .08$. The necessary number from Table 6 is $a_{\overline{9}|.08} = 6.24689$. Ms. Gonsalez must therefore donate a lump sum of

$$500(6.24689) = 3123.45,$$

or $3123.45, today.   ■

We can also use the formula above if we know the lump sum and want to find the periodic payment of the annuity. The next example shows how to do this.

Example 3   A car costs $6000. After a down payment of $1000, the balance will be paid off in 36 monthly payments with interest of 12% per year, compounded monthly. Find the amount of each payment.

A single lump sum payment of $5000 today would pay off the loan. Thus, $5000 is the present value of an annuity of 36 monthly payments with interest of $12\%/12 = 1\%$ per month. We need to find R, the amount of each payment. Start with

$$P = R \cdot a_{\overline{n}|i};$$

replace P with 5000, n with 36, and i with .01. From Table 6, $a_{\overline{36}|.01} = 30.10751$, so

$$5000 = R(30.10751)$$

$$R = 166.07,$$

or $166.07. A monthly payment of $166.07 will be needed. (Check that for 48 months, the monthly payment would be $131.67.)   ■

Without a calculator, the arithmetic in Example 3 would have been easier if we knew values of $1/a_{\overline{n}|i}$. These can be found from Table 7, which gives values of $1/s_{\overline{n}|i}$. To see how, first recall that

$$a_{\overline{n}|i} = \frac{1 - (1 + i)^{-n}}{i}, \qquad \text{so that} \qquad \frac{1}{a_{\overline{n}|i}} = \frac{i}{1 - (1 + i)^{-n}}.$$

Add $-i$ to both sides of the equation on the right.

$$\frac{1}{a_{\overline{n}|i}} - i = \frac{i}{1 - (1 + i)^{-n}} - i$$

$$= \frac{i - i + i(1 + i)^{-n}}{1 - (1 + i)^{-n}}$$

$$= \frac{i(1 + i)^{-n}}{1 - (1 + i)^{-n}}$$

Multiply numerator and denominator by $(1 + i)^n$.

$$\frac{1}{a_{\overline{n}|i}} - i = \frac{i(1+i)^{-n}(1+i)^n}{[1-(1+i)^{-n}](1+i)^n}$$

$$= \frac{i}{(1+i)^n - 1}.$$

The fraction on the right is $1/s_{\overline{n}|i}$, so that

$$\frac{1}{a_{\overline{n}|i}} - i = \frac{1}{s_{\overline{n}|i}}$$

or, finally,

$$\frac{1}{a_{\overline{n}|i}} = \frac{1}{s_{\overline{n}|i}} + i.$$

Example 4 (a) Find $1/a_{\overline{12}|.03}$.

From Table 7, $1/s_{\overline{12}|.03} = .07046$. By the formula above,

$$\frac{1}{a_{\overline{12}|.03}} = \frac{1}{s_{\overline{12}|.03}} + i$$
$$= .07046 + .03$$
$$= .10046$$

(b) Find $1/a_{\overline{20}|.06}$.

$$\frac{1}{a_{\overline{20}|.06}} = \frac{1}{s_{\overline{20}|.06}} + .06 = .08718 \qquad ∎$$

Amortization   A loan is **amortized** if both the principal and interest are paid by a sequence of equal periodic payments. In Example 3 above, a loan of $5000 at 12% interest compounded monthly could be amortized by paying $166.07 per month for 36 months, or $131.67 per month for 48 months.

Example 5   A speculator agrees to pay $15,000 for a parcel of land; this amount, with interest, will be paid over 4 years, with semiannual payments, at an interest rate of 12% compounded semiannually.

(a) Find the amount of each payment.

If the speculator were to pay $15,000 immediately, there would be no need for any payments at all. Thus, $15,000 is the present value of an annuity of R dollars, $2 \cdot 4 = 8$ periods, and $i = 12\%/2 = 6\% = .06$ per period. If P is the present value of an annuity,

$$P = R \cdot a_{\overline{n}|i}.$$

In our example, $P = 15,000$, with

$$15,000 = R \cdot a_{\overline{8}|.06}$$

or

$$R = 15,000\left(\frac{1}{a_{\overline{8}|.06}}\right).$$

By the formula above, $1/a_{\overline{8}|.06} = 1/s_{\overline{8}|.06} + .06$. From Table 7, $1/s_{\overline{8}|.06} = .10104$, so that $1/a_{\overline{8}|.06} = .10104 + .06 = .16104$, and

$$R = 15,000\,(.16104) = 2415.60,$$

or \$2415.60. Each payment is \$2415.60. (With a calculator, much of this work could have been avoided.)

(b) Find the portion of the first payment that is applied to the reduction of the debt.

   Interest is 12% per year, compounded semiannually. During the first period, the entire \$15,000 is owed. Interest on this amount for 6 months (1/2 year) is found by the formula for simple interest,

$$I = Prt$$

$$I = 15,000(.12)\left(\frac{1}{2}\right)$$

$$= 900,$$

or \$900. At the end of 6 months, the speculator makes a payment of \$2415.60; since \$900 of this represents interest, a total of

$$\$2415.60 - \$900 = \$1515.60$$

is applied to the reduction of the original debt.

(c) Find the balance due after 6 months.

   The original balance due is \$15,000. After 6 months, \$1515.60 is applied to reduction of the debt. The debt owed after 6 months is thus

$$\$15,000 - \$1515.60 = \$13,484.40.$$

(d) How much interest is owed for the second six-month period?

   A total of \$13,484.40 is owed for the second 6 months. Interest on this amount is

$$I = 13,484.40(.12)\left(\frac{1}{2}\right) = 809.06,$$

or \$809.06. A payment of \$2415.60 is made at the end of this period; a total of

$$\$2415.60 - \$809.06 = \$1606.54$$

is applied to reduction of the debt.

   By continuing this process, we can get the **amortization schedule** shown below. Note that the payment is always the same, except perhaps for a small adjustment in the final one. Payment 0 represents the original amount of the loan.   ■

Amortization Schedule

Payment number	Amount of payment	Interest for period	Portion to principal	Principal at end of period
0	------	------	------	$15,000
1	$2415.60	$900	$1515.60	$13,484.40
2	$2415.60	$809.06	$1606.54	$11,877.86
3	$2415.60	$712.67	$1702.93	$10,174.93
4	$2415.60	$610.50	$1805.10	$8369.83
5	$2415.60	$502.19	$1913.41	$6456.42
6	$2415.60	$387.39	$2028.21	$4428.21
7	$2415.60	$265.69	$2149.91	$2278.30
8	$2415.00	$136.70	$2278.30	$0

Example 6   A house is bought for $74,000, with a down payment of $16,000. Interest is charged at 10.25% per year for 30 years. Find the amount of each monthly payment to amortize the loan.

Here, the present value, P, is 58,000 (or 74,000 − 16,000). Also, $i = .1025/12 = .0085416667$, and $n = 12 \cdot 30 = 360$. We must find R. From the formula for the present value of an annuity,

$$58,000 = R \cdot a_{\overline{360}|.0085416667}$$

or

$$58,000 = R\left[\frac{1 - (1 + .0085416667)^{-360}}{.0085416667}\right].$$

We must now use a financial calculator, or a calculator with an x^y key. This gives

$$58,000 = R\left[\frac{1 - .0467967507}{.0085416667}\right]$$

$$58,000 = R\left[\frac{.9532032493}{.0085416667}\right]$$

$$58,000 = R[111.5945263]$$

or

$$R = 519.74.$$

Monthly payments of $519.74 will be required to amortize the loan.   ■

Sinking Fund   A **sinking fund** is a fund set up to receive periodic payments; these periodic payments plus the interest on them are designed to produce a given total at some time in the future. As an example, a corporation might set up a sinking fund to receive money that will be needed to pay off a loan at some time in the future (any interest payments on the loan are handled outside the sinking fund).

Example 7   The Stockdales are close to retirement. They agree to sell an antique urn to the local museum for $17,000. Their tax adviser suggests that they defer receipt of this money until they retire, 5 years in the future. (At that time, they might well be in a lower tax bracket.) The museum agrees to pay them 10% per year for 5 years, and then pay them the $17,000 in a lump sum.

(a) What annual interest payment will the museum make?
Use the formula for simple interest.

$$I = 17{,}000(.10)(1) = 1700,$$

or $1700. The museum makes annual interest payments of $1700 to the Stockdales. (This amount does not play any part in the rest of the example.)

(b) Find the amount of each payment the museum must make into a sinking fund so that it will have the necessary $17,000 in 5 years. Assume that the museum can earn 8% compounded annually on its money. Also, assume that the payments are made annually.

These payments are the periodic payments into an ordinary annuity. The annuity will amount to $17,000 in 5 years at 8% compounded annually. Thus,

$$17{,}000 = R \cdot s_{\overline{5}|.08}$$

$$R = 17{,}000\left(\frac{1}{s_{\overline{5}|.08}}\right)$$

$$= 17{,}000(.17046) \qquad \text{Table 7}$$

$$R = 2897.82,$$

or $2897.82. If the museum deposits $2897.82 at the end of each year for 5 years in an account paying 8% compounded annually, it will have the total amount that it needs. This is shown in the following table; again, the last payment differs slightly from the others.

Payment number	Amount of deposit	Interest earned	Total in account
1	$2897.82	$0	$2897.82
2	$2897.82	$231.83	$6027.47
3	$2897.82	$482.20	$9407.49
4	$2897.82	$752.60	$13,057.91
5	$2897.46	$1044.63	$17,000.00

10.4 Exercises   *Find each of the following values.*

1. $a_{\overline{15}|.06}$ **2.** $a_{\overline{10}|.03}$ **3.** $a_{\overline{18}|.04}$ **4.** $a_{\overline{30}|.015}$

5. $a_{\overline{16}|.01}$ **6.** $a_{\overline{32}|.02}$ **7.** $1/a_{\overline{6}|.05}$ **8.** $1/a_{\overline{18}|.03}$

9. $1/a_{\overline{12}|.015}$ **10.** $1/a_{\overline{32}|.01}$

Find the present value of each of the following ordinary annuities.

11. Payments of $1000 are made annually for 9 years at 8% compounded annually.

12. Payments of $5000 are made annually for 11 years at 6% compounded annually.

13. Payments of $890 are made annually for 16 years at 8% compounded annually.

14. Payments of $1400 are made annually for 8 years at 8% compounded annually.

15. Payments of $10,000 are made semiannually for 15 years at 10% compounded semiannually.

16. Payments of $50,000 are made quarterly for 10 years at 8% compounded quarterly.

17. Payments of $105,000 are made quarterly for 4 years at 12% compounded quarterly.

18. Payments of $18,500 are made every six months for 18 years at 10% compounded semiannually.

For Exercises 19–22, find the lump sum deposited today that will yield the same total amount as payments of $10,000 at the end of each year for 15 years, at each of the following interest rates. Interest is compounded annually.

19. 4%   **20.** 5%   **21.** 6%   **22.** 8%

23. In his will the late Mr. Hudspeth said that each child in his family could have an annuity of $2000 at the end of each year for 9 years, or the equivalent present value. If money can be deposited at 8% compounded annually, what is the present value?

24. In the "Million Dollar Lottery," a winner is paid a million dollars at the rate of $50,000 per year for 20 years. Assume that these payments form an ordinary annuity, and that the lottery managers can invest money at 6% compounded annually. Find the lump sum that the management must put away to pay off the "million dollar" winner.

25. Lynn Meyers buys a new car costing $6000. She agrees to make payments at the end of each monthly period for 4 years. If she pays 12% interest, compounded monthly, what is the amount of each payment?

26. Find the total amount of interest Meyers will pay. (See Exercise 25.)

27. What lump sum deposited today at 5% compounded annually for 8 years will provide the same amount as $1000 deposited at the end of each year for 8 years at 6% compounded annually?

28. What lump sum deposited today at 8% compounded quarterly for 10 years will yield the same final amount as deposits of $4000 at the end of each six month period for 10 years at 6% compounded semiannually?

Find the payments necessary to amortize each of the following loans.

29. $1000, 6% compounded annually, 9 annual payments

30. $2500, 8% compounded quarterly, 6 quarterly payments

31. $41,000, 8% compounded semiannually, 10 semiannual payments

32. $90,000, 8% compounded annually, 12 annual payments

33. $140,000, 12% compounded quarterly, 15 quarterly payments

34. $7400, 10% compounded semiannually, 18 semiannual payments

35. $5500, 18% compounded monthly, 24 monthly payments

36. $45,000, 18% compounded monthly, 36 monthly payments

Find the monthly house payment necessary to amortize the following loans. You will need a financial calculator or one with an x^y key.

37. $49,560 at 11.75% for 25 years

38. $70,892 at 12.11% for 30 years

39. $53,762 at 10.45% for 30 years

40. $96,511 at 12.57% for 25 years

For Exercises 41–48, *find the amount of each payment to be made into a sinking fund so that enough will be present to pay off the indicated loans.*

41. loan $2000, money earns 6% compounded annually, 5 annual payments

42. loan $8500, money earns 8% compounded annually, 7 annual payments

43. loan $11,000, money earns 10% compounded semiannually, 12 semiannual payments

44. loan $75,000, money earns 6% compounded semiannually, 9 semiannual payments

45. loan $50,000, money earns 6% compounded quarterly, 18 quarterly payments

46. loan $25,000, money earns 12% compounded quarterly, 14 quarterly payments

47. loan $6000, money earns 18% compounded monthly, 36 monthly payments

48. loan $9000, money earns 18% compounded monthly, 24 monthly payments

49. An insurance firm pays $4000 for a new printer for its computer. It amortizes the loan for the printer in 4 annual payments at 8% compounded annually. Prepare an amortization schedule for this machine.

50. Large semitrailer trucks cost $72,000 each. Ace Trucking buys such a truck and agrees to pay for it by a loan which will be amortized with 9 semiannual payments at 16% compounded semiannually. Prepare an amortization schedule for this truck.

51. IBM charges $1048 for a correcting Selectric typewriter. A firm of tax accountants buys 8 of these machines. They make a down payment of $1200 and agree to amortize the balance with monthly payments at 18% compounded monthly for 4 years. Prepare an amortization schedule showing the first six payments.

52. When Denise Sullivan opened her law office, she bought $14,000 worth of law books and $7200 worth of office furniture. She paid $1200 down and agreed to amortize the balance with semiannual payments for 5 years, at 8% compounded semiannually. Prepare an amortization schedule for this purchase.

53. Helen Spence sells some land in Nevada. She will be paid a lump sum of $60,000 in 7 years. Until then, the buyer pays 8% interest in quarterly payments.

   (a) Find the amount of each quarterly interest payment.
   (b) The buyer sets up a sinking fund so that enough money will be present to pay off the $60,000. The buyer wants to make semiannual payments into the sinking fund; the account pays 6% compounded semiannually. Find the amount of each payment into the fund.
   (c) Prepare a table showing the amount in the sinking fund after each deposit.

54. Jeff Reschke bought a rare stamp for his collection. He agreed to pay a lump sum of $4000 after 5 years. Until then, he pays 6% interest in semiannual payments.

   (a) Find the amount of each semiannual interest payment.
   (b) Reschke sets up a sinking fund so that enough money will be present to pay off the $4000. He wants to make annual payments into the fund. The account pays 8% compounded annually. Find the amount of each payment into the fund.
   (c) Prepare a table showing the amount in the sinking fund after each deposit.

55. When Ms. Thompson died, she left $25,000 for her husband. He deposits the money at 6% compounded annually. He wants to make annual withdrawals from the account so that the money (principal and interest) is gone in exactly 8 years.

   (a) Find the amount of each withdrawal.
   (b) Find the amount of each withdrawal if the money must last 12 years.

56. The trustees of St. Albert's College have accepted a gift of $150,000. The donor has directed the trustees to deposit the money in an account paying 6% per year, compounded semiannually. The trustees may withdraw money at the end of each 6-month period; the money must last 5 years.

(a) Find the amount of each withdrawal.

(b) Find the amount of each withdrawal if the money must last 7 years.

> *Application*   *Present Value*

The Southern Pacific Railroad, with lines running from Oregon to Louisiana, is one of the country's most profitable railroads.[1] The railroad has vast landholdings (granted by the government in the last half of the nineteenth century) and is diversified into trucking, pipelines, and data transmission.

The railroad was recently faced with a decision on the fate of an old bridge which crosses the Kings River in central California. The bridge is on a minor line which carries very little traffic. Just north of the Southern Pacific bridge is another bridge, owned by the Santa Fe Railroad. It too carries little traffic. The Southern Pacific had two alternatives. It could replace its own bridge or it could negotiate with the Santa Fe for the rights to run trains over its bridge. In the second alternative a yearly fee would be paid to the Santa Fe, and new connecting tracks would be built. The situation is shown in Figure 1.

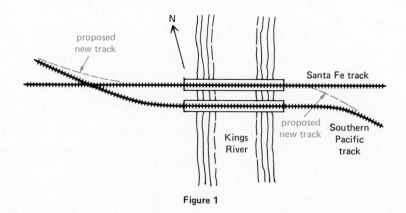

Figure 1

To find the better of these two alternatives, the railroad used the following approach.

1. Estimated expenses for each alternative were calculated.

---

[1] This section based on information supplied by The Southern Pacific Transportation Company, San Francisco.

2. Annual cash flows were calculated, in after-tax dollars. At a 48% corporate tax rate, $1 of expenses costs $.52, and $1 of revenue can bring a maximum of $.52 in profit. Cash flow for a given year is found by the following formula:

   cash flow = −.52 (operating and maintenance expenses)

   + .52 (savings and revenue) + .48 (depreciation).

3. The net present values of all cash flows for future years were calculated. The formula used was

   $$\text{net present value} = \Sigma \ (\text{cash flow in year } i)(1 + k)^{i-1}$$

   where i was a particular year in the life of the project and k the assumed annual rate of interest. (Recall: Σ indicates a sum.) The net present value was found for interest rates from 0% to 20%.

4. The interest rate that leads to a net present value of $0 was called the *rate of return* on the project.

Let us now see how these steps worked out in practice.

Alternative 1: Operate over the Santa Fe bridge

First, estimated expenses were calculated:

1980	Work done by Southern Pacific on Santa Fe track	$27,000
1980	Work by Southern Pacific on its own track	11,600
1980	Undepreciated portion of cost of current bridge	97,410
1980	Salvage value of bridge	12,690
1981	Annual maintenance of Santa Fe track	16,717
1981	Annual rental fee to Santa Fe	7,382

From these figures and others not given here, annual cash flows and net present values were calculated. The following table was then prepared.

Interest rate, %	Net present value
0	$85,731
4	67,566
8	53,332
12	42,008
16	32,875
20	25,414

Although the table does not show a net present value of $0, the interest rate that leads to that value is 44%. This is the rate of return for this alternative.

Alternative 2: Build a new bridge

Again, estimated expenses were calculated:

1980	Annual maintenance	$2,870
1980	Annual inspection	120
1980	Repair trestle	17,920

1981	Install bridge	189,943
1981	Install walks and handrails	15,060
1982	Repaint steel (every 10 years)	10,000
1982	Repair trusses (every 10 years, increases by $200)	2,000
1985	Replace ties	31,000
1992	Repair concrete (every 10 years)	400
2025	Replace ties	31,000

After cash flows and net present values were calculated, the following table was prepared.

Interest rate, %	Net present value
0	$399,577
4	96,784
7.8	0
8	−3,151
12	−43,688
16	−62,615
20	−72,126

In this alternative the net present value is $0 at 7.8%, the rate of return.

Based on this analysis, the first alternative—renting from the Santa Fe—is clearly preferable.

Exercises *Find the cash flow in each of the following years. Use the formula given above.*

1. Alternative 1, year 1981, operating expenses $6228, maintenance expenses $2976, savings $26,251, depreciation $10,778.

2. Alternative 1, year 1988, same as Exercise 1, except depreciation is only $1347.

3. Alternative 2, year 1980, maintenance $2870, operating expenses $6386, savings $26,251.

4. Alternative 2, year 1984, operating expenses $6228, maintenance expenses $2976, savings $10,618, depreciation $6736.

Key Words

interest	continuous compounding
simple interest	nominal rate
principal	stated rate
rate of interest	effective rate
time	annuity
ordinary interest	ordinary annuity
exact interest	payment period
present value	term of an annuity
simple discount	amount of an annuity
proceeds	annuity due

discounting a note   present value of an annuity
compound interest   amortization
compound amount   amortization schedule
periods   sinking fund

Chapter 10 Review Exercises

Many of these exercises will require a calculator with an x^y key.

Find the simple interest in each of the following.

1. $15,903 at 13% for 8 months

2. $4902 at 14.5% for 11 months

3. $42,368 at 12.22% for 5 months

4. $3478 at 9.4% for 88 days (assume 360 days in a year)

5. $2390 at 10.7% from May 3 to July 28 (assume 365 days in a year)

6. $69,056.12 at 11.5% from September 13 to March 25 of the following year (assume a 365 day year)

Find the present value of the following future amounts. Assume 360 days in a year; use simple interest.

7. $25,000 for 10 months, money earns 13%

8. $459.57 for 7 months, money earns 11.5%

9. $80,612 for 128 days, money earns 13.77%

Find the proceeds for the following. Assume 360 days in a year.

10. $56,882, discount rate 14%, length of loan 5 months

11. $802.34, discount rate 15.6%, length of loan 11 months

12. $12,000, discount rate 13.09%, length of loan 145 days

13. Tom Wilson owes $5800 to his mother. He has agreed to pay the money in 10 months, at an interest rate of 14%. Three months before the loan is due, the mother discounts the loan at the bank. The bank charges a 15.45% discount rate. How much money does the mother receive?

14. Larry DiCenso needs $9812 to buy new equipment for his business. The bank charges a discount of 14%. Find the amount of DiCenso's loan, if he borrows the money for 7 months.

Find the effective annual rate for the following bank discount rates. (Hint: Assume a one year loan of $1000.)

15. 14%   16. 17.5%

Find the compound amounts for the following.

17. $1000 at 8% compounded annually for 9 years

18. $2800 at 6% compounded annually for 10 years

19. $19,456.11 at 10% compounded semiannually for 7 years

20. $312.45 at 12% compounded semiannually for 16 years

21. $1900 at 8% compounded quarterly for 9 years

22. $57,809.34 at 12% compounded quarterly for 5 years

23. $2500 at 12% compounded monthly for 3 years

24. $11,702.55 at 18% compounded monthly for 4 years

Find the amount of interest earned by the following deposits.

25. $3954 at 8% compounded annually for 12 years

26. $12,699.36 at 10% compounded semiannually for 7 years

27. $7801.72 at 12% compounded quarterly for 5 years

28. $48,121.91 at 18% compounded monthly for 2 years

Find the compound amount if $5890 is invested at 10% compounded continuously for the following number of years.

29. 2   30. 5   31. 7   32. 10

Find the present value of the following amounts.

33. $5000 in 9 years if money can be invested at 8% compounded annually

34. $12,250 in 5 years if money can be invested at 12% compounded semiannually

Find the present value of $50,000 for the following number of years if the money can be deposited at 10.9% compounded continuously.

35. 4 years   36. 7 years

37. In four years, Mr. Heeren must pay a pledge of $5000 to his church's building fund. What lump sum can he invest today, at 8% compounded semiannually, so that he will have enough to pay his pledge?

38. Joann Hudspeth must make an alimony payment of $1500 in 15 months. What lump sum can she invest today, at 12% compounded monthly, so that she will have enough to pay the payment?

Find the value of each of the following annuities.

39. $500 is deposited at the end of each six month period for 8 years; money earns 6% compounded semiannually

40. $1288 is deposited at the end of each year for 14 years; money earns 8% compounded annually

41. $4000 is deposited at the end of each quarter for 7 years; money earns 12% compounded quarterly

42. $233 is deposited at the end of each month for 4 years; money earns 12% compounded monthly

43. $672 is deposited at the beginning of each quarter for 7 years; money earns 8% compounded quarterly

44. $11,900 is deposited at the beginning of each month for 13 months; money earns 12% compounded monthly

45. Georgette Dahl deposits $491 at the end of each quarter for 9 years. If the account pays 13% compounded quarterly, find the final amount in the account.

46. J. Euclid deposits $1526.38 at the end of each six-month period in an account paying 14.8% compounded semiannually. How much will be in the account after five years?

Find the present value of each of the following ordinary annuities.

47. Payments of $850 are made annually for 4 years at 8% compounded annually.

48. Payments of $1500 are made quarterly for 7 years, at 6% compounded quarterly.

49. Payments of $4210 are made semiannually for 8 years, at 11% compounded semi-annually.

50. Payments of $877.34 are made monthly for 17 months, at 14% compounded monthly.

51. Vicki Manchester borrows $20,000 from the bank. The money is to help her expand her business. She agrees to repay the money in equal payments at the end of each year for 9 years. Interest is at 15.9% compounded annually. Find the amount of each payment.

52. Ken Murrill wants to expand his pharmacy. To do this, he takes out a loan of $49,275 from the bank, and agrees to repay it at 14.2% compounded monthly, over 48 months. Find the amount of each payment necessary to amortize this loan.

Find the amount of the payment necessary to amortize the following loans.

53. $80,000 loan, 8% compounded annually, 9 annual payments

54. $3200 loan, 12% compounded quarterly, 10 quarterly payments

55. $32,000 loan, 13.4% compounded quarterly, 17 quarterly payments

56. $51,607 loan, 15.6% compounded monthly, 32 monthly payments

Find each of the following monthly house payments.

57. $56,890 at 10.74% for 25 years

58. $77,110 at 11.45% for 30 years

Find the amount of each payment to be made into a sinking fund so that enough money will be available to pay off the indicated loan.

59. $6500 loan, money earns 8% compounded annually, 6 annual payments

60. $57,000 loan, money earns 10% compounded semiannually, 17 semiannual payments

61. $233,188 loan, money earns 13.7% compounded quarterly, 31 quarterly payments

62. $1,056,788 loan, money earns 14.12% compounded monthly, 55 monthly payments

Prepare amortization schedules for the following loans.

63. $5000 at 10% compounded semiannually, for 3 years

64. $12,500 at 12% compounded quarterly, for 2 years

Appendix   Exponential and Logarithmic Functions

In this appendix we shall briefly study exponential and logarithmic functions. These functions, as we shall see, are important in many areas of mathematics and its applications. In algebra, the symbol a^m is given meaning for rational values of the exponent m. Other courses extend this to give meaning to a^m for *any* real number value of m. Based on this, we can define an exponential function.

$$f(x) = a^x, \qquad \text{where } a > 0 \text{ and } a \neq 1,$$

is called the **exponential function with base a.**

Example 1   Graph the exponential function $y = 2^x$ ($f(x) = 2^x$).

Make a table of values of x and y.

x	-3	-2	-1	0	1	2	3	4
2^x	$\frac{1}{8}$	$\frac{1}{4}$	$\frac{1}{2}$	1	2	4	8	16

If we plot these points and draw a smooth curve through them, we get the graph shown in Figure 1. This graph is typical of the graphs of exponential functions of the form $y = a^x$, where $a > 1$. The larger the value of a, the faster the graph rises.   ∎

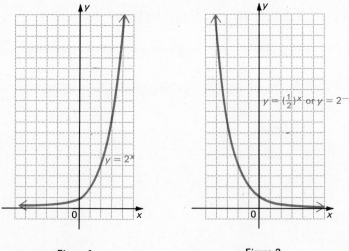

Figure 1   Figure 2

Example 2   Graph $y = (1/2)^x$.

Again, find some representative points of the graph.

x	-3	-2	-1	0	1	2	3	4
$\left(\frac{1}{2}\right)^x$	8	4	2	1	$\frac{1}{2}$	$\frac{1}{4}$	$\frac{1}{8}$	$\frac{1}{16}$

The graph, shown in Figure 2, is similar to that of $y = 2^x$ except that it goes down while the graph of $y = 2^x$ goes up. This graph is typical of the graph of any function of the form $y = a^x$, where $0 < a < 1$. ▮

Possibly the single most important exponential function is the function $y = e^x$. e is an irrational number that occurs often in practical applications. To see how the number e comes up in a practical situation, let us begin with a slight rewording of the formula for compound interest from Chapter 10: if P dollars is deposited in an account paying a rate of interest i compounded m times a year, the final amount on deposit after n years is

$$P\left(1 + \frac{i}{m}\right)^{mn}.$$

Suppose that a lucky investment produces annual interest of 100%, so that $i = 1.00$, or $i = 1$. Suppose also that you can deposit only \$1 at this rate, and for only 1 year. Then $P = 1$ and $n = 1$. Substituting into the formula for compound interest gives

$$P\left(1 + \frac{i}{m}\right)^{mn} = 1\left(1 + \frac{1}{m}\right)^{m(1)} = \left(1 + \frac{1}{m}\right)^{m}.$$

As interest is compounded more and more often, the value of this expression will increase. For annual compounding, $m = 1$, and we have

$$\left(1 + \frac{1}{m}\right)^{m} = \left(1 + \frac{1}{1}\right)^{1} = 2^1 = 2,$$

so that your \$1 will become \$2 in one year.

With a calculator or a computer, we get the results, rounded to five decimal places, shown in the following table.

m	1	2	5	10	25	50	100	500	1000	10,000	1,000,000
$\left(1 + \frac{1}{m}\right)^{m}$	2	2.25	2.48832	2.59374	2.66584	2.69159	2.70481	2.71557	2.71692	2.71815	2.71828

It appears from this table, that as m increases, getting larger and larger, the expression $(1 + 1/m)^m$ gets closer and closer to some fixed number. This fixed number is called e. By continuing the process used in the table, we can find that to nine decimal places,

$$e = 2.718281828.$$

Table 8 in this book gives various powers of e. Also, some calculators have e keys.

In Figure 3, the functions $y = 2^x$, $y = e^x$, and $y = 3^x$ are graphed for comparison. The graph of $y = e^x$ can be found using a calculator or values from Table 8 in this book.

It can be shown that in many situations involving growth or decay of a population, the amount or number present at time t can be closely approximated by a function of the form

$$y = y_0 e^{kt},$$

where y_o is the amount or number present at time $t = 0$ and k is a constant. (In other words, once the numbers y_o and k have been determined, population is a function of time.)

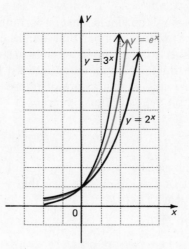

Figure 3

Example 3   Suppose the population of a midwestern city is

$$P(t) = 10,000e^{.04t},$$

where t represents time measured in years. The population at time $t = 0$ is

$$P(0) = 10,000e^{.04(0)}$$
$$= 10,000e^0$$
$$= 10,000(1)$$
$$= 10,000.$$

Thus, the population of the city is 10,000 at time $t = 0$, written $P_0 = 10,000$. The population of the city at year $t = 5$ is

$$P(5) = 10,000e^{.04(5)}$$
$$= 10,000e^{.2}$$

The number $e^{.2}$ can be found in Table 8 or by using a suitable calculator. By either of these methods, $e^{.2} \approx 1.22140$, so that

$$P(5) \approx 10,000(1.22140) = 12,214.$$

Thus, in 5 years the population of the city will be about 12,200 people.

Each exponential function a^x can be used to find a logarithmic function, written $f(x) = \log_a x$, or $y = \log_a x$. This is done using the following definition.

$$y = \log_a x \text{ if and only if } x = a^y.$$

The abbreviation log is used for logarithm. Read $\log_a x$ as "the logarithm of x to the base a." The function $y = \log_a x$ is the **logarithmic function with base a.** Figure 4 shows the graph of $y = \log_2 x$ and, for comparison, the graph of the exponential function $y = 2^x$. These two graphs are mirror images with respect to the 45° line $y = x$.

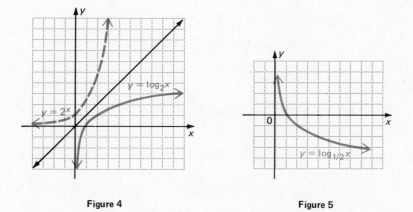

Figure 4   Figure 5

Example 4   The chart below shows several pairs of equivalent statements. The same statement is written in both exponential and logarithmic form.

Exponential form	Logarithmic form
$3^2 = 9$	$\log_3 9 = 2$
$\left(\dfrac{1}{5}\right)^{-2} = 25$	$\log_{1/5} 25 = -2$
$10^5 = 100,000$	$\log_{10} 100,000 = 5$
$4^{-3} = 1/64$	$\log_4 1/64 = -3$

Example 5   Graph $y = \log_{1/2} x$.

If we write $y = \log_{1/2} x$ in exponential form as $x = (1/2)^y$, we can find some ordered pairs which satisfy the equation. Here it is easier to choose values for y and find the corresponding values of x. Doing this, we get the following pairs.

x	$\dfrac{1}{4}$	$\dfrac{1}{2}$	1	2	4	8
y	2	1	0	-1	-2	-3

Plotting these points and connecting them with a smooth curve gives the graph of Figure 5. This graph is typical of logarithmic functions with base $0 < a < 1$. The graph of $y = \log_2 x$ in Figure 4 is typical of graphs of logarithmic functions with base $a > 1$.

Logarithms were originally important as an aid for numerical calculations, but the availability of inexpensive calculators has diminished the need for this application of logarithms. Yet the principles behind the use of logarithms for calculations are important and are based on the properties discussed in the next theorem.

Theorem 1 If x and y are any positive real numbers, r is any real number, and a is any positive real number, $a \neq 1$, then

(a) $\log_a xy = \log_a x + \log_a y$ (b) $\log_a \dfrac{x}{y} = \log_a x - \log_a y$

(c) $\log_a x^r = r \cdot \log_a x$ (d) $\log_a \sqrt[r]{x} = \dfrac{1}{r} \cdot \log_a x$

Example 6 Using the results of Theorem 1,

(a) $\log_6 7 \cdot 9 = \log_6 7 + \log_6 9$

(b) $\log_9 \dfrac{15}{7} = \log_9 15 - \log_9 7$

(c) $\log_5 \sqrt{8} = \log_5 8^{1/2} = \dfrac{1}{2} \log_5 8$

(d) $\log_a \dfrac{mnq}{p^2} = \log_a m + \log_a n + \log_a q - 2 \log_a p$

(e) $\log_a \sqrt[3]{m^2} = \dfrac{2}{3} \log_a m.$ ■

Example 7 Write $2 \log_a m - 3 \log_a n$ as a single logarithm.
By Theorem 1,

$$2 \log_a m - 3 \log_a n = \log_a m^2 - \log_a n^3$$
$$= \log_a \dfrac{m^2}{n^3}. \quad ■$$

Example 8 Assume $\log_{10} 2 = .3010$. Find the base 10 logarithms of 4 and 5.
Using the results of Theorem 1,

$$\log_{10} 4 = \log_{10} 2^2 = 2 \log_{10} 2 = 2(.3010) = .6020$$
$$\log_{10} 5 = \log_{10} \dfrac{10}{2} = \log_{10} 10 - \log_{10} 2 = 1 - .3010 = .6990. \quad ■$$

Historically, one of the main applications of logarithms has been as an aid to numerical calculation. Since our number system is base 10, logarithms to base 10 are most convenient for numerical calculations. Such logarithms are called **common logarithms.** These logarithms, however, have few applications other than numerical calculation. Most other practical applications of logarithms use the number $e \approx 2.7182818$ as base. Logarithms to base e are called **natural logarithms,** written **ln x.**

A table of natural logarithms is given in Table 9. From this table we find, for example, that

$$\ln 55 = 4.0073$$
$$\ln 1.9 = 0.6419$$
$$\ln 0.4 = -.9163.$$

Example 9   Use Table 9 to find the following logarithms.

(a) ln 83

Table 9 does not give ln 83. However, we can use the properties of logarithms in Theorem 1 to write

$$\ln 83 = \ln(8.3 \times 10)$$
$$= \ln 8.3 + \ln 10$$
$$\approx 2.1163 + 2.3026$$
$$= 4.4189.$$

(b) ln 36

Since 36 is not listed in Table 9, use the properties of logarithms.

$$\ln 36 = \ln 6^2$$
$$= 2 \cdot \ln 6$$
$$\approx 2(1.7918)$$
$$= 3.5836. \quad \blacksquare$$

Example 10   The number of years, $N(r)$, since two independently evolving languages split off from a common ancestral language is approximated by

$$N(r) = -5000 \ln r,$$

where r is the portion of words from the ancestral language common to both languages now. Find N if $r = .7$.

We want

$$N(.7) = -5000 \ln .7$$

From Table 9, or a calculator, $\ln .7 = -.3567$, with

$$N(.7) = -5000(-.3567)$$
$$= 1783.$$

Approximately 1800 years have elapsed since the two languages separated.   ∎

Appendix Exercises

1. Graph each of the following functions. Compare the graphs to that of $y = 2^x$.
   (a) $y = 2^x + 1$ (b) $y = 2^x - 4$
   (c) $y = 2^{x+1}$ (d) $y = 2^{x-4}$

2. Graph each of the following. Compare the graphs to that of $y = 3^{-x}$.
   (a) $y = 3^{-x} - 2$ (b) $y = 3^{-x} + 4$
   (c) $y = 3^{-x-2}$ (d) $y = 3^{-x+4}$

Graph each of the following functions.

3. $y = 10^{-x}$ 4. $y = 10^x$ 5. $y = e^{-x}$

6. $y = e^{2x}$ 7. $y = 2^{|x|}$ 8. $y = 2^{-|x|}$

Use a calculator to help graph each of the following.

9. $y = \dfrac{e^x + e^{-x}}{2}$ **10.** $y = \dfrac{e^x - e^{-x}}{2}$

11. $y = x \cdot e^x$ **12.** $y = (1 - x)e^x$

13. Suppose the population of a city is

$$P(t) = 1,000,000e^{.02t},$$

where t represents time measured in years. Find each of the following values.

(a) P_0 (b) $P(2)$ (c) $P(4)$ (d) $P(10)$ (e) Graph $P(t)$.

14. Suppose the quantity in grams of a radioactive substance present at time t is

$$Q(t) = 500e^{-.05t},$$

where t is measured in days. Find the quantity present at each of the following times.

(a) $t = 0$.   (b) $t = 4$ (c) $t = 8$ (d) $t = 20$ (e) Graph $Q(t)$.

15. *Escherichia coli* is a strain of bacteria that occurs naturally in many different situations. Under certain conditions, the number of these bacteria present in a colony is given by

$$E(t) = E_0 \cdot 2^{t/30},$$

where $E(t)$ is the number of bacteria present t minutes after the beginning of an experiment, and E_0 is the number present when $t = 0$. Let $E_0 = 2,400,000$, and find the number of bacteria at the following times.

(a) $t = 5$ (b) $t = 10$ (c) $t = 60$ (d) $t = 120$

16. The higher a student's grade-point average, the fewer applications that the student need send to medical schools (other things being equal). Using information given in a guidebook for prospective medical students, we constructed the function $y = 540e^{-1.3x}$ for the number of applications a student should send out. Here y is the number of applications for a student whose grade point average is x. The domain of x is $2.0 \le x \le 4.0$. Find the number of applications that should be sent out by students having a grade point average of

(a) 2.0   (b) 3.0   (c) 3.5   (d) 3.9

Change each of the following to logarithmic form.

17. $3^4 = 81$ **18.** $2^5 = 32$ **19.** $(1/2)^{-4} = 16$

20. $(2/3)^{-3} = 27/8$ **21.** $10^{-4} = .0001$ **22.** $(1/100)^{-2} = 10,000$

Change each of the following to exponential form.

23. $\log_6 36 = 2$ **24.** $\log_5 5 = 1$ **25.** $\log_{\sqrt{3}} 81 = 8$

26. $\log_4 (1/64) = -3$ **27.** $\log_m k = n$ **28.** $\log_2 r = y$

Write each of the following as a sum, difference, or product of logarithms. Simplify the result if possible.

29. $\log_3 (2/5)$ **30.** $\log_4 (6/7)$ **31.** $\log_2 \dfrac{6x}{y}$ **32.** $\log_3 \dfrac{4p}{q}$

33. $\log_5 \dfrac{5\sqrt{7}}{3}$ **34.** $\log_2 \dfrac{2\sqrt{3}}{5}$ **35.** $\log_4 (2x + 7y)$ **36.** $\log_6 (7m - 3q)$

Write each of the following expressions as a single logarithm.

37. $\log_a x + \log_a y - \log_a m$

38. $(\log_b k - \log_b m) - \log_b a$

39. $2 \log_m a - 3 \log_m b^2$

40. $\frac{1}{2} \log_y p^3 q^4 - \frac{2}{3} \log_y p^4 q^3$

41. $-\frac{3}{4} \log_x a^6 b^8 + \frac{2}{3} \log_x a^9 b^3$

42. $\log_a (pq^2) + 2 \log_a (p/q)$

Find each of the following natural logarithms. Use Table 9 and the properties of logarithms as necessary.

43. ln 20 **44.** ln 35 **45.** ln 800

46. ln 920 **47.** ln 1440 **48.** ln 490

Suppose ln .01 = −4.6052 *and* ln .001 = −6.9078. *Use Table 9 and the properties of logarithms to find each of the following.*

49. ln .08 **50.** ln. 04 **51.** ln .20

52. ln .30 **53.** ln .007 **54.** ln .009

55. The population of an animal species that is introduced in a certain area may grow rapidly at first but then grows more slowly as time goes on. A logarithmic function can provide an excellent description of such growth. Suppose that the population of foxes, $F(t)$, in an area t months after the species is introduced into the area is

$$F(t) = 500 \ln (2t + 3).$$

Find the population of foxes at the following times:

(a) when they are first released into the area (that is, when $t = 0$); (b) after 3 months; (c) after 7 months; (d) after 15 months. (e) Graph F.

56. Turnage† has shown that the pull, P, of a tracked vehicle on dry sand under certain conditions is approximated by

$$P = W\left[.2 + .16 \ln \frac{G(bl)^{3/2}}{W} \right]$$

where G is an index of sand strength, W is the load on the vehicle, b is the width of the track, and l is the length of the track. Find P if $W = 10$, $G = 5$, $b = 30.5$ cm, and $l = 61.0$ cm.

57. Carbon 14 is a radioactive isotope of carbon which has a half-life of about 5600 years. The atmosphere contains much carbon, mostly in the form of carbon dioxide, with small traces of carbon 14. Most of this atmospheric carbon is in the form of the non-radioactive isotope carbon 12. The ratio of carbon 14 to carbon 12 is virtually constant in the atmosphere. However, as a plant absorbs carbon dioxide from the air in the process of photosynthesis, the carbon 12 stays in the plant while the carbon 14 decays by conversion to nitrogen. Thus, the ratio of carbon 14 to carbon 12 is smaller in the plant than it is in the atmosphere. Even when the plant is eaten by an animal,

---

†Gerald W. Turnage, *Prediction of Track Pull Performance in a Desert Sand,* unpublished MS thesis, The Florida State University, 1971.

this ratio will continue to decrease. Based on these facts, a method of dating objects called *carbon 14 dating* has been developed. It is explained in the following problems.

(a) Suppose an Egyptian mummy is discovered in which the ratio of carbon 14 to carbon 12 is only about half the ratio found in the atmosphere. About how long ago did the Egyptian die?

(b) Let R be the (nearly constant) ratio of carbon 14 to carbon 12 found in the atmosphere, and let r be the ratio found in a fossil. It can be shown that the relationship between R and r is given by

$$\frac{R}{r} = e^{(t \ln 2)/5600},$$

where t is the age of the fossil in years. Verify the formula for $t = 0$.

(c) Verify the formula in part (b) for $t = 5600$ and then for $t = 11,200$.

(d) Suppose a specimen is found in which $r = (2/3)R$. Estimate the age of specimen. (Hint: take natural logarithms of both sides.)

58. If an object is fired vertically upward and is subject only to the force of gravity, g, and to air resistance, then the maximum height, H, attained by the object is

$$H = \frac{1}{K}\left(V_0 - \frac{g}{K} \ln \frac{g + V_0 K}{g}\right),$$

where V_0 is the initial velocity of the object and K is a constant. Find H if $K = 2.5$, $V_0 = 1000$ feet per second, and $g = 32$ feet per second per second.

Table 1 Combinations

n	$\binom{n}{0}$	$\binom{n}{1}$	$\binom{n}{2}$	$\binom{n}{3}$	$\binom{n}{4}$	$\binom{n}{5}$	$\binom{n}{6}$	$\binom{n}{7}$	$\binom{n}{8}$	$\binom{n}{9}$	$\binom{n}{10}$
0	1										
1	1	1									
2	1	2	1								
3	1	3	3	1							
4	1	4	6	4	1						
5	1	5	10	10	5	1					
6	1	6	15	20	15	6	1				
7	1	7	21	35	35	21	7	1			
8	1	8	28	56	70	56	28	8	1		
9	1	9	36	84	126	126	84	36	9	1	
10	1	10	45	120	210	252	210	120	45	10	1
11	1	11	55	165	330	462	462	330	165	55	11
12	1	12	66	220	495	792	924	792	495	220	66
13	1	13	78	286	715	1287	1716	1716	1287	715	286
14	1	14	91	364	1001	2002	3003	3432	3003	2002	1001
15	1	15	105	455	1365	3003	5005	6435	6435	5005	3003
16	1	16	120	560	1820	4368	8008	11440	12870	11440	8008
17	1	17	136	680	2380	6188	12376	19448	24310	24310	19448
18	1	18	153	816	3060	8658	18564	31824	43758	48620	43758
19	1	19	171	969	3876	11628	27132	50388	75582	92378	92378
20	1	20	190	1140	4845	15504	38760	77520	125970	167960	184756

For $r > 10$, it may be necessary to use the identity

$$\binom{n}{r} = \binom{n}{n-r}$$

$$\binom{N}{X} \cdot p^{X} \cdot (1-p)^{N-X}$$

412

Table 2 Cumulative distribution–Standard normal

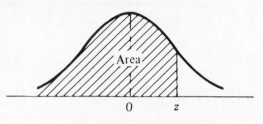

$$F_Z\ (z) = P\ [Z \leq z]$$

z	0.00	0.01	0.02	0.03	0.04	0.05	0.06	0.07	0.08	0.09
−3.4	0.0003	0.0003	0.0003	0.0003	0.0003	0.0003	0.0003	0.0003	0.0003	0.0002
−3.3	0.0005	0.0005	0.0005	0.0004	0.0004	0.0004	0.0004	0.0004	0.0004	0.0003
−3.2	0.0007	0.0007	0.0006	0.0006	0.0006	0.0006	0.0006	0.0005	0.0005	0.0005
−3.1	0.0010	0.0009	0.0009	0.0009	0.0008	0.0008	0.0008	0.0008	0.0007	0.0007
−3.0	0.0013	0.0013	0.0013	0.0012	0.0012	0.0011	0.0011	0.0011	0.0010	0.0010
−2.9	0.0019	0.0018	0.0017	0.0017	0.0016	0.0016	0.0015	0.0015	0.0014	0.0014
−2.8	0.0026	0.0025	0.0024	0.0023	0.0023	0.0022	0.0021	0.0021	0.0020	0.0019
−2.7	0.0035	0.0034	0.0033	0.0032	0.0031	0.0030	0.0029	0.0028	0.0027	0.0026
−2.6	0.0047	0.0045	0.0044	0.0043	0.0041	0.0040	0.0039	0.0038	0.0037	0.0036
−2.5	0.0062	0.0060	0.0059	0.0057	0.0055	0.0054	0.0052	0.0051	0.0049	0.0048
−2.4	0.0082	0.0080	0.0078	0.0075	0.0073	0.0071	0.0069	0.0068	0.0066	0.0064
−2.3	0.0107	0.0104	0.0102	0.0099	0.0096	0.0094	0.0091	0.0089	0.0087	0.0084
−2.2	0.0139	0.0136	0.0132	0.0129	0.0125	0.0122	0.0119	0.0116	0.0113	0.0110
−2.1	0.0179	0.0174	0.0170	0.0166	0.0162	0.0158	0.0154	0.0150	0.0146	0.0143
−2.0	0.0228	0.0222	0.0217	0.0212	0.0207	0.0202	0.0197	0.0192	0.0188	0.0183
−1.9	0.0287	0.0281	0.0274	0.0268	0.0262	0.0256	0.0250	0.0244	0.0239	0.0233
−1.8	0.0359	0.0352	0.0344	0.0336	0.0329	0.0322	0.0314	0.0307	0.0301	0.0294
−1.7	0.0446	0.0436	0.0427	0.0418	0.0409	0.0401	0.0392	0.0384	0.0375	0.0367
−1.6	0.0548	0.0537	0.0526	0.0516	0.0505	0.0495	0.0485	0.0475	0.0465	0.0455
−1.5	0.0668	0.0655	0.0643	0.0630	0.0618	0.0606	0.0594	0.0582	0.0571	0.0559
−1.4	0.0808	0.0793	0.0778	0.0764	0.0749	0.0735	0.0722	0.0708	0.0694	0.0681
−1.3	0.0968	0.0951	0.0934	0.0918	0.0901	0.0885	0.0869	0.0853	0.0838	0.0823
−1.2	0.1151	0.1131	0.1112	0.1093	0.1075	0.1056	0.1038	0.1020	0.1003	0.0985
−1.1	0.1357	0.1335	0.1314	0.1292	0.1271	0.1251	0.1230	0.1210	0.1190	0.1170
−1.0	0.1587	0.1562	0.1539	0.1515	0.1492	0.1469	0.1446	0.1423	0.1401	0.1379
−0.9	0.1841	0.1814	0.1788	0.1762	0.1736	0.1711	0.1685	0.1660	0.1635	0.1611
−0.8	0.2119	0.2090	0.2061	0.2033	0.2005	0.1977	0.1949	0.1922	0.1894	0.1867
−0.7	0.2420	0.2389	0.2358	0.2327	0.2296	0.2266	0.2236	0.2206	0.2177	0.2148
−0.6	0.2743	0.2709	0.2676	0.2643	0.2611	0.2578	0.2546	0.2514	0.2483	0.2451
−0.5	0.3085	0.3050	0.3015	0.2981	0.2946	0.2912	0.2877	0.2843	0.2810	0.2776

Reprinted with permission of Macmillan Publishing Co., Inc. from *Probability and Statistics for Engineers and Scientists*, 2/E by Ronald E. Walpole and Raymond H. Myers. Copyright © 1972, 1978 by Macmillan Publishing Co., Inc.

Table 2   Cumulative distribution–Standard normal (*continued*)

$$F_Z\ (z) = P\ [Z \leq z]$$

z	0.00	0.01	0.02	0.03	0.04	0.05	0.06	0.07	0.08	0.09
−0.4	0.3446	0.3409	0.3372	0.3336	0.3300	0.3264	0.3228	0.3192	0.3156	0.3121
−0.3	0.3821	0.3783	0.3745	0.3707	0.3669	0.3632	0.3594	0.3557	0.3520	0.3483
−0.2	0.4207	0.4168	0.4129	0.4090	0.4052	0.4013	0.3974	0.3936	0.3897	0.3859
−0.1	0.4602	0.4562	0.4522	0.4483	0.4443	0.4404	0.4364	0.4325	0.4286	0.4247
−0.0	0.5000	0.4960	0.4920	0.4880	0.4840	0.4801	0.4761	0.4721	0.4681	0.4641
0.0	0.5000	0.5040	0.5080	0.5120	0.5160	0.5199	0.5239	0.5279	0.5319	0.5359
0.1	0.5398	0.5438	0.5478	0.5517	0.5557	0.5596	0.5636	0.5675	0.5714	0.5753
0.2	0.5793	0.5832	0.5871	0.5910	0.5948	0.5987	0.6026	0.6064	0.6103	0.6141
0.3	0.6179	0.6217	0.6255	0.6293	0.6331	0.6368	0.6406	0.6443	0.6480	0.6517
0.4	0.6554	0.6591	0.6628	0.6664	0.6700	0.6736	0.6772	0.6808	0.6844	0.6879
0.5	0.6915	0.6950	0.6985	0.7019	0.7054	0.7088	0.7123	0.7157	0.7190	0.7224
0.6	0.7257	0.7291	0.7324	0.7357	0.7389	0.7422	0.7454	0.7486	0.7517	0.7549
0.7	0.7580	0.7611	0.7642	0.7673	0.7704	0.7734	0.7764	0.7794	0.7823	0.7852
0.8	0.7881	0.7910	0.7939	0.7967	0.7995	0.8023	0.8051	0.8078	0.8106	0.8133
0.9	0.8159	0.8186	0.8212	0.8238	0.8264	0.8289	0.8315	0.8340	0.8365	0.8389
1.0	0.8413	0.8438	0.8461	0.8485	0.8508	0.8531	0.8554	0.8577	0.8599	0.8621
1.1	0.8643	0.8665	0.8686	0.8708	0.8729	0.8749	0.8770	0.8790	0.8810	0.8830
1.2	0.8849	0.8869	0.8888	0.8907	0.8925	0.8944	0.8962	0.8980	0.8997	0.9015
1.3	0.9032	0.9049	0.9066	0.9082	0.9099	0.9115	0.9131	0.9147	0.9162	0.9177
1.4	0.9192	0.9207	0.9222	0.9236	0.9251	0.9265	0.9278	0.9292	0.9306	0.9319
1.5	0.9332	0.9345	0.9357	0.9370	0.9382	0.9394	0.9406	0.9418	0.9429	0.9441
1.6	0.9452	0.9463	0.9474	0.9484	0.9495	0.9505	0.9515	0.9525	0.9535	0.9545
1.7	0.9554	0.9564	0.9573	0.9582	0.9591	0.9599	0.9608	0.9616	0.9625	0.9633
1.8	0.9641	0.9649	0.9656	0.9664	0.9671	0.9678	0.9686	0.9693	0.9699	0.9706
1.9	0.9713	0.9719	0.9726	0.9732	0.9738	0.9744	0.9750	0.9756	0.9761	0.9767
2.0	0.9772	0.9778	0.9783	0.9788	0.9793	0.9798	0.9803	0.9808	0.9812	0.9817
2.1	0.9821	0.9826	0.9830	0.9834	0.9838	0.9842	0.9846	0.9850	0.9854	0.9857
2.2	0.9861	0.9864	0.9868	0.9871	0.9875	0.9878	0.9881	0.9884	0.9887	0.9890
2.3	0.9893	0.9896	0.9898	0.9901	0.9904	0.9906	0.9909	0.9911	0.9913	0.9916
2.4	0.9918	0.9920	0.9922	0.9925	0.9927	0.9929	0.9931	0.9932	0.9934	0.9936
2.5	0.9938	0.9940	0.9941	0.9943	0.9945	0.9946	0.9948	0.9949	0.9951	0.9952
2.6	0.9953	0.9955	0.9956	0.9957	0.9959	0.9960	0.9961	0.9962	0.9963	0.9964
2.7	0.9965	0.9966	0.9967	0.9968	0.9969	0.9970	0.9971	0.9972	0.9973	0.9974
2.8	0.9974	0.9975	0.9976	0.9977	0.9977	0.9978	0.9979	0.9979	0.9980	0.9981
2.9	0.9981	0.9982	0.9982	0.9983	0.9984	0.9984	0.9985	0.9985	0.9986	0.9986
3.0	0.9987	0.9987	0.9987	0.9988	0.9988	0.9989	0.9989	0.9989	0.9990	0.9990
3.1	0.9990	0.9991	0.9991	0.9991	0.9992	0.9992	0.9992	0.9992	0.9993	0.9993
3.2	0.9993	0.9993	0.9994	0.9994	0.9994	0.9994	0.9994	0.9995	0.9995	0.9995
3.3	0.9995	0.9995	0.9995	0.9996	0.9996	0.9996	0.9996	0.9996	0.9996	0.9997
3.4	0.9997	0.9997	0.9997	0.9997	0.9997	0.9997	0.9997	0.9997	0.9997	0.9998

Table 3 Compound Interest

$$(1 + i)^n$$

n \ i	1%	$1\frac{1}{2}\%$	2%	3%	4%	5%	6%	8%
1	1.01000	1.01500	1.02000	1.03000	1.04000	1.05000	1.06000	1.08000
2	1.02010	1.03023	1.04040	1.06090	1.08160	1.10250	1.12360	1.16640
3	1.03030	1.04568	1.06121	1.09273	1.12486	1.15763	1.19102	1.25971
4	1.04060	1.06136	1.08243	1.12551	1.16986	1.21551	1.26248	1.36049
5	1.05101	1.07728	1.10408	1.15927	1.21665	1.27628	1.33823	1.46933
6	1.06152	1.09344	1.12616	1.19405	1.26532	1.34010	1.41852	1.58687
7	1.07214	1.10984	1.14869	1.22987	1.31593	1.40710	1.50363	1.71382
8	1.08286	1.12649	1.17166	1.26677	1.36857	1.47746	1.59385	1.85093
9	1.09369	1.14339	1.19509	1.30477	1.42331	1.55133	1.68948	1.99900
10	1.10462	1.16054	1.21899	1.34392	1.48024	1.62889	1.79085	2.15892
11	1.11567	1.17795	1.24337	1.38423	1.53945	1.71034	1.89830	2.33164
12	1.12683	1.19562	1.26824	1.42576	1.60103	1.79586	2.01220	2.51817
13	1.13809	1.21355	1.29361	1.46853	1.66507	1.88565	2.13293	2.71962
14	1.14947	1.23176	1.31948	1.51259	1.73168	1.97993	2.26090	2.93719
15	1.16097	1.25023	1.34587	1.55797	1.80094	2.07893	2.39656	3.17217
16	1.17258	1.26899	1.37279	1.60471	1.87298	2.18287	2.54035	3.42594
17	1.18430	1.28802	1.40024	1.65285	1.94790	2.29202	2.69277	3.70002
18	1.19615	1.30734	1.42825	1.70243	2.02582	2.40662	2.85434	3.99602
19	1.20811	1.32695	1.45681	1.75351	2.10685	2.52695	3.02560	4.31570
20	1.22019	1.34686	1.48595	1.80611	2.19112	2.65330	3.20714	4.66096
21	1.23239	1.36706	1.51567	1.86029	2.27877	2.78596	3.39956	5.03383
22	1.24472	1.38756	1.54598	1.91610	2.36992	2.92526	3.60354	5.43654
23	1.25716	1.40838	1.57690	1.97359	2.46472	3.07152	3.81975	5.87146
24	1.26973	1.42950	1.60844	2.03279	2.56330	3.22510	4.04893	6.34118
25	1.28243	1.45095	1.64061	2.09378	2.66584	3.38635	4.29187	6.84848
26	1.29526	1.47271	1.67342	2.15659	2.77247	3.55567	4.54938	7.39635
27	1.30821	1.49480	1.70689	2.22129	2.88337	3.73346	4.82235	7.98806
28	1.32129	1.51722	1.74102	2.28793	2.99870	3.92013	5.11169	8.62711
29	1.33450	1.53998	1.77584	2.35657	3.11865	4.11614	5.41839	9.31727
30	1.34785	1.56308	1.81136	2.42726	3.24340	4.32194	5.74349	10.06266
31	1.36133	1.58653	1.84759	2.50008	3.37313	4.53804	6.08810	10.86767
32	1.37494	1.61032	1.88454	2.57508	3.50806	4.76494	6.45339	11.73708
33	1.38869	1.63448	1.92223	2.65234	3.64838	5.00319	6.84059	12.67605
34	1.40258	1.65900	1.96068	2.73191	3.79432	5.25335	7.25103	13.69013
35	1.41660	1.68388	1.99989	2.81386	3.94609	5.51602	7.68609	14.78534
36	1.43077	1.70914	2.03989	2.89828	4.10393	5.79182	8.14725	15.96817
37	1.44508	1.73478	2.08069	2.98523	4.26809	6.08141	8.63609	17.24563
38	1.45953	1.76080	2.12230	3.07478	4.43881	6.38548	9.15425	18.62528
39	1.47412	1.78721	2.16474	3.16703	4.61637	6.70475	9.70351	20.11530
40	1.48886	1.81402	2.20804	3.26204	4.80102	7.03999	10.28572	21.72452
41	1.50375	1.84123	2.25220	3.35990	4.99306	7.39199	10.90286	23.46248
42	1.51879	1.86885	2.29724	3.46070	5.19278	7.76159	11.55703	25.33948
43	1.53398	1.89688	2.34319	3.56452	5.40050	8.14967	12.25045	27.36664
44	1.54932	1.92533	2.39005	3.67145	5.61652	8.55715	12.98548	29.55597
45	1.56481	1.95421	2.43785	3.78160	5.84118	8.98501	13.76461	31.92045
46	1.58046	1.98353	2.48661	3.89504	6.07482	9.43426	14.59049	34.47409
47	1.59626	2.01328	2.53634	4.01190	6.31782	9.90597	15.46592	37.23201
48	1.61223	2.04348	2.58707	4.13225	6.57053	10.40127	16.39387	40.21057
49	1.62835	2.07413	2.63881	4.25622	6.83335	10.92133	17.37750	43.42742
50	1.64463	2.10524	2.69159	4.38391	7.10668	11.46740	18.42015	46.90161

Table 4 Present Value of a Dollar

$$(1 + i)^{-n}$$

n \ i	1%	$1\frac{1}{2}$%	2%	3%	4%	5%	6%	8%
1	0.99010	0.98522	0.98039	0.97087	0.96154	0.95238	0.94340	0.92593
2	0.98030	0.97066	0.96117	0.94260	0.92456	0.90703	0.89000	0.85734
3	0.97059	0.95632	0.94232	0.91514	0.88900	0.86384	0.83962	0.79383
4	0.96098	0.94218	0.92385	0.88849	0.85480	0.82270	0.79209	0.73503
5	0.95147	0.92826	0.90573	0.86261	0.82193	0.78353	0.74726	0.68058
6	0.94205	0.91454	0.88797	0.83748	0.79031	0.74622	0.70496	0.63017
7	0.93272	0.90103	0.87056	0.81309	0.75992	0.71068	0.66506	0.58349
8	0.92348	0.88771	0.85349	0.78941	0.73069	0.67684	0.62741	0.54027
9	0.91434	0.87459	0.83676	0.76642	0.70259	0.64461	0.59190	0.50025
10	0.90529	0.86167	0.82035	0.74409	0.67556	0.61391	0.55839	0.46319
11	0.89632	0.84893	0.80426	0.72242	0.64958	0.58468	0.52679	0.42888
12	0.88745	0.83639	0.78849	0.70138	0.62460	0.55684	0.49697	0.39711
13	0.87866	0.82403	0.77303	0.68095	0.60057	0.53032	0.46884	0.36770
14	0.86996	0.81185	0.75788	0.66112	0.57748	0.50507	0.44230	0.34046
15	0.86135	0.79985	0.74301	0.64186	0.55526	0.48102	0.41727	0.31524
16	0.85282	0.78803	0.72845	0.62317	0.53391	0.45811	0.39365	0.29189
17	0.84438	0.77639	0.71416	0.60502	0.51337	0.43630	0.37136	0.27027
18	0.83602	0.76491	0.70016	0.58739	0.49363	0.41552	0.35034	0.25025
19	0.82774	0.75361	0.68643	0.57029	0.47464	0.39573	0.33051	0.23171
20	0.81954	0.74247	0.67297	0.55368	0.45639	0.37689	0.31180	0.21455
21	0.81143	0.73150	0.65978	0.53755	0.43883	0.35894	0.29416	0.19866
22	0.80340	0.72069	0.64684	0.52189	0.42196	0.34185	0.27751	0.18394
23	0.79544	0.71004	0.63416	0.50669	0.40573	0.32557	0.26180	0.17032
24	0.78757	0.69954	0.62172	0.49193	0.39012	0.31007	0.24698	0.15770
25	0.77977	0.68921	0.60953	0.47761	0.37512	0.29530	0.23300	0.14602
26	0.77205	0.67902	0.59758	0.46369	0.36069	0.28124	0.21981	0.13520
27	0.76440	0.66899	0.58586	0.45019	0.34682	0.26785	0.20737	0.12519
28	0.75684	0.65910	0.57437	0.43708	0.33348	0.25509	0.19563	0.11591
29	0.74934	0.64936	0.56311	0.42435	0.32065	0.24295	0.18456	0.10733
30	0.74192	0.63976	0.55207	0.41199	0.30832	0.23138	0.17411	0.09938
31	0.73458	0.63031	0.54125	0.39999	0.29646	0.22036	0.16425	0.09202
32	0.72730	0.62099	0.53063	0.38834	0.28506	0.20987	0.15496	0.08520
33	0.72010	0.61182	0.52023	0.37703	0.27409	0.19987	0.14619	0.07889
34	0.71297	0.60277	0.51003	0.36604	0.26355	0.19035	0.13791	0.07305
35	0.70591	0.59387	0.50003	0.35538	0.25342	0.18129	0.13011	0.06763
36	0.69892	0.58509	0.49022	0.34503	0.24367	0.17266	0.12274	0.06262
37	0.69200	0.57644	0.48061	0.33498	0.23430	0.16444	0.11579	0.05799
38	0.68515	0.56792	0.47119	0.32523	0.22529	0.15661	0.10924	0.05369
39	0.67837	0.55953	0.46195	0.31575	0.21662	0.14915	0.10306	0.04971
40	0.67165	0.55126	0.45289	0.30656	0.20829	0.14205	0.09722	0.04603
41	0.66500	0.54312	0.44401	0.29763	0.20028	0.13528	0.09172	0.04262
42	0.65842	0.53509	0.43530	0.28896	0.19257	0.12884	0.08653	0.03946
43	0.65190	0.52718	0.42677	0.28054	0.18517	0.12270	0.08163	0.03654
44	0.64545	0.51939	0.41840	0.27237	0.17805	0.11686	0.07701	0.03383
45	0.63905	0.51171	0.41020	0.26444	0.17120	0.11130	0.07265	0.03133
46	0.63273	0.50415	0.40215	0.25674	0.16461	0.10600	0.06854	0.02901
47	0.62646	0.49670	0.39427	0.24926	0.15828	0.10095	0.06466	0.02686
48	0.62026	0.48936	0.38654	0.24200	0.15219	0.09614	0.06100	0.02487
49	0.61412	0.48213	0.37896	0.23495	0.14634	0.09156	0.05755	0.02303
50	0.60804	0.47500	0.37153	0.22811	0.14071	0.08720	0.05429	0.02132

(handwritten)

$$Z = \frac{X - \mu}{\sigma}$$

μ = expected value

σ^2 = VAR

σ = SD

Table 5   Amount of an Annuity

$$s_{\overline{n}|}\,i = \frac{(1+i)^n - 1}{i}$$

i n	1%	$1\frac{1}{2}\%$	2%	3%	4%	5%	6%	8%
1	1.00000	1.00000	1.00000	1.00000	1.00000	1.00000	1.00000	1.00000
2	2.01000	2.01500	2.02000	2.03000	2.04000	2.05000	2.06000	2.08000
3	3.03010	3.04523	3.06040	3.09090	3.12160	3.15250	3.18360	3.24640
4	4.06040	4.09090	4.12161	4.18363	4.24646	4.31013	4.37462	4.50611
5	5.10101	5.15227	5.20404	5.30914	5.41632	5.52563	5.63709	5.86660
6	6.15202	6.22955	6.30812	6.46841	6.63298	6.80191	6.97532	7.33593
7	7.21354	7.32299	7.43428	7.66246	7.89829	8.14201	8.39384	8.92280
8	8.28567	8.43284	8.58297	8.89234	9.21423	9.54911	9.89747	10.63663
9	9.36853	9.55933	9.75463	10.15911	10.58280	11.02656	11.49132	12.48756
10	10.46221	10.70272	10.94972	11.46388	12.00611	12.57789	13.18079	14.48656
11	11.56683	11.86326	12.16872	12.80780	13.48635	14.20679	14.97164	16.64549
12	12.68250	13.04121	13.41209	14.19203	15.02581	15.91713	16.86994	18.97713
13	13.80933	14.23683	14.68033	15.61779	16.62684	17.71298	18.88214	21.49530
14	14.94742	15.45038	15.97394	17.08632	18.29191	19.59863	21.01507	24.21492
15	16.09690	16.68214	17.29342	18.59891	20.02359	21.57856	23.27597	27.15211
16	17.25786	17.93237	18.63929	20.15688	21.82453	23.65749	25.67253	30.32428
17	18.43044	19.20136	20.01207	21.76159	23.69751	25.84037	28.21288	33.75023
18	19.61475	20.48938	21.41231	23.41444	25.64541	28.13238	30.90565	37.45024
19	20.81090	21.79672	22.84056	25.11687	27.67123	30.53900	33.75999	41.44626
20	22.01900	23.12367	24.29737	26.87037	29.77808	33.06595	36.78559	45.76196
21	23.23919	24.47052	25.78332	28.67649	31.96920	35.71925	39.99273	50.42292
22	24.47159	25.83758	27.29898	30.53678	34.24797	38.50521	43.39229	55.45676
23	25.71630	27.22514	28.84496	32.45288	36.61789	41.43048	46.99583	60.89330
24	26.97346	28.63352	30.42186	34.42647	39.08260	44.50200	50.81558	66.76476
25	28.24320	30.06302	32.03030	36.45926	41.64591	47.72710	54.86451	73.10594
26	29.52563	31.51397	33.67091	38.55304	44.31174	51.11345	59.15638	79.95442
27	30.82089	32.98668	35.34432	40.70963	47.08421	54.66913	63.70577	87.35077
28	32.12910	34.48148	37.05121	42.93092	49.96758	58.40258	68.52811	95.33883
29	33.45039	35.99870	38.79223	45.21885	52.96629	62.32271	73.63980	103.96594
30	34.78489	37.53868	40.56808	47.57542	56.08494	66.43885	79.05819	113.28321
31	36.13274	39.10176	42.37944	50.00268	59.32834	70.76079	84.80168	123.34587
32	37.49407	40.68829	44.22703	52.50276	62.70147	75.29883	90.88978	134.21354
33	38.86901	42.29861	46.11157	55.07784	66.20953	80.06377	97.34316	145.95062
34	40.25770	43.93309	48.03380	57.73018	69.85791	85.06696	104.18375	158.62667
35	41.66028	45.59209	49.99448	60.46208	73.65222	90.32031	111.43478	172.31680
36	43.07688	47.27597	51.99437	63.27594	77.59831	95.83632	119.12087	187.10215
37	44.50765	48.98511	54.03425	66.17422	81.70225	101.62814	127.26812	203.07032
38	45.95272	50.71989	56.11494	69.15945	85.97034	107.70955	135.90421	220.31595
39	47.41225	52.48068	58.23724	72.23423	90.40915	114.09502	145.05846	238.94122
40	48.88637	54.26789	60.40198	75.40126	95.02552	120.79977	154.76197	259.05652
41	50.37524	56.08191	62.61002	78.66330	99.82654	127.83976	165.04768	280.78104
42	51.87899	57.92314	64.86222	82.02320	104.81960	135.23175	175.95054	304.24352
43	53.39778	59.79199	67.15947	85.48389	110.01238	142.99334	187.50758	329.58301
44	54.93176	61.68887	69.50266	89.04841	115.41288	151.14301	199.75803	356.94965
45	56.48107	63.61420	71.89271	92.71986	121.02939	159.70016	212.74351	386.50562
46	58.04589	65.56841	74.33056	96.50146	126.87057	168.68516	226.50812	418.42607
47	59.62634	67.55194	76.81718	100.39650	132.94539	178.11942	241.09861	452.90015
48	61.22261	69.56522	79.35352	104.40840	139.26321	188.02539	256.56453	490.13216
49	62.83483	71.60870	81.94059	108.54065	145.83373	198.42666	272.95840	530.34274
50	64.46318	73.68283	84.57940	112.79687	152.66708	209.34800	290.33590	573.77016

Table 6 Present Value of an Annuity

$$a_{\overline{n}|\,i} = \frac{1 - (1 + i)^{-n}}{i}$$

n \ i	1%	$1\frac{1}{2}$%	2%	3%	4%	5%	6%	8%
1	0.99010	0.98522	0.98039	0.97087	0.96154	0.95238	0.94340	0.92593
2	1.97040	1.95588	1.94156	1.91347	1.88609	1.85941	1.83339	1.78326
3	2.94099	2.91220	2.88388	2.82861	2.77509	2.72325	2.67301	2.57710
4	3.90197	3.85438	3.80773	3.71710	3.62990	3.54595	3.46511	3.31213
5	4.85343	4.78264	4.71346	4.57971	4.45182	4.32948	4.21236	3.99271
6	5.79548	5.69719	5.60143	5.41719	5.24214	5.07569	4.91732	4.62288
7	6.72819	6.59821	6.47199	6.23028	6.00205	5.78637	5.58238	5.20637
8	7.65168	7.48593	7.32548	7.01969	6.73274	6.46321	6.20979	5.74664
9	8.56602	8.36052	8.16224	7.78611	7.43533	7.10782	6.80169	6.24689
10	9.47130	9.22218	8.98259	8.53020	8.11090	7.72173	7.36009	6.71008
11	10.36763	10.07112	9.78685	9.25262	8.76048	8.30641	7.88687	7.13896
12	11.25508	10.90751	10.57534	9.95400	9.38507	8.86325	8.38384	7.53608
13	12.13374	11.73153	11.34837	10.63496	9.98565	9.39357	8.85268	7.90378
14	13.00370	12.54338	12.10625	11.29607	10.56312	9.89864	9.29498	8.24424
15	13.86505	13.34323	12.84926	11.93794	11.11839	10.37966	9.71225	8.55948
16	14.71787	14.13126	13.57771	12.56110	11.65230	10.83777	10.10590	8.85137
17	15.56225	14.90765	14.29187	13.16612	12.16567	11.27407	10.47726	9.12164
18	16.39827	15.67256	14.99203	13.75351	12.65930	11.68959	10.82760	9.37189
19	17.22601	16.42617	15.67846	14.32380	13.13394	12.08532	11.15812	9.60360
20	18.04555	17.16864	16.35143	14.87747	13.59033	12.46221	11.46992	9.81815
21	18.85698	17.90014	17.01121	15.41502	14.02916	12.82115	11.76408	10.01680
22	19.66038	18.62082	17.65805	15.93692	14.45112	13.16300	12.04158	10.20074
23	20.45582	19.33086	18.29220	16.44361	14.85684	13.48857	12.30338	10.37106
24	21.24339	20.03041	18.91393	16.93554	15.24696	13.79864	12.55036	10.52876
25	22.02316	20.71961	19.52346	17.41315	15.62208	14.09394	12.78336	10.67478
26	22.79520	21.39863	20.12104	17.87684	15.98277	14.37519	13.00317	10.80998
27	23.55961	22.06762	20.70690	18.32703	16.32959	14.64303	13.21053	10.93516
28	24.31644	22.72672	21.28127	18.76411	16.66306	14.89813	13.40616	11.05108
29	25.06579	23.37608	21.84438	19.18845	16.98371	15.14107	13.59072	11.15841
30	25.80771	24.01584	22.39646	19.60044	17.29203	15.37245	13.76483	11.25778
31	26.54229	24.64615	22.93770	20.00043	17.58849	15.59281	13.92909	11.34980
32	27.26959	25.26714	23.46833	20.38877	17.87355	15.80268	14.08404	11.43500
33	27.98969	25.87895	23.98856	20.76579	18.14765	16.00255	14.23023	11.51389
34	28.70267	26.48173	24.49859	21.13184	18.41120	16.19290	14.36814	11.58693
35	29.40858	27.07559	24.99862	21.48722	18.66461	16.37419	14.49825	11.65457
36	30.10751	27.66068	25.48884	21.83225	18.90828	16.54685	14.62099	11.71719
37	30.79951	28.23713	25.96945	22.16724	19.14258	16.71129	14.73678	11.77518
38	31.48466	28.80505	26.44064	22.49246	19.36786	16.86789	14.84602	11.82887
39	32.16303	29.36458	26.90259	22.80822	19.58448	17.01704	14.94907	11.87858
40	32.83469	29.91585	27.35548	23.11477	19.79277	17.15909	15.04630	11.92461
41	33.49969	30.45896	27.79949	23.41240	19.99305	17.29437	15.13802	11.96723
42	34.15811	30.99405	28.23479	23.70136	20.18563	17.42321	15.22454	12.00670
43	34.81001	31.52123	28.66156	23.98190	20.37079	17.54591	15.30617	12.04324
44	35.45545	32.04062	29.07996	24.25427	20.54884	17.66277	15.38318	12.07707
45	36.09451	32.55234	29.49016	24.51871	20.72004	17.77407	15.45583	12.10840
46	36.72724	33.05649	29.89231	24.77545	20.88465	17.88007	15.52437	12.13741
47	37.35370	33.55319	30.28658	25.02471	21.04294	17.98102	15.58903	12.16427
48	37.97396	34.04255	30.67312	25.26671	21.19513	18.07716	15.65003	12.18914
49	38.58808	34.52468	31.05208	25.50166	21.34147	18.16872	15.70757	12.21216
50	39.19612	34.99969	31.42361	25.72976	21.48218	18.25593	15.76186	12.23348

Table 7 $1/s_{\overline{n}|i}$

$$\frac{1}{s_{\overline{n}|i}} = \frac{i}{(1+i)^n - 1}$$

n \ i	1%	$1\frac{1}{2}$%	2%	3%	4%	5%	6%	8%
1	1.00000	1.00000	1.00000	1.00000	1.00000	1.00000	1.00000	1.00000
2	0.49751	0.49628	0.49505	0.49261	0.49020	0.48780	0.48544	0.48077
3	0.33002	0.32838	0.32675	0.32353	0.32035	0.31721	0.31411	0.30803
4	0.24628	0.24444	0.24262	0.23903	0.23549	0.23201	0.22859	0.22192
5	0.19604	0.19409	0.19216	0.18835	0.18463	0.18097	0.17740	0.17046
6	0.16255	0.16053	0.15853	0.15460	0.15076	0.14702	0.14336	0.13632
7	0.13863	0.13656	0.13451	0.13051	0.12661	0.12282	0.11914	0.11207
8	0.12069	0.11858	0.11651	0.11246	0.10853	0.10472	0.10104	0.09401
9	0.10674	0.10461	0.10252	0.09843	0.09449	0.09069	0.08702	0.08008
10	0.09558	0.09343	0.09133	0.08723	0.08329	0.07950	0.07587	0.06903
11	0.08645	0.08429	0.08218	0.07808	0.07415	0.07039	0.06679	0.06008
12	0.07885	0.07668	0.07456	0.07046	0.06655	0.06283	0.05928	0.05270
13	0.07241	0.07024	0.06812	0.06403	0.06014	0.05646	0.05296	0.04652
14	0.06690	0.06472	0.06260	0.05853	0.05467	0.05102	0.04758	0.04130
15	0.06212	0.05994	0.05783	0.05377	0.04994	0.04634	0.04296	0.03683
16	0.05794	0.05577	0.05365	0.04961	0.04582	0.04227	0.03895	0.03298
17	0.05426	0.05208	0.04997	0.04595	0.04220	0.03870	0.03544	0.02963
18	0.05098	0.04881	0.04670	0.04271	0.03899	0.03555	0.03236	0.02670
19	0.04805	0.04588	0.04378	0.03981	0.03614	0.03275	0.02962	0.02413
20	0.04542	0.04325	0.04116	0.03722	0.03358	0.03024	0.02718	0.02185
21	0.04303	0.04087	0.03878	0.03487	0.03128	0.02800	0.02500	0.01983
22	0.04086	0.03870	0.03663	0.03275	0.02920	0.02597	0.02305	0.01803
23	0.03889	0.03673	0.03467	0.03081	0.02731	0.02414	0.02128	0.01642
24	0.03707	0.03492	0.03287	0.02905	0.02559	0.02247	0.01968	0.01498
25	0.03541	0.03326	0.03122	0.02743	0.02401	0.02095	0.01823	0.01368
26	0.03387	0.03173	0.02970	0.02594	0.02257	0.01956	0.01690	0.01251
27	0.03245	0.03032	0.02829	0.02456	0.02124	0.01829	0.01570	0.01145
28	0.03112	0.02900	0.02699	0.02329	0.02001	0.01712	0.01459	0.01049
29	0.02990	0.02778	0.02578	0.02211	0.01888	0.01605	0.01358	0.00962
30	0.02875	0.02664	0.02465	0.02102	0.01783	0.01505	0.01265	0.00883
31	0.02768	0.02557	0.02360	0.02000	0.01686	0.01413	0.01179	0.00811
32	0.02667	0.02458	0.02261	0.01905	0.01595	0.01328	0.01100	0.00745
33	0.02573	0.02364	0.02169	0.01816	0.01510	0.01249	0.01027	0.00685
34	0.02484	0.02276	0.02082	0.01732	0.01431	0.01176	0.00960	0.00630
35	0.02400	0.02193	0.02000	0.01654	0.01358	0.01107	0.00897	0.00580
36	0.02321	0.02115	0.01923	0.01580	0.01289	0.01043	0.00839	0.00534
37	0.02247	0.02041	0.01851	0.01511	0.01224	0.00984	0.00786	0.00492
38	0.02176	0.01972	0.01782	0.01446	0.01163	0.00928	0.00736	0.00454
39	0.02109	0.01905	0.01717	0.01384	0.01106	0.00876	0.00689	0.00419
40	0.02046	0.01843	0.01656	0.01326	0.01052	0.00828	0.00646	0.00386
41	0.01985	0.01783	0.01597	0.01271	0.01002	0.00782	0.00606	0.00356
42	0.01928	0.01726	0.01542	0.01219	0.00954	0.00739	0.00568	0.00329
43	0.01873	0.01672	0.01489	0.01170	0.00909	0.00699	0.00533	0.00303
44	0.01820	0.01621	0.01439	0.01123	0.00866	0.00662	0.00501	0.00280
45	0.01771	0.01572	0.01391	0.01079	0.00826	0.00626	0.00470	0.00259
46	0.01723	0.01525	0.01345	0.01036	0.00788	0.00593	0.00441	0.00239
47	0.01677	0.01480	0.01302	0.00996	0.00752	0.00561	0.00415	0.00221
48	0.01633	0.01437	0.01260	0.00958	0.00718	0.00532	0.00390	0.00204
49	0.01591	0.01396	0.01220	0.00921	0.00686	0.00504	0.00366	0.00189
50	0.01551	0.01357	0.01182	0.00887	0.00655	0.00478	0.00344	0.00174

Table 8 Powers of e

x	e^x	e^{-x}	x	e^x	e^{-x}
0.00	1.00000	1.00000			
0.01	1.01005	0.99004	1.60	4.95302	0.20189
0.02	1.02020	0.98019	1.70	5.47394	0.18268
0.03	1.03045	0.97044	1.80	6.04964	0.16529
0.04	1.04081	0.96078	1.90	6.68589	0.14956
0.05	1.05127	0.95122	2.00	7.38905	0.13533
0.06	1.06183	0.94176			
0.07	1.07250	0.93239	2.10	8.16616	0.12245
0.08	1.08328	0.92311	2.20	9.02500	0.11080
0.09	1.09417	0.91393	2.30	9.97417	0.10025
0.10	1.10517	0.90483	2.40	11.02316	0.09071
			2.50	12.18248	0.08208
0.11	1.11628	0.89583	2.60	13.46372	0.07427
0.12	1.12750	0.88692	2.70	14.87971	0.06720
0.13	1.13883	0.87810	2.80	16.44463	0.06081
0.14	1.15027	0.86936	2.90	18.17412	0.05502
0.15	1.16183	0.86071	3.00	20.08551	0.04978
0.16	1.17351	0.85214			
0.17	1.18530	0.84366	3.50	33.11545	0.03020
0.18	1.19722	0.83527	4.00	54.59815	0.01832
0.19	1.20925	0.82696	4.50	90.01713	0.01111
0.20	1.22140	0.81873	5.00	148.41316	0.00674
0.30	1.34985	0.74081	5.50	224.69193	0.00409
0.40	1.49182	0.67032			
0.50	1.64872	0.60653	6.00	403.42879	0.00248
0.60	1.82211	0.54881	6.50	665.14163	0.00150
0.70	2.01375	0.49658	7.00	1096.63316	0.00091
0.80	2.22554	0.44932	7.50	1808.04241	0.00055
0.90	2.45960	0.40656			
1.00	2.71828	0.36787	8.00	2980.95799	0.00034
			8.50	4914.76884	0.00020
1.10	3.00416	0.33287			
1.20	3.32011	0.30119	9.00	8103.08392	0.00012
1.30	3.66929	0.27253	9.50	13359.72683	0.00007
1.40	4.05519	0.24659			
1.50	4.48168	0.22313	10.00	22026.46579	0.00005

Table 9 Natural Logarithms

x	$\ln x$	x	$\ln x$	x	$\ln x$
0.0		4.5	1.5041	9.0	2.1972
0.1	−2.3026	4.6	1.5261	9.1	2.2083
0.2	−1.6094	4.7	1.5476	9.2	2.2192
0.3	−1.2040	4.8	1.5686	9.3	2.2300
0.4	−0.9163	4.9	1.5892	9.4	2.2407
0.5	−0.6931	5.0	1.6094	9.5	2.2513
0.6	−0.5108	5.1	1.6292	9.6	2.2618
0.7	−0.3567	5.2	1.6487	9.7	2.2721
0.8	−0.2231	5.3	1.6677	9.8	2.2824
0.9	−0.1054	5.4	1.6864	9.9	2.2925
1.0	0.0000	5.5	1.7047	10	2.3026
1.1	0.0953	5.6	1.7228	11	2.3979
1.2	0.1823	5.7	1.7405	12	2.4849
1.3	0.2624	5.8	1.7579	13	2.5649
1.4	0.3365	5.9	1.7750	14	2.6391
1.5	0.4055	6.0	1.7918	15	2.7081
1.6	0.4700	6.1	1.8083	16	2.7726
1.7	0.5306	6.2	1.8245	17	2.8332
1.8	0.5878	6.3	1.8405	18	2.8904
1.9	0.6419	6.4	1.8563	19	2.9444
2.0	0.6931	6.5	1.8718	20	2.9957
2.1	0.7419	6.6	1.8871		
2.2	0.7885	6.7	1.9021	25	3.2189
2.3	0.8329	6.8	1.9169	30	3.4012
2.4	0.8755	6.9	1.9315	35	3.5553
				40	3.6889
2.5	0.9163	7.0	1.9459		
2.6	0.9555	7.1	1.9601	45	3.8067
2.7	0.9933	7.2	1.9741	50	3.9120
2.8	1.0296	7.3	1.9879	55	4.0073
2.9	1.0647	7.4	2.0015	60	4.0943
				65	4.1744
3.0	1.0986	7.5	2.0149		
3.1	1.1314	7.6	2.0281	70	4.2485
3.2	1.1632	7.7	2.0412	75	4.3175
3.3	1.1939	7.8	2.0541	80	4.3820
3.4	1.2238	7.9	2.0669	85	4.4427
				90	4.4998
3.5	1.2528	8.0	2.0794		
3.6	1.2809	8.1	2.0919	95	4.5539
3.7	1.3083	8.2	2.1041	100	4.6052
3.8	1.3350	8.3	2.1163		
3.9	1.3610	8.4	2.1281		
4.0	1.3863	8.5	2.1401		
4.1	1.4110	8.6	2.1518		
4.2	1.4351	8.7	2.1633		
4.3	1.4586	8.8	2.1748		
4.4	1.4816	8.9	2.1861		

Answers to Selected Exercises

CHAPTER 1

Section 1.1   (page 5)

1. $(-2, -3), (-1, -2), (0, -1), (1, 0), (2, 1), (3, 2)$; range: $\{-3, -2, -1, 0, 1, 2\}$

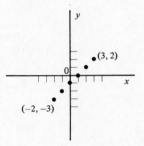

3. $(-2, 17), (-1, 13), (0, 9), (1, 5), (2, 1), (3, -3)$; range: $\{17, 13, 9, 5, 1, -3\}$

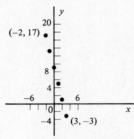

5. $(-2, -3), (-1, -4), (0, -5), (1, -6), (2, -7), (3, -8)$; range: $\{-3, -4, -5, -6, -7, -8\}$

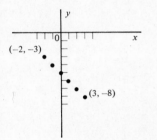

7. $(-2, 13), (-1, 11), (0, 9), (1, 7), (2, 5), (3, 3)$; range: $\{13, 11, 9, 7, 5, 3\}$

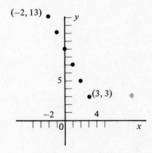

9. $(-2, 3/2), (-1, 2), (0, 5/2), (1, 3), (2, 7/2), (3, 4)$; range: $\{3/2, 2, 5/2, 3, 7/2, 4\}$

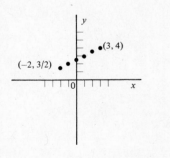

11. $(-2, 2), (-1, 0), (0, 0), (1, 2), (2, 6), (3, 12)$; range: $\{0, 2, 6, 12\}$

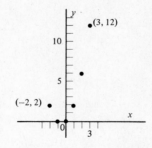

13. $(-2, 4), (-1, 1), (0, 0), (1, 1), (2, 4), (3, 9)$; range: $\{4, 1, 0, 9\}$

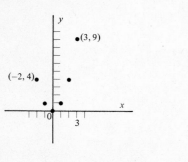

15. $(-2, -13), (-1, -1), (0, 3), (1, -1), (2, -13), (3, -33)$; range: $\{-13, -1, 3, -33\}$

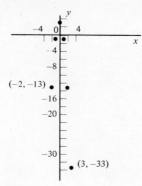

17. $(-2, 1), (-1, 1/2), (0, 1/3), (1, 1/4), (2, 1/5), (3, 1/6)$; range: $\{1, 1/2, 1/3, 1/4, 1/5, 1/6\}$

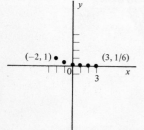

19. $(-2, -3), (-1, -3/2), (0, -3/5), (1, 0), (2, 3/7), (3, 3/4)$; range: $\{-3, -3/2, -3/5, 0, 3/7, 3/4\}$

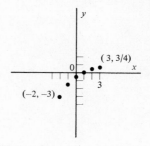

21. $(-2, 4), (-1, 4), (0, 4), (1, 4), (2, 4), (3, 4)$; range: $\{4\}$

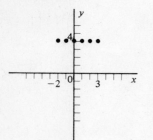

23. Function   **25.** Not a function   **27.** Function   **29.** (a) 14   (b) -7 (c) 2   (d) $3a + 2$
31. (a) -12 (b) 2   (c) -4 (d) $-2a - 4$ **33.** (a) 6   (b) 6   (c) 6   (d) 6
35. (a) 48   (b) 6   (c) 0   (d) $2a^2 + 4a$ **37.** (a) 5   (b) -23 (c) 1   (d) $-a^2 + 5a + 1$
39. (a) 30   (b) 2   (c) 2   (d) $(a + 1)(a + 2)$, or $a^2 + 3a + 2$ **41.** (a) -3 (b) -5 (c) -15
(d) 5   (e) $2a - 3$ (f) $-2r - 3$ (g) $2m + 3$ (h) $2p - 7$ (i) -1 (j) -21 **43.** (a) \$11
(b) \$11   (c) \$18   (d) \$32   (e) \$32   (f) \$39   (g) \$39

Section 1.2   (page 14)

1.

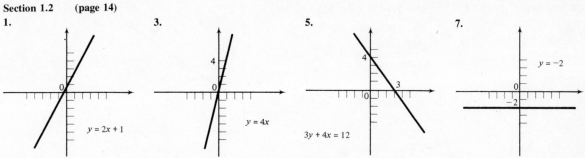

$y = 2x + 1$

3.

$y = 4x$

5.

$3y + 4x = 12$

7.

$y = -2$

9.

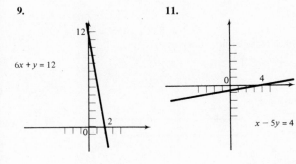

$6x + y = 12$

11.

$x - 5y = 4$

13.

$x + 5 = 0$

not a function

15.

$5y - 3x = 12$

17.

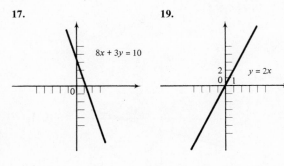

$8x + 3y = 10$

19.

$y = 2x$

21.

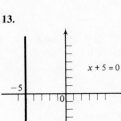

$y = -4x$

23.

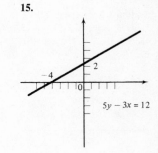

$x + 4y = 0$

25. (a) 16   (b) 11   (c) 6
(d) 8   (e) 4   (f) 0
(g)

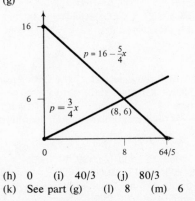

$p = 16 - \frac{5}{4}x$

$p = \frac{3}{4}x$

$(8, 6)$

(h) 0   (i) 40/3   (j) 80/3
(k) See part (g)   (l) 8   (m) 6

27. (a)

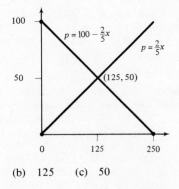

$p = 100 - \frac{2}{5}x$

$p = \frac{2}{5}x$

$(125, 50)$

(b) 125   (c) 50

29. (a) $135   (b) $205   (c) $275   (d) $345   (e)

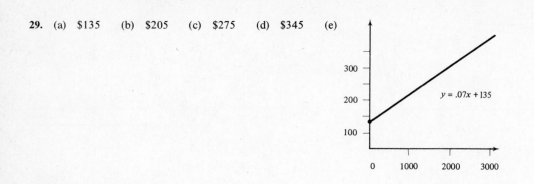

$$y = .07x + 135$$

Section 1.3   (page 22)
1. $-1/5$ **3.** $2/3$ **5.** $-3/2$ **7.** No slope   **9.** 0 **11.** $3; 4$ **13.** $-4; 8$ **15.** $-3/4; 5/4$
17. $-3; 0$ **19.** $-2/5; 0$ **21.** $0; 8$ **23.** $0; -2$ **25.** No slope; no y-intercept
27.   **29.**   **31.**

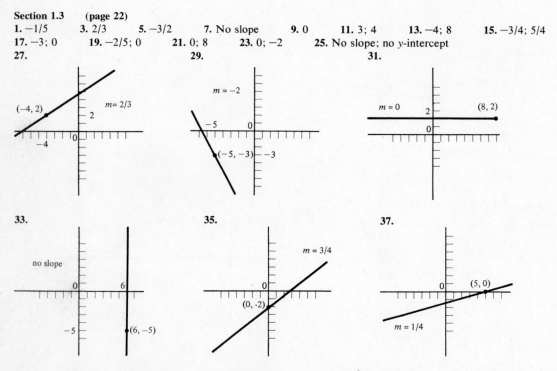

39. $4y = -3x + 16$ **41.** $2y = -x - 4$ **43.** $4y = 6x + 5$ **45.** $y = 2x + 9$ **47.** $y = -3x + 3$
49. $4y = x + 5$ **51.** $3y = 4x + 7$ **53.** $3y = -2x$ **55.** $x = -8$ **57.** $y = 3$ **59.** $y = 640x + 1100; m = 640$
61. $y = -1000x + 40,000; m = -1000$ **63.** $y = 2.5x - 80; m = 2.5$ **65.** (a) $h = 3.5r + 83$ (b) About
163.5 cm; about 177.5 cm   (c) About 25 cm

Section 1.4   (page 29)
1. (a) 2600   (b) 2900   (c) 3200   (d) 2000   (e) 300   **3.** (a) 100 thousand   (b) 70 thousand
(c) 0   (d) -5 thousand; the number is decreasing   **5.** (a) $y = 82,500x + 850,000$ (b) $1,097,500
(c) about $1,510,000   **7.** (a) 480   (b) 360   (c) 120   (d) June 29   (e) -20 **9.** If $C(x)$ is the
cost of renting a saw for x hours, then $C(x) = 12 + x$.   **11.** If $P(x)$ is the cost (in cents) of parking for x half-hours,
then $P(x) = 35x + 50$.   **13.** $C(x) = 30x + 100$ **15.** $C(x) = 25x + 1000$ **17.** $C(x) = 50x + 500$
19. $C(x) = 90x + 2500$ **21.** (a) $97   (b) $97.097   (c) $.097, or 9.7¢   (d) $.097, or 9.7¢
23. (a) $100   (b) $36   (c) $24   **25.** $2000; $32,000   **27.** $2000; $52,000   **29.** $150,000;
$600,000   **31.** (a) $20,000 per year   (b) $80,000   **33.** (a) $4000; $3000; $2000; $1000   (b) $2500
35. $5090.91; $3563.64   **37.** $4171.43; $2085.71   **39.** 500 units; $30,000   **41.** Break-even point is 45 units;
don't produce   **43.** Break-even point is -50 units; impossible to make a profit here

45.   (a)   (b)   (c) $y = 8x + 50$

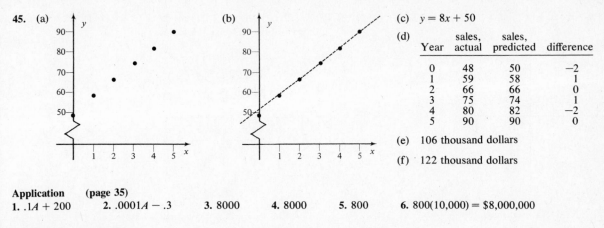

(d)

Year	sales, actual	sales, predicted	difference
0	48	50	−2
1	59	58	1
2	66	66	0
3	75	74	1
4	80	82	−2
5	90	90	0

(e)   106 thousand dollars

(f)   122 thousand dollars

Application   (page 35)

1. $.1A + 200$ **2.** $.0001A − .3$ **3.** 8000   **4.** 8000   **5.** 800   **6.** $800(10,000) = \$8,000,000$

Application   (page 36)

1. 4.8 million units   **2.** Portion of a straight line going through (3.1, 10.50) and (5.7, 10.67)   **3.** In the interval under discussion (3.1 to 5.7 million units), the marginal cost always exceeds the selling price.   **4.**   (a)   9.87; 10.22   (b)   portion of a straight line through (3.1, 9.87) and (5.7, 10.22)   (c)   .83 million units, which is not in the interval under discussion

Section 1.5   (page 41)

1.   (a)   \$60,000   (b)   \$40,000   (c)   \$26,667   (d)   \$17,778

3.

Depreciation in year	Straight-line	Double declining	Sum-of-years' digits
1	\$375	\$933	\$563
2	\$375	\$192*	\$375
3	\$375	\$0	\$188
Totals	\$1125	\$1125	\$1126

*Only \$192 of depreciation may be taken, since the total may not exceed the net cost of \$1125.

5.   (a)   \$7740   (b)   \$6460   (c)   \$5500   (d)   \$4540   (e)   About 50,900   (f)   About 44,600
(g)   About 42,100   (h)   About 37,750

Application   (page 44)

1. 16   **2.** 30   **3.** 335   **4.** 14   **5.** 2.0%   **6.** 13.7%   **7.** 52.0%   **8.** 97.3%

Chapter 1   Review Exercises   (page 44)

1. $(-3, -16/5), (-2, -14/5), (-1, -12/5), (0, -2), (1, -8/5)$,
$(2, -6/5), (3, -4/5)$;
range: $\{-16/5, -14/5, -12/5, -2, -8/5, -6/5, -4/5\}$

3. $(-3, 20), (-2, 9), (-1, 2), (0, -1), (1, 0), (2, 5), (3, 14)$;
range: $\{-1, 0, 2, 5, 9, 14, 20\}$

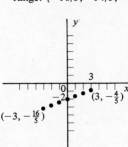

5. $(-3, 7)$, $(-2, 2)$, $(-1, -1)$, $(0, -2)$, $(1, -1)$, $(2, 2)$, $(3, 7)$; range: $\{-2, -1, 2, 7\}$

7. $(-3, 1/5)$, $(-2, 2/5)$, $(-1, 1)$, $(0, 2)$, $(1, 1)$, $(2, 2/5)$, $(3, 1/5)$; range: $\{1/5, 2/5, 1, 2\}$

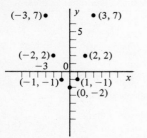

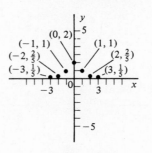

9. $(-3, -1)$, $(-2, -1)$, $(-1, -1)$, $(0, -1)$, $(1, -1)$, $(2, -1)$, $(3, -1)$; range: $\{-1\}$

11. (a) 23 (b) -9 (c) -17 (d) $4r + 3$
13. (a) -28 (b) -12 (c) -28 (d) $-r^2 - 3$

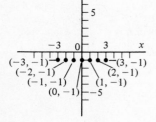

15. (a) -13 (b) 3 (c) -32 (d) 22 (e) $-k^2 - 4k$ (f) $-9m^2 + 12m$ (g) $-k^2 + 14k - 45$
(h) $12 - 5p$ (i) -28 (j) -21
17.

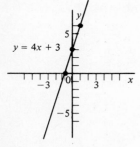

19.

21.

23.

25. (a) $7/6$; $9/2$ (b) 2; 2 (c) $5/2$, $1/2$
(d)   (e) 15
(f) 2; 2

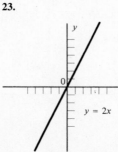

27. $1/3$ **29.** $-2/11$ **31.** $-2/3$ **33.** No slope   **35.** $3y = 2x - 13$ **37.** $5x + 4y = 17$ **39.** $x = -1$
41. $C(x) = 30x + 60$ **43.** $C(x) = 30x + 85$ **45.** (a) 5 units   (b) $\$200$

47.

Year	Straight-line	Double	Sum
1	$17,000	$39,500	$27,200
2	$17,000	$19,750	$20,400
3	$17,000	$8750	$13,600
4	$17,000	$0	$6800

CHAPTER 2

Section 2.1   (page 51)

1. False, not all corresponding elements equal   **3.** True   **5.** True   **7.** 2×2, square   **9.** 3×4
11. 2×1, column   **13.** $x = 2, y = 4, z = 8$ **15.** $x = -15, y = 5, k = 3$ **17.** $z = 18, r = 3, s = 3, p = 3,$
$a = 3/4$ **19.** $\begin{bmatrix} 9 & 12 & 0 & 2 \\ 1 & -1 & 2 & -4 \end{bmatrix}$ **21.** $\begin{bmatrix} 5 & 13 & 0 \\ 3 & 1 & 8 \end{bmatrix}$ **23.** Not possible   **25.** $\begin{bmatrix} 1 & 5 & 6 & -9 \\ 5 & 7 & 2 & 1 \\ -7 & 2 & 2 & -7 \end{bmatrix}$

27. $\begin{bmatrix} -12x + 8y & -x + y \\ x & 8x - y \end{bmatrix}$ **29.** $\begin{bmatrix} x & y \\ z & w \end{bmatrix} + \begin{bmatrix} r & s \\ t & u \end{bmatrix} = \begin{bmatrix} x + r & y + s \\ z + t & w + u \end{bmatrix}$ (a 2×2 matrix)

31. $\begin{bmatrix} x + (r + m) & y + (s + n) \\ z + (t + p) & w + (u + q) \end{bmatrix} = \begin{bmatrix} (x + r) + m & (y + s) + n \\ (z + t) + p & (w + u) + q \end{bmatrix}$ **33.** $\begin{bmatrix} m + 0 & n + 0 \\ p + 0 & q + 0 \end{bmatrix} = \begin{bmatrix} m & n \\ p & q \end{bmatrix}$

35. $\begin{bmatrix} 7 & 2 \\ 9 & 0 \\ 8 & 6 \end{bmatrix}; \begin{bmatrix} 7 & 9 & 8 \\ 2 & 0 & 6 \end{bmatrix}$ **37.** (a) $\begin{bmatrix} 2 & 1 & 2 & 1 \\ 3 & 2 & 2 & 1 \\ 4 & 3 & 2 & 1 \end{bmatrix}$ (b) $\begin{bmatrix} 5 & 0 & 7 \\ 0 & 10 & 1 \\ 0 & 15 & 2 \\ 10 & 12 & 8 \end{bmatrix}$ (c) $\begin{bmatrix} 8 \\ 4 \\ 5 \end{bmatrix}$

Section 2.2   (page 59)

1. $2 \times 2; 2 \times 2$ **3.** $4 \times 4; 2 \times 2$ **5.** $3 \times 2; BA$ does not exist   **7.** AB does not exist; 3×2
9. $\begin{bmatrix} -4 & 8 \\ 0 & 6 \end{bmatrix}$ **11.** $\begin{bmatrix} 24 & -8 \\ -16 & 0 \end{bmatrix}$ **13.** $\begin{bmatrix} -22 & -6 \\ 20 & -12 \end{bmatrix}$ **15.** $\begin{bmatrix} 13 \\ 25 \end{bmatrix}$ **17.** $\begin{bmatrix} -2 & 10 \\ 0 & 8 \end{bmatrix}$ **19.** $\begin{bmatrix} 13 & 5 \\ 25 & 15 \end{bmatrix}$
21. $\begin{bmatrix} 13 \\ 29 \end{bmatrix}$ **23.** $\begin{bmatrix} 110 \\ 40 \\ -50 \end{bmatrix}$ **25.** $\begin{bmatrix} 22 & -8 \\ 11 & -4 \end{bmatrix}$ **27.** (a) $\begin{bmatrix} 16 & 22 \\ 7 & 19 \end{bmatrix}$ (b) $\begin{bmatrix} 5 & -5 \\ 0 & 30 \end{bmatrix}$ (c) No; no   (d) No

33. (a) P, P, X (b) T (c) I maintains the identity of any 2×2 matrix under multiplication.
35. (a) $\begin{bmatrix} 20 & 52 & 27 \\ 25 & 62 & 35 \\ 30 & 72 & 43 \end{bmatrix}$ The rows represent the amounts of fat, carbohydrate, and protein, respectively in each of the daily meals.

(b) $\begin{bmatrix} 75 \\ 45 \\ 70 \\ 168 \end{bmatrix}$ The rows give the number of calories in one exchange of each of the food groups.

Application   (page 62)

1. (a) 3   (b) 3   (c) 5   (d) 3   **2.** (a) 21   (b) 25
3. (a) $B = \begin{bmatrix} 0 & 2 & 3 \\ 2 & 0 & 4 \\ 3 & 4 & 0 \end{bmatrix}$ (b) $\begin{bmatrix} 13 & 12 & 8 \\ 12 & 20 & 6 \\ 8 & 6 & 25 \end{bmatrix}$ (c) 12   (d) 14

4. (a)

	S	J	NO	H
S	0	1	2	1
J	1	0	1	0
NO	2	1	0	1
H	1	0	1	0

(b) 2
(c) 2
(d) 2

5. (a)

	dogs	rats	cats	mice
dogs	0	1	1	1
rats	0	0	0	1
cats	0	1	0	1
mice	0	0	0	0

$$C = \begin{bmatrix} 0 & 1 & 1 & 1 \\ 0 & 0 & 0 & 1 \\ 0 & 1 & 0 & 1 \\ 0 & 0 & 0 & 0 \end{bmatrix}$$

(b)

$$C^2 = \begin{bmatrix} 0 & 1 & 0 & 2 \\ 0 & 0 & 0 & 0 \\ 0 & 0 & 0 & 1 \\ 0 & 0 & 0 & 0 \end{bmatrix}$$

C^2 gives the number of food sources once removed from the feeder. Thus, since dogs eat rats and rats eat mice, mice are an indirect as well as a direct food source.

Application   (page 64)

1.
$$PQ = \begin{bmatrix} 1 & 2 & 0 & 2 & 1 & 1 \\ 0 & 1 & 0 & 1 & 0 & 0 \\ 1 & 1 & 0 & 1 & 2 & 1 \end{bmatrix}$$
2. none   **3.** yes, the third person

4. The second and fourth persons in the third group each had four contacts in all.

Section 2.3   (page 74)

1. (3, 6)   **3.** (−1, 4)   **5.** (−2, 0)   **7.** (1, 3)   **9.** (4, −2)   **11.** (2, −2)   **13.** ∅   **15.** Same line
17. (12, 6)   **19.** (7, −2)   **21.** (1, 2, −1)   **23.** (2, 0, 3)   **25.** ∅   **27.** (0, 2, 4)   **29.** (1, 2, 3)
31. (−1, 2, 1)   **33.** (4, 1, 2)   **35.** x arbitrary, $y = x + 5$, $z = -2x + 1$ **37.** x arbitrary, $y = x + 1$, $z = -x + 3$
39. x arbitrary, $y = 4x - 7$, $z = 3x + 7$, $w = -x - 3$ **41.** (3, −4)   **43.** ∅   **45.** ∅   **47.** Wife 40 days;
husband 32 days   **49.** 5 model 201; 8 model 301   **51.** $10,000 at 16%; $7000 at 20%; $8000 at 18%
53. $k = 3$; solution is (−1, 0, 2)

Section 2.4   (page 82)

1. $\begin{bmatrix} 2 & 3 & | & 11 \\ 1 & 2 & | & 8 \end{bmatrix}$ **3.** $\begin{bmatrix} 1 & 5 & | & 6 \\ 0 & 1 & | & 1 \end{bmatrix}$ **5.** $\begin{bmatrix} 2 & 1 & 1 & | & 3 \\ 3 & -4 & 2 & | & -7 \\ 1 & 1 & 1 & | & 2 \end{bmatrix}$ **7.** $\begin{bmatrix} 1 & 1 & 0 & | & 2 \\ 0 & 2 & 1 & | & -4 \\ 0 & 0 & 1 & | & 2 \end{bmatrix}$ **9.** $\begin{bmatrix} 1 & 0 & 0 & | & 5 \\ 0 & 1 & 0 & | & -2 \\ 0 & 0 & 1 & | & 3 \end{bmatrix}$

11. $x = 2$ **13.** $2x + y = 1$ **15.** $x = 2$
$\quad\ \ y = 3$ $\qquad 3x - 2y = -9$ $\qquad y = 3$
$\qquad\qquad\qquad\qquad\qquad\qquad\qquad\qquad z = -2$

17. (2, 3)   **19.** (−3, 0)   **21.** (7/2, −1)   **23.** (5/2, −1)   **25.** ∅   **27.** Same line   **29.** (−2, 1, 3)
31. (−1, 23, 16)   **33.** (3, 2, −4)   **35.** ∅   **37.** (0, 2, −2, 1) The answers are given in the order x, y, z, w.
39. (a) $\begin{bmatrix} 1 & 0 & 0 & 1 & | & 1000 \\ 1 & 1 & 0 & 0 & | & 1100 \\ 0 & 1 & 1 & 0 & | & 700 \\ 0 & 0 & 1 & 1 & | & 600 \end{bmatrix}$, $\begin{bmatrix} 1 & 0 & 0 & 1 & | & 1000 \\ 0 & 1 & 0 & -1 & | & 100 \\ 0 & 0 & 1 & 1 & | & 600 \\ 0 & 0 & 0 & 0 & | & 0 \end{bmatrix}$
(b) $x_1 + x_4 = 1000$; $x_2 - x_4 = 100$; $x_3 + x_4 = 600$
(c) $x_4 = 1000 - x_1$; $x_4 = x_2 - 100$; $x_4 = 600 - x_3$
(d) 1000; 1000   (e) 100   (f) 600; 600
(g) $x_1 = 1000$; $x_2 = 700$; $x_3 = 600$; $x_4 = 600$

Section 2.5   (page 90)

1. Yes   **3.** No   **5.** No   **7.** Yes   **9.** $\begin{bmatrix} 0 & 1/2 \\ -1 & 1/2 \end{bmatrix}$ **11.** $\begin{bmatrix} 2 & 1 \\ 5 & 3 \end{bmatrix}$ **13.** None   **15.** $\begin{bmatrix} 1 & 0 & 0 \\ 0 & -1 & 0 \\ -1 & 0 & 1 \end{bmatrix}$

17. $\begin{bmatrix} 15 & 4 & -5 \\ -12 & -3 & 4 \\ -4 & -1 & 1 \end{bmatrix}$ **19.** None   **21.** $\begin{bmatrix} 7/4 & 5/2 & 3 \\ -1/4 & -1/2 & 0 \\ -1/4 & -1/2 & -1 \end{bmatrix}$ **23.** $\begin{bmatrix} 1/2 & 1/2 & -1/4 & 1/2 \\ -1 & 4 & -1/2 & -2 \\ -1/2 & 5/2 & -1/4 & -3/2 \\ 1/2 & -1/2 & 1/4 & 1/2 \end{bmatrix}$

25. (−1, 4)   **27.** (2, 1)   **29.** (2, 3)   **31.** Line $-x - 8y = 12$ **33.** (−8, 6, 1)   **35.** (15, −5, −1)
37. (−31, 24, −4)   **39.** No inverse, no solution for system   **41.** (−7, −34, −19, 7)
49. (a) $\begin{bmatrix} 72 \\ 48 \\ 60 \end{bmatrix}$ (b) $\begin{bmatrix} 2 & 4 & 2 \\ 2 & 1 & 2 \\ 2 & 1 & 3 \end{bmatrix} \begin{bmatrix} x_1 \\ x_2 \\ x_3 \end{bmatrix} = \begin{bmatrix} 72 \\ 48 \\ 60 \end{bmatrix}$ (c) 8, 8, 12

Application   (page 94)

1. (a) $\begin{bmatrix} 47 \\ 56 \end{bmatrix}$ $\begin{bmatrix} 0 \\ 130 \end{bmatrix}$ $\begin{bmatrix} 107 \\ 60 \end{bmatrix}$ $\begin{bmatrix} 53 \\ 202 \end{bmatrix}$ $\begin{bmatrix} 72 \\ 88 \end{bmatrix}$ $\begin{bmatrix} -7 \\ 172 \end{bmatrix}$ $\begin{bmatrix} 11 \\ 12 \end{bmatrix}$ $\begin{bmatrix} 83 \\ 74 \end{bmatrix}$ (b) $M^{-1} = \begin{bmatrix} 3/13 & 1/26 \\ -1/13 & 2/13 \end{bmatrix}$

2. $\begin{bmatrix} 39 \\ -98 \end{bmatrix}$ $\begin{bmatrix} -18 \\ 35 \end{bmatrix}$ $\begin{bmatrix} 19 \\ -49 \end{bmatrix}$ $\begin{bmatrix} -25 \\ 49 \end{bmatrix}$ $\begin{bmatrix} 34 \\ -95 \end{bmatrix}$ $\begin{bmatrix} -2 \\ 3 \end{bmatrix}$ $\begin{bmatrix} 5 \\ -24 \end{bmatrix}$ $\begin{bmatrix} 15 \\ -51 \end{bmatrix}$ $\begin{bmatrix} 10 \\ -32 \end{bmatrix}$ $\begin{bmatrix} 33 \\ -85 \end{bmatrix}$ $\begin{bmatrix} 35 \\ -97 \end{bmatrix}$

$\begin{bmatrix} 10 \\ -35 \end{bmatrix}$ $\begin{bmatrix} 39 \\ -105 \end{bmatrix}$ $\begin{bmatrix} 27 \\ -69 \end{bmatrix}$ $\begin{bmatrix} -4 \\ 4 \end{bmatrix}$

3. Santa Claus is fat

4. $\begin{bmatrix} 76 \\ 77 \\ 96 \end{bmatrix}$ $\begin{bmatrix} 62 \\ 67 \\ 75 \end{bmatrix}$ $\begin{bmatrix} 88 \\ 108 \\ 97 \end{bmatrix}$ $\begin{bmatrix} 141 \\ 160 \\ 168 \end{bmatrix}$ $\begin{bmatrix} 147 \\ 166 \\ 174 \end{bmatrix}$ $\begin{bmatrix} 105 \\ 120 \\ 123 \end{bmatrix}$ $\begin{bmatrix} 111 \\ 131 \\ 119 \end{bmatrix}$ $\begin{bmatrix} 92 \\ 119 \\ 94 \end{bmatrix}$ $\begin{bmatrix} 75 \\ 93 \\ 79 \end{bmatrix}$ $\begin{bmatrix} 181 \\ 208 \\ 208 \end{bmatrix}$

Section 2.6   (page 98)

1. $\begin{bmatrix} 32/3 \\ 25/3 \end{bmatrix}$ **3.** $\begin{bmatrix} 23,000/3579 \\ 93,500/3579 \end{bmatrix}$ or $\begin{bmatrix} 6.43 \\ 26.12 \end{bmatrix}$ **5.** $\begin{bmatrix} 20/3 \\ 20 \\ 10 \end{bmatrix}$

7. 1079 metric tons of wheat; 1428 metric tons of oil   **9.** 1285 units of agriculture; 1455 units of manufacturing and 1202 units of transportation   **11.** 3077 units of agriculture, 2564 units of manufacturing, and 3179 units of transportation   **13.** (a) 7/4 bushels of yams; 15/8 of a pig   (b) 167.5 bushels of yams; 153.75 pigs **15.** 33:47:23

Application   (page 103)

1. (a) $A = \begin{bmatrix} 0.245 & 0.102 & 0.051 \\ 0.099 & 0.291 & 0.279 \\ 0.433 & 0.372 & 0.011 \end{bmatrix}$ $D = \begin{bmatrix} 2.88 \\ 31.45 \\ 30.91 \end{bmatrix}$ $X = \begin{bmatrix} x_1 \\ x_2 \\ x_3 \end{bmatrix}$

(b) $I - A = \begin{bmatrix} 0.755 & -0.102 & -0.051 \\ -0.099 & 0.709 & -0.279 \\ -0.433 & -0.372 & 0.989 \end{bmatrix}$ (d) $\begin{bmatrix} 18.2 \\ 73.2 \\ 66.8 \end{bmatrix}$ (e) \$18.2 billion of agriculture, \$73.2 billion of manufacturing, and \$66.8 billion of household would be required (rounded to three significant digits).

2. (a) $A = \begin{bmatrix} 0.293 & 0 & 0 \\ 0.014 & 0.207 & 0.017 \\ 0.044 & 0.010 & 0.216 \end{bmatrix}$ $D = \begin{bmatrix} 138,213 \\ 17,597 \\ 1,786 \end{bmatrix}$ (b) $I - A = \begin{bmatrix} .707 & 0 & 0 \\ -0.014 & 0.793 & -0.017 \\ -0.044 & -0.010 & 0.784 \end{bmatrix}$

(d) Agriculture 195,000 million pounds; manufacture 26,000 million pounds; energy 13,600 million pounds

Chapter 2   Review Exercises   (page 104)

1. 2×2; $a = 2$; $b = 3$; $c = 5$; $q = 9$; square   **3.** 1×4; $m = 12$; $k = 4$; $z = -8$; $r = -1$; row

5. $\begin{bmatrix} 8 & 8 & 8 \\ 10 & 5 & 9 \\ 7 & 10 & 7 \\ 8 & 9 & 7 \end{bmatrix}$ **7.** $\begin{bmatrix} 9 & 10 \\ -3 & 0 \\ 10 & 16 \end{bmatrix}$ **9.** $\begin{bmatrix} 23 & 20 \\ -7 & 3 \\ 24 & 39 \end{bmatrix}$ **11.** $\begin{bmatrix} -17 & 20 \\ 1 & -21 \\ -8 & -17 \end{bmatrix}$

13. Not possible   **15.** [9]   **17.** [−26 −35]   **19.** No inverse   **21.** (−4, 6)   **23.** (−1, 2, 3) **25.** 8 thousand standard; 6 thousand extra large   **27.** 5 blankets; 3 rugs; 8 skirts   **29.** (−9, 3) **31.** (7, −9, −1)   **33.** $x = 6 - (7/3)z$; $y = 1 + (1/3)z$, z arbitrary   **35.** (34, −9) **37.** $\begin{bmatrix} 3 & -1 \\ -5 & 2 \end{bmatrix}$ **39.** $\begin{bmatrix} 1/2 & 0 \\ 1/10 & 1/5 \end{bmatrix}$ **41.** $\begin{bmatrix} 2/3 & 0 & -1/3 \\ 1/3 & 0 & -2/3 \\ -2/3 & 1 & 1/3 \end{bmatrix}$ **43.** No inverse   **45.** $\begin{bmatrix} 218.1 \\ 318.3 \end{bmatrix}$

47. (a) $\begin{bmatrix} 0 & 1/2 \\ 2/3 & 0 \end{bmatrix}$ (b) 1200 units of cheese; 1600 units of goats

49. $X = \begin{bmatrix} 3 \\ 4 \end{bmatrix}$ **51.** $X = \begin{bmatrix} 6 \\ 15 \\ 16 \end{bmatrix}$

CHAPTER 3

Section 3.1 (page 112)

1.

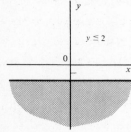

$x + y \leq 2$

3.

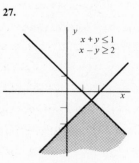

$x \geq 3 + y$

5.

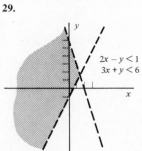

$4x - y < 6$

7.

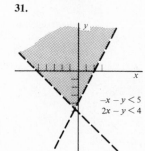

$3x + y < 6$

9.

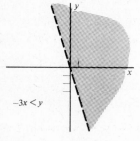

$x + 3y \geq -2$

11.

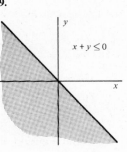

$4x + 3y > -3$

13.

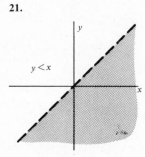

$2x - 4y < 3$

15.

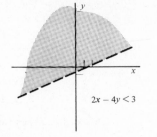

$x \leq 5y$

17.

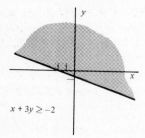

$-3x < y$

19.

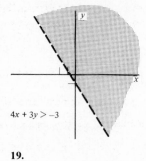

$x + y \leq 0$

21.

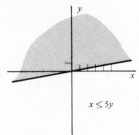

$y < x$

23.

$x < 4$

25.

$y \leq 2$

0

27.

$x + y \leq 1$
$x - y \geq 2$

29.

$2x - y < 1$
$3x + y < 6$

31.

$-x - y < 5$
$2x - y < 4$

33.   **35.**   **37.**   **39.**

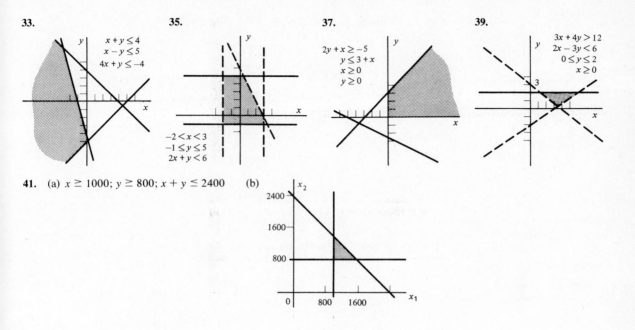

41.   (a) $x \geq 1000$; $y \geq 800$; $x + y \leq 2400$ (b)

Section 3.2   **(page 119)**
1. Maximum of 65 at (5, 10); minimum of 8 at (1, 1)   **3.** Maximum of 9 at (0, 12); minimum of 0 at (0, 0)
5. $x = 6/5$; $y = 6/5$ **7.** $x = 17/3$; $y = 5$ **9.** $x = 105/8$; $y = 25/8$ **11.** (a) Maximum of 204; $x = 18$; $y = 2$
(b) Maximum of 117 3/5; $x = 12/5$; $y = 39/5$ (c) Maximum of 102; $x = 0$; $y = 17/2$ **13.** 6.4 million gallons of
gasoline and 3.2 million gallons of fuel oil, for maximum revenue of $11,200,000   **15.** 1200 Type 1 and 2400 Type 2
for maximum revenue of $408   **17.** Ship 20 to A and 80 to B; $1040   **19.** 150 kg half-and-half mix; 75 kg other
21. 3 3/4 servings of A; 1 7/8 servings of B

Section 3.3   **(page 125)**
1. (0, 4, 2); (1, 7/2, 0); (0, 4, 0); (0, 0, 0); (0, 0, 10); (10/3, 0, 0); maximum of 14 at (0, 4, 2); minimum of 0 at (0, 0, 0)
3. (0, 0, 0, 12); (2, 4, 6, 0); (6, 0, 6, 0); (0, 0, 0, 0); maximum of 72 at (0, 0, 0, 12); minimum of 0 at (0, 0, 0, 0)
5. 160 kg. of Supreme and none of the others for a maximum profit of $320   **7.** 30 on first shift, 30 on second
shift, and 340/13 on third shift, for a maximum profit of $258.46   **9.** Plant 60 acres of potatoes, no corn, and no
cabbage, for a maximum profit of $7200.

Chapter 3   Review Exercises   **(page 126)**
1.   **3.**   **5.**   **7.**

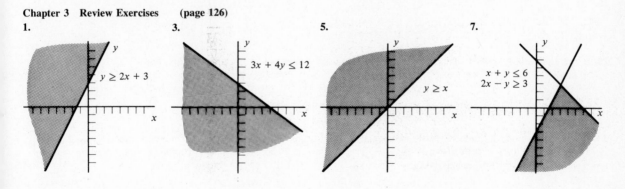

9.   **11.**   **13.**

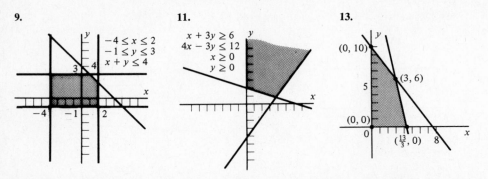

13. Let x = number of batches of cakes and y = number of batches of cookies. Then $x \geq 0$, $y \geq 0$, and $2x + (3/2)y \leq 15$
$$3x + (2/3)y \leq 13.$$

15. Minimum of 8 at (2, 1); maximum of 40 at (6, 7)   **17.** Maximum of 24 at (0, 6)   **19.** Minimum of 40 at any point on the segment connecting (0, 20) and (10/3, 40/3)   **21.** Make 3 batches of cakes, and 6 of cookies, for a maximum profit of \$210   **23.** Maximum of 210 at (18, 6, 0); minimum of 150 at (10, 10, 0)

CHAPTER 4

Section 4.1   (page 136)

1. $x_1 + 2x_2 + x_3 = 6$ **3.** $2x_1 + 4x_2 + 3x_3 + x_4 = 100$ **5.** (a) 3   (c) $4x_1 + 2x_2 + x_3 = 20$
(b) x_3, x_4, x_5 $5x_1 + x_2 + x_4 = 50$
$2x_1 + 3x_2 + x_5 = 25$

7. (a) 2   (c) $7x_1 + 6x_2 + 8x_3 + x_4 = 118$
(b) x_4, x_5 $4x_1 + 5x_2 + 10x_3 + x_5 = 220$

9. (0, 0, 20, 0, 15)   **11.** (0, 0, 8, 0, 6, 7)

13. (a)
$$\begin{array}{cccccc} x_1 & x_2 & x_3 & x_4 & x_5 & \end{array}$$
$$\left[\begin{array}{ccccc|c} -1 & 0 & 3 & 1 & -1 & 16 \\ 1 & 1 & 1/2 & 0 & 1/2 & 20 \end{array}\right]$$

(b) (0, 20, 0, 16, 0)

15. (a)
$$\begin{array}{cccccc} x_1 & x_2 & x_3 & x_4 & x_5 & x_6 \end{array}$$
$$\left[\begin{array}{cccccc|c} 2 & 2 & 1 & 1 & 0 & 0 & 12 \\ -5 & -4 & 0 & -3 & 1 & 0 & 9 \\ 1 & -1 & 0 & -1 & 0 & 1 & 8 \end{array}\right]$$

(b) (0, 0, 12, 0, 9, 8)

17. (a)
$$\begin{array}{cccccc} x_1 & x_2 & x_3 & x_4 & x_5 & x_6 \end{array}$$
$$\left[\begin{array}{cccccc|c} -1/2 & 1/2 & 0 & 1/2 & 1 & -1/2 & 0 \\ 3/2 & 1/2 & 1 & 1/2 & 0 & 1/2 & 50 \end{array}\right]$$

(b) (0, 0, 50, 0, 0, 0)

19.
$$\begin{array}{cccc} x_1 & x_2 & x_3 & x_4 \end{array}$$
$$\left[\begin{array}{cccc|c} 2 & 3 & 1 & 0 & 6 \\ 4 & 1 & 0 & 1 & 6 \end{array}\right]$$

21.
$$\begin{array}{ccccc} x_1 & x_2 & x_3 & x_4 & x_5 \end{array}$$
$$\left[\begin{array}{ccccc|c} 1 & 1 & 1 & 0 & 0 & 10 \\ 5 & 2 & 0 & 1 & 0 & 20 \\ 1 & 2 & 0 & 0 & 1 & 36 \end{array}\right]$$

23.
$$\begin{array}{cccc} x_1 & x_2 & x_3 & x_4 \end{array}$$
$$\left[\begin{array}{cccc|c} 3 & 1 & 1 & 0 & 12 \\ 1 & 1 & 0 & 1 & 15 \end{array}\right]$$

25. If x_1 is the number of kg of half-and-half mix and x_2 is the number of kg of the other mix, find $x_1 \geq 0$, $x_2 \geq 0$, $x_3 \geq 0$, $x_4 \geq 0$ so that $(1/2)x_1 + (1/3)x_2 + x_3 = 100$; $(1/2)x_1 + (2/3)x_2 + x_4 = 125$, and $z = 6x_1 + 4.8x_2$ is maximized.
$$\begin{array}{cccc} x_1 & x_2 & x_3 & x_4 \end{array}$$
$$\left[\begin{array}{cccc|c} \frac{1}{2} & \frac{1}{3} & 1 & 0 & 100 \\ \frac{1}{2} & \frac{2}{3} & 0 & 1 & 125 \end{array}\right]$$

27. If x_1 is the number of prams, x_2 is the number of runabouts, and x_3 is the number of trimarans, find $x_1 \geq 0$, $x_2 \geq 0$, $x_3 \geq 0$, $x_4 \geq 0$, $x_5 \geq 0$, $x_6 \geq 0$ so that $x_1 + 2x_2 + 3x_3 + x_4 = 6240$, $2x_1 + 5x_2 + 4x_3 + x_5 = 10,800$, $x_1 + x_2 + x_3 + x_6 = 3000$, and $z = 75x_1 + 90x_2 + 100x_3$ is maximized.
$$\begin{array}{cccccc} x_1 & x_2 & x_3 & x_4 & x_5 & x_6 \end{array}$$
$$\left[\begin{array}{cccccc|c} 1 & 2 & 3 & 1 & 0 & 0 & 6240 \\ 2 & 5 & 4 & 0 & 1 & 0 & 10,800 \\ 1 & 1 & 1 & 0 & 0 & 1 & 3000 \end{array}\right]$$

29. If x_1 is the number of Siamese cats and x_2 is the number of Persian cats, find $x_1 \geq 0$, $x_2 \geq 0$, $x_3 \geq 0$, $x_4 \geq 0$, $x_5 \geq 0$ so that $2x_1 + x_2 + x_3 = 90$, $x_1 + 2x_2 + x_4 = 80$, $x_1 + x_2 + x_5 = 50$, and $z = 12x_1 + 10x_2$ is maximized.

$$\begin{array}{ccccc} x_1 & x_2 & x_3 & x_4 & x_5 \end{array}$$
$$\left[\begin{array}{ccccc|c} 2 & 1 & 1 & 0 & 0 & 90 \\ 1 & 2 & 0 & 1 & 0 & 80 \\ 1 & 1 & 0 & 0 & 1 & 50 \end{array}\right]$$

Section 4.2   (page 145)

1. (0, 4, 0, 0, 2); maximum is 20   **3.** (4, 0, 8, 2, 0); maximum is 8   **5.** (16, 4, 0, 0, 16); maximum is 264
7. (0, 10, 0, 40, 4); maximum is 120   **9.** (118, 0, 0, 0, 102); maximum is 944   **11.** (0, 0, 0, 50, 0, 50); maximum
is 250   **13.** 163.6 kg of food P; none of Q; 1090.9 kg of R; 145.5 kg of S; maximum is 87,454.5   **15.** 150 kg of
the half-and-half mix; 75 kg of the other; maximum revenue is $1260   **17.** Make no 1-speed or 3-speed bicycles;
make 2700 10-speed bicycles; maximum profit is $59,400

Section 4.3   (page 152)

1. $2x_1 + 3x_2 + x_3 = 8$ **3.** $x_1 + x_2 + x_3 + x_4 = 100$ **5.** Change the objective function to maximize
    $x_1 + 4x_2 - x_4 = 7$ $x_1 + x_2 + x_3 - x_5 = 75$ $z = -4x_1 - 3x_2 - 2x_3$.   **7.** Change the objective
            $x_1 + x_2 - x_6 = 27$ function to maximize $z = -x_1 - 2x_2 - x_3 - 5x_4$.
9. (40, 0, 16, 0, 0); maximum is 480   **11.** (0, 150, 0, 50, 0, 0); maximum is 750   **13.** (0, 100, 50, 0, 10, 0);
maximum is 300   **15.** (10, 0, 0, 50); minimum is 40   **17.** (0, 100, 0, 0, 50); minimum is 100
19. 800,000 for whole tomatoes and 80,000 for sauce for a minimum cost of $3,460,000   **21.** 3 of pill #1 and 2 of
pill #2 for a minimum cost of 70¢   **23.** Buy 1000 small and 500 large for a minimum cost of $210   **25.** 2 2/3
units of I, none of II, and 4 of III for a minimum cost of $30.67

Application   (page 156)

1.

From	To			
	A	B	C	D
A	0	10	200	1000
B	20	0	30	15
C	300	0	0	50
D	800	0	75	0

2. 135,552   **3.** 756,970   **4.** 2,756,984   **5.** 5,886,972
6. 732,652

Application   (page 157)

1. 1: .95; 2: .83; 3: .75; 4: 1.00; 5: .87; 6: .94   **2.** $x_1 = 100$; $x_2 = 0$; $x_3 = 0$; $x_4 = 90$; $x_5 = 0$; $x_6 = 210$

Chapter 4   Review Exercises   (page 159)

1. (a) Let $x_1 =$ number of item A, $x_2 =$ number of item B, and $x_3 =$ number of item C she should buy
(b) $z = 4x_1 + 3x_2 + 3x_3$ (c) $5x_1 + 3x_2 + 6x_3 \leq 1200$
                $x_1 + 2x_2 + 2x_3 \leq 800$
                $2x_1 + x_2 + 5x_3 \leq 500$
3. (a) Let $x_1 =$ number of gallons of fruity wine and $x_2 =$ number of gallons of crystal wine to be made
    (b) $z = 12x_1 + 15x_2$ (c) $2x_1 + x_2 \leq 110$
                $2x_1 + 3x_2 \leq 125$
                $2x_1 + x_2 \leq 90$
5. (a) $2x_1 + 5x_2 + x_3 = 50$;   (b)
    $x_1 + 3x_2 + x_4 = 25$;
    $4x_1 + x_2 + x_5 = 18$;
    $x_1 + x_2 + x_6 = 12$

$$\begin{array}{cccccc} x_1 & x_2 & x_3 & x_4 & x_5 & x_6 \end{array}$$
$$\left[\begin{array}{cccccc|c} 2 & 5 & 1 & 0 & 0 & 0 & 50 \\ 1 & 3 & 0 & 1 & 0 & 0 & 25 \\ 4 & 1 & 0 & 0 & 1 & 0 & 18 \\ 1 & 1 & 0 & 0 & 0 & 1 & 12 \\ \hline -5 & -3 & 0 & 0 & 0 & 0 & 0 \end{array}\right]$$

7. (a) $x_1 + x_2 + x_3 + x_4 = 90;$
$2x_1 + 5x_2 + x_3 + x_5 = 120;$
$x_1 + 3x_2 - x_6 = 80$

(b)
$$\begin{array}{cccccc} x_1 & x_2 & x_3 & x_4 & x_5 & x_6 \\ \end{array}$$
$$\left[\begin{array}{cccccc|c} 1 & 1 & 1 & 1 & 0 & 0 & 90 \\ 2 & 5 & 1 & 0 & 1 & 0 & 120 \\ 1 & 3 & 0 & 0 & 0 & -1 & 80 \\ \hline -5 & -8 & -6 & 0 & 0 & 0 & 0 \end{array}\right]$$

9. (68/5, 0, 24/5, 0, 0,); maximum is 412/5; or (13.6, 0, 4.8, 0, 0); maximum is 82.4 **11.** (20/3, 0, 65/3, 35, 0, 0); maximum is 76 2/3 **13.** Change the objective function to maximize $z = -10x_1 - 15x_2$ **15.** Change the objective function to maximize $z = -7x_1 - 2x_2 - 3x_3$ **17.** (8, 12, 0, 1, 0, 2); minimum is -62 **19.** None of A; 400 of B; none of C; maximum profit is $1200 **21.** Produce 36.25 gallons of fruity and 17.5 gallons of crystal for a maximum profit of $697.50

CHAPTER 5

Section 5.1 (page 166)
1. False **3.** True **5.** True **7.** True **9.** False **11.** False **13.** False **15.** True
17. False **19.** True **21.** True **23.** False **25.** True **27.** True **29.** False **31.** False
33. False **35.** 8 **37.** 32 **39.** 1 **41.** 32 **43.** 4 **45.** 3 **47.** 0 **49.** (a) all except $\emptyset$,
{s}, {1}, {m}, {s, 1}, {s, m} (b) all except $\emptyset$, {s}, {1}

Section 5.2 (page 169)
1. True **3.** False **5.** True **7.** False **9.** True **11.** True **13.** {3, 5}
15. U, or {2, 3, 4, 5, 7, 9} **17.** U, or {2, 3, 4, 5, 7, 9} **19.** {7, 9} **21.** $\emptyset$ **23.** $\emptyset$
25. U, or {2, 3, 4, 5, 7, 9} **27.** All students in this school not taking this course **29.** All students in this school taking accounting and zoology **31.** All students in this school taking this course or zoology **33.** 6 **35.** 5
37. 1 **39.** 5 **41.** $B' = \{$stocks having a price to earnings ratio of less than 10$\}$; $B' = \{$ATT, GE, Hershey, Mobil, RCA$\}$ **43.** $B \cap C = \{$stocks having a price to earnings ratio of 10 or more, *and* having a positive price change$\}$; $B \cap C = \emptyset$ **45.** $(A \cap B)'$ is made up of the stocks that are *not* in the set {stock with a dividend greater than $3 and a price to earnings ratio of at least 10}; $(A \cap B)' = \{$ATT, GE, Hershey, Mobil, RCA$\}$
47. (a) {s, d, c, g, i, m, h} (b) {s, d, c} (c) {i, m, h} (d) {g} (e) {s, d, c, g, i, m, h} (f) {s, d, c}

Section 5.3 (page 174)
1. **3.** **5.** **7.** $\emptyset$

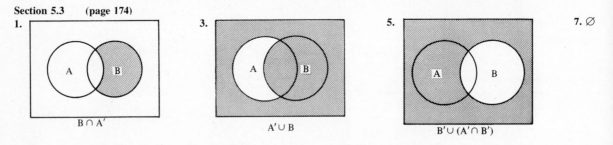

$B \cap A'$ $A' \cup B$ $B' \cup (A' \cap B')$

9. **11.** **13.** **15.**

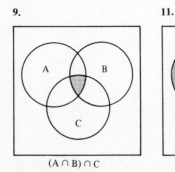 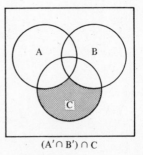 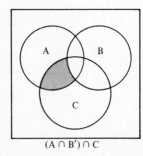

$(A \cap B) \cap C$ $A \cap (B \cup C')$ $(A' \cap B') \cap C$ $(A \cap B') \cap C$

17. 9   **19.** 16   **21.** Yes; his data add up to 142 people   **23.** (a) 37   (b) 22   (c) 50   (d) 11
(e) 25   (f) 11   **25.** (a) 50   (b) 2   (c) 14   **27.** (a) 54   (b) 17   (c) 10   (d) 7
(e) 15   (f) 3   (g) 12   (h) 1   **29.** (a) 40   (b) 30   (c) 95   (d) 110   (e) 160   (f) 65

Section 5.4   (page 184)
1. 12   **3.** 56   **5.** 8   **7.** 24   **9.** 792   **11.** 156   **13.** $\binom{52}{2} = 1326$ **15.** 10
17. (a) $5^7 = 78{,}125$ (b) $9 \cdot 10^5 \cdot 1 = 900{,}000$ (c) $9 \cdot 10^4 \cdot 1^2 = 90{,}000$ (d) $1^3 \cdot 10^4 = 10{,}000$
(e) $9 \cdot 9 \cdot 8 \cdot 7 \cdot 6 \cdot 5 \cdot 4 = 544{,}320$ **19.** (a) $\binom{8}{5} = 56$ (b) $\binom{11}{5} = 462$ (c) $\binom{8}{3} \cdot \binom{11}{2} = 56 \cdot 55 = 3080$
21. $\binom{5}{2} \cdot \binom{4}{1} + \binom{5}{3} \cdot \binom{4}{0} = 40 + 10 = 50$ **23.** (a) $\binom{5}{3} = 10$ (b) 0   (c) $\binom{3}{3} = 1$ (d) $\binom{5}{2} \cdot \binom{4}{1} = 10$
(e) $\binom{5}{2} \cdot \binom{3}{1} = 30$ (f) $\binom{5}{1} \cdot \binom{3}{2} = 15$ (g) 0   **25.** $5 \cdot 3 \cdot 2 = 30$ **27.** (a) $2 \cdot 25 \cdot 24 \cdot 23 = 27{,}600$
(b) $2 \cdot 26 \cdot 26 \cdot 26 = 35{,}152$ (c) $2 \cdot 24 \cdot 23 \cdot 1 = 1104$ **29.** $\binom{4}{2} = 6$ **31.** (a) $\binom{7}{1} = 21$
(b) $1 \cdot \binom{6}{1} = 6$ (c) $\binom{2}{1} \cdot \binom{5}{1} + \binom{2}{2} \cdot \binom{5}{0} = 11$ **33.** (a) $\binom{12}{3} = 220$ (b) $\binom{12}{9} = 220$ **35.** $6! = 720$
37. $5! = 120$ **39.** $2 \cdot 26 \cdot 25 \cdot 24 \cdot 10 \cdot 9 \cdot 8 = 22{,}464{,}000$ **41.** (a) $\binom{9}{3} = 84$ (b) $\binom{5}{3} = 10$
(c) $\binom{5}{2} \cdot \binom{4}{1} = 40$ (d) $1 \cdot \binom{8}{2} = 28$

Section 5.5   (page 189)
1. $m^4 + 4m^3n + 6m^2n^2 + 4mn^3 + n^4$ **3.** $729x^6 - 2916x^5y + 4860x^4y^2 - 4320x^3y^3 + 2160x^2y^4 - 576xy^5 + 64y^6$
5. $m^5/32 - 15m^4n/16 + 45m^3n^2/4 - 135m^2n^3/2 + 405mn^4/2 - 243n^5$ **7.** $p^{10} + 10p^9q + 45p^8q^2 + 120p^7q^3$
9. $a^{15} + 30a^{14}b + 420a^{13}b^2 + 3640a^{12}b^3$ **11.** $2^{10} = 1024$ **13.** $2^6 = 64$ (This result includes a pizza made with
none of the six toppings.)   **15.** $\binom{20}{2} + \binom{20}{3} + \binom{20}{4} + \binom{20}{5} + \binom{20}{6} = 60{,}439$ **17.** 64

Chapter 5   Review Exercises   (page 190)
1. False   **3.** False   **5.** True   **7.** False   **9.** False   **11.** $\{6, 7\}$; 2   **13.** $\{1, 2, 3, 4\}$; 4
15. $2^4 = 16$ **17.** $\{a, b, e\}$ **19.** $\{c, d, g\}$ **21.** $\{a, b, e, f\}$ **23.** U **25.** All employees in the
accounting department who also have at least 10 years with the company   **27.** All employees who are in the
accounting department or who have MBA degrees   **29.** All employees who are not in the sales department and
who have worked less than 10 years with the company
31.   **33.**

$A \cup B'$ $(A \cap B) \cup C$

35. 52   **37.** 12   **39.** $6! = 720$ **41.** $5 \cdot 4 = 20$ **43.** $\binom{8}{3} \cdot \binom{6}{2} = 840$ **45.** $32m^5 + 80m^4n + 80m^3n^2 +$
$40m^2n^3 + 10mn^4 + n^5$ **47.** $x^{20} - 10x^{19}y + 95x^{18}y^2/2 - 285x^{17}y^3/2$ **49.** $2^8 - 8 - 1 = 247$

CHAPTER 6

Section 6.1   (page 197)
1. {January, February, March, $\cdots$, December}   **3.** $\{0, 1, 2, \cdots, 80\}$ **5.** {go ahead, cancel}
7. $\{0, 1, 2, 3, \cdots, 5000\}$ **9.** {hhhh, hhht, hhth, hthh, thhh, hhtt, htth, tthh, thth, thht, htht, httt, thtt, ttht, ttth, tttt}
11. {h, th, tth, ttth, tttth, $\cdots$}   **13.** (a) {(1, 1), (1, 2), (1, 3), (1, 4), (1, 5), (1, 6), (2, 1), (2, 2), (2, 3), (2, 4), (2, 5),
(2, 6), (3, 1), (3, 2), (3, 3), (3, 4), (3, 5), (3, 6), (4, 1), (4, 2), (4, 3), (4, 4), (4, 5), (4, 6), (5, 1), (5, 2), (5, 3), (5, 4), (5, 5),
(5, 6), (6, 1), (6, 2), (6, 3), (6, 4), (6, 5), (6, 6)}   (b) $F = \{(3, 1), (3, 2), (3, 3), (3, 4), (3, 5), (3, 6)\}$
(c) $G = \{(2, 6), (3, 5), (4, 4), (5, 3), (6, 2)\}$ (d) $H = \varnothing$ **15.** (a) {hh, hth, thh, htth, thth, tthh, tthh, tthth,
thtth, htttt, htttt, thttt, tthtt, ttthht, tttth, tttttt}   (b) $E = \{hh\}$ (c) $F = \{hth, thh\}$ (d) $G = \{tttttt, ttttth, tttht,$
tthtt, thttt, htttt}   **17.** (a) worker is male   (b) worker has worked five years or more   (c) worker is female
and has worked less than five years   (d) worker has worked less than five years or has contributed to a voluntary
retirement plan   (e) worker is female or does not contribute to a voluntary retirement plan   (f) worker has
worked five years or more and does not contribute to a voluntary retirement plan   **19.** yes   **21.** no   **23.** yes
25. yes

27. no   **29.** (a) E' (b) $E \cap F$ (c) $E' \cap F'$ or $(E \cup F)'$ (d) $E \cap F'$ (e) $(E \cup F) \cap (E \cap F)'$
(f) $E \cup F$ **31.** {r, s, t}; {r, s}; {r, t}; {s, t}; {r}; {s}; {t}; $\varnothing$

Section 6.2   (page 206)
1. 1/6   **3.** 2/3   **5.** 1/13   **7.** 1/26   **9.** 1/52   **11.** 1 to 5   **13.** 2 to 1   **15.** {2}; {4}; {6}
17. {up}; {down}; {stays the same}   **19.** feasible   **21.** not feasible; sum of probabilities is less than 1
23. not feasible; $P(s_2) < 0$ **25.** .12   **27.** .50   **29.** .77   **31.** 6 to 19   **33.** .15   **35.** .85
37. .70   **39.** 1/5   **41.** 4/15   **43.** 8 to 7   **45.** 11 to 4   **47.** 4/11   **49.** .41   **51.** .21
53. .79   **55.** Not subjective   **57.** Subjective   **59.** Subjective   **61.** Not subjective
63. Not subjective   **65.** "Odds *in favor* of a direct hit . . ."

Application   (page 209)
1. .715; .569; .410; .321; .271

Section 6.3   (page 212)
1. 1/36   **3.** 1/9   **5.** 5/36   **7.** 1/12   **9.** 5/18   **11.** 11/36   **13.** 5/12   **15.** 2/13   **17.** 3/26
19. 3/13   **21.** 7/13   **23.** 1/2   **25.** 3/10   **27.** 7/10   **29.** 1/10   **31.** 2/5   **33.** 7/20   **35.** .88
37. .25   **39.** .38   **41.** .89   **43.** .50   **45.** .09   **47.** .39   **49.** .77   **51.** .951   **53.** .473
55. .007   **57.** 3/4   **59.** 1/4   **61.** 1/4   **63.** .23   **65.** 2/3

Section 6.4   (page 219)
1. $\binom{7}{1}/\binom{9}{1} = 7/9$ **3.** $\binom{7}{3}/\binom{9}{3} = 35/84$ **5.** .424   **7.** $\binom{6}{3}/\binom{10}{3} = 1/6$ **9.** $\binom{4}{2}\binom{6}{1}/\binom{10}{3} = 3/10$ **11.** $\binom{52}{2} = 1326$
13. $(4 \cdot 48 + 6)/\binom{52}{2} = 33/221$ **15.** $4 \cdot \binom{13}{2}/\binom{52}{2} = 52/221$ **17.** $\binom{40}{2}/\binom{52}{2} = 130/221$ **19.** $(1/26)^5$
21. $1 \cdot 25 \cdot 24 \cdot 23 \cdot 22/(26)^4 = 18{,}975/28{,}561$ **23.** $4/\binom{52}{5} = 1/649{,}740$ **25.** $13 \cdot \binom{4}{4}\binom{48}{1}/\binom{52}{5} = 1/4165$
27. $1/\binom{52}{13}$ **29.** $\binom{3}{3}\binom{4}{3}\binom{44}{7}/\binom{52}{13}$ **31.** $1 - P(365, 41)/(365)^{41}$ **33.** 1   **35.** 3/8; 1/4
37. $120/343 \approx .3498$ **39.** $3! = 6$ **41.** $9! = 362{,}880$ **43.** 1/3

Section 6.5   (page 229)
1. 0   **3.** 1   **5.** 1/6   **7.** 4/17   **9.** 25/51   **11.** $\frac{13}{52} \cdot \frac{12}{51} \cdot \frac{11}{50} \cdot \frac{10}{49} \cdot \frac{9}{48} \approx .00050$ **13.** $9/48 = .1875$
15. $4\left(\frac{13}{52} \cdot \frac{12}{51} \cdot \frac{11}{50} \cdot \frac{10}{49} \cdot \frac{9}{48}\right) \approx .00198$ **17.** $(2/3)^3 \approx .296$ **19.** 1/10   **21.** 0   **23.** 2/7   **25.** 2/7
27. The probability of a customer cashing a check given that the customer made a deposit is 5/7.   **29.** The
probability of a customer not cashing a check given that the customer did not make a deposit is 1/4.
31. The probability of a customer not both cashing a check and making a deposit is 6/11   **33.** 1/6   **35.** 0
37. 2/3   **39.** .06   **41.** 1/4   **43.** 1/4   **45.** 1/7   **47.** .049   **49.** .534   **51.** 42/527 or .080
53. Yes   **55.** .05   **57.** .25   **59.** not very reasonable   **61.** $1 - .000015 = .999985$
63. $1/2000 = .0005$ **65.** $(1999/2000)^a$ **67.** $(1999/2000)^{Nc}$ **69.** $1 - .741 = .259$ **71.** True
73. True   **75.** True   **77.** False   **79.** 0

Section 6.6   (page 238)
1. 1/3   **3.** 2/41   **5.** 21/41   **7.** 8/17   **9.** .146   **11.** .082   **13.** $119/131 \approx .908$ **15.** .824
17. $1/176 \approx .006$ **19.** 1/11   **21.** 5/9   **23.** 5/26   **25.** 72/73   **27.** 165/343

Application   (page 242)
1. .076   **2.** .542   **3.** .051

Section 6.7   (page 246)
1. $\binom{5}{2}\left(\frac{1}{2}\right)^2\left(\frac{1}{2}\right)^3 = 5/16$ **3.** $\binom{5}{0}\left(\frac{1}{2}\right)^0\left(\frac{1}{2}\right)^5 = 1/32$ **5.** $\binom{5}{4}\left(\frac{1}{2}\right)^4\left(\frac{1}{2}\right)^1 + \binom{5}{5}\left(\frac{1}{2}\right)^5\left(\frac{1}{2}\right)^0 = 3/16$ **7.** $\binom{5}{0}\left(\frac{1}{2}\right)^0\left(\frac{1}{2}\right)^5 +$

$\binom{5}{1}\left(\frac{1}{2}\right)^1\left(\frac{1}{2}\right)^4 + \binom{5}{2}\left(\frac{1}{2}\right)^2\left(\frac{1}{2}\right)^3 + \binom{5}{3}\left(\frac{1}{2}\right)^3\left(\frac{1}{2}\right)^2 = 13/16$ **9.** $\binom{12}{12}\left(\frac{1}{6}\right)^{12}\left(\frac{5}{6}\right)^0 \approx .0000000005$ **11.** $\binom{12}{1}\left(\frac{1}{6}\right)^1\left(\frac{5}{6}\right)^{11} \approx .269$

13. $\binom{12}{0}\left(\frac{1}{6}\right)^0\left(\frac{5}{6}\right)^{12} + \binom{12}{1}\left(\frac{1}{6}\right)^1\left(\frac{5}{6}\right)^{11} + \binom{12}{2}\left(\frac{1}{6}\right)^2\left(\frac{5}{6}\right)^{10} + \binom{12}{3}\left(\frac{1}{6}\right)^3\left(\frac{5}{6}\right)^9 \approx .875$ **15.** $\binom{5}{5}\left(\frac{1}{2}\right)^5\left(\frac{1}{2}\right)^0 = 1/32$

17. $\binom{5}{0}\left(\frac{1}{2}\right)^0\left(\frac{1}{2}\right)^5 + \binom{5}{1}\left(\frac{1}{2}\right)^1\left(\frac{1}{2}\right)^4 + \binom{5}{2}\left(\frac{1}{2}\right)^2\left(\frac{1}{2}\right)^3 + \binom{5}{3}\left(\frac{1}{2}\right)^3\left(\frac{1}{2}\right)^2 = 13/16$ **19.** $\binom{20}{0}(.05)^0(.95)^{20} \approx .358$

21. $\binom{6}{2}\left(\frac{1}{5}\right)^2\left(\frac{4}{5}\right)^4 \approx .246$ **23.** $\binom{6}{4}\left(\frac{1}{5}\right)^4\left(\frac{4}{5}\right)^2 + \binom{6}{5}\left(\frac{1}{5}\right)^5\left(\frac{4}{5}\right)^1 + \binom{6}{6}\left(\frac{1}{5}\right)^6\left(\frac{4}{5}\right)^0 \approx .017$ **25.** $\binom{3}{1}\left(\frac{5}{50}\right)^1\left(\frac{45}{50}\right)^2 \approx .243$

27. $\binom{10}{5}(.20)^5(.80)^5 \approx .026$ **29.** .999   **31.** $\binom{20}{17}(.70)^{17}(.30)^3 \approx .072$ **33.** .035

35. $\binom{10}{7}\left(\frac{1}{5}\right)^7\left(\frac{4}{5}\right)^3 \approx .00079$ **37.** $\approx .999922$ **39.** $\binom{3}{1}(.80)^1(.20)^2 = .096$

41. $\binom{3}{0}(.80)^0(.20)^3 + \binom{3}{1}(.80)^1(.20)^2 = .104$ **43.** .185   **45.** .256   **47.** $1 - (.99999975)^{10,000} \approx .0025$

49. $\binom{12}{4}(.2)^4(.8)^8 \approx .133$ **51.** .795   **53.** 3/16   **55.** 21/256   **57.** 125/23,328   **59.** $(165(5)^8)/6^{12}$

61. 6 teams   **63.** 10 teams

Chapter 6   Review Exercises   (page 249)
1. $\{1, 2, 3, 4, 5, 6\}$ **3.** $\{0, .5, 1, 1.5, 2, \cdots, 299.5, 300\}$ **5.** $\{(3, R), (3, G), (5, R), (5, G), (7, R), (7, G),$
$(9, R), (9, G), (11, R), (11, G)\}$ **7.** $\{(3, G), (5, G), (7, G), (9, G), (11, G)\}$ **9.** $E' \cap F'$ **11.** 1/4
13. 3/13   **15.** 1/2   **17.** 1   **19.** 1 to 25   **21.** .86   **23.**   **25.** 1/2

	N_2	T_2
N_1	N_1N_2	N_1T_2
T_1	T_1N_2	T_1T_2

27. 5/36   **29.** 1/6   **31.** 1/6   **33.** 2/11   **35.** .66   **37.** .71   **39.** $\binom{4}{3}/\binom{11}{3} = 4/165$
41. $\binom{4}{2}\binom{5}{1}/\binom{11}{3} = 2/11$ **43.** $\binom{5}{2}\binom{2}{1}/\binom{11}{3} = 4/33$ **45.** 25/102   **47.** 15/34   **49.** 4/17   **51.** 3
53. 2/5 or .4   **55.** 3/4 or .75   **57.** 19/22   **59.** 1/3   **61.** 1/7   **63.** 5/16   **65.** 11/32
67. $\binom{20}{4}(.01)^4(.99)^{16} \approx .00004$ **69.** $.81791 + .16523 + .01586 + .00096 + .00004 = 1.0000$ **71.**   (a)   8 wells
(b)   10 wells

CHAPTER 7

Section 7.1   (page 259)
1. No   **3.** Yes   **5.** No   **7.** Yes   **9.** No   **11.** Yes   **13.** No   **15.** No   **17.** Yes
19. $A^1 = \begin{bmatrix} 1 & 0 \\ .8 & .2 \end{bmatrix}$; $A^2 = \begin{bmatrix} 1 & 0 \\ .96 & .04 \end{bmatrix}$; $A^3 = \begin{bmatrix} 1 & 0 \\ .992 & .008 \end{bmatrix}$; 0   **21.** $C^1 = \begin{bmatrix} .5 & .5 \\ .72 & .28 \end{bmatrix}$; $C^2 = \begin{bmatrix} .61 & .39 \\ .5616 & .4384 \end{bmatrix}$;

$C^3 = \begin{bmatrix} .5858 & .4142 \\ .596448 & .403552 \end{bmatrix}$; .4142   **23.** $E^1 = \begin{bmatrix} .8 & .1 & .1 \\ .3 & .6 & .1 \\ 0 & 1 & 0 \end{bmatrix}$; $E^2 = \begin{bmatrix} .67 & .24 & .09 \\ .42 & .49 & .09 \\ .3 & .6 & .1 \end{bmatrix}$; $E^3 = \begin{bmatrix} .608 & .301 & .091 \\ .483 & .426 & .091 \\ .42 & .49 & .09 \end{bmatrix}$; .301

25.   (a)
	Small	Large
Small	.80	.20
Large	.60	.40

(b) $[.10 \quad .90]$ (c) $[.724 \quad .276]$ (d) $[.7448 \quad .2552]$
(e) $[.7490 \quad .2510]$ (f) $[.7498 \quad .2502]$ **27.**   (a) $[.53 \quad .47]$ (b) $[.5885 \quad .4115]$
(c) $[.6148 \quad .3852]$ (d) $[.62666 \quad .37334]$ **29.**   (a)   42,500; 5000; 2500
(b)   36,125; 8250; 5625   (c)   30,706; 10,213; 9081   (d)   26,100; 11,241; 12,659
31.   (a) $[.257 \quad .597 \quad .146]$ (b) $[.255 \quad .594 \quad .151]$ (c) $[.254 \quad .590 \quad .156]$
33.   (a)
	single	multiple
single	.90	.10
multiple	.05	.95

(b) $[.75 \quad .25]$
(c) $[68.8\% \quad 31.3\%]$
(d) $[63.4\% \quad 36.6\%]$

Section 7.2   (page 266)
1. Regular   **3.** Not regular   **5.** Regular   **7.** $[2/5 \quad 3/5]$ **9.** $[4/11 \quad 7/11]$
11. $[14/83 \quad 19/83 \quad 50/83]$ **13.** $[170/563 \quad 197/563 \quad 196/563]$ **15.** $[3/4 \quad 1/4]$
17. $[0 \quad 0 \quad 1]$ **19.** $[2/17 \quad 11/17 \quad 4/17]$ **21.** $[.244 \quad .529 \quad .227]$ **23.** 16/17
25. $[51/209 \quad 88/209 \quad 70/209]$ **27.** $[1/3 \quad 1/3 \quad 1/3]$ **29.** 1/2

Section 7.3   (page 274)
1. State 2 is absorbing; matrix is that of an absorbing Markov chain   **3.** State 2 is absorbing; matrix is not that of an absorbing Markov chain   **5.** States 2 and 4 are absorbing; matrix is that of an absorbing Markov chain

7. States 2 and 4 are absorbing; matrix is that of an absorbing Markov chain   **9.** $F = [2]$; $FR = [.4 \quad .6]$

11. $F = [5]$; $FR = [3/4 \quad 1/4]$ **13.** $F = [3/2]$; $FR = [1/2 \quad 1/2]$ **15.** $F = \begin{bmatrix} 1 & 0 \\ 1/3 & 4/3 \end{bmatrix}$;

$FR = \begin{bmatrix} 1/3 & 2/3 \\ 4/9 & 5/9 \end{bmatrix}$ **17.** $F = \begin{bmatrix} 25/17 & 5/17 \\ 5/34 & 35/34 \end{bmatrix}$; $FR = \begin{bmatrix} 4/17 & 15/34 & 11/34 \\ 11/34 & 37/68 & 9/68 \end{bmatrix}$ **19.** (a) $F = \begin{bmatrix} 3/2 & 1 & 1/2 \\ 1 & 2 & 1 \\ 1/2 & 1 & 3/2 \end{bmatrix}$;

$FR = \begin{bmatrix} 3/4 & 1/4 \\ 1/2 & 1/2 \\ 1/4 & 3/4 \end{bmatrix}$ (b) 3/4   (c) 1/4   **21.** (a) $F = \begin{bmatrix} 20/17 & 240/289 \\ 0 & 20/17 \end{bmatrix}$; $FR = \begin{bmatrix} 145/289 & 144/289 \\ 5/17 & 12/17 \end{bmatrix}$

(b) 144/289   **23.** .8756

25.

p	.1	.2	.3	.4	.5	.6	.7	.8	.9
x_a	.9999999997	.99999905	.99979	.98295	.5	.017046	.000209	.00000095	.0000000003

Chapter 7   Review Exercises   (page 276)

1. Yes   **3.** Yes   **5.** (a) $C^1 = \begin{bmatrix} .6 & .4 \\ 1 & 0 \end{bmatrix}$; $C^2 = \begin{bmatrix} .76 & .24 \\ .6 & .4 \end{bmatrix}$; $C^3 = \begin{bmatrix} .696 & .304 \\ .76 & .24 \end{bmatrix}$ (b) .76

7. (a) $E^1 = \begin{bmatrix} .2 & .5 & .3 \\ .1 & .8 & .1 \\ 0 & 1 & 0 \end{bmatrix}$; $E^2 = \begin{bmatrix} .09 & .8 & .11 \\ .1 & .79 & .11 \\ .1 & .8 & .1 \end{bmatrix}$; $E^3 = \begin{bmatrix} .098 & .795 & .107 \\ .099 & .792 & .109 \\ .1 & .79 & .11 \end{bmatrix}$ (b) .099

9. $[.453 \quad .547]$; $[5/11 \quad 6/11]$ or $[.455 \quad .545]$ **11.** $[.48 \quad .28 \quad .24]$; $[47/95 \quad 26/95 \quad 22/95]$ or $[.495 \quad .274 \quad .232]$
13. (a) $[.54 \quad .46]$ (b) $[.6464 \quad .3536]$ **15.** $[.428 \quad .322 \quad .25]$ **17.** $[.431 \quad .284 \quad .285]$ **19.** .2
21. .196   **23.** .28   **25.** $[.195 \quad .555 \quad .25]$ **27.** $[.194 \quad .556 \quad .25]$ **29.** Regular
31. Not regular   **33.** State 2 is absorbing; matrix is not that of an absorbing Markov chain

35. $F = [5/4]$; $FR = [5/8 \quad 3/8]$ **37.** $F = \begin{bmatrix} 7/4 & 4/5 \\ 1 & 8/5 \end{bmatrix}$; $FR = \begin{bmatrix} .55 & .45 \\ .6 & .4 \end{bmatrix}$ **39.** State 1, 3, and 6

41. $F = \begin{bmatrix} 5/2 & 1 & 1/2 \\ 1 & 2 & 1 \\ 1/2 & 1 & 5/2 \end{bmatrix}$; $FR = \begin{bmatrix} 11/16 & 1/8 & 3/16 \\ 3/8 & 1/4 & 3/8 \\ 3/16 & 1/8 & 11/16 \end{bmatrix}$

CHAPTER 8

Section 8.1   (page 286)
1.
(a)

Number	0	1	2	3	4	5
Probability	0	0	.1	.3	.4	.2

(b)

3.
(a)

Number	0	1	2	3	4	5	6
Probability	0	.04	0	.16	.40	.32	.08

(b)

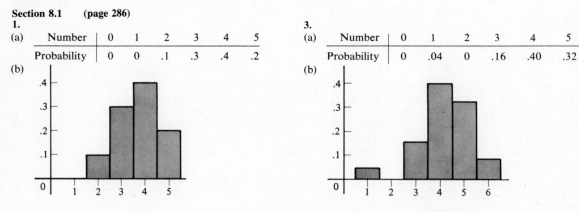

5. (a)

Number	0	1	2	3	4	5
Probability	.15	.25	.3	.15	.1	.05

(b)

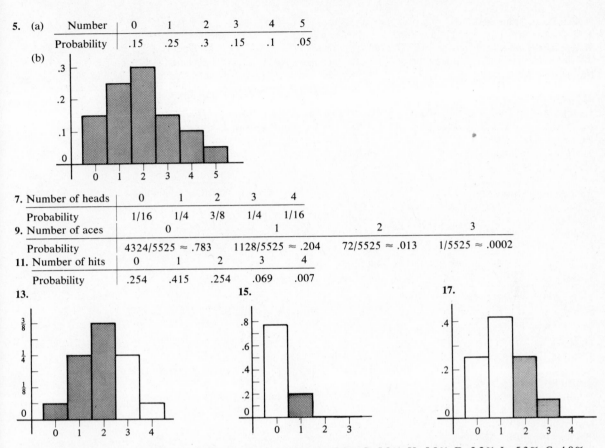

7.

Number of heads	0	1	2	3	4
Probability	1/16	1/4	3/8	1/4	1/16

9.

Number of aces	0	1	2	3
Probability	$4324/5525 \approx .783$	$1128/5525 \approx .204$	$72/5525 \approx .013$	$1/5525 \approx .0002$

11.

Number of hits	0	1	2	3	4
Probability	.254	.415	.254	.069	.007

13.   **15.**   **17.**

19. E: 18.1%; T: 8.8%; A: 6.2%; O: 4.4%; N: 7.5%; I: 7.1%; R: 5.8%; S: 5.8%; H: 5.8%; D: 2.2%; L: 5.3%; C: 4.9%; U: 4.0%; M: 1.8%; F: 4.0%; P: 1.8%; Y: 1.3%; W: 1.8%; G: 4.0%; B: 0%; V: .4%; K: 0%; X: 0%; J: 0%; Q: 1.3%; Z: 0%

Section 8.2   (page 293)
1. 3.6   **3.** 14.64   **5.** 2.7   **7.** 18   **9.** −$.64; no   **11.** 9/7 or 1.3   **13.** (a) 5/3 or 1.67
(b) 4/3 or 1.33   **15.** 1   **17.** No, the expected value is about −21¢.   **19.** −2.7¢   **21.** −20¢
23. (a) Yes, the probability of a match is still 1/2.   (b) 40¢   (c) −40¢   **25.** 2.5   **27.** $4500
29. 118   **31.** 3.51   **33.**

Account	EV	Total	Class
3	$2000	$22,000	C
4	1000	51,000	B
5	25,000	30,000	C
6	60,000	60,000	A
7	16,000	46,000	B

35. 1.58

37. (a) $94.0 million for seeding; 116.0 for not seeding   (b) seed

Application   (page 299)
1. (a) $69.01   (b) $79.64   (c) $58.96   (d) $82.54   (e) $62.88   (f) $64.00   **2.** Stock only
part 3 on the truck   **4.** 2^n

Application   (page 301)
1. .9886 million dollars   **2.** −.9714 million dollars

incorrect

Section 8.3 **(page 305)**
1. $\sqrt{407.4} = 20.2$ **3.** $\sqrt{215.3} = 14.7$ **5.** $\sqrt{59.4} = 7.7$ **7.** ~~Variance is .81; $\sigma = .9$~~
9. Variance is .000124; $\sigma = 0.11$ **11.** .700 **13.** .745 **15.** (b) **17.** 2.41 **19.** (a) at least 3/4
(b) at least 15/16 (c) at least 24/25 **21.** (a) mean is 320; standard deviation is 170.8 (b) 6 (c) 6
(d) at least 3/4 **23.** (a) 12.5 (b) −3.0 (c) 4.3 (d) 3.7 (e) 15.5 (f) 7.55 to 23.45

Section 8.4 **(page 315)**
1. 49.38% **3.** 17.36% **5.** 45.64% **7.** 49.91% **9.** 7.7% **11.** 47.35% **13.** 92.42%
15. 32.56% **17.** −1.64 or −1.65 **19.** 1.04 **21.** 5000 **23.** 4332 **25.** 642 **27.** 9918
29. 19 **31.** 15.87% **33.** .62% **35.** 84.13% **37.** 37.79% **39.** 2.27% **41.** 99.38%
43. 189 **45.** .0062 **47.** .4325 **49.** $38.62 and $25.89 **51.** .0823 **53.** 6.68% **55.** 38.3%
57. 82 **59.** 70

Application **(page 319)**
1. No; yes **2.** No; yes

Section 8.5 **(page 325)**
1. (a)

x	0	1	2	3	4	5	6
P(x)	.335	.402	.201	.054	.008	.001	.000

(b) 1.00 (c) .91

3. (a)

x	0	1	2	3
P(x)	.941	.058	.001	.000

(b) .06 (c) .24

5. (a)

x	0	1	2	3	4
P(x)	.0081	.0756	.2646	.4116	.2401

(b) 2.8 (c) .92 **7.** 12.5; 3.49 **9.** 51.2; 3.2

11. .1974 **13.** .1210 **15.** .0240 **17.** .9032 **19.** .8665 **21.** .0956 **23.** .0760 **25.** .6443
27. .0146 **29.** .1974 **31.** .0092 **33.** .0001 **35.** .9945

Section 8.6 **(page 329)**
1. (a) $y' = .3x + 1.5$ (b) $r = .20$ (c) 2.4 **3.** (a) $y' = 3.35x - 78.2$ (b) 123 (c) 156
(d) $r = .66$ **5.** (a) 26,920; 23,340; 19,770 (b) 29,370; 25,790; 22,210 (c) $42
7. $y' = 8.06x + 49.52$; $r = .996$ **9.** (a) $y' = 1.02x - 135$
(b) $r = .74$ (c) $375,000

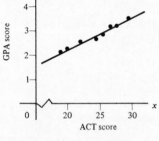

Chapter 8 **Review Exercises** **(page 333)**
1. (a)

x	1	2	3	4	5
Probability	.125	.292	.375	.125	.083

(b)

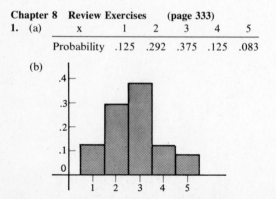

3. (a)

Number	0	1	2	3
Probability	$\frac{1}{8} = .125$	$\frac{3}{8} = .375$	$\frac{3}{8} = .375$	$\frac{1}{8} = .125$

(b)

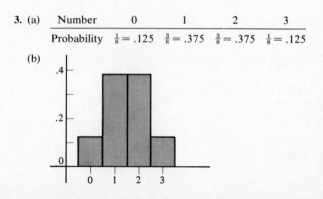

5. (a)

Number	0	1	2	3	4	5
Probability	.1	.1	.2	.3	.3	0

(b)

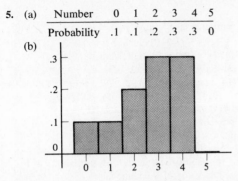

7. .6   **9.** −28¢   **11.** $8500   **13.** 11.87   **15.** 2.6   **17.** $1.29   **19.** Variance is 1.64; standard deviation is 1.28   **21.** (a) diet A   (b) diet B   **23.** (a) 98.76%   (b) Chebyshev's inequality would give "at least 84%"   **25.** $z = 1.41$

27. (a)

x	0	1	2	3	4
Probability	.9801	.0197	.0001485	.0000004975	6×10^{-10}

(b) mean = .02, $\sigma = .141$

29. (a) .1019   (b) .0008   **31.** (a) 25.14%   (b) 28.10%   (c) 22.92%   (d) 56.25%
33. $y' = 3.9x - 7.9$ **35.** .998

CHAPTER 9

Section 9.1   (page 339)
1. (a) Buy speculative   (b) buy blue-chip   (c) buy speculative; $24,300   (d) buy blue-chip
3. (a) Set up in the stadium   (b) set up in the gym   (c) set up both; $1010
5. (a)

	New product better	Not better
Market new product	50,000	−25,000
Don't	−40,000	−10,000

(b) $5000 if they market new product, −$22,000 if they don't; market the new product
7. (a)

	Strike	None
Bid $30,000	−5500	4500
Bid $40,000	4500	0

(b) bid $40,000   **9.** Environment; 14.25

Application   (page 345)
1. $E_1 = 18.61M$, $E_2 = 2.898M - 42.055$, $E_3 = .56M - 48.6$ **2.** $E_1 = 19.7325M$, $E_2 = .054M - 47.973$, $E_3 = .108M - 49.73$ **3.** $E_1 = 19M$, $E_2 = 1.7M - 45$, $E_3 = .6M - 48.5$

Section 9.2   (page 350)
1. $6 from B to A   **3.** $2 from A to B   **5.** $1 from A to B   **7.** Yes   **9.** $\begin{bmatrix} -2 & 8 \\ -1 & -9 \end{bmatrix}$ **11.** $\begin{bmatrix} 4 & -1 \\ 3 & 5 \end{bmatrix}$

13. $\begin{bmatrix} 8 & -7 \\ -2 & 4 \end{bmatrix}$ **15.** (1, 1); 3; strictly determined   **17.** No saddle point; not strictly determined   **19.** (3, 1);
3; strictly determined   **21.** (1, 3); 1; strictly determined   **23.** No saddle point; not strictly determined
25. (2, 3); 6   **27.**

	stone	scissors	paper
stone	0	1	−1
scissors	−1	0	1
paper	1	−1	0

; no   **29.**

		B		
		1	2	3
	1	15	−2	6
A	2	7	15	9
	3	3	−3	15

; no

Section 9.3 **(page 359)**
1. (a) −1 (b) −.28 (c) −1.54 (d) −.46 **3.** Player A: 1: 1/5, 2: 4/5; player B: 1: 3/5, 2: 2/5; value 17/5 **5.** Player A: 1: 7/9, 2: 2/9; player B: 1: 4/9, 2: 5/9; value −8/9 **7.** Player A: 1: 8/15, 2: 7/15; player B: 1: 2/3, 2: 1/3; value 5/3 **9.** Player A: 1: 6/11, 2: 5/11; player B: 1: 7/11, 2: 4/11; value −12/11 **11.** Strictly determined game; saddle point at (2, 2); value is −5/12 **13.** Player A: 1: 2/5, 2: 3/5; player B: 1: 1/5, 2: 4/5; value 7/5 **15.** Player A: 1: 1/14, 2: 0, 3: 13/14; player B: 1: 1/7, 2: 6/7; value 50/7 **17.** Player A: 1: 2/3, 2: 1/3; player B: 1: 0, 2: 1/9, 3: 8/9; value: 10/3 **19.** Player A: 1: 0; 2: 3/4; 3: 1/4; player B: 1: 0, 2: 1/12, 3: 11/12; value 33/4 **21.** Allied should select strategy 1 with probability 10/27 and strategy 2 with probability 17/27. The value of the game is 1/18, which represents increased sales of $55,556. **23.** The doctor should prescribe medicine 1 about 5/6 of the time and medicine 2 about 1/6 of the time. The effectiveness will be about 50%. **25.** (a) Number of fingers (b) For both players A and B: choose 1 with probability 3/4 and 2 with probability 1/4.

$$\begin{array}{cc} & \begin{array}{cc} 0 & 2 \end{array} \\ \begin{array}{c} \text{Number} \quad 0 \\ \text{of fingers} \quad 2 \end{array} & \begin{bmatrix} 0 & -2 \\ -2 & 4 \end{bmatrix} \end{array}$$

The value of the game is −1/2.

27. He should invest in rainy day goods about 5/9 of the time and sunny day goods about 4/9 of the time for a steady profit of $72.22.

Application **(page 366)**
1. $\begin{bmatrix} 2 & 2 \\ 1 & 3 \end{bmatrix}$ **2.** Both (1, 1) and (1, 2) are saddle points.

Chapter 9 Review Exercises (page 366)
1. Hostile to A **3.** Hostile; $785 **5.** Oppose **7.** Oppose; 2700 **9.** $2 from A to B **11.** $7 from B **13.** Row 3 dominates row 1; column 1 dominates column 4 **15.** $\begin{bmatrix} -11 & 6 \\ -10 & -12 \end{bmatrix}$ **17.** $\begin{bmatrix} -2 & 4 \\ 3 & 2 \\ 0 & 3 \end{bmatrix}$ **19.** (1, 1); value is −2 **21.** (2, 2); value is 0; fair game **23.** (2, 3); value is −3 **25.** Player A: 1: 5/6, 2: 1/6; player B: 1: 1/2, 2: 1/2; value 1/2 **27.** Player A: 1: 1/9, 2: 8/9; player B: 1: 5/9, 2: 4/9; value 5/9 **29.** Player A: 1: 1/5, 2: 4/5; player B: 1: 3/5, 2: 0; 3: 2/5; value −12/5 **31.** Player A: 1: 3/4, 2: 1/4, 3: 0; player B: 1: 1/2, 2: 1/2; value 1/2

CHAPTER 10

Section 10.1 **(page 372)**
1. $120 **3.** $2437.50 **5.** $128.97 **7.** $336 **9.** $96.20 **11.** $438.90 **13.** $329.10 **15.** $108.74 **17.** $14,062.50 **19.** $5043.65 **21.** $14,850.27 **23.** $6363.50 **25.** $330.52 **27.** $29,081.38 **29.** $1732.90 **31.** $6053.59 **33.** $3598; 20.1% **35.** $7246.80

Section 10.2 **(page 378)**
Answers in the remainder of this chapter may differ by a few cents, depending on whether tables or a calculator are used.
1. $1368.57 **3.** $14,432.13 **5.** $955.41 **7.** $82,379.10 **9.** $15,299.18 **11.** $13,213.14 **13.** $5105.58 **15.** $60,484.66 **17.** $818.86 **19.** $2663.55 **21.** $2217.26 **23.** $916.22 **25.** $4872.38 **27.** $1000 now **29.** $21,665.60 **31.** $44,510.80 **33.** $25,424.98 **35.** $13,693.34 **37.** $7743.93 **39.** 4.04% **41.** 8.16% **43.** 12.36% **45.** About 18 years **47.** About 12 years **49.** About 12 years **51.** $142,886.40 **53.** $123,506.50

Section 10.3 **(page 383)**
1. 15.91713 **3.** 21.82453 **5.** 22.01900 **7.** .04994 **9.** .02962 **11.** $1318.08 **13.** $305,390.00 **15.** $671,994.62 **17.** $4,180,929.88 **19.** $35,601.60 **21.** $516,397.05 **23.** $6294.79 **25.** $158,456.00 **27.** $13,486.56 **29.** $5303.43 **31.** $527 **33.** $952.50 **35.** $137,895.84 **37.** $34,186.91 **39.** $226.10 **41.** $522.90

Section 10.4 **(page 391)**
1. 9.71225 **3.** 12.65930 **5.** 14.71787 **7.** .19702 **9.** .09168 **11.** $6246.89 **13.** $7877.72 **15.** $153,724.50 **17.** $1,318,915.50 **19.** $111,183.90 **21.** $97,122.50 **23.** $12,493.78

25. $158.00   **27.** $6699.00   **29.** $147.02   **31.** $5054.93   **33.** $11,727.32   **35.** $274.58
37. $512.85   **39.** $489.77   **41.** $354.80   **43.** $691.08   **45.** $2440.29   **47.** $126.91

49.

Payment number	Amount of payment	Interest for period	Portion to principal	Principal at end of period
0	–	–	–	$4000
1	$1207.68	$320.00	$887.68	$3112.32
2	$1207.68	$248.99	$958.69	$2153.63
3	$1207.68	$172.29	$1035.39	$1118.24
4	$1207.70	$89.46	$1118.24	$0

51.

Payment number	Amount of payment	Interest for period	Portion to principal	Principal at end of period
0	–	–	–	$7184
1	$211.03	$107.76	$103.27	$7080.73
2	$211.03	$106.21	$104.82	$6975.91
3	$211.03	$104.64	$106.39	$6869.52
4	$211.03	$103.04	$107.99	$6761.53
5	$211.03	$101.42	$109.61	$6651.92
6	$211.03	$99.78	$111.25	$6540.67

53. (a) $1200   (b) $3511.58   (c)

Payment number	Amount of deposit	Interest earned	Total
1	$3511.58	$0	$3511.58
2	$3511.58	$105.35	$7128.51
3	$3511.58	$213.86	$10,853.95
4	$3511.58	$325.62	$14,691.15
5	$3511.58	$440.73	$18,643.46
6	$3511.58	$559.30	$22,714.34
7	$3511.58	$681.43	$26,907.35
8	$3511.58	$807.22	$31,226.15
9	$3511.58	$936.78	$35,674.51
10	$3511.58	$1070.24	$40,256.33
11	$3511.58	$1207.69	$44,975.60
12	$3511.58	$1349.27	$49,836.45
13	$3511.58	$1495.09	$54,843.12
14	$3511.59	$1645.29	$60,000.00

55. (a) $4025.90   (b) $2981.93

Application   (page 396)
1. $14,038   **2.** $9511   **3.** $8837   **4.** $3968

Chapter 10   Review Exercises   (page 397)
1. $1378.26   **3.** $2157.24   **5.** $60.25   **7.** $22,556.39   **9.** $76,849.45   **11.** $687.61
13. $6226.51   **15.** 16.3%   **17.** $1999.00   **19.** $38,521.74   **21.** $3875.79   **23.** $3576.93
25. $6002.84   **27.** $6289.04   **29.** $7194.05   **31.** $11,860.99   **33.** $2501.25   **35.** $32,330.89
37. $3653.45   **39.** $10,078.44   **41.** $171,723.68   **43.** $25,396.38   **45.** $32,671.66   **47.** $2815.31
49. $44,045.70   **51.** $4326.54   **53.** $12,806.37   **55.** $2499.49   **57.** $546.93   **59.** $886.05

61. $4339.46   **63.**

Payment number	Amount of payment	Interest for period	Portion to principal	Principal at end of period
0	–	–	–	$5000
1	$985.09	$250.00	$735.09	$4264.91
2	$985.09	$213.25	$771.84	$3493.07
3	$985.09	$174.65	$810.44	$2682.63
4	$985.09	$134.13	$850.96	$1831.67
5	$985.09	$91.58	$893.51	$938.16
6	$985.07	$46.91	$938.16	$0

APPENDIX EXERCISES   (page 407)

1. (a)

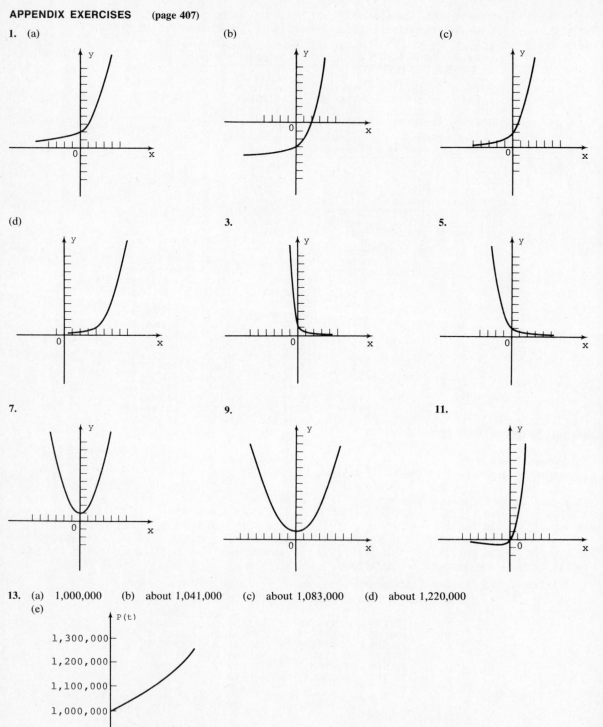

(b)

(c)

(d)

3.

5.

7.

9.

11.

13. (a) 1,000,000   (b) about 1,041,000   (c) about 1,083,000   (d) about 1,220,000
(e)

15. (a) about 2,690,000   (b) about 3,024,000   (c) about 9,600,000   (d) about 38,400,000
17. $\log_3 81 = 4$ **19.** $\log_{1/2} 16 = -4$ **21.** $\log_{10} .0001 = -4$ or $\log .0001 = -4$ **23.** $6^2 = 36$
25. $(\sqrt{3})^8 = 81$ **27.** $m^n = k$ **29.** $\log_3 2 - \log_3 5$ **31.** $\log_2 6 + \log_2 x - \log_2 y$
33. $1 + (1/2) \log_5 7 - \log_5 3$ **35.** Cannot be simplified using the properties   **37.** $\log_a (xy)/m$ **39.** $\log_m (a^2/b^6)$
41. $\log_x a^{3/2}b^{-4}$ or $\log_x (a^{3/2}/b^4)$ **43.** 2.9957   **45.** 6.6846   **47.** 7.2724   **49.** −2.5258   **51.** −1.6095
53. −4.9619   **55.** (a) about 549   (b) about 1100   (c) about 1420   (d) about 1750

(e)

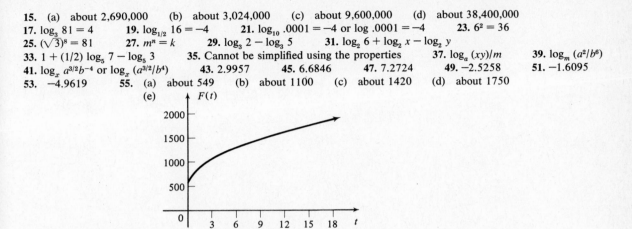

57. (a) about 5600 years ago   (d) about 3280 years

Index

Notes

Notes

Notes

Notes

Notes

Notes

Notes

Notes

We would appreciate it if you would take a few minutes to answer these questions. Then cut the page out, fold it, seal it, and mail it. No postage is required.

Which chapters did you cover?
(circle)   1   2   3   4   5   6   7   8   9   10   Appendix   All

Which helped most?
Explanations _____   Examples _____   Exercises _____
All three _____

Does the book have enough worked-out examples? Yes _____   No _____

Does the book have enough exercises? Yes _____   No _____

Were the answers in the back of the book helpful? Yes _____   No _____

Were the applications in the text helpful? Yes _____   No _____

Did your instructor cover the applications in class? Yes _____   No _____

Did you use the *Study Guide with Computer Problems?*
Yes _____   No _____   Did not know of it _____

If yes, was the *Study Guide* helpful?
Yes _____   For some topics _____   No _____

For you, was the course elective _____   required by _____

Do you plan to take more mathematics courses? Yes _____   No _____

If yes, which ones?
Statistics _____   College Algebra _____
Analytic Geometry _____   Trigonometry _____
Calculus _____   Other _____

How much algebra did you have before this course?
Years in high school   (circle)   0   ½   1   1½   2   more
Courses in college   0   1   2   3

If you had algebra before, how long ago?
Last 2 years _____   3-5 years _____   5 years or more _____

What is your major or your career goal? _____   Your age? _____

We would appreciate knowing of any errors you found in the book. (Please supply page numbers.)

What did you like most about the book?

FOLD HERE
··

What did you like least about the book?

College _____ State _____

FOLD HERE
··

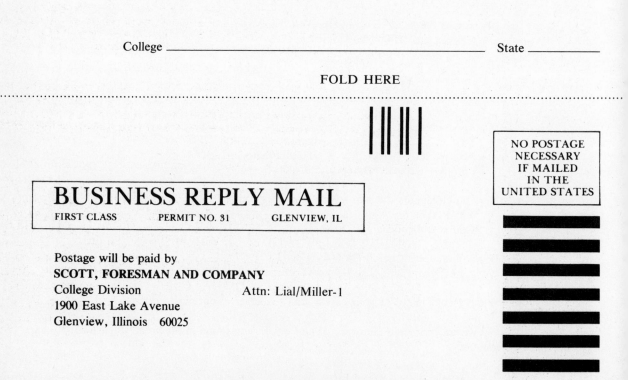

BUSINESS REPLY MAIL
FIRST CLASS   PERMIT NO. 31   GLENVIEW, IL

Postage will be paid by
SCOTT, FORESMAN AND COMPANY
College Division   Attn: Lial/Miller-1
1900 East Lake Avenue
Glenview, Illinois 60025

NO POSTAGE
NECESSARY
IF MAILED
IN THE
UNITED STATES